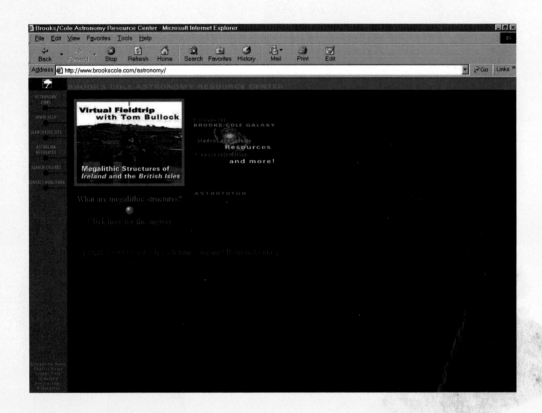

www.brookscole.com/astronomy

Discover the Brooks/Cole Galaxy is the heart of our Astronomy Resource Center. It includes a wealth of information and links to all Brooks/Cole astronomy titles. At brookscole.com/astronomy, you can find out about supplements, demonstration software, and student resources. You can also send e-mail to authors and preview new publications and exciting new technologies.

The Solar System

About the Author

Mike Seeds is the John W. Wetzel Professor of Astronomy at Franklin and Marshall College as well as Director of the College's Joseph R. Grundy Observatory. His research interests focus on peculiar variable stars and the automation of astronomical photometry. He is the Principal Astronomer in charge of the Phoenix 10, the first fully robotic telescope located in southern Arizona. In 1989 he received the Christian R. and Mary F. Lindback Award for Distinguished Teaching. In addition to writing textbooks, Seeds pursues an active interest in astronomical research, creates education tools for use in computer-smart classrooms, and continues to develop his upper-level courses in Archaeoastronomy and in the History of Astronomy. He has also published educational software for preliterate toddlers! Seeds was Senior Consultant in the creation of the 26-episode telecourse *Universe: The Infinite Frontier.* He is author of four other books published by Brooks/Cole, *Horizons: Exploring the Universe,* Sixth Edition; *Astronomy: The Solar System and Beyond,* Second Edition; *Foundations of Astronomy,* Sixth Edition; and *Stars and Galaxies,* Second Edition.

The Solar System

Second Edition

Michael A. Seeds

Joseph R. Grundy Observatory
Franklin and Marshall College

Brooks/Cole
Thomson Learning™

Australia • Canada • Mexico • Singapore • Spain • United Kingdom • United States

For Gary and Karen Seeds

Sponsoring Editors: *Gary Carlson and Jennifer Huber*
Marketing Manager: *Steve Catalano*
Project Development Editor: *Marie Carigma-Sambilay*
Marketing Communications: *Laura Hubrich*
Marketing Assistant: *Christina DeVito*
Editorial Assistant: *Samuel Subity*
Production Editor: *Kirk Bomont*
Production Service: *Heckman & Pinette*

Manuscript Editor: *Margaret Pinette*
Interior Design: *Jeanne Calabrese*
Cover Design: *Vernon T. Boes and Bob Western*
Interior Illustration: *Precision Graphics*
Print Buyer: *Vena Dyer*
Typesetting: *Thompson Type*
Printing and Binding: *Transcontinental Printing Inc.*

For more information, contact:
BROOKS/COLE
511 Forest Lodge Road
Pacific Grove, CA 93950
USA
www.brookscole.com

For permission to use material from this work, contact us by
Web: www.thomsonrights.com
fax: 1-800-730-2215
phone: 1-800-730-2214

Printed in Canada

10 9 8 7 6 5 4 3 2

Library of Congress Cataloging-in-Publication Data
Seeds, Michael A.
    The solar system / Michael A. Seeds.—2nd ed.
      p.   cm.
    Includes index.
    ISBN 0-534-38050-6
    1. Solar system.   I. Seeds, Michael A. Foundations of astronomy.   II. Title.
QB501.S43 2001
523.2—dc21   00-040396

Brief Contents

Contents

A Note to the Student

From Mike Seeds

Astronomy is about us. Although an astronomy course covers planets, and stars, and galaxies, it is really about you and me. Astronomy helps us answer the ultimate question of human existence. What are we? Great minds like Plato, Beethoven, and Hemingway have tried to give their personal answers, but we must each find our own answer. Astronomy helps us understand the meaning of our own existence.

Of course, part of our personal answer must include philosophical and cultural issues. We define ourselves by what we create, what we worship, what we admire, and what we expect of each other. But a major part of our answer is imbedded in the physical universe. We cannot hope to find a complete answer until we understand how the birth of the universe, the origin of galaxies, the evolution of stars, and the formation of planets created and brought together the atoms of which we are made. A basic understanding of astronomy is fundamental to knowing what we are. Astronomy is exciting because it is about us.

Part of the excitement of astronomy is the discovery of new things. Astronomers expect to be astonished. That excitement is reflected in this book in the form of new images, new discoveries, and new understandings that take us, in an introductory course, to the frontier of human knowledge. Here we explore the new evidence of an ancient ocean on Mars and a modern ocean on Jupiter's moon Europa. We will study the newest images of dying stars, and we will struggle to understand new evidence that the expansion of the universe is speeding up. The Hubble Space Telescope, spacecraft orbiting Mars and Jupiter, the new Chandra X-ray observatory, and giant Earth-based telescopes on remote mountaintops provide us with a daily dose of excitement that goes far beyond sensationalism. These new discoveries in astronomy are exciting because they are about us. They tell us more and more about what we are.

As a teacher, my quest is simple. I want you to understand your place in the universe—not just your location in space, but your location in the unfolding history of the physical universe. Not only do I want you to know where you are and what you are in the universe, but I want you to understand how we know. By the end of this book, I want you to know that the uni-verse is very big, but that it is described by a small set of rules and that we have found a way to figure out the rules—a method called science.

The most important concept in introductory astronomy is the process of science, the process by which scientists ask questions of nature and gradually puzzle out the beautiful secrets of the physical world. You should not memorize facts or believe principles because some astronomer says it is so. Of course not! To be a participant in the adventure and not a mere observer, you need to understand how evidence and hypotheses interact to create new understanding. Those who understand how science works are part of the adventure. Those who do not are baggage.

Another reason you need to understand how science works is that you live in a technological world that is changing rapidly through the practical application of new scientific discoveries. To control your life in such a world, to guide and care for others, you need to understand the power and the limitation of scientific study.

Science is based on the interplay of evidence and hypothesis, and that interplay is the principal organizing theme for this book at every level—from the individual sentences to the order of the chapters. We will deal with each topic by surveying basic observational facts, synthesizing hypotheses, and then discussing further observations as evidence that supports or contradicts those hypotheses.

A good example is how a survey of the solar system in Chapter 20 leads to theories to explain the origin of the solar system, which is followed by further evidence in Chapters 21 through 26 that helps us test and generalize our theories. At a different level, the paragraphs that discuss Jupiter's rings show how basic observations allow us to propose theories for the origin of the ring particles.

One big advantage of emphasizing evidence and hypothesis is that it unifies the mass of facts that discourage so many nonscience readers. A good example of how the ordering of chapters also follows this theme can be seen with Chapter 20, "The Origin of the Solar System." Its placement at the beginning of the section on the solar system allows you to use the following chapters on the planets as further tests of the "origins" hypothesis.

I want you to understand logical arguments. My approach makes the facts easier to remember, but, more importantly, the facts make sense because they fit into a logical framework. Understanding nature is the heart of science, and the interplay of evidence and hypothesis is the tool science uses to unlock the secrets of the universe. Once you begin to understand that exciting story, the facts will fall into place.

To help you grasp the logic of astronomy, I have created a number of features within the book.

- **Windows on Science** provide a parallel commentary on how all of science works. For example, the Windows point out where we are using statistical evidence, where we are reasoning by analogy, and where we are building a scientific model in place of a scientific hypothesis.

- **Guideposts** on the opening page of each chapter help you see the organization of the book. The Guidepost connects the chapter with preceding and following chapters to provide an overall organizational guide.

- **Review/Critical Inquiry** at the end of each text section is a carefully designed question to help you review and synthesize concepts and understand the scientific procedures being used in the section. A short answer follows to show how scientists construct logical arguments (from observations, evidence, theories, and natural laws) that lead to a conclusion. A further question then gives you a chance to construct your own argument on a related issue.

- **End of Chapter Review Questions** are designed to help you review and test your understanding of the material.

- **End of Chapter Discussion Questions** go beyond the text and invite you to think critically and creatively about scientific questions. You can think about these questions yourself or discuss them in class.

- **Critical Inquiries for the Web** conclude each chapter by challenging you to use the World Wide Web to explore further and to think creatively and analytically about astronomy.

- **Exploring The Sky** are experiments you can perform with the software *The Sky,* experiments that will allow you to see the sky and celestial bodies in new ways on your own computer screen.

As additional aids, be aware that this book also includes, free of charge, the following electronic enhancements:

- **The Sky *Student Edition CD-ROM.*** With this CD-ROM, a personal computer becomes a powerful personal planetarium. Loaded with data on 118,000 stars and 13,000 deep-sky objects with images, it allows you to view the universe at any point in time from 4000 years ago to 8000 years in the future, to see the sky in motion, to view constellations, to print star charts, and much more.

- **InfoTrac® *College Edition.*** You receive free unlimited access for one academic term to this online database of complete articles and images from more than 700 popular and scholarly periodicals of the past four years, updated daily. An online student guide gives you tips for using InfoTrac and correlates each chapter in this book to InfoTrac articles. (Check with your instructor for availability outside North America.)

- **The Brooks/Cole Astronomy Resource Center (www.brookscole.com/astronomy).** Thousands of students and instructors visit this free Web site every week. The Astronomy Resource Center offers—among other things—online study aids, virtual field trips, and links to other sites.

Do not be humble. Astronomy tells us that the universe is vast and powerful, but it also tells us that we are astonishing creatures. We humans are the parts of the universe that think. The human brain is the most complex piece of matter known, so as you explore the universe, remember that it is your human brain that is capable of understanding the depth and beauty of the cosmos.

To appreciate your role in this beautiful universe, you must learn more than just the facts of astronomy. You must understand what we are and how we know. Every page of this book reflects that ideal.

Mike Seeds
m_seeds@acad.fandm.edu

Acknowledgments

I started writing astronomy textbooks in 1973, and over the years I have had the guidance of a great many people who care about astronomy and teaching. I would like to thank all of the students and teachers who have responded so enthusiastically to *Foundations of Astronomy*. Their comments and suggestions have been very helpful in shaping this book. I would especially like to thank the many reviewers whose careful analysis and thoughtful suggestions have been invaluable in completing this new edition.

The following institutions have provided the many photos and diagrams in this book: Alcoa, American Image Inc., Anglo-Australian Telescope Board, Arecibo Observatory, Association of Universities for Research in Astronomy, The Astronomical Journal, The Astrophysical Journal, A.T.&T. Archives, Arizona State University Computer Science and Geography Departments, Brookhaven National Laboratories, Calar Alto Observatory, California Association for Research in Astronomy, California Institute of Technology, Carnegie Institute of Washington, Celestron International, Cerro Tololo Inter-American Observatory, Chandra X-Ray Center, Cornell University, European Southern Observatory, Galileo Imaging Team, Gemini 8-m Telescope Project, Goddard Space Flight Center, Granger Collection, Grundy Observatory, Hobby-Eberly Telescope Project, Hubble Heritage Team, Infrared Processing and Analysis Center, Institute for Astronomy, Japan/US Yohkoh Team, Johns Hopkins Applied Physics Lab, Joint Astronomy Center, JPL, Lawrence Livermore Laboratory, Lick Observatory, Los Alamos National Laboratory, Lowell Observatory, Lunar and Planetary Institute, Max-Planck Institute for Extraterrestrial Physics, McDonald Observatory, NASA, National Astronomy and Ionospheric Center, National Geophysical Data Center, National Museum of Natural History, National Oceanic and Atmospheric Administration, National Optical Astronomy Observatory, National Radio Astronomy Observatory, New England Meteorical Services, North Museum Lancaster PA, Ozone Processing Team/Goddard Space Flight Center, Rhodes College, Sacramento Peak Observatory, SETI Institute, Smithsonian Astrophysical Observatory, The Solar and Heliospheric Observatory, Sovfoto, Space Telescope Science Institute, Steward Observatory, University of Hawaii, U.S. Geological Survey, William Keck Observatory, Yerkes Observatory.

I would especially like to thank the following individuals who went out of their way to help me locate materials for *Foundations of Astronomy:* H. A. Abt, David P. Anderson, Andy Aperala, Anthony Aveni, Dana Backman, Sam Barden, Thomas G. Barns, Ginny Beal, Allen Beechel, Roger Bell, Dana Berry, David Bradstreet, Michael Briley, Marc Buie, Jack Burns, Adam Burrows, G. Byrd, Bruce Campbell, Duncan Chesley, Chip Clark, Cheryl Davis, Don Davis, Frank Drake, R. J. Dufour, John Eddy, Richard Fluck, Sidney W. Fox, Henry Freudenreich, Bruce Fryxell, James Gelb, Margaret J. Geller, Dan Gezari, Bronagh Glaser, Bruce A. Goldberg, Daniel Good, Bradford Greeley, Randall Grubbs, Calvin Hamilton, F. R. Harnden, William Hartmann, John Hayes, Floyd Herbert, Helen Horstman, John Hughes, Vincent Icke, George Jacoby, Rebecca Johnson, William Keel, Charles Keller, R. Kempton, Dean Ketelsen, Stephen Larson, Tod Lauer, Kathy Lestition, Deborah Levine, Kenneth Libbrecht, William Livingston, Kwok-Yung Lo, Bruce Macintosh, John W. Mackenty, David Malin, Jordan Marché, Jack Marling, Ursula B. Marvin, Alfred S. McEwen, Maureen McLean, Richard McPeters, David Miller, Stanley Miller, Reinhard Mundt, Lynne Nakauchi, Harold Nations, Seven Ostro, David Parker, Timothy Pearson, Richard Perley, C. M. Pieters, Carolyn C. Porco, Kathleen Powell, Vera Rubin, Laurence Rudnick, Anneila Sargent, Rudolph Schild, Maartin Schmidt, J. William Schopf, François Schweizer, M. Seldner, Nigel Sharp, Seth Shostak, L. D. Simon, Damon Simonelli, Martin Slade, T. Scott Smith, Steve Snowden, Hyron Spinrad, Alan Stockton, J. Trümper, Joseph Ververka, Steven Vogt, Steve Wampler, Mike Werner, Michael J. West, Gart Westerhout, Richard E. White, Simon D. M. White, Robin Witmore, Russell Wooldridge, Dennis Zaritsky, Benn Zellner.

I'd also like to thank the following manuscript reviewers: Gene Byrd, University of Alabama; R. Kent Clark, University of South Alabama; Geoffrey Clayton, Louisiana State University; Stephen Gottesman, University of Florida; Bob Hamilton, Solano College; Bruce Hanna, Old Dominion University; Russell Harkay, Keene State College; Gary Mechler, Pima Community College, West Campus; Harrison Prosper, Florida State University; Peter Shull, Oklahoma State University; Peter Strine, Bloomsburg University, and David Weinberg, Ohio State University.

It has been a great pleasure to work with and learn from all of the Brooks/Cole team. Special thanks go to all of the people who have contributed to the project, including Vernon Boes, Kirk Bomont, Jeanne Calabrese, Marie Carigma-Sambilay, Steve Catalano, Christina DeVito, Laura Hubrich, Sam Subity, and Pat Waldo. I have enjoyed working with Margaret Pinette and Bill Heckman of Heckman & Pinette on production, and I appreciate their understanding of my multifaceted failings. I would especially like to thank my editor, Gary Carlson, for all his help and advice and guidance on all of my books, and for his efforts to develop the entire astronomy list at Brooks/Cole. I would also like to thank my new editor, Jennifer Huber, for her help throughout production of this new edition.

Most of all, I would like to thank my wife, Janet, and my daughter, Kate, for putting up with "the books." They know all too well that textbooks are made of time.

Mike Seeds

The Solar System

Exploring *The Sky*

1. Install and open *The Sky*—a program that turns your computer screen into a window through which you view a virtual sky. Take this opportunity to familiarize yourself with it. *The Sky* can show your local sky at any time and date (well, within limits . . .) as well as many objects you can only see with binoculars and telescopes. You can face different directions by clicking on the **Look North, Look East, Look South, Look West, Look Up** buttons on your **Orientation Toolbar.** Or you can simply drag the slider on the bottom of your screen. Also experiment with changing between the **Daytime Sky Mode** and **Nighttime Sky Mode** by clicking on the second button in the **View Toolbar.**

   Before you experiment much further, it is useful to define some settings.

   a. First, set *The Sky* for your location, i.e., your latitude and longitude. Click on **Data,** then on **Site Information.** The easiest way to input your location is simply to choose a large city that is closest to you—for our purpose, that will most likely be close enough to your actual location. Also check to make sure that the chosen time at which you'll view *The Sky* is the computer clock (the default)—check to see that a check mark appears next to **Use Computer's Clock** within the **Site Information** window.

   b. It will also be useful to change the **Status Bar** at the bottom of your screen to show date and time. To do this, click on **View,** then **Status Bar...,** then check **date** and **time** in **Status Bar window.**

CHAPTER ONE

The Scale of the Cosmos

The longest journey begins with a

single step.

Confucius

Guidepost

How can we study something so big it includes everything, including us? How do we start thinking about the universe? Perhaps the best way is to grab a quick impression as we zoom from things our own size up to the largest things in the entire universe.

That cosmic zoom, the subject of this chapter, gives us our first glimpse of the subject of the rest of this book—the universe. In the next chapter, we will return to Earth to think about the appearance of the sky, and subsequent chapters will discuss other aspects of the universe around us.

As we study the universe, we must also observe the process by which we learn. That process, science, gives us a powerful way to understand not only the universe but also ourselves.

We are going on a voyage out to the end of the universe. Marco Polo journeyed east, Columbus west, but we will travel away from our home on Earth, out past the moon, sun, and other planets, past the stars we see in the sky, and past billions more that we cannot see without the aid of the largest telescopes. We will journey through great whirlpools of stars to the most distant galaxies visible from Earth—and then we will continue on, carried only by experience and imagination—looking for the structure of the universe itself.

Besides journeying through space, we will also travel in time. We will explore the past to see the sun and planets form and search for the formation of the first stars and the origin of the universe. We will also explore forward in time to watch the sun die and Earth wither. Our imagination will become a scientific time machine searching for the ultimate end of the universe.

Though we may find an end to the universe, a time when it will cease to exist, we will not discover an edge. It is possible that our universe is infinite and extends in all directions without limit. Such vastness dwarfs our human dimensions but not our intelligence or imagination.

Astronomy—this imagined voyage—is more than the study of stars and planets. It is the study of the universe in which we exist. Our personal lives are confined to a small part of a small planet circling a small sun drifting through the universe, but astronomy can take us out of ourselves and thus help us understand what we are.

Our study of astronomy introduces us to sizes, distances, and times far beyond our common experience. The comparisons in this chapter are designed to help us grasp their meaning.

We live small lives on our planet, but our study of astronomy introduces us to sizes, distances, and times far beyond our usual experience. Our task in this chapter is to grasp the meaning of these unfamiliar sizes, distances, and times. The solution lies in a single word *scale*. We will compare objects of different sizes to grasp the scale of the universe.

We begin with a human figure whose size we recognize. Figure 1-1 shows a region about 52 feet across occupied by a human being, a sidewalk, and a few trees—all objects whose size we can understand. Each successive picture in this chapter will show a region of the universe that is 100 times wider than the preceding picture. That is, each step will widen our view by a factor of 100.

In Figure 1-2 we widen our field of view by a factor of 100 and see an area 1 mile in diameter. The arrow points to the scene shown in the preceding photograph. People, trees, and sidewalks have vanished, but now we can see a college campus and the surrounding streets and houses. The dimensions of houses and streets are familiar to us. We have been in houses, crossed streets, and walked or run a mile. This is the familiar world around us, and we can relate such objects to the scale of our bodies.

We have begun our adventure using feet and miles, but we should use the metric system of units. Not only is it used by all scientists around the world, but it makes

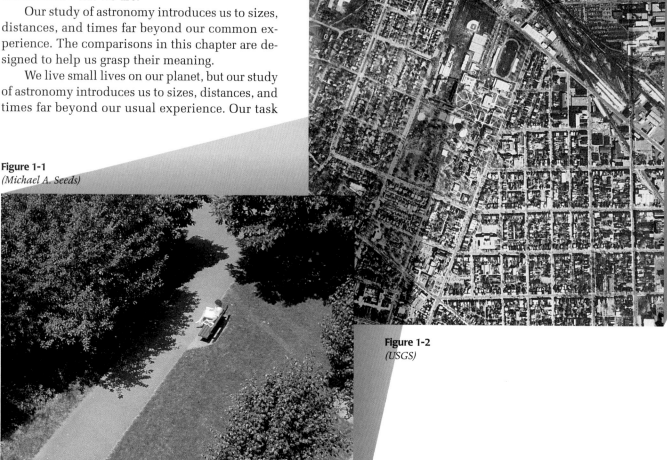

Figure 1-1
(Michael A. Seeds)

Figure 1-2
(USGS)

Figure 1-3
(NASA infrared photograph)

see that it, like Earth's surface, is evolving. Change is the norm in the universe.

Look again at Figure 1-3, and note the red color. This photo is an infrared photograph in which healthy green leaves and crops show up as red. Our eyes are only sensitive to a narrow range of colors. As we explore the universe, we must learn to use a wide range of "colors," from X rays to radio waves, to reveal sights invisible to our unaided eyes.

At the next step in our journey, we see our entire planet (Figure 1-4), which is 12,756 km in diam-

calculations much easier. For example, just calculating the number of inches in a mile is a chore. We must multiply 5280 feet per mile by 12 inches per foot. It is much easier to calculate the number of centimeters in 1 kilometer. We just move the decimal place three places to the right to get 1000 meters per kilometer and then move it two more places to get 100,000 centimeters per kilometer. If you are not already familiar with the metric system, or need a review, study Appendix A before reading on.

The photo in Figure 1-2 is 1 mile in diameter. A mile equals 1.609 kilometers, so we can see in the photo that a kilometer is a bit over two thirds of a mile—a short walk across a neighborhood.

Our view in Figure 1-3 spans 160 kilometers. Green foliage of different kinds appears as shades of red in this infrared photo. The college campus is now invisible. The gray blotches are small cities and the suburbs of Philadelphia are visible at the lower right. At this scale, we see the natural features of Earth's surface. The Allegheny Mountains of southern Pennsylvania cross the image in the upper left, and the Susquehanna River flows southeast into Chesapeake Bay. What appear as white bumps are a few puffs of clouds.

These features remind us that we live on the surface of a changing planet. Forces in Earth's crust pushed the mountain ranges up into parallel folds, like a rug wrinkled on a polished floor. The clouds remind us that Earth's atmosphere is rich in water, which falls as rain and erodes the mountains, washing material down the rivers and into the sea. The mountains and valleys that we know are only temporary features; they are constantly changing. As we explore the universe, we will

Figure 1-4
(NASA)

eter. Earth rotates on its axis once a day, exposing half of its surface to daylight at any particular moment. The photo shows most of the daylight side of the planet. The blurriness at the extreme right is the sunset line. The rotation of Earth carries us eastward, and as we cross the sunset line into darkness, we say the sun has set. Thus, the rotation of our planet causes the cycle of day and night.

We know that Earth's interior is made mostly of iron and nickel and that its crust is mostly silicate rocks. Only a thin layer of water makes up the oceans, and the atmosphere is only a few hundred kilometers deep. On the scale of this photograph, the depth of the atmosphere on which our lives depend is less than the thickness of a piece of thread.

Again we enlarge our field of view by a factor of 100, and we see a region 1,600,000 km wide (Figure 1-5). Earth is the small white dot in the center, and the moon, whose diameter is only one-fourth that of Earth, is an even smaller dot along its orbit 380,000 km from Earth.

These numbers are so large that it is inconvenient to write them out. Astronomy is the science of big numbers, and we will use numbers much larger than these to discuss the universe. Rather than writing out these numbers as in the previous paragraph, it is convenient to write them in **scientific notation.** This is nothing more than a simple way to write numbers without writing a great many zeros. In scientific notation, we write 380,000 as 3.8×10^5. The 5 tells us to begin with 3.8 and move the decimal point five places to the right to get 380,000. In the same way, we write 1,600,000 as 1.6×10^6. Notice that we could also write this as 16×10^5 or 0.16×10^7. All represent the same quantity. If you are not familiar with scientific notation, read the section on Powers of 10 Notation in the Appendix. The universe is too big to discuss without using scientific notation.

When we once again enlarge our field of view by a factor of 100 (Figure 1-6), Earth, the moon, and the moon's orbit all lie in the small red box at lower left. But now we see the sun and two other planets that are part of our solar system. Our **solar system** consists of the sun, its family of plan-

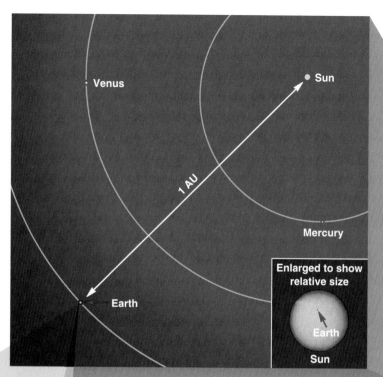

Figure 1-6

ets, and some smaller bodies such as moons and comets.

Like the Earth, Venus and Mercury are **planets,** small, nonluminous bodies that shine by reflected light. Venus is about the size of Earth, and Mercury is a bit larger than our moon. On this diagram, they are both too small to be seen as anything but tiny dots. The sun is a **star,** a self-luminous ball of hot gas that generates its own energy. The sun is 109 times larger in diameter than Earth (inset), but it too is nothing more than a dot in this diagram.

This diagram has a diameter of 1.6×10^8 km. One way astronomers deal with large numbers is to define new units. The average distance from the Earth to the Sun is a unit of distance called the **astronomical unit (AU),** a distance of 1.5×10^{11} m. Using this unit we can say that the average distance from Venus to the sun is about 0.7 AU. The average distance from Mercury to the sun is about 0.39 AU.

The orbits of the planets are not perfect circles, and this is particularly apparent for Mercury. Its orbit carries it as close to the sun as 0.307 AU and as far away as 0.467 AU. Earth's orbit is more circular, and its distance from the sun varies by only a few percent.

Our first field of view was only 52 feet (about 16 m) in width. After only six steps of enlarging by a factor of 100, we now see the entire solar system (Figure 1-7). Our field of view is 1 trillion (10^{12}) times wider than in our first view. The details of

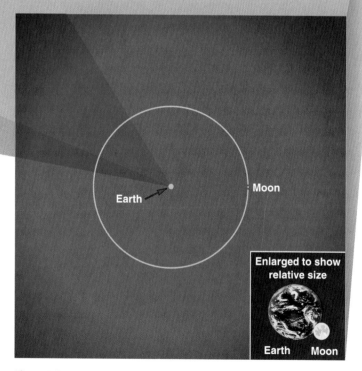

Figure 1-5

the preceding figure are now lost in the red square at the center of this diagram. We see only the brighter, more widely separated objects as we enlarge our view. The sun, Mercury, Venus, and Earth lie so close together that we cannot separate them at this scale. Mars, the next outward planet, lies only 1.5 AU from the sun. In contrast, Jupiter, Saturn, Uranus, Neptune, and Pluto are so far from the sun that they are easy to place in this diagram. These are cold worlds far from the sun's warmth. Light from the sun reaches Earth in only 8 minutes, but it takes over 4 hours to reach Neptune. Notice that Pluto's orbit is so elliptical that Pluto can come closer to the sun than Neptune does, as Pluto did between 1979 and 1999.

When we again enlarge our field of view by a factor of 100, our solar system vanishes (Figure 1-8). The sun is only a point of light, and all the planets and their orbits are now crowded into the small, red square at the

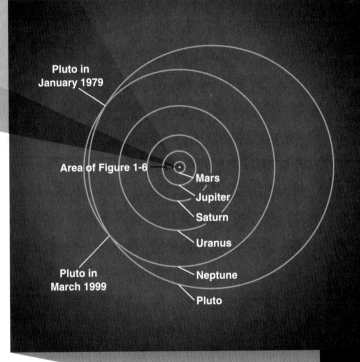

Figure 1-7

center. The planets are too small and reflect too little light to be visible so near the brilliance of the sun.

Nor are any stars visible except for the sun. The sun is a fairly typical star, and it seems to be located in a fairly average neighborhood in the universe. Although there are many billions of stars like the sun, none are close enough to be visible in this diagram, which shows an area only 11,000 AU in diameter. The stars are typi-

cally separated by distances about 10 times larger than the distance represented in this diagram. We will see stars in our next field of view, but, except for the sun at the center, this diagram is empty.

It is difficult to grasp the isolation of the stars. If the sun were represented by a golf ball in New York City, the nearest star would be another golf ball in Chicago. Except for the widely scattered stars and a few atoms of gas drifting between the stars, the universe is nearly empty.

In Figure 1-9, our field of view has expanded to a diameter a bit over 1 million AU. The sun is at the center, and we see a few of the nearest stars. These stars are so distant that it is not reasonable to give their distances in astronomical units. We must define a new unit of distance, the light-year. One **light-year (ly)** is the distance that light travels in one year, roughly 10^{13} km or 63,000 AU. The diameter of our field of view is, in our new unit, 17 ly. The nearest star to the sun, Proxima Centauri, is 4.2 ly from Earth. In other words, light from Proxima Centauri takes 4.2 years to reach Earth.

Although these stars are roughly the same size as the sun, they are so far away that we cannot see them as anything but points of light. Even with the largest telescopes on Earth, we still see only points of light when we look at stars. In Figure 1-9, the sizes of the dots represent not the size of the stars but their brightness. This is the custom in astronomical diagrams, and it is also how star images are recorded on photographs. Bright stars make larger spots on a photograph than faint stars. Thus, the size of a star image in a photograph tells us not how big the star is but only how bright it looks.

Figure 1-8

Sun

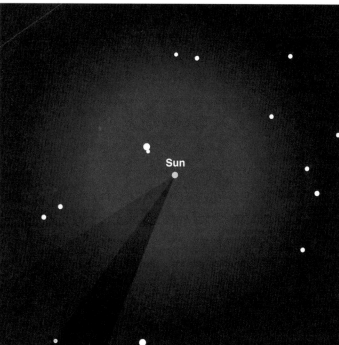

Figure 1-9

In Figure 1-10, we expand our field of view by another factor of 100, and the sun and its neighboring stars vanish into the background of thousands of stars. The field of view is now 1700 ly in diameter. Of course, no one has ever journeyed thousands of light-years from Earth to look back and photograph the solar neighborhood, so we use a representative photo of the sky. The sun is a relatively faint star that would not be easily located in a photo at this scale. Looking at this photograph, we notice a loose cluster of stars in the lower left quarter of the image. We will discover that many stars are born in clusters and that both old and young clusters exist in the sky. Star clusters are forming right now.

What we do not see is critically important. We do not see the thin gas that fills the spaces between the stars. Although those clouds of gas are thinner than the best vacuum on Earth, it is those clouds that give birth to new stars. Our sun formed from such a cloud about 5 billion years ago. We will see more star formation in our next view.

If we expand our field of view by a factor of 100, we see our galaxy (Figure 1-11). A **galaxy** is a great cloud of stars, gas, and dust bound together by the combined gravity of all the matter. Galaxies range from 1500 to over 300,000 ly in diameter and can contain over 100 billion stars. In the night sky, we see our galaxy as a great, cloudy wheel of stars ringing the sky as the **Milky Way,** and we refer to our galaxy as the **Milky Way Galaxy.** Of course, no one can journey far enough

into space to look back and photograph our galaxy, so the photo in Figure 1-11 shows a galaxy similar to our own. Our sun would be invisible in such a photo, but if we could see it, we would find it in the outer parts of the disk of the galaxy.

Ours is a fairly large galaxy. Only a century ago astronomers thought it was the entire universe—an island universe of stars in an otherwise empty vastness. Now we know that our galaxy is not unique. Indeed ours is only one of many billions of galaxies scattered throughout the universe.

As we expand our field of view by another factor of 100, our galaxy appears as a tiny lu-

Figure 1-10
This box ∎ represents the relative size of the previous frame. *(NOAO)*

minous speck surrounded by other specks (Figure 1-12). This diagram includes a region 17 million ly in diameter, and each of the dots represents a galaxy. Notice that our galaxy is part of a cluster of a few dozen galaxies. We will find that galaxies are commonly grouped together in clusters. Some of these galaxies have beautiful spiral patterns like our own galaxy, but others do not. Some are strangely distorted. One of the mysteries we will try to solve is what produces these differences among the galaxies.

If we again expand our field of view, we see that the clusters of galaxies are connected in a vast network

Figure 1-13
This box ■ represents the relative size of the previous frame. *(Detail from galaxy map from M. Seldner, B. L. Siebers, E. J. Groth, and P. J. E. Peebles,* Astronomical Journal *82 [1977])*

of the billions of galaxies contains billions of stars. Most of those stars probably have families of planets like our solar system, and on some of those billions of planets liquid-water oceans and protective atmospheres may have sheltered the spark of life. It is possible that some other planets are inhabited by intelligent creatures who share our curiosity and our wonder at the scale of the cosmos.

Summary

Our goal in this chapter is to preview the scale of astronomical objects. To do so, we journey outward from a familiar campus scene by expanding our field of view by factors of 100. Only 12 such steps take us to the largest structures in the universe.

The numbers in astronomy are so large it is not convenient to express them in the usual way. Instead, we use the metric system to simplify our calculations and scientific notation to simplify the writing of big numbers. The metric system and scientific notation are discussed in Appendix A.

We live on the rotating planet Earth, which orbits a rather typical star we call the sun. A unit of distance convenient for

(Figure 1-13). Clusters are grouped into super-clusters—clusters of clusters—and the superclusters are linked to form long filaments and walls outlining voids that seem nearly empty of galaxies. These appear to be the largest structures in the universe. Were we to expand our field of view another time, we would probably see a uniform fog of filaments and voids. When we puzzle over the origin of these structures, we are at the frontier of human knowledge.

Our problem in studying astronomy is to keep a proper sense of scale. Remember that each

Figure 1-11
(© Anglo-Australian Telescope Board)

Milky Way Galaxy →

Figure 1-12

discussing our solar system, the astronomical unit (AU), equals the average distance from Earth to the sun. Of the other planets in our solar system, Mercury is closest to the sun, only 0.39 AU away, and Neptune is about 30 AU.

The sun, like most stars, is very far from its neighboring stars, and this leads us to use another unit of distance, the light-year. A light-year is the distance light travels in one year. The star nearest to the sun is Proxima Centauri at a distance of 4.2 ly.

As we enlarge our field of view, we discover that the sun is only one of about 100 billion stars in our galaxy and that our galaxy is only one of many billions of galaxies in the universe. Galaxies appear to be grouped together in clusters, superclusters, and filaments, the largest structures known.

As we explore, we note that the universe is evolving. Earth's surface is evolving, and so are stars. Stars form from the gas in space, grow old, and eventually die. We do not yet understand how galaxies form or evolve.

Among the billions of stars in each of the billions of galaxies, many probably have planets, but detecting planets orbiting other stars is very difficult, and only a few have been found. We suppose that among the many planets in the universe, there must be some that are like Earth. We wonder if a few are inhabited by intelligent beings like ourselves.

New Terms

scientific notation	light-year (ly)
solar system	galaxy
planet	Milky Way
star	Milky Way Galaxy
astronomical unit (AU)	

Review Questions

1. What is the largest dimension you have personal knowledge of? Have you run a mile? Hiked 10 miles? Run a marathon?

2. Why are astronomical units more convenient than miles or kilometers for measuring some astronomical distances?

3. In what ways is our planet changing?

4. What is the difference between our solar system, our galaxy, and the universe?

5. Why do all stars, except for the sun, look like points of light as seen from Earth?

6. Why are light-years—rather than miles or kilometers—more convenient units for measuring some astronomical distances?

7. Why is it difficult to see planets orbiting other stars?

8. Which is the outermost planet in our solar system? Why does that change?

9. Why can't we measure the diameters of stars from the size of the star images on photographs? What does the diameter of a star image really tell us about the star?

10. How long does it take light to cross the diameter of our solar system or of our galaxy?

11. What are the largest known structures in the universe?

12. How many planets inhabited by intelligent life do you think the universe contains? Explain your answer.

Problems

1. If 1 mile equals 1.609 km and the moon is 2160 miles in diameter, what is its diameter in kilometers? in meters?

2. If sunlight takes 8 minutes to reach Earth, how long does moonlight take?

3. How many suns would it take, laid edge to edge, to reach the nearest star?

4. How many kilometers are there in a light-minute? (*Hint:* The speed of light is 3×10^5 km/s.)

5. How many galaxies like our own, laid edge to edge, would it take to reach the nearest large galaxy (which is 2×10^6 ly away)?

Exploring *The Sky*

1. Locate and center one example of each of three different types of objects:

   a. A planet, such as Saturn. Find its rising and setting time. Such objects have distances measured in **astronomical units (AU).**

   How to proceed: Decide on the object you want to locate. Then find and center the object by pressing the **Find** button on the **Object Toolbar.** The second method is to press the **F** key. The third is to click **Edit,** then **Find.** Once you have the **Object Information** window, press the **center** button.

   b. A star. All stars in *The Sky* belong to our Milky Way Galaxy. Give the star's name, its magnitude, and its distance in light-years.

   How to proceed: Click on any star, which brings up an **Object Information** window containing a variety of information about the star.

   c. A galaxy; give its name and/or its designation.

   How to proceed: To show galaxies, click on the **Galaxies button** in the **Object Toolbar,** then click on any galaxy. Distances to galaxies are on the order of millions and billions of light years.

 Go to the Brooks/Cole Astronomy Resource Center (www. brookscole.com/astronomy) for critical thinking exercises, articles, and additional readings from InfoTrac College Edition, Brooks/Cole's online student library.

CHAPTER TWO

The Sky

The Southern Cross I saw every night abeam. The sun every morning came up astern; every evening it went down ahead. I wished for no other compass to guide me, for these were true.

Captain Joshua Slocum
Sailing Alone Around the World

Guidepost

Our goal in studying astronomy is to learn about ourselves. We search for an answer to the question "What are we?" The quick answer is that we are thinking creatures living on a planet that circles a star we call the sun. In this chapter, we begin trying to understand that answer. What does it mean to live on a planet?

The preceding chapter gave us a quick overview of the universe, and chapters later in the book will discuss details. This chapter and the next help us understand what the universe looks like seen from the surface of our spinning planet.

We will see in Chapter 4 how difficult it has been for humanity to understand what we see in the sky every day. In fact, we will discover that science was born when people tried to understand the appearance of the sky.

The night sky is the rest of the universe as seen from our planet. When we look up at the stars, we look out through a layer of air only a few hundred kilometers deep. Beyond that, space is nearly empty, with the stars scattered light-years apart. We begin our search for the natural laws that govern the universe by trying to understand what that universe looks like.

What we see in the sky is influenced dramatically by a simple fact. We live on a planet. The stars are scattered into the void all around us, most very distant and some closer. Our planet rotates on its axis once a day, so from our viewpoint the sky appears to rotate around us each day. Not only does the sun rise in the east and set in the west, but so also do the stars. That apparent daily motion of the sky is caused by the rotation of our planet. As you read this chapter and learn about the sky, keep in mind that you are also learning what it means to live on a planet.

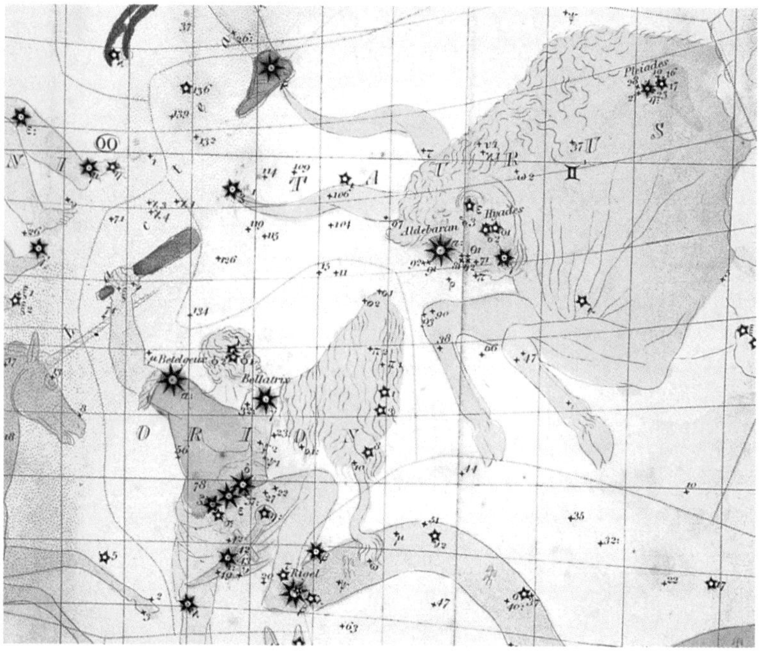

Figure 2-1
The ancient constellations Orion and Taurus celebrate the conflict between a great hunter and a powerful bull. *(From Duncan Bradford,* Wonders of the Heavens, *Boston: John B. Russell, 1837.)*

2-1 The Stars

Constellations

All around the world, ancient cultures celebrated heroes, gods, and mythical beasts by naming groups of stars—**constellations** (Figure 2-1). We should not be surprised that the star patterns do not look like the creatures they represent any more than Columbus, Ohio, looks like Christopher Columbus. The constellations "celebrate" the most important mythical figures in each culture. The constellations we know in Western culture originated in Mesopotamia over 5000 years ago, with other constellations added by Babylonian, Egyptian, and Greek astronomers during the classical age. Of these ancient constellations, 48 are still in use.

Different cultures grouped stars and named constellations differently. The constellation we know as Orion was known as Al Jabbar, the giant, to the ancient Syrians, as the White Tiger to the Chinese, and as Prajapati in the form of a stag in ancient India. The Pawnee Indians knew the constellation Scorpius as two groupings. The long tail of the scorpion was the Snake, and the two bright stars at the tip of the scorpion's tail were the Two Swimming Ducks.

Many ancient cultures around the world, including the Greeks, northern Asians, and Native Americans, associated the stars of the Big Dipper with a bear. The celestial bear may have crossed the land bridge into North America with the first Americans over 10,000 years ago. Some of the constellations we know today may be among the oldest surviving traces of human culture.

To the ancients, a constellation was a loose grouping of stars. Many of the fainter stars were not included in any constellation, and regions of the southern sky not visible to the ancient astronomers of northern latitudes were not identified with constellations. Constellation boundaries, when they were defined at all, were only approximate (Figure 2-2a), so a star like Alpheratz could be thought of as part of Pegasus or part of Andromeda. In recent centuries astronomers have added 40 modern constellations to fill gaps, and in 1928 the International Astronomical Union established 88 official constellations with clearly defined boundaries (Figure 2-2b). Thus a constellation now represents not a group of stars but an area of the sky, and any star within the region belongs to one and only one constellation.

In addition to the 88 official constellations, the sky contains a number of less formally defined groupings called **asterisms.** The Big Dipper, for example, is a well-known asterism that is part of the constellation Ursa Major (the Great Bear). Another asterism is the Great Square of Pegasus (Figure 2-2b), which includes three stars from Pegasus and one from Andromeda. The star charts at the end of this book will introduce you to the brighter constellations and asterisms.

Although we name constellations and asterisms, most are made up of stars that are not physically associated with one another. Some stars may be many times farther away than others and moving through space in different directions. The only thing they have in common is that they lie in approximately the same direction from Earth (Figure 2-3).

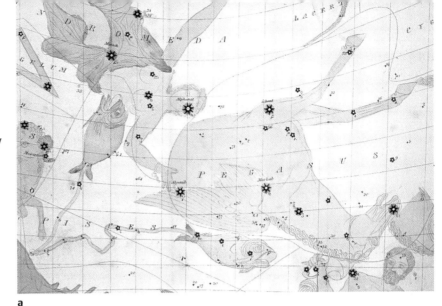

Figure 2-2
(a) In antiquity, constellation boundaries were poorly defined, as shown on this map by the curving dotted lines that separate Pegasus from Andromeda. *(From Duncan Bradford,* Wonders of the Heavens, *Boston: John B. Russell, 1837)*
(b) Modern constellation boundaries are precisely defined by international agreement.

The Names of the Stars

The brightest stars were named thousands of years ago, and these names, along with the names of the constellations, found their way into the first catalogs of stars. We take our constellation names from Greek versions translated into Latin—the language of science from the fall of Rome to the 19th century—but most star names come from ancient Arabic. Names such as Sirius (the Scorched One), Capella (the Little She Goat), and Aldebaran (the Follower of the Pleiades) are beautiful additions to the mythology of the sky.

Giving the stars individual names is not very helpful, because we see thousands of stars, and these names do not help us locate the star in the sky. In which constellation is Antares, for example? In 1603, Bavarian lawyer Johann Bayer published an atlas of the sky called

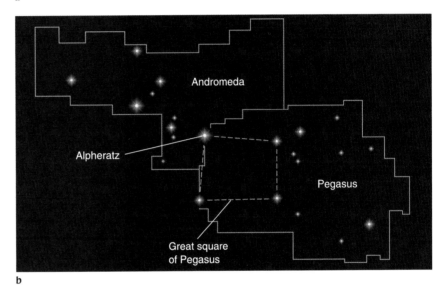

Uranometria in which he assigned lowercase Greek letters to the brighter stars of each constellation in approximate order of brightness. Astronomers have used those Greek letters ever since. (See the Appendix table with the Greek alphabet.) Thus, the brightest star is usually designated α (alpha), the second-brightest β (beta), and so on (Figure 2-4). To identify a star in this way, we give the Greek letter followed by the possessive form of the constellation name, such as α Scorpii (for Antares). Now we know that Antares is in the constellation Scorpius and that it is probably the brightest star in that constellation.

This method of identifying a star's brightness, however, is only approximate. To be precise, we must

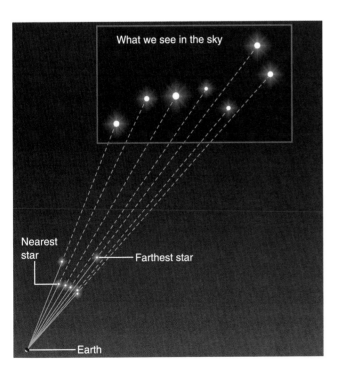

Figure 2-3
The stars we see in the Big Dipper—the brighter stars of the constellation Ursa Major, the Great Bear—are not at the same distance from Earth. We see the stars in a group in the sky because they lie in the same general direction as seen from Earth. The size of the star dots in the star chart represents the apparent brightness of the stars.

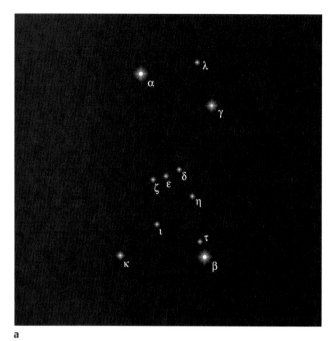

a

Figure 2-4

The brighter stars in the constellation Orion are designated by Greek letters. Although most of the stars are lettered in order of brightness, β is brighter than α and κ is brighter than η. (b) A long-exposure photograph reveals the many faint stars that lie within Orion's boundaries. These are members of the constellation, but they do not have Greek-letter designations. Note the differences in the colors of the stars. The crosses through the bright stars are caused by an optical effect in the telescope. *(Photo courtesy William Hartmann)*

b

have an accurate way of referring to the brightness of stars, and for that we must consult one of the first great astronomers.

The Brightness of Stars

Astronomers measure the brightness of stars using the **magnitude scale,** a system that had its start over two millennia ago when the Greek astronomer Hipparchus (160–127 BC) (Figure 2-5) divided the stars into six classes. The brightest he called first-class stars, and those that were fainter, second-class. His scale continued down to sixth-class stars, the faintest visible to the human eye. Thus, the larger the magnitude number, the fainter the star. This makes sense if we think of the brightest stars as first-class stars and the faintest stars visible as sixth-class stars.

Hipparchus used his brightness scale in a great catalog of the stars, and successive generations of astronomers have continued to use his system. Later astronomers had to revise the ancient magnitude system to include very faint and very bright stars. Telescopes reveal many stars fainter than those the eye can detect, so the magnitude scale was extended to include these faint stars. The Hubble Space Telescope, currently the most sensitive astronomical telescope, can detect stars

fainter than +28 magnitude. In contrast, the brightest stars are actually brighter than the first brightness class in Hipparchus's system. Thus, modern astronomers extended the magnitude system into negative numbers to account for brighter objects. For instance, Vega (alpha Lyrae) is almost zero magnitude at 0.04, and Sirius, the

Figure 2-5

Hipparchus (2nd century BC) was the first great observational astronomer. Among other things, he constructed a catalog listing 1080 of the brightest stars by position and dividing them into six brightness classes now known as magnitudes. He is honored here on a Greek stamp, which also shows one of his observing instruments.

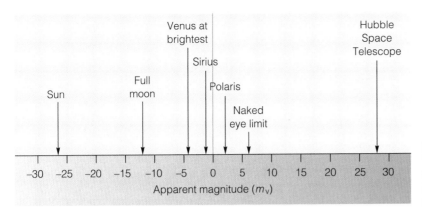

Figure 2-6
The scale of apparent visual magnitudes extends into negative numbers to represent the brightest objects and to positive numbers larger than 6 to represent objects fainter than the human eye can see.

brightest star in the sky, has a magnitude of −1.42. We can even place the sun on this scale at −26.5 and the moon at −12.5 (Figure 2-6).

These numbers are known as **apparent visual magnitudes (m_v)**, and they describe how the stars look to human eyes seen from Earth. A star that is a million times more luminous than the sun might appear very faint if it is far away, and a star that is much less luminous might look bright if it is nearby. Look at the brightness and distance of the stars in Figure 2-3. In Chapter 9, we will develop a magnitude scale that tells us how bright the stars really are. Apparent visual magnitudes tell us only how bright the stars appear in our sky.

Magnitude and Intensity

Nearly every star catalog from today back to the time of Hipparchus uses his magnitude scale of stellar brightness. However, brightness is subjective depending on such things as the physiology of the eye and the psychology of perception. To be accurate in measurements and calculations astronomers use the more precise term *intensity*—a measure of the light energy from a star that hits 1 square meter in 1 second. Thus astronomers need a way to convert between magnitudes and intensities.

The human eye senses the brightness of objects by comparing the ratios of their intensities. If two stars have intensities I_A and I_B, we can compare them by writing the ratio of their intensities, I_A/I_B. The magnitude system is based on a constant intensity ratio of about 2.5 for each magnitude. That is, if two stars differ by 1 magnitude, then they have an intensity ratio of about 2.5. If they differ by 2 magnitudes, their ratio is 2.5 × 2.5, and so on.

When 19th-century astronomers began measuring starlight, they realized they needed to define a mathematical magnitude system that was precise, but for convenience they wanted a system that agreed at least roughly with that of Hipparchus. The stars that Hipparchus classified as first and sixth differ by 5 magnitudes and have an intensity ratio of almost exactly 100, so the modern system of magnitudes specifies that a magnitude difference of 5 magnitudes corresponds to an intensity ratio of 100. That means that 1 magnitude corresponds to an intensity ratio of precisely 2.512, the fifth root of 100. That is, $100 = (2.512)^5$.

Now we can compare the light we receive from two stars. The light from a first-magnitude star is 2.512 times more intense than that from a second-magnitude star. The light from a third-magnitude star is $(2.512)^2$ times more intense than the light from a fifth-magnitude star. In general, the intensity ratio equals 2.512 raised to the power of the magnitude difference. That is:

$$I_A/I_B = (2.512)^{(m_B - m_A)}$$

For instance, if two stars differ by 6.32 magnitudes, we can calculate the ratio of their intensities as $2.512^{6.32}$. A pocket calculator tells us that the ratio is 337.

This magnitude system has some advantages. It compresses a tremendous range of intensity into a small range of magnitudes (Table 2-1) and makes it possible

Table 2-1 Magnitude and Intensity

Magnitude Difference	Intensity Ratio
0	1
1	2.5
2	6.3
3	16
4	40
5	100
6	250
7	630
8	1600
9	4000
10	10,000
⋮	⋮
15	1,000,000
20	100,000,000
25	10,000,000,000
⋮	⋮

In everyday language, we use the word *model* in various ways—fashion model, model airplane, model student—but scientists use the word in a very specific way. A **scientific model** is a carefully devised mental conception of how something works, a framework that helps scientists think about some aspect of nature. For example, astronomers use the celestial sphere as a way to think about the rotations of the sky, sun, moon, and stars.

Although we will think of a scientific model as a mental conception, it can take many forms. Some models are quite abstract—the psychologist's model of how the human mind processes visual information into images, for instance. But other models are so specific that they can be expressed as a set of mathematical equa-

tions. For example, an astronomer might use a set of equations to describe in detail how gas falls into a black hole. We could refer to such a calculation as a model. Of course, we could use metal and plastic to build a celestial globe, but the thing we build wouldn't really be the model any more than the equations are a model. A scientific model is a mental conception, an idea in our heads, that helps us think about nature.

On the other hand, a model is not meant to be a statement of truth. The celestial sphere is not real; we know the stars are scattered through space at various distances, but we can imagine a celestial sphere and use it to help us think about the sky. A scientific model does not have to be true to be useful. Chemists, for

example, think about the atoms in molecules by visualizing them as balls joined together by bonds. This model of a molecule is not fully correct, but it is a helpful way to think about molecules; it gives chemists a framework within which to organize their ideas.

Because scientific models are not meant to be totally correct, we must always remember the assumptions on which they are based. If we begin to think a model is true, it can mislead us instead of helping us. The celestial sphere, for instance, can help us think about the sky, but we must remember that it is only a mental crutch. The universe is much larger and much more interesting than this ancient scientific model of the heavens.

for modern astronomers to compare their measurements with all the recorded measurements of the past, right back to the first star catalog assembled by Hipparchus over 2000 years ago.

REVIEW Critical Inquiry

Nonastronomers sometimes complain that the magnitude scale is awkward. Why would they think it is awkward, and how did it get that way?

Two things might make the magnitude scale seem awkward. First, it is backwards; the bigger the magnitude number the fainter the star. Of course, that arose because Hipparchus was not measuring the brightness of stars but rather classifying them, and first-class stars would be brighter than second-class stars. The second awkward feature of the magnitude scale is its mathematical relation to intensity. If two stars differ by one magnitude, one is about 2.5 times brighter than the other. But if they differ by two magnitudes, one is 2.5×2.5 times brighter. This mathematical relationship arises because of the way we perceive brightness as ratios of intensity.

If the magnitude scale is so awkward, why do you suppose astronomers have used it for over two millennia?

In this section, we have found a way to identify stars by name and by brightness. Next we must look at the sky as a whole and notice its motion.

2-2 The Celestial Sphere

Ancient astronomers thought the sky was a great sphere with the stars stuck on the inside like thumbtacks in the ceiling. They imagined that Earth was very small and located at the center of this vast **celestial sphere.** Of course, modern astronomers know the celestial sphere is not real. The stars are not attached to a sphere but are scattered at different distances through space. Nevertheless, the celestial sphere is still a useful model to help us think about the motions of the sky (Window on Science 2-1).

A Model of the Sky

The sky we see above us at any moment is the half of the celestial sphere above the **horizon,** the circular boundary between Earth and sky. The other half of the sphere is hidden below the horizon. The very top of the sky, the **zenith,** is directly overhead (Figure 2-7).

If we watch the night sky for a few hours, we can see movement (Figure 2-8a). As the rotation of Earth carries us eastward, stars move across the sky and set in the west as new stars rise in the east.

Figure 2-8b shows how stars in the northern sky appear to revolve around a point called the **north celestial pole,** the point on the sky directly above Earth's North Pole. The star Polaris happens to lie very near the north celestial pole and thus hardly moves as Earth rotates. At any time of night, in any season of the year,

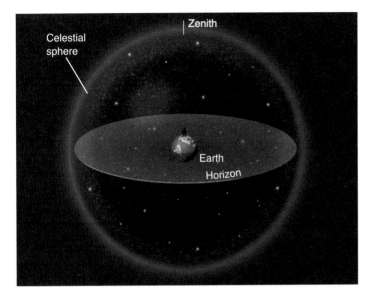

Figure 2-7
The horizon is the dividing line between the half of the celestial sphere we can see and the half we cannot see. The zenith is the point on the sky directly above our heads. Notice that people in different parts of the world have different horizons and zeniths.

Polaris stands above our northern horizon and is consequently known as the North Star. (Locate Polaris on the star charts at the end of the book.) Obviously, the name Polaris is associated with the location of this star. The Pawnee Indian name for Polaris means "Star That Does Not Walk Around."

The **south celestial pole** is the corresponding point directly above Earth's South Pole. No easily visible star marks the south celestial pole. The south celestial pole lies below the southern horizon as seen from the United States. Residents of the United States can see stars in the southern sky following curved paths westward as they circle the south celestial pole (Figure 2-8d).

Earth's rotation defines the celestial poles and the cardinal points around the horizon. The **north point** on the horizon is the point directly below

Figure 2-8
Motion in the sky. (a) Stars rising in the northeast make streaks in this time-exposure photograph taken from a road near an observatory dome. *(NOAO)* (b) Looking north, we would see stars circling the north celestial pole. (c) In the eastern sky, stars rise along straight lines, and (d) in the southern sky, the stars follow arcs as they circle the south celestial pole hidden below the southern horizon.

the north celestial pole, and the **south point** is directly opposite the north point. The **east point** and the **west point** are halfway between the north and south points. Thus the directions we use every day are defined astronomically.

The **celestial equator** is an imaginary line directly above Earth's equator. It touches the horizon at the east point and the west point and divides the celestial sphere into a two equal hemispheres (Figure 2-9).

Angles on the Sky

One way we might use the celestial sphere is to describe the location of objects. Just as we might tell friends that we will meet them 10 kilometers north of Flagstaff, Arizona, we might want to tell our friends to look for an interesting star a certain distance north of the bright star Sirius. However, we can't measure distances in the sky in kilometers. Rather, we need to measure angles on the celestial sphere.

We measure angles in degrees, minutes of arc, and seconds of arc. There are 360° in a circle and 90° in a right angle. Each degree is divided into 60 **minutes of arc** (sometimes abbreviated 60′). If you view a 25¢ piece from the length of a football field, it is about 1 minute of arc in diameter. Each minute of arc is divided into 60 **seconds of arc** (sometimes abbreviated 60″). The dot on this letter i seen from the length of a football field is about 1 second of arc in diameter.

Astronomers use angles to describe distance across the sky. They might say, for instance, that the **angular distance** between the moon and the bright star Spica is 8°, meaning that if we point one arm at the moon and the other arm at Spica, the angle between our arms is 8° (Figure 2-10). We know the star is hundreds of light-years away, and we know the moon is much closer, so the true distance between them is immense. But if we imagine them painted on the celestial sphere, we can think of their separation as an "angle on the sky." Thus we can discuss the angular distance between two objects even when we don't know their true distance from each other.

The **angular diameter** of an object is the angular distance across the object. For example, if we point at the left edge of the full moon with one arm and the right edge with the other arm, the angle between our arms is the angular diameter of the moon, about 0.5°. The sun also has an angular diameter of about 0.5°. The inset in Figure 2-10 shows how we can use our hand at arm's length to estimate angular distances.

Being able to measure angles on the sky is a great help as we discuss the appearance of the sky as seen from different locations on Earth.

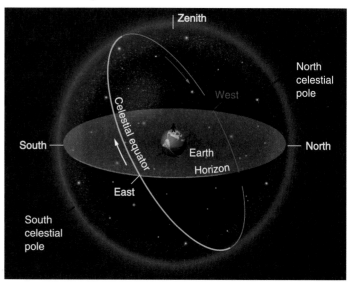

Figure 2-9
The celestial equator circles the celestial sphere directly above Earth's equator and touches the horizon at the east point and the west point (yellow). Earth's rotation carries us eastward and makes stars rise in the east and set in the west. Earth's axis of rotation (red) extended out to the celestial sphere defines the celestial poles (red), the pivots around which the sky appears to rotate.

Latitude and the Sky

Our discussion so far has assumed we live at a latitude representative of North America, that is, about 40° north of Earth's equator. If we live at a different latitude, then our zenith and horizon will be quite different. Notice in Figure 2-7 how someone in southern Africa would have a different zenith and horizon.

What we see in the sky depends on our latitude. If we live in Australia, the south celestial pole would be above our horizon, and the north celestial pole would be hidden below the northern horizon. To be precise, the angular distance from the horizon to the north celestial pole equals the latitude of the observer.

Figure 2-11 illustrates the effect of latitude on what we see in the sky by taking us on an imaginary journey southward from Earth's North Pole. We begin in the ice and snow at Earth's North Pole where our latitude is 90° N, and the north celestial pole is directly overhead, 90° from the horizon. As we walk southward, our latitude decreases, and the north celestial pole sinks closer to the northern horizon. As we cross Earth's equator, as shown in Figure 2-11d, our latitude is 0°, and the north celestial pole lies on our northern horizon. If we continue our journey south of Earth's equator, our latitude becomes negative, and the north celestial pole sinks below the northern horizon. That is, its angular distance "above" the horizon becomes negative.

This relationship between latitude and the distance from the horizon to the north celestial pole is a great

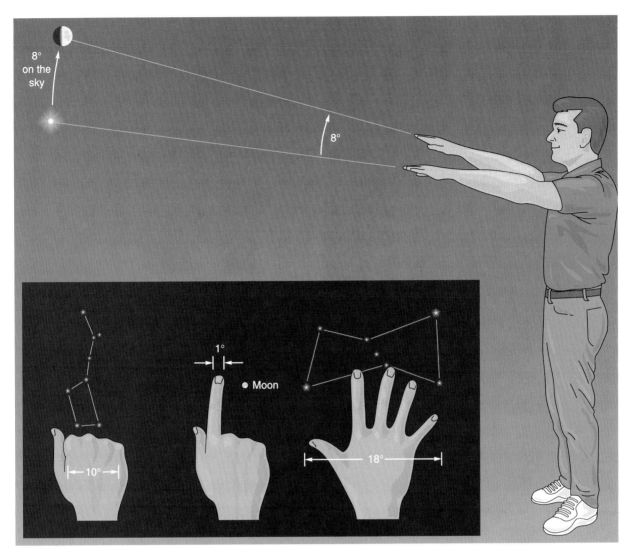

Figure 2-10
The angular separation between two objects is the angle your arms would make if you pointed to the two objects. Your hand held at arm's length makes a convenient measuring tool. Your fist is about 10° across, and your index finger is about 1° wide. Your spread fingers span about 18°.

aid to sailors. So long as we can see Polaris, the north star, we can find our latitude by measuring its distance above the northern horizon. (For precision, we would want to use navigation tables to correct for the fact that Polaris is a little less than a degree from the north celestial pole.) Of course, if we were voyaging through southern oceans, we could never see Polaris, and then we would have to observe the south celestial pole, which is more difficult because of the lack of a bright star as a marker.

Our latitude also determines what constellations we can see. From northern latitudes, constellations near the north celestial pole never set below the horizon. These **north circumpolar constellations,** such as Ursa Minor and Cassiopeia, are always above the hori-

zon (Figure 2-12). Constellations near the south celestial pole never rise above the horizon for an observer in Earth's northern hemisphere. These **south circumpolar constellations** are not visible from the United States and similar latitudes. How many constellations are circumpolar depends on your latitude; if you live on Earth's equator, you have no circumpolar constellations at all.

Many of the technical terms that describe the celestial sphere are not common in our daily lives, but we do not learn them here for their own sake. Merely naming things is not the same as understanding (Window on Science 2-2). Our goal in science is to understand nature, so we learn these new terms so we can discuss nature without confusion. Now that we know some basic vocabulary we are ready to analyze the motions of Earth and heavenly bodies.

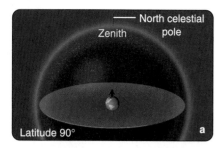

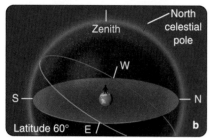

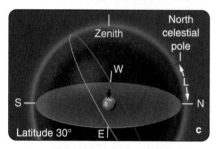

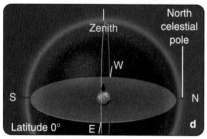

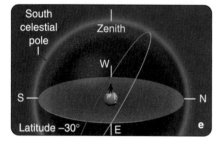

Figure 2-11

A journey beginning at Earth's North Pole (a) shows how our latitude L determines what we see in the sky. As we travel south, our latitude decreases, and the north celestial pole moves down toward the northern horizon. The angular distance from the northern horizon up to the north celestial pole always equals our latitude. When we cross Earth's equator, our latitude becomes negative, the north celestial pole sinks below the horizon, and the south celestial pole rises above our southern horizon.

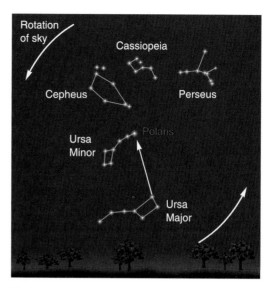

Figure 2-12

The brighter north circumpolar constellations visible from the latitude of the United States never set. The pointer stars at the front of the cup of the Big Dipper always point toward Polaris.

Precession

Over 2000 years ago, Hipparchus compared a few of his star positions with those made nearly two centuries before and realized that the celestial poles and equator were slowly moving across the sky. Later astronomers understood that this motion is caused by Earth's top-like motion.

If you have ever played with a gyroscope or top, you have seen how the spinning mass resists any change in the direction of its axis of rotation. The more massive the top and the more rapidly it spins, the more difficult it is to change the direction of its axis of rotation. But you probably recall that the axis of even the most rapidly spinning top sweeps around in a conical motion. The weight of the top tends to make it tip over, and this combines with its rapid rotation to make its axis sweep around in a conical motion called **precession** (Figure 2-13a).

Earth spins like a giant top, but it does not spin upright in its orbit; it is inclined 23.5° from vertical. Earth's large mass and rapid rotation keep its axis of rotation pointed toward a spot near the star Polaris, and the axis would not wander if Earth were a perfect sphere. However, Earth, because of its rotation, has a slight bulge around its middle, and the gravity of the sun and moon pull on this bulge, tending to twist Earth upright in its orbit. The combination of these forces and Earth's rotation causes Earth's axis to precess in a conical motion, taking about 26,000 years for one cycle (Figure 2-13b).

Because the celestial poles and equator are defined by Earth's rotational axis, precession moves these reference marks. We notice no change at all from night

One of the fascinations of science is that it can reveal things we normally do not sense, such as a slow drift in the direction of Earth's axis. Science can take our imagination into realms beyond our experience, from the inside of an atom to the inside of a star. Because these experiences are so unfamiliar, we need a vocabulary of technical terms just to talk about them, and that leads to a common confusion between naming things and understanding things.

The first step in understanding something is naming the parts, but it is only the beginning. The scientist's true goal is not naming but understanding. Thus, biolo-gists studying plants go beyond memorizing the names of the flowers' colors. They search for evidence to test their hypothesis that specific plants have developed specific flower colors to attract specific kinds of insects for efficient pollination. Naming the colors is only the beginning.

Some people complain about jargon in science, but the terminology is unfamiliar only because it describes phenomena we don't notice in everyday life. We don't learn this specialized vocabulary because we want to be pretentious or confusing. We learn proper terminology so we can discuss nature with precision.

We must know the proper words, but real understanding comes from being able to use that vocabulary to discuss the way nature works. We must be able to tell the stories that scientists call theories and hy-potheses. We must be able to use technical terms precisely to cite the evidence that gives us confidence that stories are true. That goes far beyond merely memorizing terminology; that is real understanding.

In folklore, naming a thing gives us power over it—recall the story of Rumpel-stiltskin. In science, naming things is the first step toward understanding.

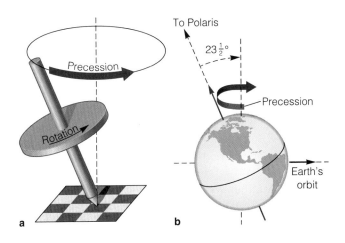

a   b

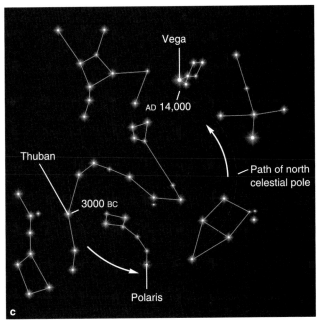

c

to night or year to year, but precise measurements reveal the precessional motion of the celestial poles and equator.

Over centuries, precession has dramatic effects. Egyptian records show that 4800 years ago the north celestial pole was near the star Thuban (α Draconis). The pole is now approaching Polaris and will be closest to it about 2100. In about 12,000 years, the pole will have moved to within 5° of Vega (α Lyrae). Figure 2-13c shows the path followed by the north celestial pole.

REVIEW Critical Inquiry

Does everyone see the same circumpolar constellations?

Here we can use the celestial sphere as a convenient model of the sky (Figure 2-9). A circumpolar constellation is one that does not set or rise. Which constellations are circumpolar depends on our latitude. If we live on Earth's equator, we see all the constellations rising and setting, and there are no circumpolar constellations at all. If we live at Earth's North Pole, all the constellations north of the celestial equator never set, and all the constellations south of the celestial equator never rise. In that case, every constellation is circumpolar.

Figure 2-13
Precession. (a) A spinning top precesses in a conical motion around the perpendicular to the floor because its weight tends to make it fall over. (b) Earth precesses around the perpendicular to its orbit because the gravity of the sun and moon tend to twist it upright. (c) Precession causes the north celestial pole to drift among the stars, completing a circle in 26,000 years.

At intermediate latitudes, the circumpolar regions are caps whose angular radius equals the latitude of the observer. If we live in Iceland, the caps are very large, and if we live in Egypt, near the equator, the caps are much smaller.

Locate Ursa Major and Orion on the star charts at the end of this book. For people in Canada, Ursa Major is circumpolar, but people in Mexico see most of this constellation slip below the horizon. From much of the United States, some of the stars of Ursa Major set, and some do not. In contrast, Orion rises and sets as seen from nearly everywhere on Earth. Explorers at Earth's poles, however, never see Orion rise or set. How would you improve the definition of a circumpolar constellation to clarify the status of Ursa Major? Would your definition help in the case of Orion?

The motions of the celestial sphere may seem to have little to do with our lives, but at the end of this chapter we will see how precession may be one of the causes of the ice ages. Before we can think about ice ages, however, we must consider Earth's orbital motion around the sun and the resulting apparent motion of the sun.

2-3 The Cycles of the Sun

Perhaps the most obvious cycles in the sky are those of the sun, but those apparent motions are produced by Earth's rotation and revolution. **Rotation** is the turning of a body about an axis through its center. Thus we say that Earth rotates on its axis once a day. **Revolution** is the motion of a body around a point located outside the body, and we say that Earth revolves around the sun once a year.

The cycle of day and night is caused by the rotation of Earth on its axis. In Figure 2-14 we see that four people in different places on Earth have different times of day, and that Earth's rotation carries us across the daylight side and then across the night side of Earth, producing the cycle of day and night.

Earth's annual revolution around the sun produces a cycle in the sky that we experience as the seasons. To understand that motion, we must imagine that we can make the sun fainter.

The Annual Motion of the Sun

If the sun were fainter, it would not illuminate the atmosphere and prevent us from seeing the stars, and we would notice that day by day the sun was moving slowly eastward against the background of stars. This motion is caused by Earth's motion as it revolves around the

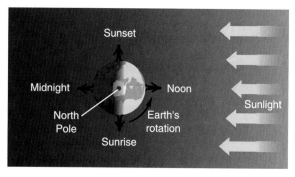

Figure 2-14
Looking down on Earth from above the North Pole shows how the time of day or night depends on our location on Earth.

sun. In Figure 2-15, we can see how the sun would appear in front of the stars of the constellation Sagittarius on January 1, and how Earth's motion along its orbit would cause the sun to appear to move eastward against the constellations. By March 1 the sun appears to be in Aquarius.

Of course, it is not quite correct to say that the sun is "in Aquarius." The sun is only 93 million miles away, and the stars of Aquarius are at least a million times further away. But in March of each year, we would see the sun against the background of the stars in Aquarius, and thus we could say, "The sun is in Aquarius."

If we continue watching the sun against the background of stars throughout the year, we could plot its path on a star chart. After one full year, we would see the sun begin to retrace this line as it continued its annual cycle of motion around the sky. This line, the apparent path of the sun around the sky, is called the **ecliptic.** Another way to define the ecliptic is to say it is the projection of Earth's orbit on the sky. If the sky were a great screen illuminated by the sun at the center, then the shadow cast by Earth's orbit would be the ecliptic. Yet a third way to define the ecliptic is to refer to it as the plane of Earth's orbit. These three definitions of the ecliptic are equivalent, and it is worth considering them all because the ecliptic is one of the most important reference lines on the sky. We will use it, for instance, to discuss the seasons.

The eastward motion of the sun along the ecliptic is a consequence of Earth's moving around its orbit. Earth completely circles the sun in 365.25 days, and consequently the sun circles the sky along the ecliptic in the same number of days. This means the sun travels 360° around the ecliptic in 365.25 days, which is about 1° eastward each day. The sun is about 0.5° in angular diameter, so it travels twice its own angular diameter each day.

The sun's annual motion around the sky is complicated by the fact that Earth rotates about an axis that is tipped 23.5° from the perpendicular to its orbit. Like a spinning top, it holds its axis fixed in space as it revolves around the sun (Figure 2-16). We saw in a previ-

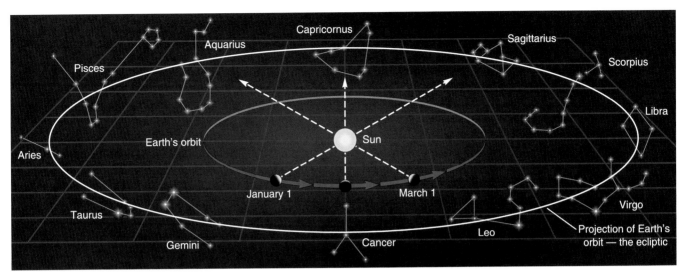

Figure 2-15
Earth's motion around the sun makes the sun appear to move against the background of the stars. Earth's circular orbit is thus projected on the sky as the circular path of the sun, the ecliptic.

ous section that Earth's axis always points toward a spot on the sky very near Polaris, the North Star. Although precession causes Earth's axis to drift, the change is very slow and does not alter the angle between the axis of rotation and the perpendicular.

Because Earth is tipped 23.5° in its orbit, the ecliptic is tipped 23.5° from the celestial equator. Recall that the celestial equator is the projection of Earth's equator and that the ecliptic is the projection of Earth's orbit. Because Earth is tipped 23.5°, its equator is tipped 23.5° from the plane of its orbit. When we project these two lines onto the sky, we find that the ecliptic and celestial equator meet at an angle of 23.5°. (Find the celestial equator and ecliptic on the star charts at the end of this book.)

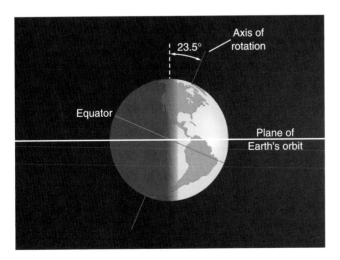

Figure 2-16
Earth's axis is inclined 23.5° from the perpendicular to the plane of its orbit. As it moves along its orbit, its axis of rotation remains fixed, pointing at the North Star.

The ecliptic and the celestial sphere give us four more reference marks on the sky. As we see in Figure 2-17, the ecliptic and celestial equator cross at two places called equinoxes. (Notice that they cross at an angle of 23.5°.) The **vernal equinox** is the place where the sun crosses the celestial equator moving northward, and the **autumnal equinox** is the place where it crosses moving southward. The sun crosses the vernal equinox on or about March 21, and it crosses the autumnal equinox on or about September 22. The exact dates of the equinoxes can vary by a day or two because of leap year and other factors. Figure 2-17 can help us identify two other reference marks on the ecliptic, the two points where the sun is farthest from the celestial equator. About June 22, the sun is farthest north at the point called the **summer solstice.** The **winter solstice** is the point where the sun is farthest south, about December 22.

Note that the equinoxes and solstices are points on the sky, but the same words refer to the times when the sun crosses those points. You might hear someone say, "This year the vernal equinox occurs at 3:02 AM on March 21." Whether we think of them as places or times, the equinoxes and solstices are important because they mark the beginning of each of the seasons.

The Seasons

The seasonal temperature depends on the amount of heat we receive from the sun. The motion of the sun around the ecliptic tips the balance between the amount of heat a particular place on Earth gains and the amount that is radiated to space. The balance is tipped one way in summer and the opposite way in winter. Because the ecliptic is at an angle to the celestial equator (Figure 2-17), the sun spends half the year in the northern

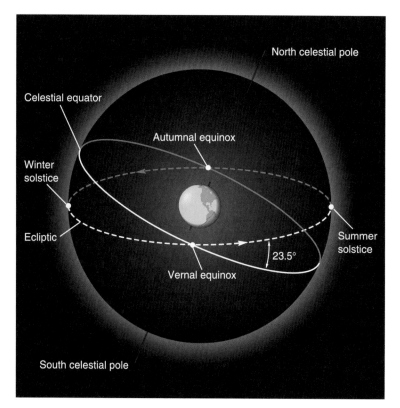

Figure 2-17
The ecliptic crosses the celestial equator at the equinoxes and reaches its most northerly point at the summer solstice and its most southerly point at the winter solstice.

above our horizon longer in summer, we receive more energy each day.

Second, the sun stands high in the sky at noon on a summer day. It shines almost straight down, as shown by our small shadows. On a winter day, however, the noon sun is low in the southern sky. Each square meter of ground gains little heat from the winter sun because the sunlight strikes the ground at an angle and spreads out (Figure 2-19). We can see that the winter sunlight is spread out by looking at our long shadows, as shown in Figure 2-18. These two effects work together to tip the heat balance and produce the seasons.

We mark the beginning of the seasons by the position of the sun. Spring begins at the moment the sun crosses the celestial equator going north (the vernal equinox). Summer begins at the moment the sun reaches its most northerly point (the summer solstice), and autumn begins when the sun crosses the celestial equator going south

celestial hemisphere and half the year in the southern celestial hemisphere. When the sun is in the northern sky, Earth's northern half receives more direct sunlight—and therefore more heat—than the southern half (Figure 2-16). This makes North American, Europe, and Asia warmer.

The seasons are reversed in Earth's southern half. While the sun is in the northern celestial hemisphere warming North America, South America becomes cooler. Chile has warm weather on New Year's Day and cold in July.

To see how the sun can give us more heat in summer, think about the path the sun takes across the sky between sunrise and sunset. Figure 2-18 shows these paths when the sun is at the summer solstice and at the winter solstice as seen by a person living at latitude 40°, a good average latitude for the United States. Notice that at the summer solstice, the sun rises in the northeast, moves high across the sky, and sets in the northwest (Figure 2-18a). But at the winter solstice, the sun rises in the southeast, moves low across the sky, and sets in the southwest (Figure 2-18b). Two features of these paths tip the heat balance.

First, the summer sun is above the horizon for more hours of each day than the winter sun. Summer days are long, and winter days are short. Because the sun is

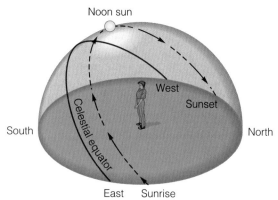

a At summer solstice

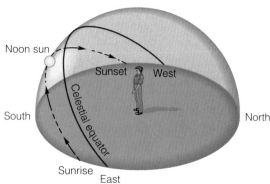

b At winter solstice

Figure 2-18
The daily paths of the sun at latitude 40°. Notice the person's shadows. (a) The summer sun is above the horizon for more than 12 hours and stands high in the sky at noon. (b) The winter sun is above the horizon for less than 12 hours and never stands high in the sky, even at noon.

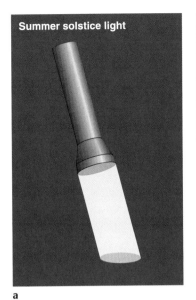
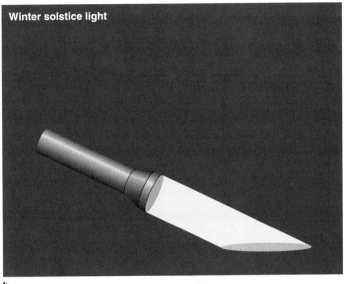

Figure 2-19
The spreading of light. (a) At noon on the day of the summer solstice, the sun shines from nearly overhead at the average latitude of the United States. Like the light from a flashlight shining nearly straight downward, the summer sunlight is not spread out very much. (b) On the day of the winter solstice, sunlight strikes the ground at a steep angle and spreads out. Thus, the ground receives less energy per square meter from the winter sun than from the summer sun.

(the autumnal equinox). We mark the official beginning of winter when the sun reaches its most southerly position (the winter solstice).

Of course, the weather does not turn warm the instant spring begins. The ground, air, and oceans are still cool from winter, and they take a while to warm up. Likewise, in the autumn, the ground, air, and oceans slowly release the heat stored through the summer. Due to this thermal lag, the average daily temperatures lag behind the solstices by about one month. Although the sun crosses the summer solstice on about June 22, the hottest months at northern latitudes are July and August. The coldest months are January and February, even though the sun passes the winter solstice earlier, about December 22.

The seasons are not related to variations in Earth–sun distance. The average distance from Earth to the sun is one astronomical unit (1 AU), 1.5×10^8 km. Earth's orbit is slightly elliptical, however, so its distance varies. About January 4, Earth reaches **perihelion,** its closest point to the sun; it reaches **aphelion,** its farthest point, about July 4. The total variation is only about 3 percent, and it has only a tiny influence on the seasons.

To summarize, notice that Earth's axis remains fixed in space pointing at Polaris as it circles the sun, and that its distance from the sun does not change significantly. Rather, the seasons occur because the Earth is inclined toward the sun on one side of Earth's orbit and away from the sun on the opposite side (Figure 2-20).

In ancient times, the solstices and equinoxes were celebrated with rituals and festivals. Shakespeare's play *A Midsummer Night's Dream* describes the enchantment of the summer solstice night. (In Shakespeare's

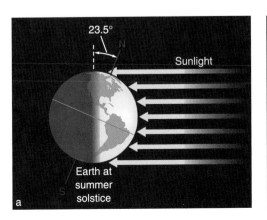
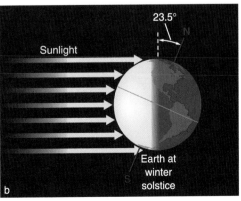

Figure 2-20
(a) In late June, at the time of the summer solstice, northern latitudes receive sunlight from nearly overhead, and southern latitudes receive sunlight at a steep angle. (b) In late December, on the other side of Earth's orbit, Earth is at winter solstice and receives less energy in the northern hemisphere and more energy in the southern hemisphere.

time, the equinoxes and solstices were taken to mark the midpoint of the seasons.) Many North American Indians marked the summer solstice with ceremonies and dances. Early church officials placed Christmas day in late December to coincide with an earlier pagan celebration of the winter solstice.

REVIEW Critical Inquiry

If Earth had a significantly elliptical orbit, how would its seasons be different?

Suppose Earth had an elliptical orbit so that perihelion occurred in July and aphelion in January. At perihelion, Earth would be closer to the sun, and the entire surface of Earth would be a bit warmer. If that happened in July, it would be summer in the northern hemisphere and winter in the southern hemisphere, and both would be warmer than they now are. It could be a dreadfully hot summer in Canada, and it might not snow at all in southern Argentina. Six months later, aphelion would occur in January, which means Earth would be farther from the sun. Winter in northern latitudes would be frigid, and summers in Argentina would be cool.

Of course, this doesn't happen. Earth's orbit is nearly circular, and the seasons are caused not by a variation in the distance of Earth from the sun but by the inclination of Earth in its orbit.

Nevertheless, Earth's orbit is slightly elliptical. Earth passes perihelion about January 4th and aphelion about July 4th. Although Earth's oceans tend to store heat and reduce the importance of this effect, this very slight variation in distance does affect the seasons. Does it make your winters warmer or cooler?

The sun is not the only body that moves along the ecliptic. The moon and planets maintain constant traffic along the sun's highway.

2-4 The Motion of the Planets

The ecliptic is important to us because it is the path of the sun around the sky, and that motion gives rise to the seasons. In a later chapter, we will consider the moon's cycle around the ecliptic, but here it is worthwhile to mention the motion of the planets. Not only are most of the planets visible to the naked eye, but their motion is the basis for one of the longest-living superstitions in human history—that our fate is written in the stars.

The Moving Planets

Most of the planets of our solar system are visible to the unaided eye, though they produce no light of their own.

We see them by reflected sunlight. Mercury, Venus, Mars, Jupiter, and Saturn are all visible to the naked eye, but Uranus is usually too faint to be seen (magnitude 5.6 at its brightest) and Neptune is never bright enough. Pluto is even fainter, and we need a large telescope to find it. Although Uranus, Neptune, and Pluto are usually too faint to see, their motions are the same as those of the other planets.

All the planets of our solar system move in nearly circular orbits around the sun. If we were looking down on the solar system from the north celestial pole, we would see the planets moving in the same counterclockwise direction around their orbits (Chapter 1). The farther they are from the sun, the more slowly the planets move.

When we look for planets in the sky, we always find them near the ecliptic, because their orbits lie in nearly the same plane as the orbit of Earth. As they orbit the sun, they appear to move generally eastward along the ecliptic.* In fact, the word *planet* comes from a Greek word meaning "wanderer." Mars moves completely around the ecliptic in slightly less than two years, but Saturn, being farther from the sun, takes nearly 30 years.

As seen from Earth, Venus and Mercury can never move far from the sun because their orbits are inside Earth's orbit. They sometimes appear near the western horizon just after sunset or near the eastern horizon just before sunrise. Venus is easier to locate because its larger orbit carries it higher above the horizon than Mercury (Figure 2-21). Mercury's orbit is so small that it can never get farther than about 28° from the sun. Consequently, it is usually hard to see against the sun's glare and is often hidden in the clouds and haze near the horizon. At certain times when it is farthest from the sun, however, Mercury shines brightly and can be located near the horizon in the evening or morning sky. (See the Appendix for the best times to observe Venus and Mercury.)

By tradition, any planet visible in the evening sky is called an **evening star,** although planets are not stars. Any planet visible in the sky shortly before sunrise is called a **morning star.** Perhaps the most beautiful is Venus, which can become as bright as minus fourth magnitude. As Venus moves around its orbit, it can dominate the western sky each evening for about half the year, but eventually its orbit carries it back toward the sun and it is lost in the haze near the horizon. In a few weeks it reappears in the dawn sky as a brilliant morning star.

Astrology

Seen from Earth, the planets move gradually eastward along the ecliptic, but they don't follow the ecliptic exactly. Also, each travels at its own pace and seems to

---

*We will discuss occasional exceptions to this eastward motion in Chapter 4.

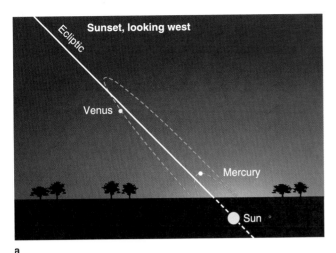

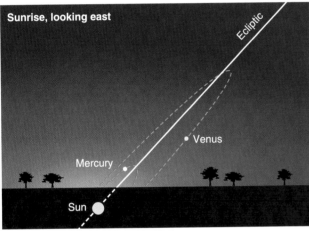

Figure 2-21
Mercury and Venus are sometimes visible in the western sky just after sunset (a) or in the eastern sky just before sunrise (b).

speed up and slow down at various times. To the ancients, this complex motion reflected the moods of the sky gods, and astrology was born.

Ancient astrologers defined a **zodiac,** a band 18° wide centered on the ecliptic, as the highway the planets follow. They divided this band into 12 segments named for the constellations along the ecliptic—the signs of the zodiac. A **horoscope** shows the location of the sun, moon, and planets among the zodiacal signs with respect to the horizon at the moment of a person's birth as seen from that longitude and latitude. Even if astrology worked, the generalized horoscopes published in newspapers and tabloids can't have been calculated accurately for the readers.

Astrology buffs argue that a person's personality, life history, and fate are revealed in his or her horoscope, but the evidence contradicts this belief. Astrology has been tested many times over the centuries, and it just doesn't work. Believers, however, don't give up on it. Thus, astrology is a superstition that depends on faith and not a science that depends on evidence (Window on Science 2-3).

One reason astronomers find astrology irritating is that it has no link to the physical world. For example, precession has moved the constellations so that they no longer match the zodiacal signs. Whatever sign you were "born under," the sun was probably in the previous zodiacal constellation. In fact, if you were born on or between November 30th and December 17th, the sun was passing through a corner of the nonzodiacal constellation Ophiuchus, and you have no official zodiacal sign.* Furthermore, there is no mechanism by which the planets could influence us. The gravitational field of a doctor who is delivering a baby is many times more powerful than the gravitational field of the planets.

Astrology makes sense only when we think of the world as the ancients did. They believed in multiple sky gods whose moods altered events on Earth. They believed in mystical influences between natural events and human events. The ancients did not understand natural forces such as gravity as the mechanisms that cause things to happen, and thus they did not believe in cause and effect as we do. If a house burned, they might conclude that it caught fire because it was cursed and not because someone was careless with an oil lamp.

Modern science left astrology behind centuries ago, but it survives as a fascinating part of human history—an early attempt to understand the meaning of the sky.

REVIEW Critical Inquiry

Planets like Mars can sometimes be seen rising in the east as the sun sets. Why is the same not true for Mercury or Venus?

Mars has an orbit outside the orbit of Earth, and that means it can reach a location opposite the sun in the sky. As the sun sets in the west, Mars rises in the east. But Mercury and Venus follow orbits that are smaller than the orbit of Earth, so they can never reach a point opposite the sun in the sky. As they follow their orbits around the sun, we on Earth see them gradually swing out on the east side of the sun, reach a maximum distance from the sun, swing back toward the sun, move out to the west side of the sun, reach a maximum distance west, and then move back toward the sun again. Because they never get very far from the sun in the sky, they can never be seen rising in the east as the sun sets.

The planetary motions we see in the sky are produced by the orbital motions of the planets around the sun. What would we see if the planets followed orbits that were all in exactly the same plane as the orbit of Earth?

---

*The author of this book was born on December 14th and thus has no astrological sign. An astronomer friend claims that the author must therefore have no personality.

Astronomers have a low opinion of astrology not so much because it is groundless but because it pretends to be a science. It is a pseudoscience, from the prefix *pseudo*, meaning false. There are many examples of pseudoscience, and it is illuminating to consider the difference between pseudoscience and science.

A pseudoscience is a set of beliefs that appear to be based on scientific ideas but that fail to obey the most basic rules of science. For example, some years ago a fad arose in which people placed objects under pyramids made of paper, plastic, wire, and so on. The claim was that the pyramidal shape would focus cosmic forces on anything inside and so preserve fruit, sharpen razor blades, and do other miraculous things. Many books promoted this idea, but simple experiments showed that any shape would protect a piece of fruit from airborne spores and allow it to dry without rotting. Likewise, any shape would allow oxidation to improve the cut-

ting edge of a razor blade. In short, experimental evidence contradicted the claim. Nevertheless, supporters of the theory declined to abandon or revise their claims. Thus, the fad of pyramid power was a pseudoscience.

One characteristic of a pseudoscience is that it appeals to our needs and desires. Thus, some pseudoscientific claims are self-fulfilling. For example, some people bought pyramidal tents to put over their beds and thus improve their rest. While there is no logical mechanism by which such a tent could affect a sleeper, because people wanted and expected the claim to be true they slept more soundly. Many pseudoscientific claims involve medical cures, ranging from copper bracelets and crystals to focus spiritual power to astonishingly expensive and illegal treatments for cancer. Logic is a stranger to pseudoscience, but human fears and needs are not.

Astrology is a pseudoscience. Over the centuries, astrology has been tested re-

peatedly, and no correlation has been found. But it survives, and its supporters disregard any evidence that it doesn't work. Like all pseudosciences, astrology is not open to revision in the face of contradictory evidence. Furthermore, astrology fulfills our human need to believe that there is order and meaning to our lives. It may comfort us to believe that our sweetheart has rejected us because of the motions of the planets rather than to admit that we behaved badly on our last date. Comfort aside, astrology is a poor basis for life decisions.

Human nature and human needs probably ensure that pseudoscientific beliefs will continue to plague us like emotional viruses propagating from person to person. But if we recognize them for what they are, we can more easily guide our lives by rational principles and not by giving credit for our successes and blame for our failings to the stars.

Modern astronomers have a low tolerance for astrological superstition, yet the motions of the heavenly bodies do affect our lives. The motion of the sun produces the seasons, and, as we will see in Chapter 3, the moon governs the tides. In addition, small changes in Earth's orbit may partially control global climate.

2-5 Astronomical Influences on Earth's Climate

Weather is what happens today; climate is the average of what happens over decades and centuries. We know that Earth has gone through past episodes, called ice ages, when the worldwide climate was cooler and dryer and thick layers of ice covered northern latitudes. The earliest known ice age occurred about 570 million years ago, and the next about 280 million years ago. The most recent ice age began only about 3 million years ago and is still going on. We are living in one of the periodic episodes when the glaciers melt and Earth grows slightly warmer. The current warm period began about 20,000 years ago.

Ice ages seem to occur with a period of roughly 250 million years, and cycles of glaciation within ice ages

occur with a period of about 40,000 years. Many scientists now believe that these cyclic changes have an astronomical origin.

The Hypothesis

Sometimes a theory or hypothesis is proposed long before scientists can find the critical evidence to test it. That situation happened in 1920 when Yugoslavian meteorologist Milutin Milankovitch proposed what became known as the **Milankovitch hypothesis**—that changes in the shape of Earth's orbit, in precession, and in inclination affect Earth's climate and trigger ice ages. We will examine each of these three motions in turn.

First, astronomers know that the elliptical shape of Earth's orbit varies slightly over a period of about 100,000 years. At present, Earth's orbit carries it 1.7 percent closer than average to the sun during northern-hemisphere winters and 1.7 percent farther away in northern-hemisphere summers. This makes the northern climate warmer, and that is critical—most of the land mass where ice can accumulate is in the northern hemisphere. If Earth's orbit became more elliptical, Milankovitch suggested, northern winters might be warm enough to prevent the accumulation of snow and ice from forming glaciers, and Earth's climate would warm.

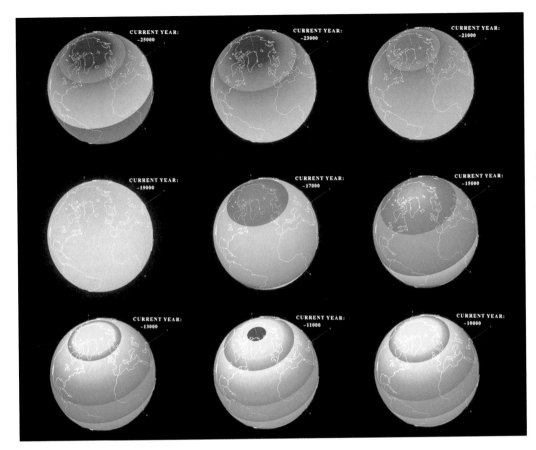

Figure 2-22
Color-coded globes show the importance of the Milanko-vich effect. These globes show Earth at the summer solstice from 25,000 years ago (upper left), during the last glaciation, to only 10,000 years ago, after the end of the glaciation. Red, yellow, and green regions receive more solar energy than average; blue and violet regions receive less. Changes in Earth's rotational axis and orbital shape affect the total amount of solar energy received and thus alter the climate. *(Courtesy Arizona State University, Computer Science and Geography Departments)*

A second factor is also at work. Precession causes Earth's axis to sweep around a cone with a period of about 26,000 years, and that changes the location of the seasons around Earth's orbit. Northern winters now occur when Earth is 1.7 percent closer to the sun, but in 13,000 years northern winters will occur on the other side of Earth's orbit where Earth is farther from the sun. Northern winters will be colder, and glaciers may grow.

The third factor is the inclination of Earth's equator to its orbit. Currently at 23.5°, this angle varies from 22° to 24° with a period of roughly 41,000 years. When the inclination is greater, seasons are more severe.

In 1920, Milankovitch proposed that these three factors cycled against each other to produce complex periodic variations in Earth's climate and the advance and retreat of glaciers (Figure 2-22). But no evidence was available to test the theory in 1920, and scientists treated it with skepticism. Many thought it was laughable.

The Evidence

By the middle 1970s, Earth scientists could collect the data that Milankovitch needed. Oceanographers could drill deep into the seafloor and collect samples, and geologists could determine the age of the samples from the natural radioactive atoms they contained. From all this, scientists constructed a history of ocean temperatures that convincingly matched the predictions of the Milankovitch hypothesis (Figure 2-23).

The evidence seemed very strong, and by the 1980s, the Milankovitch hypothesis was widely discussed as

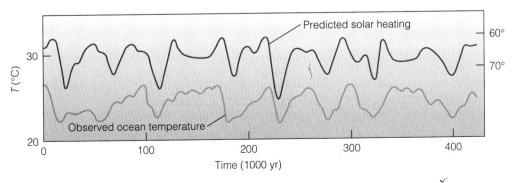

Figure 2-23
The Milankovich theory predicts periodic changes in solar heating (shown here as the equivalent summer latitude of the sun). Over the last 400,000 years, these changes seem to have varied in step with ocean temperatures measured from fossils in sediment layers. *(Adapted from Cesare Emiliani)*

Science is based on evidence. Every theory and conclusion must be supported by evidence obtained from experiments or from observation. If a theory is supported by many pieces of evidence but is clearly contradicted by a single experiment or observation, scientists quickly abandon it. For a theory to be true, there can be no contradictory evidence.

Of course, scientists argue about the significance of particular evidence and often disagree on the interpretation of evidence. Some observations may seem significant at first glance, but upon closer examination we may find that the procedure was flawed and so the piece of evidence is not important. Or we might conclude that the observational fact is correct but is being misinterpreted. The observation may not mean what it seems. Some of the most famous disagreements in science, such as those surrounding Galileo and Darwin, have arisen over the interpretation of well-established factual evidence.

Furthermore, scientists are not allowed to be selective in considering evidence. A lawyer in court can call a certain witness and intentionally fail to ask a critical question that would reveal evidence harmful to the lawyer's case. But a scientist may not ignore any known evidence. The difference in their methods is revealing. The lawyer is attempting to prove only one side of the case and rightly may ignore contradictory evidence. The scientist, however, is searching for the truth and so must test any theory against all available evidence. In a sense, the scientist, in dealing with evidence, must act as both the prosecution and the defense.

As you read about any science, look for the evidence in the form of measurements or observations. Every theory or conclusion should have supporting evidence. If you can find and understand the evidence, the science will make sense. All scientists, from astronomers to zoologists, demand evidence. You should, too.

---

the leading hypothesis. But science follows a mostly unstated set of rules that holds that a hypothesis must be tested over and over against all available evidence (Window on Science 2-4). In 1988, scientists discovered contradictory evidence.

While scuba diving in a water-filled crack in Nevada called Devil's Hole, scientists drilled out samples of calcite, a mineral that contains oxygen atoms. For 500,000 years, layers of calcite have built up in Devil's Hole, recording in their oxygen atoms the temperature of the atmosphere when rain fell there. Finding the ages of the mineral samples was difficult, but the results seemed to show that the previous ice age ended thousands of years too early to have been caused by Earth's motions.

These contradictory findings are irritating because we naturally prefer certainty, but such circumstances are common in science. The disagreement between ocean floor samples and Devil's Hole samples triggered a scramble to understand the problem. Were the ages of one or the other set of samples wrong? Were the ancient temperatures wrong? Or were scientists misunderstanding the significance of the evidence?

In 1997, a new study of the ages of the samples confirmed that those from the ocean floor are correctly dated. This seems to give scientists renewed confidence in the Milankovitch hypothesis. But the same study found that the ages of the Devil's Hole samples are also correct. Many now believe the temperatures at Devil's Hole tell us about local climate changes in the region that became the southwestern United States. The ocean floor samples, which agree with the Milankovitch hypothesis, seem to tell us about global climate. Thus Milutin Milan-kovitch's hypothesis, first proposed in 1920, is still being tested as we try to understand the world we live on.

REVIEW Critical Inquiry

How do precession and the shape of Earth's orbit interact to affect the earth's climate?

One technique that is useful in the critical analysis of an idea is exaggeration. If we exaggerate the variation in the shape of Earth's orbit, we can see dramatically the influence of precession. At present, Earth reaches perihelion during winter in the northern hemisphere and aphelion during summer. The variation in distance is only about 1.7 percent, and that difference doesn't cause much change in the severity of the seasons. But if Earth's orbit were much more elliptical, then winter in the northern hemisphere would be much warmer, and summer would be much cooler.

Now we can see the importance of precession. As Earth's axis precesses, it points gradually in different directions, and the seasons occur at different places in Earth's orbit. In 13,000 years, northern winter will occur at aphelion, and if Earth's orbit were highly elliptical, northern winter would be terrible. Similarly, summer would occur at perihelion, and the heat would be awful. Such extremes might deposit large amounts of ice in the winter but then melt it away in the hot summer, thus preventing the accumulation of glaciers.

Continue this analysis by exaggeration. What effect would precession have if Earth's orbit were more circular?

This chapter is about the sky, but it has told us a great deal about Earth. We have discovered that Earth's orbital motion produces the apparent motion of the sun along the ecliptic, causes the seasons, and may even cause the cycle of the ice ages. As is often the case, astronomy helps us understand what we are and where we are. In this case, our study of the sky has helped us understand what it means to live on a planet. However, we have omitted one of the most dramatic objects in the sky—the moon. We will repair that omission in the next chapter.

Summary

Astronomers divide the sky into 88 areas called constellations. Although the ancient constellations originated in Greek mythology, the names are Latin. Even the modern constellations, added to fill in the areas between the ancient figures, have Latin names. The names of stars usually come from ancient Arabic, though modern astronomers often refer to a star by constellation and Greek letters assigned according to brightness within each constellation.

The magnitude system is the astronomers' brightness scale. First-magnitude stars are brighter than second-magnitude stars, which are brighter than third-magnitude stars, and so on. The magnitude we see when we look at a star in the sky is its apparent visual magnitude.

The celestial sphere is a model of the sky, carrying the celestial objects around Earth. Because Earth rotates eastward, the celestial sphere appears to rotate westward on its axis. The northern and southern celestial poles are the pivots on which the sky appears to rotate. The celestial equator, an imaginary line around the sky above Earth's equator, divides the sky in half.

Because Earth orbits the sun, the sun appears to move eastward around the sky following the ecliptic. Because the ecliptic is tipped 23.5° to the celestial equator, the sun spends half the year in the northern celestial hemisphere and half the year in the southern celestial hemisphere, producing the seasons. The seasons are reversed south of Earth's equator. That is, while the northern hemisphere is experiencing a warm season, the southern hemisphere is experiencing a cold season.

Of the nine planets in our solar system, Mercury, Venus, Mars, Jupiter, and Saturn are visible to the naked eye. Their orbital motion around the sun carries them along the zodiac, a band 18° wide centered on the ecliptic. The positions of the sun, the moon, and these five planets form the basis of astrology, an ancient superstition that originated in Babylonia about 1000 BC.

Earth's motion may change in ways that can affect the climate. Changes in orbital shape, in precession, and in axial tilt can alter the planet's heat balance and may be responsible for the ice ages and glacial periods.

New Terms

constellation

asterism

magnitude scale

apparent visual magnitude (m_v)

celestial sphere

scientific model

horizon

zenith

north celestial pole

south celestial pole

north point

south point

east point

west point

celestial equator

minute of arc

second of arc

angular distance

angular diameter

north circumpolar constellation

south circumpolar constellation

precession

rotation

revolution

ecliptic

vernal equinox

autumnal equinox

summer solstice

winter solstice

perihelion

aphelion

evening star

morning star

zodiac

horoscope

Milankovitch hypothesis

Review Questions

1. Why are most modern constellations composed of faint stars or located in the southern sky?

2. What does a star's Greek-letter designation tell us that its ancient Arabic name does not?

3. From your knowledge of star names and constellations, which of the following stars in each group is the brighter and which is the fainter? Explain your answers.

   a. α Ursae Majoris; θ Ursae Majoris

   b. λ Scorpii; β Pegasus

   c. β Telescopium; β Orionis

4. Give two reasons why the magnitude scale might be confusing.

5. Why do modern astronomers continue to use the celestial sphere when they know that stars are not all at the same distance?

6. How do we define the celestial poles and the celestial equator?

7. From what locations on Earth is the north celestial pole not visible? the south celestial pole? the celestial equator?

8. If Earth did not turn on its axis, could we still define an ecliptic? Why or why not?

9. Give two reasons why winter days are colder than summer days.

10. How do the seasons in Earth's southern hemisphere differ from those in the northern hemisphere?

11. Why don't the planets move exactly along the ecliptic?

12. Why can we be sure the astrological predictions printed in newspapers and magazines are not based on true calculations of the positions of celestial bodies in a horoscope?

13. Why should the eccentricity of Earth's orbit make winter in Earth's northern hemisphere different from winter in the southern hemisphere?

14. How might small changes in the inclination of Earth's axis to the plane of its orbit affect the growth of glaciers?

Discussion Questions

1. Have you thought of the sky as a ceiling? as a dome overhead? as a sphere around Earth? as a limitless void?

2. How would the seasons be different if Earth were inclined 90° instead of 23.5°? 0° instead of 23.5°?

Problems

1. If one star is 6.3 times brighter than another star, how many magnitudes brighter is it?

2. If one star is 40 times brighter than another star, how many magnitudes brighter is it?

3. If two stars differ by 7 magnitudes, what is their intensity ratio?

4. If two stars differ by 8.6 magnitudes, what is their intensity ratio?

5. If star A is third magnitude and star B is fifth magnitude, which is brighter and by what factor?

6. If star A is magnitude 4 and star B is magnitude 9.6, which is brighter and by what factor?

7. By what factor is the sun brighter than the full moon? (*Hint:* See Figure 2-6.)

8. What is the angular distance from the north celestial pole to the summer solstice? to the winter solstice?

9. As seen from your latitude, what is the angle between the north celestial pole and the northern horizon? between the southern horizon and the noon sun at the summer solstice?

10. Draw a diagram like that in Figure 2-18 to show the path of the sun across the sky at the time of the vernal equinox.

Critical Inquiries for the Web

1. Can you see the figure of a hunter in the constellation Orion? Orion's central location on the celestial sphere means that most cultures have noticed this familiar pattern of stars. Search the Internet for information on the mythologies associated with this star pattern. What stories do you find? Do the stories have common threads?

2. Would the stars of a familiar constellation such as Pegasus or Orion look the same if we lived on planets orbiting other stars? Find a table of distance data for the bright stars in a familiar constellation, and construct a diagram that visualizes the positions of these stars in space. (See Figure 2-3.)

3. Is astrology a science or a pseudoscience? The Internet is full of astrological Web sites. Read the Windows on Science essays for this chapter, and examine several astrology-related Web pages. Look for evidence that a scientific approach is being applied. What points can you make to indicate that the information presented has not been generated through scientific processes?

Exploring *The Sky*

1. As discussed in this chapter, Earth's rotation about its own axis gives us the impression that the whole sky rotates around the **north celestial pole** in a period of one day. This apparent motion of the **celestial sphere** is difficult to notice because it happens so slowly. However, *The Sky* makes it possible to simulate this motion at a pace that is easy to observe by using a feature called **Time Skip.** Observe and describe the apparent motion of the sky as you see it looking north, east, south, and west.
How to proceed: Set the **Time Skip Increment** (a dropdown menu on the **Time Skip Toolbar**) to 1 minute, and click on the **Go Forward** button to begin the simulation. View the sky from the four cardinal directions, due north, south, east, and west. (You'll find **Time Skip** under the **Tools** menu as well.)

2. In the course of a year, the sun appears to move through twelve constellations called the **zodiac.** Using *The Sky,*

a. Determine in which constellation of the zodiac the sun appears today (or some other chosen date).
How to proceed: Locate the sun on your screen using **Find** (remember the button on the toolbar or the keystroke **F**). To actually bring the sun into the center of your screen, click on the **center** button in the **Object Information** window. Then click on the **Constellation Lines** button as well as the **Constellation Boundaries** button located on the **Objects Toolbar.**

b. Determine the constellation the sun is in on June 21, December 21, March 21, and September 21. These are the approximate dates of the summer and winter solstices and vernal and autumnal equinoxes.
How to proceed: To set the date and time, click on the **Data** button, then on **Site Information.**

 Go to the Brooks/Cole Astronomy Resource Center (www. brookscole.com/astronomy) for critical thinking exercises, articles, and additional readings from InfoTrac College Edition, Brooks/Cole's online student library.

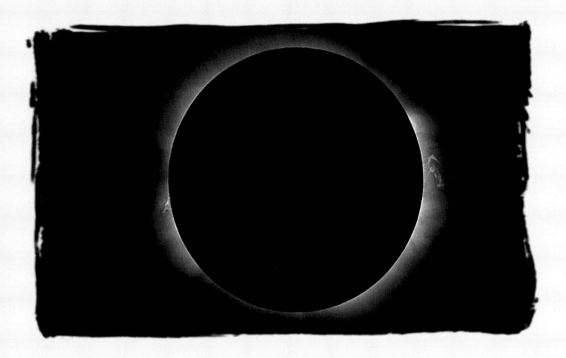

CHAPTER THREE

The Cycles of the Moon

Even a man who is pure in heart

And says his prayers by night

May become a wolf when the

wolfbane blooms

And the moon shines full and bright.

Proverb from old Wolfman movies

Guidepost

In the preceding chapter, we saw how the sun dominates our sky and determines the seasons. The moon is not as bright as the sun, but the moon passes through dramatic phases and occasionally participates in eclipses. The sun dominates the daytime sky, but the moon rules the night.

As we try to understand the appearance and motions of the moon in the sky, we discover that what we see is a product of light and shadow. To understand the appearance of the universe, we must understand light. Later chapters will show that much of astronomy hinges on the behavior of light.

In the next chapter, we will see how Renaissance astronomers found a new way to describe the appearance of the sky and the motions of the sun, moon, and planets.

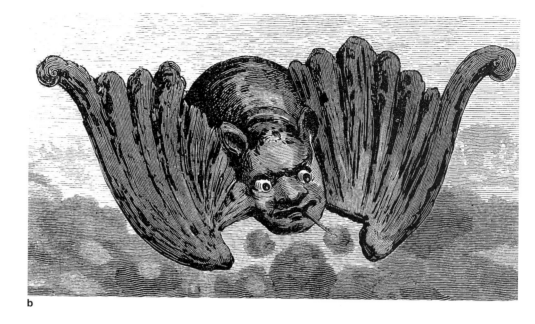

a

b

c

Figure 3-1

(a) A 12th-century Mayan symbol believed to represent a solar eclipse. The black-and-white sun symbol hangs from a rectangular sky symbol, and a voracious serpent approaches from below. (b) Chinese representation of a solar eclipse as a dragon flying in front of the sun. *(From the collection of Yerkes Observatory)* (c) A wall carving from the ruins of a temple at Vijayanagara in southern India. It symbolizes a solar eclipse as two snakes approaching the disk of the sun. *(T. Scott Smith)*

"Don't stare at the moon—you'll go crazy," more than one child has been warned.* The word *lunatic* comes from a time when even doctors thought that the insane were "moonstruck." A "mooncalf" is someone who has been crazy since birth, and the word is probably related to the belief that moonlight can harm unborn children. Everyone "knows" that people act less rationally when the moon is bright, but everyone is wrong. Careful studies of hospital and police records show that there is no real correlation between the moon and erratic behavior. The moon is so bright and its cycles through the sky are so dramatic we *expect* it to influence us, and many people are disappointed to learn the moon does not influence us. Some simply refuse to believe the evidence.

---

* When I was very small, my grandmother told me if I gazed at the moon, I might go crazy. But it was too beautiful, and I ignored her warning. I secretly watched the moon from my window, became fascinated by the sky, and became an astronomer.

The cycles of the moon are indeed dramatic. Many different cultures around the world explain eclipses of the sun as invisible monsters devouring the sun (Figure 3-1). A Chinese story tells of two astronomers Hsi and Ho who were too drunk to predict the solar eclipse of October 22, 2137 BC or perhaps failed to conduct the proper ceremonies to scare away the dragon snacking on the sun. When the emperor recovered from the terror of the eclipse, he had the two astronomers beheaded. The cycles of the moon provide a wealth of beauty in the sky and an important part of the cultural beliefs of Earth's peoples.

This chapter discusses the lunar cycles of phases, tides, and eclipses. Studying these events will help us understand what we see in the night sky. They will also introduce us to some of the most basic concepts of astronomy. We will discover how gravity produces tides, how light and shadow move through space, how the size and distance of an object affect what we see, and how cyclic events can be analyzed by searching for their

Isn't it weird that Isaac Newton is said to have "discovered" gravity in the late 17th century—as if people didn't have gravity before that, as if people in the 16th century floated around holding on to tree branches? Newton's accomplishment was that he realized that gravity was universal. The same force that makes an apple fall holds the moon in its orbit, guides the planets around the sun, and shapes the motion of the universe. We will discuss Newton's accomplishment in its proper historical place in Chapter 5, but here we need to recognize gravity for what it is, the master of the universe.

Newton realized that gravity was a property of matter, and consequently every object made of matter has to exert a gravitational attraction on every other object. Earth attracts the moon and the moon attracts Earth, but they also attract you and your book. Further, you and your book exert a gravitational force on each other and on every other object in the universe.

The size of an object's gravitational effect depends on the amount of matter it contains. A small amount of matter has a small gravitational effect, which is why you don't have satellites orbiting around you. Most objects in astronomy contain more matter, and their gravitational effects are larger. Moons, planets, stars, and galaxies each exert their gravitational attraction in proportion to the amount of matter they contain.

Of course, objects that are far away have less of an effect than objects that are nearby. That is another important point that Newton recognized. Earth's gravitational attraction presses us to our chairs, and the gravitational attraction of a distant galaxy containing much more matter than Earth is undetectably small because the galaxy is so far away.

The universe is filled with these gravitational influences, and every object in the universe moves under their guidance. The moon orbits Earth, and Earth orbits the sun. The sun orbits our galaxy, which moves under the influence of the gravitation of all the other galaxies in the universe. The universe is a swirling waltz of matter dancing to the music of gravity.

patterns. Most of all, this chapter forces us to begin thinking of our home as a world in space.

3-1 The Phases of the Moon

Starting this evening, look for the moon in the sky. If it is a cloudy night or if the moon is in the wrong part of its orbit, you may not see it, but keep trying on successive evenings, and within a week or two you will see the moon. Then watch for the moon on following evenings, and you will see it following its orbit around Earth and cycling through its phases as it has done for billions of years (Window on Science 3-1).

The Motion of the Moon

When we watch the moon night after night, we notice two things about its motion. First, we see it moving eastward against the background of stars; second, we notice that the markings on its face don't change. These two observations help us understand the motion of the moon and the origin of the moon's phases.

The moon moves rapidly among the constellations. If you watch the moon for just an hour, you can see it move eastward against the background of stars by slightly more than its angular diameter. In the previous chapter, we discovered that the moon is about 0.5° in angular diameter, so it moves eastward a bit more than 0.5° per hour. In 24 hours, it moves 13°. Each night when we look at the moon, we see it about 13° eastward of its location the night before. This eastward movement is the result of the motion of the moon along its orbit around Earth.

The moon orbits around Earth at an average distance of 384,400 km, about 30 times Earth's diameter (Figure 3-2a). The orbit is slightly elliptical, so its distance from Earth can vary by about 6 percent. Seen from far above Earth's North Pole, the moon orbits counterclockwise (eastward) with a period of 27.321661 days. This is called the **sidereal period,** meaning that it is measured in relation to the stars. The moon takes 27.321661 days to circle the sky once and return to the same place among the stars.

In its eastward motion, the moon stays near the ecliptic. Its orbit is tipped by 5°9′ to the plane of Earth's orbit, so its path around the sky is inclined to the ecliptic by the same angle. It can never wander farther than 5°9′ north or south of the ecliptic. Thus, it follows the zodiac around the sky.

Whenever we look at the moon, we see the same markings on its face because it rotates on its axis to keep one side facing Earth. If it did not rotate, that is, if it remained facing the same direction with respect to the stars, then we would see different sides of the moon in different parts of its orbit. But because the moon rotates, a mountain on the moon facing Earth will always face Earth as the moon follows its orbit (Figure 3-2b). Thus, we can talk about the near side of the moon (the side we see) and the back side of the moon (the side we can never see from Earth). We will discover later in this section why the moon rotates in this way.

Earth   Moon

a

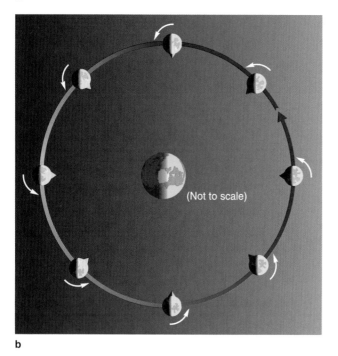

(Not to scale)

b

Figure 3-2

(a) The Earth–moon system to scale. The moon is about $\frac{1}{4}$ the diameter of Earth and about 30 Earth diameters away. (b) As the moon orbits Earth counterclockwise, as seen from the north, it rotates in the same direction and keeps the same side (marked by a mountain in this diagram) facing Earth.

The Phase Cycle

Because the moon, like the planets, does not produce visible light of its own, it is visible only by the sunlight it reflects, and we can see only that portion illuminated by the sun. As the moon moves around the sky, the sun illuminates different amounts of the side of the moon facing Earth, and so the moon passes through a sequence of phases.

The top panel of Figure 3-3 shows the phases of our moon. We can understand these by looking at the location of Earth's moon relative to the sun. For example, the *new moon* in the second panel of Figure 3-3 is located roughly between Earth and the sun, and the side of the moon facing Earth is in darkness. We see no moon at new moon. A few days after new moon, it has moved far enough along its orbit to allow the sun to illuminate a small sliver of the side toward us, and we see a thin crescent. Night by night this crescent moon waxes (grows), until, about a week later, we see half of the side toward us illuminated by sunlight and refer to it as *first quarter*. The moon continues to wax, becoming gibbous

(from the Latin for humpbacked), and then, when it is nearly opposite the sun, the side toward Earth is fully illuminated, and we see a *full moon.*

The second half of the lunar cycle reverses the first half. After reaching full, the moon wanes (shrinks) through gibbous phase to *third quarter,* then through crescent back to new moon. To distinguish between the gibbous and crescent phases of the first and second half of the cycle, we refer to *crescent waxing* and *gibbous waxing* when the moon is growing, and to *gibbous waning* and *crescent waning* when it is shrinking.

The cycle of lunar phases takes 29.53 days, the synodic period of the moon, or about four weeks. Thus new moon, first quarter, full moon, and third quarter occur at roughly one-week intervals. In general, an object's **synodic period** is its orbital period relative to the sun. (*Synodic* comes from the Greek words meaning "together" and "path.") To see why the moon's synodic period is longer than its sidereal period, imagine we begin observing at new moon—that is, when the moon is near the sun in the sky. After 27.322 days, the moon's sidereal period, it has circled the sky and returned to the same place among the stars where it was last new, but the sun has moved about 27° eastward along the ecliptic. The moon needs slightly more than two days to catch up with the sun and reach new moon again (Figure 3-4). Thus the moon's synodic period is longer than its sidereal period.

You can make a moon-phase dial from Figure 3-3 by covering the lower half of the moon's orbit with a sheet of paper, aligning the edge of the paper to pass through the word "Full" at the left and the word "New" at the right. Push a pin through the edge of the paper at Earth's North Pole to make a pivot, and under the word "Full," write on the paper, "Eastern Horizon." Under the word "New," write "Western Horizon." The paper now represents the horizon.

You can set your moon-phase dial for a given time by rotating the diagram behind the horizon-paper. Set the dial to sunset by turning the diagram until the human figure labeled "Sunset" is standing at the top of the Earth globe; the dial shows, for example, that the full moon at sunset would be at the eastern horizon. Compare the times of moonrise and moonset that you determine from your moon-phase dial with those given in Table 3-1.

Table 3-1 Times of Moonrise and Moonset

Phase	Moonrise	Moonset
New	Dawn	Sunset
First quarter	Noon	Midnight
Full	Sunset	Dawn
Third quarter	Midnight	Noon

As seen from Earth

New | Waxing crescent | First quarter | Waxing gibbous | Full | Waning gibbous | Third quarter | Waning crescent

a

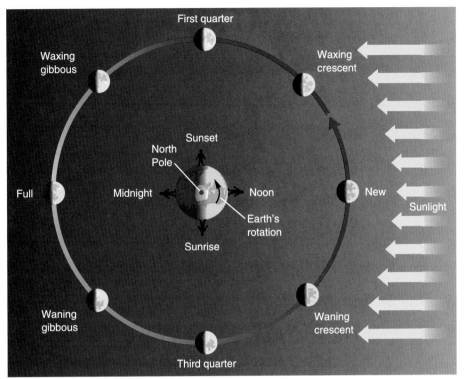

b

Figure 3-3
The phases of the moon are caused by the varying amounts of sunlight that reach the side of the moon facing Earth. At first quarter, for example, only half of the near side is illuminated. At full moon, all of the near side is illuminated. To locate a specific lunar phase in the sky at a given time of day or night, rotate the diagram until the corresponding human figure is at the top of the Earth globe; the upper half of the moon's orbit will show the location of the lunar phases above the horizon with east at the left and west at the right.

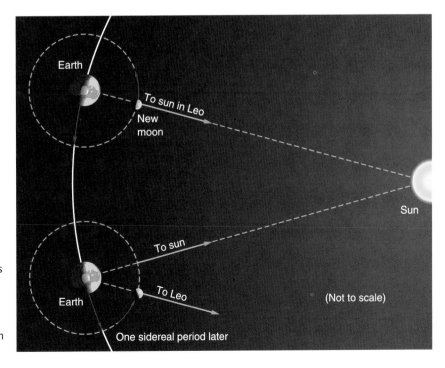

Figure 3-4
One sidereal period after new moon, the moon returns to the same place among the stars—the constellation Leo in this diagram—but the sun has moved eastward because of Earth's orbital motion. Thus, the moon must travel a bit over two more days to catch up with the sun and return to new. The period of the lunar phases (synodic period) is slightly more than two days longer than the moon's orbital period (sidereal period).

We can now follow the cycle of the moon around the sky, and we begin at sunset. If the moon were new, it would be invisible near the sun on the western horizon. A few days after new moon, we would see the waxing crescent moon moving upward in the sunset sky (Figure 3-5a); and a week after new moon, we would find the first-quarter moon standing high in the sky at sunset. As the days passed, we would see the moon wax gibbous as it moved eastward, reaching full moon at the eastern horizon about two weeks after new moon.

Lunar phases after full moon are not visible in the sunset sky, so we must shift to the sunrise sky to continue. At sunrise, the full moon would be at the western horizon, and it would wane through gibbous each day until it reached third quarter high in the sunrise sky about three weeks after new moon. Each morning we would see the moon farther eastward as it waned through crescent phase (Figure 3-5b) until it reached new moon again at the eastern horizon in the sunrise sky. New moon to new moon takes just over four weeks.

Almost everyone is familiar with the changing phases of the moon, but those who live near the seashore are probably familiar with another phenomenon related to the lunar cycle—the periodic advance and retreat of the ocean tides.

Figure 3-5
(a) When the moon is at waxing crescent phase, it is just above the western horizon soon after sunset. (b) When the moon is at waning crescent phase, it is above the eastern horizon just before dawn.

R E V I E W Critical Inquiry

Why do we sometimes see the moon in the daytime?

The full moon rises at sunset and sets at sunrise, so it is visible in the night sky but never in the daytime sky. But at other phases, it is possible to see the moon in the daytime. For example, when the moon is a waxing gibbous moon, it rises a few hours before sunset. If you look in the right spot in the sky, you can see it in the late afternoon in the southeast sky. It looks pale and washed out because of sunlight in our atmosphere, but it is quite visible once you notice it. You can also locate the waning gibbous moon in the morning sky. It sets a few hours after sunrise, so you would look for it in the southwestern sky in the morning.

If you look in the right place, you might even see the first- or third-quarter moon in the daytime sky, but you have probably never seen a crescent moon in the daytime. Why not? Where would you have to look?

The phases of the moon don't affect us directly—we don't act crazier than usual at full moon. But the moon does have an important influence on Earth.

3-2 The Tides

Anyone who lives near the sea is familiar with a dramatic lunar effect—the ebb and flow of the tides. These periodic changes in the ocean are caused by the moon's gravity.

The Cause of the Tides

We feel Earth's gravity drawing us downward with a force we refer to as our weight, but the moon also exerts a gravitational force on us. Because the moon is less massive and more distant, its gravitational force acting on an object on Earth's surface is only about 0.0003 percent of Earth's gravitational force. That is a tiny force, but it is enough to cause tides in Earth's oceans.

Tides are produced by a *difference* between the gravitational force acting on different parts of an object. The side of Earth facing the moon is about 6400 km (4000 miles) closer to the moon than is Earth's center, and the moon's gravity pulls more strongly on the oceans on the near side than on Earth's center. The difference is small, only about 3 percent of the moon's total gravitational force on Earth, but it is enough to make the ocean waters flow into a bulge on the side of Earth facing the moon.

A bulge also forms on the side of Earth facing away from the moon. Earth's far side is about 6400 km farther from the moon than is Earth's center, and the moon's gravity pulls on it less strongly than it does on Earth's center. Thus, relative to Earth's center, a small force makes the ocean waters on Earth's far side flow away from the moon (Figure 3-6).

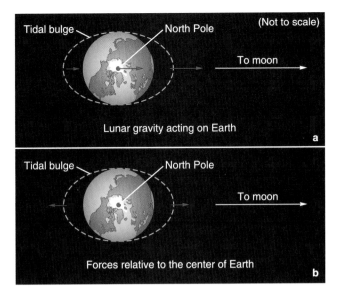

a

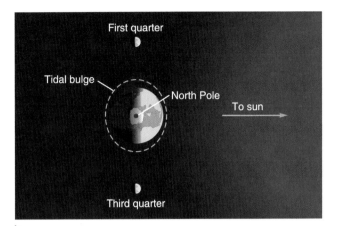

b

Figure 3-6
(a) Because one side of Earth is closer to the moon than the other side, the moon's gravity does not attract all parts of Earth with the same force. (b) Relative to the center of Earth, small differences in the forces cause tidal bulges (much exaggerated here) on the sides of Earth facing toward and away from the moon.

Figure 3-7
Looking down on Earth from above the North Pole, we can draw the shape of the tidal bulges (exaggerated here for clarity). (a) When the moon and sun pull along the same line, their tidal forces combine, and the tidal bulges are larger. (b) When the moon and sun pull at right angles, their tidal forces do not combine, and the tidal bulges are smaller.

We can see dramatic evidence of tidal forces if we watch the ocean shore for a few hours. Although Earth rotates on its axis, the tidal bulges remain fixed along the Earth–moon line. As the turning Earth carries us into a tidal bulge, the ocean water deepens, and the tide crawls up the beach. To be precise, the tide does not "come in." Rather, we are "carried into" the tidal bulge. Because there are two bulges on opposite sides of Earth, the tides rise and fall twice a day, and the times of high and low tide depend on the phase of the moon.

In reality, the tide cycle at any given location can be quite complex, depending on the latitude of the site, the shape of the shore, the north–south location of the moon, and so on. Tides in the Bay of Fundy occur twice a day and can exceed 12 m. The northern coast of the Gulf of Mexico has only one tidal cycle of roughly 30 cm each day.

The sun, too, produces tides on Earth. The sun is roughly 27 million times more massive than the moon, but it lies almost 400 times farther from Earth. Consequently, tides on Earth caused by the sun are only about half those caused by the moon. At new moon and full moon, the moon and sun produce tidal bulges that join together to cause extreme tidal changes (Figure 3-7a); high tide is very high, and low tide is very low. Known as **spring tides,** these occur at every new and full moon. **Neap tides** occur at first- and third-quarter moon, when the moon and sun pull at right angles to each other. Then the smaller tide caused by the sun slightly reduces the tide caused by the moon, and the resulting high and low tides are less extreme (Figure 3-7b).

Though the ocean has been used as an example, you should note that tides would occur even if Earth had no oceans. The difference in the gravitational forces acting on different parts of Earth causes slight bulges in the rocky shape of Earth itself. We do not feel the mountains and plains rising and falling by a few centimeters; the changes in the fluid oceans are much more obvious.

Tidal Effects

Tidal forces can have surprising effects on both rotation and orbital motion. The friction of Earth's ocean waters against the seabeds slows Earth's rotation, and our days are getting longer by 0.001 second per century. Some marine animals deposit layers in their shells in phase with the tides and the day–night cycle, and fossils of these shells confirm that only 400 million years ago Earth's day was 22 hours long. Tides are slowing Earth's rotation by about 0.002 second per day. This adds almost a second to the length of the year. You may have heard end-of-year news stories that astronomers at the U.S. Naval Observatory were adding a second to the length of the year. Tidal forces will make the 21st century almost a minute longer than the 20th.

In addition, Earth's gravitational field exerts tidal forces on the moon, and although there are no oceans on the moon, tides do flex its rocky bulk. The resulting friction in the rock has slowed the moon's rotation. It once rotated much faster, but the tides caused by Earth

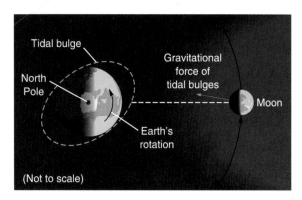

Figure 3-8
Earth's rotation drags the tidal bulges ahead of the Earth–moon line (exaggerated here). The gravitational attraction of these masses of water does not pull directly toward Earth's center but rather pulls the moon slightly forward in its orbit (green arrow), forcing its orbit to grow larger.

have slowed the moon until it now rotates to keep one side permanently facing Earth. It is thus tidally locked to Earth. We will discover that most of the moons in the solar system are tidally locked to their planets.

Tidal forces can also affect orbital motion. For example, friction with the rotating Earth drags the tidal bulges eastward out of a direct Earth–moon line (Figure 3-8). These tidal bulges contain a large amount of mass, and the gravitation of the bulge near the moon pulls the moon slightly forward in its orbit. The bulge on Earth's far side pulls the moon slightly backward in its orbit, but because it is farther from the moon, that bulge is less effective. Thus, the net effect is to drag the moon forward in its orbit. As a result, the moon's orbit is growing larger and the moon is receding from Earth at about 3.8 cm per year, an effect that astronomers can measure by bouncing laser beams off reflectors left on the lunar surface by the Apollo astronauts. Thus, we have clear evidence that tides are changing the moon's orbit.

REVIEW Critical Inquiry
Would Earth have tides if it had no moon?

Yes, there would be tides, but they would be much smaller. Don't forget that the sun also produces tides on Earth. The solar tides are less extreme than the lunar tides, but we would still see the advance and retreat of the oceans if the moon did not exist.

There would be an interesting difference, however, if the only tides were solar tides. The sun rises every 24 hours, but the moon moves rapidly eastward in the sky and rises, on average, every 24.8 hours. How many hours would separate high tides caused by the sun compared with high tides caused by the moon?

Tides and tidal forces are important in many areas of astronomy. In later chapters, we will see how tidal forces can pull gas away from stars, rip galaxies apart, and melt the interiors of satellites orbiting near massive planets. Now, however, we must consider yet another kind of lunar phenomenon—eclipses.

3-3 Lunar Eclipses

A lunar eclipse occurs at full moon when the moon moves through Earth's shadow. Because the moon shines only by reflected sunlight, it gradually darkens as it enters the shadow.

Earth's Shadow

Earth's shadow consists of two parts. The **umbra** is the region of total shadow. If we were floating in space in the umbra of Earth's shadow, we would see no portion

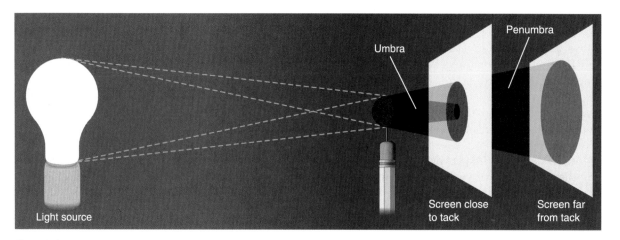

Figure 3-9
The shadows cast by a map tack resemble those of Earth and the moon. The umbra is the region of total shadow; the penumbra is the region of partial shadow.

Much of science is explained in diagrams; one particular type of diagram can be confusing. When artists draw three-dimensional diagrams on flat sheets of paper, they use lots of artistic clues that have been discovered over the centuries. Perspective, shading, color, and shadows help us see the three-dimensional figure if the drawing is familiar—a house, for example. But when a drawing shows something that is unfamiliar, as is often the case in science, the artist's clues don't work as well. When we look at drawings of a molecule, a nerve cell, layers of rock under the ocean, or Earth and its shadows, we must pay special attention to the three-dimensional nature of the figure.

When you see a three-dimensional diagram in any science book, it helps to decide what the point of view is. For example, if the diagram were a photograph, where was the camera? In Figure 3-10, the camera would have to have been in space looking back at Earth and the moon. Of course, astronauts have never been that far from Earth, but we can make such a voyage in our imagination, and that helps us understand the geometry of eclipses. Other three-dimensional diagrams have other points of view, so always be sure to first determine the point of view when you look at any three-dimensional diagram.

of the sun. However, if we moved into the **penumbra,** we would be in partial shadow and would see part of the sun peeking around Earth's edge. In the penumbra, the sunlight is dimmed but not extinguished.

We can construct a model of this by pressing a map tack into the eraser of a pencil and holding the tack between a lightbulb a few feet away and a white cardboard screen (Figure 3-9). The lightbulb represents the sun, and the map tack represents Earth. When we hold the screen close to the tack, we see that the umbra is nearly as large as the tack and that the penumbra is only slightly larger. However, as we move the screen away from the tack, the umbra shrinks and the penumbra expands. Beyond a certain point, the shadow has no dark core at all, indicating that the screen is beyond the end of the umbra.

The umbra of Earth's shadow is about 1.4 million km (860,000 miles) long and points directly away from the sun. A giant screen placed in the shadow at the average distance of the moon would reveal a dark umbra about 9000 km (5700 miles) in diameter, and the faint outer edges of the penumbra would mark a circle about 16,000 km (10,000 miles) in diameter. For comparison, the moon's diameter is only 3476 km (2160 miles). Thus, when the moon's orbit carries it through the umbra, it has plenty of room to become completely immersed in shadow.

Total Lunar Eclipses

A lunar eclipse occurs when the moon passes through Earth's shadow and grows dark. If the moon passes through the umbra and no part of the moon remains outside the umbra in the partial sunlight of the penumbra, we say the eclipse is a **total lunar eclipse.** Figure 3-10 shows the three dimensional arrangement of Earth, the moon, and shadows during a total lunar eclipse (Window on Science 3-2).

A total lunar eclipse occurs in stages as the moon, following its orbit, passes through the different parts of

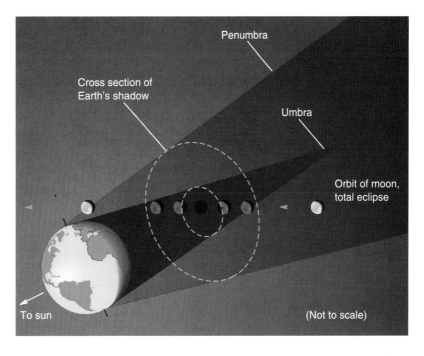

Figure 3-10
During a total lunar eclipse, the moon's orbit carries it through the penumbra and completely into the umbra. Compare the cross section of Earth's shadow with Figure 3-11.

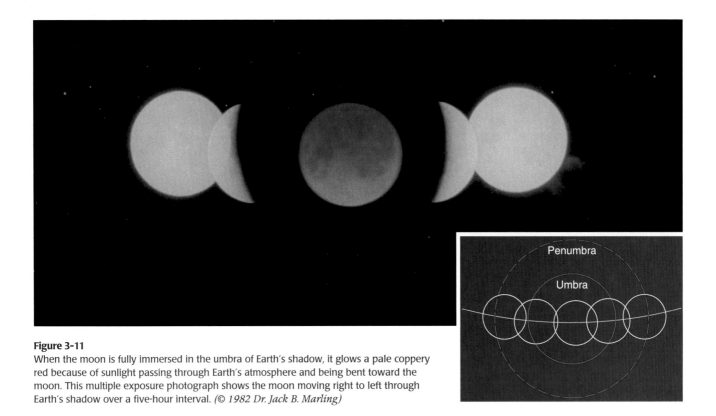

Figure 3-11
When the moon is fully immersed in the umbra of Earth's shadow, it glows a pale coppery red because of sunlight passing through Earth's atmosphere and being bent toward the moon. This multiple exposure photograph shows the moon moving right to left through Earth's shadow over a five-hour interval. *(© 1982 Dr. Jack B. Marling)*

Earth's shadow (Figure 3-11). As the moon begins to enter the penumbra it is only slightly dimmed, and a casual observer may not notice anything odd. After an hour, the moon is deeper in the penumbra and dimmer, and once it begins to enter the umbra, we see a dark bite on the edge of the lunar disk. The moon travels its own diameter in an hour, so it takes about an hour to enter the umbra completely and become totally eclipsed.

Even when the moon is totally eclipsed, it does not disappear completely. Sunlight, bent by Earth's atmosphere, leaks into the umbra and bathes the moon in a faint glow. Because blue light is scattered by Earth's atmosphere more easily than red light, it is red light that penetrates to illuminate the moon in a coppery glow (Figure 3-12). If we were on the moon during totality and looked back at Earth, we would not see any part of the sun because it would be entirely hidden behind Earth. However, we would see Earth's atmosphere illuminated from behind by the sun in a spectacular sunset completely ringing Earth. It is the red glow from this sunset that gives the totally eclipsed moon its reddish color.

How dim the totally eclipsed moon becomes depends on a number of things. If Earth's atmosphere is especially cloudy in those regions that must bend light into the umbra, the moon will be darker than usual. An unusual amount of dust in Earth's atmosphere (from volcanic eruptions, for instance) also causes a dark eclipse. Also, total lunar eclipses tend to be darkest when the moon's orbit carries it through the center of the umbra.

As the moon moves through Earth's umbral shadow, we can see that the shadow is circular. From this the Greek philosopher Aristotle (384–322 BC) concluded

Figure 3-12
During a total lunar eclipse, the moon turns coppery red as sunlight refracted by Earth's atmosphere illuminates the moon in a red, sunset glow. An astronaut on the moon during such an eclipse would see this red light coming from a sunset completely encircling Earth. *(Celestron International)*

that Earth had to be a sphere, because only a sphere could cast a shadow that was always circular.

Depending on the geometry of the eclipse, the moon can take as long as 1 hour 40 minutes to cross the umbra and another hour to emerge into the penumbra. Still another hour passes as it emerges into full sunlight. A total eclipse of the moon, including the penumbral stage, can take almost six hours from start to finish.

Partial and Penumbral Lunar Eclipses

Not all eclipses of the moon are total. Because the moon's orbit is inclined a bit over 5° to the plane of Earth's orbit, the moon might not pass through the center of the umbra.

If the moon's orbit carries the full moon too far north or south of the umbra, the moon may only partially enter the umbra. The resulting **partial lunar eclipse** is usually not as dramatic as a total lunar eclipse. Because part of the moon remains outside the umbra, it receives some sunlight and looks bright in contrast with the dark part of the moon inside the umbra. Unless the moon almost completely enters the umbra, the glare from the illuminated part of the moon drowns out the fainter red glow inside the umbra. Partial lunar eclipses are interesting because part of the full moon is darkened, but they are not as beautiful as a total lunar eclipse.

If the orbit of the moon carries the moon far enough north or south of the umbra, the moon may only pass through the penumbra and never reach the umbra. Such **penumbral eclipses** are not dramatic at all. In the partial shadow of the penumbra, the moon is only partially dimmed. Most people glancing at a penumbral eclipse would not notice any difference from a full moon.

Total, partial, or penumbral, lunar eclipses are interesting events in the night sky and are not difficult to observe. When the moon passes through Earth's shadow, the eclipse is visible from anywhere on Earth's dark side. One or two lunar eclipses occur in most years. Consult Table 3-2 to find the next lunar eclipse visible in your part of the world.

Table 3-2 Total and Partial Eclipses of the Moon, 2000 to 2011

Year	Date	Time* of Mideclipse (GMT)	Length of Totality (Hours:Min)	Length of Eclipse† (Hours:Min)
2000	Jan. 21	4:45	1:16	3:22
2000	July 16	13:57	1:46	3:56
2001	Jan. 9	20:22	1:00	3:16
2001	July 5	14:57	Partial	2:38
2003	May 16	3:41	0:52	3:14
2003	Nov. 9	1:20	0:22	3:30
2004	May 4	20:32	1:16	3:22
2004	Oct. 28	3:05	1:20	3:38
2005	Oct. 17	12:04	Partial	0:56
2006	Sept. 7	18:52	Partial	1:30
2007	Mar. 3	23:22	1:14	3:40
2007	Aug. 28	10:38	1:30	3:32
2008	Feb. 21	3:27	0:50	3:24
2008	Aug. 16	21:11	Partial	3:08
2009	Dec. 31	19:24	Partial	1:00
2010	June 26	11:40	Partial	2:42
2010	Dec. 21	8:18	1:12	3:28
2011	June 15	20:13	1:40	3:38
2011	Dec. 10	14:33	0:50	3:32

* Times are Greenwich Mean Time. Subtract 5 hours for Eastern Standard Time, 6 hours for Central Standard Time, 7 hours for Mountain Standard Time, and 8 hours for Pacific Standard Time. From your time zone, lunar eclipses that occur between sunset and sunrise will be visible, and those that occur at midnight will be best placed.

†Does not include penumbral phase.

a partial eclipse is almost total, so that only a small sliver of moon extends out of the shadow into sunlight, we can sometimes detect the red glow in the shadow.

Of course, this red glow does not happen for every planet–moon combination in the universe. Suppose a planet had a moon but no atmosphere. Would the moon glow red during a total eclipse? Why not?

Lunar eclipses are slow and stately. For drama and excitement, there is nothing like a solar eclipse.

REVIEW Critical Inquiry

Why doesn't Earth's shadow on the moon look red during a partial lunar eclipse?

During a partial lunar eclipse, part of the moon protrudes from Earth's umbral shadow into sunlight. This part of the moon is very bright compared to the fainter red light inside Earth's shadow, and the glare of the reflected sunlight makes it difficult to see the red glow. If

3-4 Solar Eclipses

A solar eclipse occurs when the moon moves between Earth and the sun. If the moon covers the disk of the sun completely, the eclipse is a **total solar eclipse.** If the moon covers only part of the sun, the eclipse is a **partial solar eclipse.** During a particular solar eclipse, people in one place on Earth may see a total eclipse,

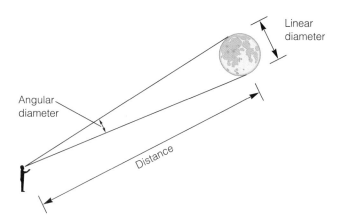

Linear
diameter

Angular
diameter

Distance

Figure 3-13
The small-angle formula relates angular diameter, linear diameter, and distance. Angular diameter is the angle formed by lines extending from our eye to opposite sides of the object—in this figure, the moon. Linear diameter and distance are typically measured in kilometers or meters.

while people only a few hundred kilometers away see only a partial eclipse.

These spectacular sights are possible because we on Earth are very lucky. Our moon has the same angular diameter as our sun, so it can cover the sun almost exactly. That lucky coincidence allows us to see total solar eclipses.

The Angular Diameter of the Sun and Moon

We discussed the angular diameter of an object in Chapter 2; now we need to think carefully about how the size and distance of an object like the moon determine its angular diameter (Figure 3-13). This is the key to understanding solar eclipses.

Linear diameter is simply the distance between an object's opposite sides. We use linear diameter when we order a 16-inch pizza—the pizza is 16 inches in diameter. The linear diameter of the moon is 3476 km. The angular diameter of an object is the angle formed by lines extending toward us from opposite sides of the object and meeting at our eye. Clearly, the farther away an object is, the smaller its angular diameter.

The **small-angle formula** gives us a way to figure out the angular diameter of any object, whether it is a pizza, the moon, or a galaxy. In the small-angle formula, we always express angular diameter in seconds of arc,* and we always use the same units for distance and linear diameter:

$$\frac{\text{angular diameter}}{206{,}265''} = \frac{\text{linear diameter}}{\text{distance}}$$

---

* The number 206,265″ is the number of seconds of arc in a radian. When we divide by 206,265″, we convert the angle from seconds of arc to radians.

Of course, we can use this formula to find any one of these three quantities if we know the other two; here we are interested in finding the angular diameter of the moon.

The moon has a linear diameter of 3476 km and a distance from Earth of about 384,000 km. What is its angular diameter? The moon's linear diameter and distance are both given in the same units, so we can put them directly into the small-angle formula:

$$\frac{\text{angular diameter}}{206{,}265''} = \frac{3476 \text{ km}}{384{,}000 \text{ km}}$$

To solve for angular diameter, we multiply both sides by 206,265 and find that the angular diameter is 1870 seconds of arc. If we divide by 60, we get 31 minutes of arc or, dividing by 60 again, about 0.5°. The moon's orbit is slightly elliptical, so it can sometimes look a bit larger or smaller, but its angular diameter is always close to 0.5°.

We can repeat this calculation for the angular diameter of the sun. The sun is 1.39×10^6 km in linear diameter and 1.50×10^8 km from Earth. Using the small-angle formula, we discover that the sun has an angular diameter of 1900 seconds of arc, which is 32 minutes of arc or about 0.5°. Earth's orbit is slightly elliptical, and thus the sun can sometimes look slightly larger or smaller, but it, like the moon, is always close to 0.5° in angular diameter.

By fantastic good luck, we live on a planet with a moon that is almost exactly the same angular diameter as our sun. When the moon passes in front of the sun, it is almost exactly the right size to block the brilliant surface of the sun. Then we see the most exciting sight in astronomy—a total solar eclipse. There are few other worlds where this can happen, because the angular diameters of the sun and a satellite rarely match so closely. To see this beautiful sight, all we have to do is arrange to be in the moon's shadow when the moon crosses in front of the sun.

The Moon's Shadow

Like Earth's shadow, the moon's shadow consists of a central umbra of total shadow and a penumbra of partial shadow. What we see when the moon crosses in front of the sun depends on where we are in the moon's shadow. The moon's umbral shadow produces a spot of darkness roughly 269 km (167 miles) in diameter on Earth's surface (Figure 3-14). (The exact size of the umbral shadow depends on the location of the moon in its elliptical orbit and the angle at which the shadow strikes the earth.) If we are in this spot of total shadow, we see a total solar eclipse. If we are just outside the umbral shadow but in the penumbra, we see part of the sun peeking around the moon, and the eclipse is partial. Of course, if we are outside the penumbra, we see no eclipse at all. Because of the orbital motion of the

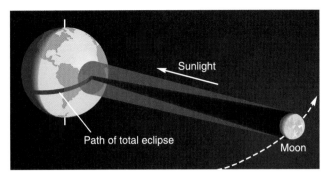

Figure 3-14
Observers in the path of totality see a total solar eclipse when the umbral shadow sweeps over them. Those in the penumbra see a partial eclipse.

moon, its shadow sweeps across Earth at speeds of at least 1700 km/h (1060 mph). To be sure of seeing a total solar eclipse, we must select an appropriate eclipse (Table 3-3), plan far in advance, and place ourselves in the **path of totality,** the path swept out by the umbral spot.

Total Solar Eclipses

A total solar eclipse begins when we first see the edge of the moon encroaching on the sun. This is the moment when the edge of the penumbra sweeps over our location.

During the partial phase, part of the sun remains visible, and it is hazardous to look at the eclipse without protection. Dense filters and exposed film do not necessarily provide protection, because some do not block the invisible heat radiation (infrared) that can burn the retina of our eyes. This has led officials to warn the public not to look at solar eclipses and has even frightened some people into locking themselves and their children into windowless rooms during eclipses. In fact, the sun is a bit less dangerous than usual during an eclipse because part of the bright surface is covered by the moon. But an eclipse is dangerous in that it can tempt us to look at the sun directly and burn our eyes.

The safest and simplest way to observe the partial phases of a solar eclipse is to use pinhole projection. Poke a small pinhole in a sheet of cardboard. Hold the sheet with the hole in the sunlight, and allow light to pass through the hole to a second sheet of cardboard (Figure 3-15). On a day when there is no eclipse, the result is a small, round spot of light that is an image of the sun. During the partial phases of a solar eclipse, the image shows the dark silhouette of the moon obscuring part of the sun. These pinhole images of the partially eclipsed sun can also be seen in the shadows of trees as the sunlight peeks through the tiny openings between the leaves and branches. This can produce an eerie effect just before totality as the remaining sliver of sun produces thin crescents of light on the ground under trees.

Table 3-3 Total and Annular Eclipses of the Sun, 2000 to 2009*

Date	Time of Mideclipse[†] (GMT)	Total/ Annular (T/A)	Max. Length of Total or Annular Phase (Min:Sec)	Area of Visibility
2001 June 21	12^h	T	4:56	Atlantic, S. Africa, Madagascar
2001 Dec. 14	21^h	A	3:54	Pacific, Central America
2002 June 10	24^h	A	1:13	Pacific
2002 Dec. 4	8^h	T	2:04	S. Africa, Indian Ocean, Australia
2003 May 31	4^h	A	3:37	Iceland, Arctic
2003 Nov. 23	23^h	T	1:57	Antarctica
2005 Apr. 8	21^h	A, T	0:42	Pacific, N. of S. America
2005 Oct. 3	11^h	A	4:32	Atlantic, Spain, Africa
2006 Mar. 29	10^h	T	4:07	Atlantic, Africa, Turkey
2006 Sept. 22	12^h	A	7:09	N.E. of S. America, Atlantic
2008 Feb. 7	4^h	A	2:14	S. Pacific, Antarctica
2008 Aug. 1	10^h	T	2:28	N. Canada, Arctic, Siberia
2009 Jan. 26	8^h	A	7:56	S. Atlantic, Indian Ocean
2009 July 22	3^h	T	6:40	Asia, Pacific

The next total solar eclipse visible from the United States will occur August 21, 2017.

*No total or annular eclipses of the sun occur in 2000.

†Times are Greenwich Mean Time. Subtract 5 hours for Eastern Standard Time, 6 hours for Central Standard Time, 7 hours for Mountain Standard Time, and 8 hours for Pacific Standard Time.

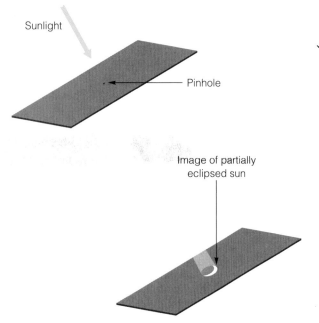

Figure 3-15
A safe way to view the partial phases of a solar eclipse. Use a pinhole in a card to project an image of the sun on a second card. The greater the distance between the cards, the larger (and fainter) the image will be.

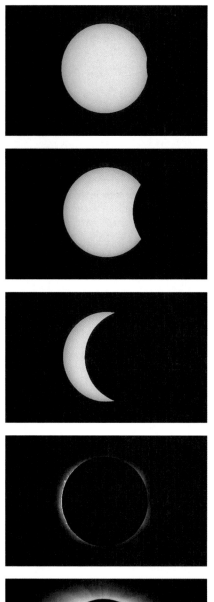

Throughout the partial phases of a solar eclipse, the moon gradually covers the bright disk of the sun (Figure 3-16). Totality begins as the last sliver of the sun's bright surface disappears behind the moon. This is the moment when the edge of the umbra sweeps over our location. So long as any of the sun is visible, the countryside is bright, but as the last of the sun disappears, dark falls in a few seconds. Automatic streetlights come on, car drivers switch on their headlights, and birds go to roost. The darkness of totality depends on a number of factors, including the weather at the observing site, but it is usually dark enough to make it difficult to read the settings on cameras.

The totally eclipsed sun is a spectacular sight. With the moon covering the bright disk of the sun, called the **photosphere,*** we can see the sun's faint outer atmosphere, the **corona,** glowing with a pale, white light so faint we can safely look at it directly. This corona is made of low-density, hot gas, which is given a wispy appearance by the solar magnetic field as shown in the last frame of Figure 3-16. Also visible just above the photosphere is a thin layer of bright gas called the **chromosphere.** The chromosphere is often marked by eruptions on the solar surface called **prominences** (Figure 3-17a), which glow with a clear, pink color due to the high temperature of the gases involved. The small-

Figure 3-16
A sequence of photographs shows the first half of a total solar eclipse. The brilliant surface of the sun is gradually covered by the moon moving from lower right to upper left in this sequence. Once the photosphere is covered, a much longer exposure is needed to photograph the fainter corona. *(Daniel Good)*

angle formula tells us that a large prominence is about 3.5 times the diameter of Earth.

Totality cannot last longer than 7.5 minutes under any circumstances, and the average is only 2 to 3 min-

---

*The photosphere, corona, chromosphere, and prominences will be discussed in detail in Chapter 8. Here the terms are used as the names of features we see during a total solar eclipse.

a   b

Figure 3-17
(a) During a total solar eclipse, the moon covers the photosphere, and the white corona and pink prominences are visible. Note the streamers in the corona caused by the sun's magnetic field. *(Daniel Good)* (b) The diamond ring effect can sometimes occur momentarily at the beginning or end of totality if a small segment of the photosphere peeks out through a valley at the edge of the lunar disk. *(National Optical Astronomy Observatory)*

utes. Totality ends when the sun's bright surface reappears at the trailing edge of the moon. This corresponds to the moment when the trailing edge of the moon's umbra sweeps over the observer.

Just as totality begins or ends, a small part of the photosphere can peek out from behind the moon through a valley at the edge of the lunar disk. Although it is intensely bright, such a small part of the photosphere does not completely drown out the fainter corona, which forms a silvery ring of light with the brilliant spot of photosphere gleaming like a diamond (Figure 3-17b). This **diamond ring effect** is one of the most spectacular of astronomical sights, but it is not visible during every solar eclipse. Its occurrence depends on the exact orientation and motion of the moon.

Once totality is over, daylight returns quickly, and the corona and chromosphere vanish. Astronomers travel great distances to place their instruments in the path of totality to study the faint outer corona and make other measurements possible only during the few minutes of a total solar eclipse.

Sometimes when the moon crosses in front of the sun, it is too small to fully cover the sun, and we see an **annular eclipse,** a solar eclipse in which a ring (or annulus) of the photosphere is visible around the disk of the moon (Figure 3-18). With a portion of the photosphere visible, the eclipse never becomes total; it never quite gets dark; and we can't see the prominences, chromosphere, and corona. Annular eclipses occur because

the moon follows a slightly elliptical orbit around Earth, and thus its angular diameter can vary (Figure 3-19). When it is at **perigee,** its point of closest approach to Earth, it looks significantly larger than when it is at **apogee,** the most distant point in its orbit. Furthermore, Earth's orbit is slightly elliptical, so the Earth–sun distance varies slightly, and thus the diameter of the solar disk varies slightly. If the moon is in the farther part of its orbit during totality, its angular diameter will be less than the angular diameter of the sun, and thus we see the annular eclipse. Such an annular eclipse of the sun swept across the United States on May 10, 1994.

R E V I E W Critical Inquiry

If people on Earth were seeing a total solar eclipse, what would astronauts on the moon see when they looked at Earth?

Astronauts on the moon could see Earth only if they were on the side that faces Earth. Because solar eclipses always happen at new moon, the near side of the moon would be in darkness, and the far side of the moon would be in full sunlight. The astronauts would be standing in darkness, and they would be looking at the fully illuminated side of Earth. They would see a "full Earth." The moon's shadow would be crossing Earth, and if the

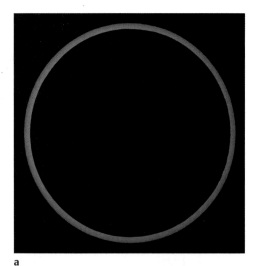

a

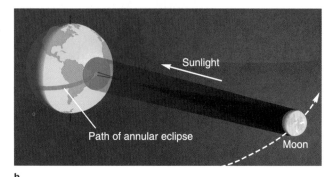

b

Figure 3-18

(a) The annular eclipse of 1994. A bright ring of photosphere remains visible around the moon, and the corona and prominences are not visible. *(Daniel Good)* (b) An annular eclipse occurs when the moon is in the farther part of its orbit and its umbral shadow does not reach Earth. From Earth, we see an annular eclipse because the moon's angular diameter is smaller than the angular diameter of the sun.

astronauts looked closely, they might be able to see the spot of darkness where the moon's umbral shadow touched Earth. It would take hours for the shadow to cross Earth.

Standing on the moon and watching the moon's umbral shadow sweep across the earth would be a cold, tedious assignment. Perhaps it would be more interesting for astronauts on the moon to watch Earth while people on Earth were seeing a total lunar eclipse. What would the astronauts see then?

Eclipses of the sun and moon are often dramatic and mysterious. Ancient astronomers studied them and found ways to predict the coming of an eclipse. In the section that follows, we will see how eclipses can be predicted from a basic understanding of the motion of the sun and the moon.

3-5 Predicting Eclipses

To make exact eclipse predictions, you would need to calculate the precise motions of the sun and moon, and that requires a computer and proper software. Such software is available for desktop computers, but it isn't necessary if you are satisfied with making less exact predictions. In fact, many primitive peoples, such as the builders of Stonehenge and the ancient Maya, are believed to have made eclipse predictions.

We examine eclipse prediction for three reasons. First, it is an important part of the history of science. Second, it illustrates how apparently complex phenomena can be analyzed in terms of cycles. Third, eclipse prediction exercises our mental muscles and forces us to see Earth, the moon, and the sun as objects moving through space.

Conditions for an Eclipse

We can predict eclipses by understanding the conditions that make them possible. As we begin to think about these conditions, we must be sure we understand our point of view. (See Window on Science 3-2.) Later we will change our point of view, but to begin we will imagine that we can look up into the sky from our home on Earth and see the sun moving along the ecliptic and the moon moving along its orbit.

The orbit of the moon is tipped 5°8′43″ to the plane of Earth's orbit, so we see the moon follow a path tipped by that angle to

Figure 3-19

The angular diameter of the moon (left) varies dramatically because its orbit is elliptical and its distance varies from perigee (closest) to apogee (farthest). The angular diameter of the sun (right) varies by a smaller amount because Earth's orbit is nearly circular.

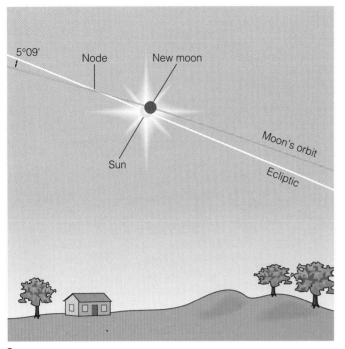

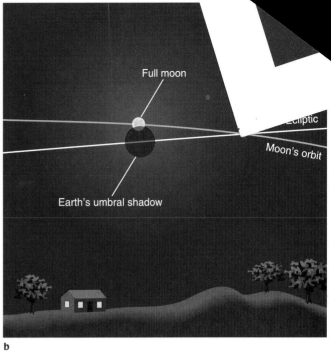

a

b

Figure 3-20
Eclipses can occur only near the nodes of the moon's orbit. (a) A solar eclipse occurs when the moon meets the sun near a node. (b) A lunar eclipse occurs when the sun and moon are near opposite nodes. Partial eclipses are shown here for clarity.

the ecliptic. Each month, the moon crosses the ecliptic at two points called **nodes.** At one node it crosses going southward, and two weeks later it crosses at the other node going northward.

Eclipses can only occur when the sun is near one of the nodes of the moon's orbit. A solar eclipse is caused by the moon passing in front of the sun. Most new moons pass too far north or too far south of the sun to cause an eclipse. Only when the sun is near a node in the moon's orbit can the moon cross in front of the sun, as shown in Figure 3-20a. A lunar eclipse doesn't happen at every full moon because most full moons pass too far north or too far south of the ecliptic and miss Earth's shadow. The moon can enter Earth's shadow only when the shadow is near a node in the moon's orbit, and that means the sun must be near the other node. This is shown in Figure 3-20b.

Thus, there are two conditions for an eclipse: The sun must be crossing a node, and the moon must be crossing either the same node (solar eclipse) or the other node (lunar eclipse). Clearly, solar eclipses can occur only when the moon is new, and lunar eclipses can occur only when the moon is full.

An **eclipse season** is the period during which the sun is close enough to a node for an eclipse to occur. For solar eclipses, an eclipse season is about 32 days. Any new moon during this period will produce a solar eclipse. For lunar eclipses, the eclipse season is a bit shorter, about 22 days. Any full moon in this period will be eclipsed.

This makes eclipse prediction easy. We simply keep track of where the moon crosses the ecliptic, and when the sun is near one of these nodes we predict that the nearest new moon will cause a solar eclipse and the nearest full moon will cause a lunar eclipse. This system works fairly well, and ancient astronomers such as the Maya may have used such a system. But we can do better if we change our point of view.

The View from Space

Let us change our point of view and imagine that we are looking at the orbits of Earth and the moon from a point far away in space. We see the moon's orbit as a smaller disk tipped at an angle to the larger disk of Earth's orbit. As Earth orbits the sun, the moon's orbit remains fixed in direction. The nodes of the moon's orbit are the points where it passes through the plane of Earth's orbit; an eclipse season occurs each time the line connecting these nodes, the **line of nodes,** points toward the sun (Figure 3-21).

The shadows of Earth and moon, seen from space, are very long and thin (Figure 3-22). Only at the time of an eclipse season, when the line of nodes points toward the sun, do the shadows produce eclipses.

From our point of view in space, we would see the orbit of the moon precess like a hubcap spinning on the

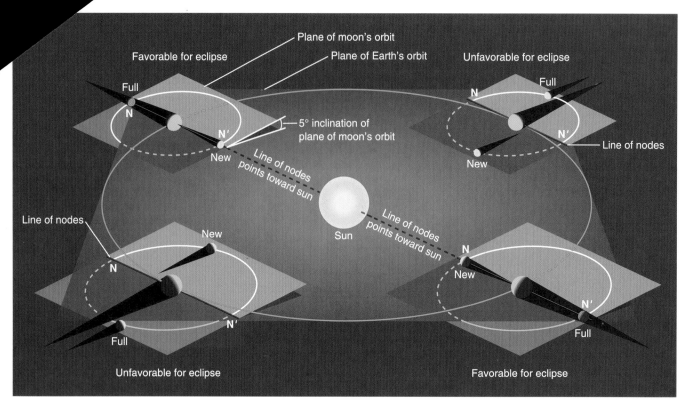

Figure 3-21

The moon's orbit is tipped about 5° to Earth's orbit. The nodes N and N' are the points where the moon passes through the plane of Earth's orbit. If the line of nodes does not point at the sun, the shadows miss, and there are no eclipses at new moon and full moon. At those parts of Earth's orbit where the line of nodes points toward the sun, eclipses are possible at new moon and full moon.

ground. This precession is caused mostly by the gravitational influence of the sun, and it makes the line of nodes rotate once every 18.6 years. People back on Earth see the nodes slipping westward along the ecliptic 19.4° per year, and the sun takes only 346.62 days (an **eclipse year**) to return to a node. This means that, according to our calendar, the eclipse seasons begin about 19 days earlier every year (Figure 3-23).

The cyclic pattern of eclipses shown in Figure 3-23 makes eclipse prediction simple. We know that the eclipse seasons occur 19 days earlier each year because the moon's orbit precesses. Any new moon during an eclipse season will cross in front of the sun and cause

a solar eclipse, and any full moon will enter Earth's shadow and cause a lunar eclipse.

The Saros Cycle

Ancient astronomers could predict eclipses in a crude way using the eclipse seasons, but they could have been much more accurate if they recognized that eclipses occur following certain patterns. The most important of these is the **saros cycle** (sometimes referred to simply as the saros). After one saros cycle of 18 years $11\frac{1}{3}$ days, the pattern of eclipses repeats. In fact, *saros* comes from a Greek word that means "repetition."

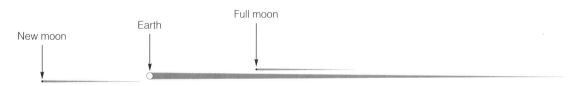

Figure 3-22

Umbral shadows of Earth and the moon. Notice how easy it is for the shadows to miss their mark at full moon and at new moon and fail to produce eclipses. That is, it is easy for the moon to reach full phase and not enter Earth's shadow. It is also easy for the moon to reach new phase and not cast its shadow on Earth. (The diameters of Earth and the moon are exaggerated by a factor of 2 for clarity.)

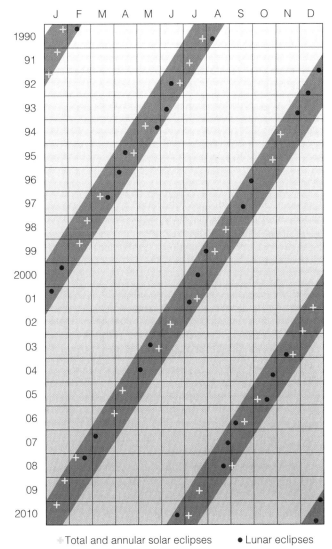

JFMAMJJASOND

1990
91
92
93
94
95
96
97
98
99
2000
01
02
03
04
05
06
07
08
09
2010

✦Total and annular solar eclipses   • Lunar eclipses

Figure 3-23
A calendar of eclipse seasons. Each year, the eclipse seasons begin about 19 days earlier. Any new moon or full moon that occurs during an eclipse season results in an eclipse. Not all eclipses are shown here.

The eclipses repeat because, after one saros cycle, the moon and the nodes of its orbit return to the same place with respect to the sun. One saros contains 6585.321 days, which is equal to 223 lunar months. Therefore, after one saros cycle the moon is back to the same phase it had when the cycle began. But one saros is nearly equal to 19 eclipse years. After one saros cycle, the sun has returned to the same place it occupied with respect to the nodes of the moon's orbit when the cycle began. If an eclipse occurs on a given day, then 18 years $11\frac{1}{3}$ days later the sun, the moon, and the nodes of the moon's orbit return to nearly the same relationship, and the eclipse occurs all over again.

Although the eclipse repeats almost exactly, it is not visible from the same place on Earth. The saros

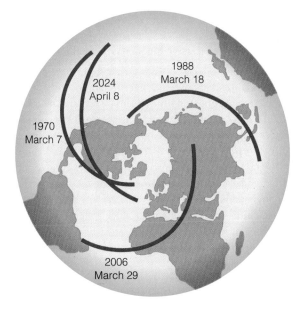

Figure 3-24
The saros cycle at work. The total solar eclipse of March 7, 1970, recurred after 18 years $11\frac{1}{3}$ days over the Pacific Ocean. After another interval of 18 years $11\frac{1}{3}$ days, the same eclipse will be visible from Asia and Africa. After a similar interval, the eclipse will again be visible from the United States.

cycle is one-third of a day longer than 18 years 11 days. When the eclipse recurs, Earth will have rotated one-third of a turn farther east, and the eclipse will occur eight hours of longitude west of its earlier location (Figure 3-24). Thus, after three saros cycles—a period of 54 years 1 month—the same eclipse occurs in the same part of Earth.

One of the most famous predictors of eclipses was Thales of Miletus (about 640–546 BC), who supposedly learned of the saros cycle from the Chaldeans, who had discovered it. No one knows which eclipse Thales predicted, but some scholars suspect the eclipse of May 28, 585 BC. In any case, the eclipse occurred at the height of a battle between the Lydians and the Medes, and the mysterious darkness in midafternoon so startled the two factions that they concluded a truce.

In fact, many historians doubt that Thales actually predicted the eclipse. It would have been very difficult to gather enough information about past eclipses, because total solar eclipses are rare. If you stay in one city, you will see a total solar eclipse about once in 360 years. Also, 585 BC is very early for the Greeks to have known of the saros cycle. The important point is not that Thales did it, but that he could have done it. If he had had records of past eclipses of the sun visible from the area, he could have discovered that they tended to recur with a period of 54 years 1 month (three saros cycles). Indeed, he could have predicted the eclipse without ever understanding what the sun and moon were or how they moved.

Why can't two successive full moons be totally eclipsed?

A total lunar eclipse occurs when the moon passes through Earth's shadow, and that can happen only when the sun is near one node and the moon crosses the other node. Earth's shadow always points toward the ecliptic exactly opposite the sun, so most of the time the moon will pass north or south of Earth's shadow, and there will be no eclipse. Only when the shadow is near one of the nodes of the moon's orbit can the moon enter the shadow and be eclipsed.

An eclipse season for a total lunar eclipse is only 22 days long. If the moon crosses a node more than 11 days before or after the sun crosses the other node, there will be no eclipse. The moon takes 29.5 days to go from one full moon to the next, so if one full moon is totally eclipsed, the next full moon 29.5 days later will occur too late for an eclipse.

Use your knowledge of the cycles of the sun and moon to explain why the sun can be eclipsed by two successive new moons.

Predicting eclipses isn't very hard, and many ancient peoples were probably familiar enough with the cycles of the sun and moon to know when eclipses were likely. We might call those first predictions early science, but science involves understanding nature, not just predicting events. In the next chapter, we will see how modern science was born out of astronomers' attempts to understand the cycles they saw in the sky.

Summary

Because we see the moon by reflected sunlight, its shape appears to change as it orbits Earth. The lunar phases wax from new moon to first quarter to full moon and wane from full moon to third quarter to new moon. A complete cycle of lunar phases takes 29.53 days.

The moon's gravitational field exerts tidal forces on Earth that pull the ocean waters up into two bulges, one on the side of Earth facing the moon and the other on the side away from the moon. As the rotating Earth carries the continents through these bulges of deeper water, the tides ebb and flow. Friction with the seabeds slows Earth's rotation, and the gravitational force the bulges exert on the moon forces its orbit to grow larger.

A lunar eclipse occurs when the moon enters Earth's shadow. If the moon becomes completely immersed in the umbra, the central shadow, the eclipse is termed total. The moon glows coppery red due to sunlight bent as it passes through Earth's atmosphere. If the moon enters the umbral shadow only partly, the eclipse is termed partial, and the reddish glow is not visible. A penumbral eclipse occurs when the moon passes through the penumbra, the region of partial shadow. Penumbral eclipses are not very noticeable.

A solar eclipse occurs when Earth passes through the moon's shadow. To see a total solar eclipse, observers must place themselves in the path of totality, the path swept by the umbra of the moon. As the umbra sweeps over the observers, they see the bright surface of the sun, the photosphere, blotted out by the moon. Then the fainter chromosphere and corona, higher layers of the sun's atmosphere, become visible. Eruptions on the solar surface, called prominences, may be visible peeking around the edge of the moon.

The corona, chromosphere, and prominences are not visible to observers outside the path of totality. Observers in the path of the penumbra of the moon will see a partial solar eclipse, but the photosphere will never be completely hidden.

If an eclipse of the sun occurs when the moon is not in the nearer part of its orbit, the moon's umbra does not reach Earth's surface. In these cases, the eclipse is not total. The moon's angular diameter is too small to cover the sun completely. Even at maximum eclipse, therefore, a bright ring, or annulus, of the photosphere is visible around the edge of the moon. This annular eclipse is not as dramatic as a total eclipse.

Ancient astronomers could predict eclipses because they occur not randomly but in a pattern. Two conditions must be met if an eclipse is to occur. First, the moon must be on or near the ecliptic. The two points where the moon crosses the ecliptic are called the nodes of the moon's orbit. The second condition is that the sun must be at or near one of the nodes. This means that eclipses can occur only at new moon or full moon during two eclipse seasons that are about 32 days long and that occur almost 6 months apart.

Because the moon's orbit precesses, the nodes slip westward along the ecliptic, and the eclipse seasons begin 19 days earlier each year. The moon, sun, and nodes return to the same relative positions every 18 years $11\frac{1}{3}$ days in what is called the saros cycle. After the passage of a saros cycle the pattern of eclipses begins to repeat. This means that ancient astronomers could predict eclipses just by examining the date of previous eclipses. The Chaldeans discovered the saros cycle, and the ancient Greeks used it. Many other primitive cultures also used this method.

New Terms

sidereal period	corona
synodic period	chromosphere
spring tides	prominences
neap tides	diamond ring effect
umbra	annular eclipse
penumbra	perigee
total eclipse (lunar or solar)	apogee
partial eclipse (lunar or solar)	node
	eclipse season
penumbral eclipse	line of nodes
small-angle formula	eclipse year
path of totality	saros cycle
photosphere	

Review Questions

1. Which lunar phases would be visible in the sky at dawn? at midnight?

2. If you looked back at Earth from the moon, what phase would you see when the moon was full? new? a first-quarter moon? a waxing crescent?

3. Give examples to show how tides can alter the rotation and revolution of celestial bodies. (*Hint:* Recall from Chapter 2 the difference between rotation and revolution.)

4. Could a solar-powered spacecraft generate any electricity while passing through Earth's umbral shadow? the penumbral shadow?

5. Draw the umbral and penumbral shadows onto Figure 3-3. Explain why lunar eclipses can occur only at full moon and solar eclipses can occur only at new moon.

6. How did lunar eclipses lead Aristotle to conclude that Earth was a sphere?

7. Why isn't the corona visible during partial or annular solar eclipses?

8. Why can't the moon be eclipsed when it is halfway between the nodes of its orbit?

9. Why aren't eclipses separated by one saros cycle visible from the same location on Earth?

10. How could Thales of Miletus have predicted the date of a solar eclipse without observing the location of the moon in the sky?

Discussion Questions

1. If the moon were closer to Earth such that it had an orbital period of 24 hours, what would the tides be like?

2. How would eclipses be different if the moon's orbit were not tipped with respect to the plane of Earth's orbit?

3. Are there other planets in our solar system from whose surface we could see a lunar eclipse? a total solar eclipse?

Problems

1. Identify the phases of the moon if on March 21 the moon is located at (a) the vernal equinox, (b) the autumnal equinox, (c) the summer solstice, (d) the winter solstice.

2. Identify the phases of the moon if at sunset the moon is (a) near the eastern horizon, (b) high in the southern sky, (c) in the southeastern sky, (d) in the southwestern sky.

3. About how many days must elapse between first-quarter moon and third-quarter moon?

4. How many hours would elapse between successive high tides if tides at a given location were caused only by the sun's gravity? only by the moon's gravity? Why is there a difference?

5. How many times larger than the moon is the diameter of Earth's umbral shadow at the moon's distance? (*Hint:* See Figure 3-11.)

6. Use the small-angle formula to calculate the angular diameter of Earth as seen from the moon.

7. During solar eclipses, large solar prominences are often seen extending 5 minutes of arc from the edge of the sun's disk. How far is this in kilometers? in Earth diameters?

8. If a solar eclipse occurs on October 3, why can't there be a lunar eclipse on October 13? Why can't there be a solar eclipse on December 28?

9. A total eclipse of the sun was visible from Canada on July 10, 1972. When did this eclipse occur next? From what part of Earth was it total?

10. When will the eclipse described in Problem 9 next be total as seen from Canada?

11. When will the eclipse seasons occur during the current year? What eclipse(s) will occur?

Critical Inquiries for the Web

1. Search the Web for myths about the moon. What common themes do different cultures ascribe to the moon and its cycle of phases?

2. A total solar eclipse swept across Europe and into Asia in the summer of 1999. Search for Web pages showing photos and observations of the eclipse. Can you find Web pages that show cultural responses to the eclipse such as celebrations or religious ceremonies?

3. Most people see more total lunar eclipses in a lifetime than total solar eclipses. Why is this so? Compare the regions of visibility for a number of past and upcoming eclipses, and determine which future events will be visible from your area.

4. How do the tides on Earth vary with the phases of the moon? Use the Internet to explore the range of tides for a particular location through the next few weeks. Also, look up moon phase information for that same interval. How do the tidal amplitudes correlate with moon phase during that period?

 Go to the Brooks/Cole Astronomy Resource Center (www. brookscole.com/astronomy) for critical thinking exercises, articles, and additional readings from InfoTrac College Edition, Brooks/Cole's online student library.

CHAPTER FOUR

The Origin of Modern Astronomy

Oh, my dear Kepler, how I wish

that we could have one hearty laugh

together!

From a letter by Galileo Galilei

Guidepost

The sun, moon, and planets sweep out a beautiful and complex dance across the heavens. Previous chapters have described that dance; this chapter describes how astronomers learned to understand what they saw in the sky and how that changed humanity's understanding of what we are.

In learning to interpret what they saw, Renaissance astronomers invented a new way of knowing about nature, a way of knowing that we recognize today as a modern science.

This chapter tells the story of heavenly motion from a cultural perspective. In the next chapter, we will give meaning to the motions in the sky by adding the ingredient Renaissance astronomers were missing—gravity.

The history of astronomy is more like a soap opera than a situation comedy. In a simple half-hour sitcom, only a few characters carry the story, but the plot of a soap opera is so complex, so filled with characters and subplots only tenuously related to one another, that the action jumps from character to character, from subplot to subplot, with almost no connection. The history of astronomy is similarly complex, filled with brilliant people who lived at different times and worked in different parts of the world.

Two subplots twine through our story. One is the human quest to understand the place of the Earth. That plot will lead us from Aristotle to Copernicus and finally to the trial of Galileo before the Inquisition. But the second subplot, the puzzle of planetary motion, is equally important to our story. That plot will involve an astronomer with a false nose and another whose mother was tried for witchcraft. We will see that the true place of the Earth could be understood only when the puzzle of planetary motion was solved.

Our story goes far beyond the birth of astronomy. The quest for the place of the Earth and the puzzle of planetary motion became, in the 16th and 17th centuries, the center of a storm of controversy over the best way to understand our world and ourselves. That conflict is now known as the scientific revolution. Thus, our story describes not only the birth of astronomy but also the birth of modern science.

Of course, astronomy had its beginnings long before the 16th century. It began when the first semihuman creature looked up at the moon and stars and wondered what they were. Since then, hundreds of generations of talented astronomers have lived and worked on this planet and left almost no record of their accomplishments. Only through archaeology can we get a hint of their insights and triumphs. Other cultures, such as ancient Greece, have left written records from which we can reconstruct the sophisticated astronomy of the ancient world.

4-1 The Roots of Astronomy

Astronomy has its origin in that most noble of all human traits, curiosity. Just as modern children ask their parents what the stars are and why the moon changes, so did ancient humans ask themselves these questions. The answers, often couched in mythical or religious terms, reveal a great reverence for the order of the heavens.

Archaeoastronomy

The study of the astronomy of ancient peoples, **archaeoastronomy,** came to public attention in 1965 when Gerald Hawkins published a book called *Stonehenge Decoded*. He reported that Stonehenge, the prehistoric ring of stones, was a sophisticated astronomical observatory.

Stonehenge, standing on Salisbury Plain in southern England, was built in stages from about 3000 BC to about 1800 BC, a period extending from the late Stone Age into the Bronze Age. Though the public is most familiar with the massive stones of Stonehenge (Figure 4-1), those were added late in its history. In its first stages, Stonehenge consisted of a circular ditch slightly larger in diameter than the length of a football field, with a concentric bank just inside the ditch and a long avenue leading away toward the northeast. A massive stone, the Heelstone, stood then, as it does now, outside the ditch in the opening of the avenue.

As early as AD 1740, the English scholar W. Stukely suggested that the avenue pointed toward the rising sun at the summer solstice, but few accepted the idea. More recently, astronomers have recognized significant astronomical alignments at Stonehenge. Seen from the center of the monument, the summer-solstice sun rises behind the Heelstone. Other sight lines point toward the most northerly and most southerly risings of the moon.

The significance of these alignments has been debated. Hawkins claims that the Stone Age people who built Stonehenge were using it as a device to predict lunar eclipses. Others have had more conservative notions, but the truth may never be known. The builders of Stonehenge had no written language and left no records of their intentions. Nevertheless, the presence of solar and lunar alignments at Stonehenge and at many other Stone Age monuments dotting England and continental Europe shows that so-called primitive peoples were paying detailed attention to the sky. The roots of astronomy lie not in sophisticated science and logic but in human curiosity and wonder.

The early inhabitants of North America were also interested in astronomy. The Big Horn Medicine Wheel,* located on a 3000-m shoulder of the Big Horn Mountains of Wyoming, was used by the Plains Indians about AD 1500–1750. This arrangement of rocks marks a 28-spoke wheel about 27 m in diameter. Piles of rocks, called cairns, mark the center and six locations on the circumference. Astronomer John Eddy has discovered that these cairns mark sight lines toward a number of important points on the eastern horizon and could have been used as a calendar to help schedule hunting, planting, harvesting, and celebrations (Figure 4-2).

Many other American Indian sites have astronomical alignments. More than three dozen medicine wheels are known, although most do not have the sophistication of the Big Horn Medicine Wheel. The Moose Mountain Wheel in Saskatchewan, Canada, for instance, seems to have been in use as early as AD 100. Alignments at some Mound Builder sites of the midwest show that these peoples were familiar with the sky. In the southwest, the Anasazi Indian ruins in Chaco Canyon, New Mexico, among others, have alignments that point to-

---

* *Medicine* is used here to mean magical power.

a

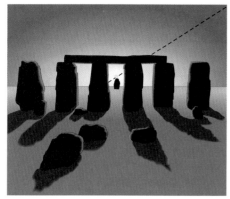

b

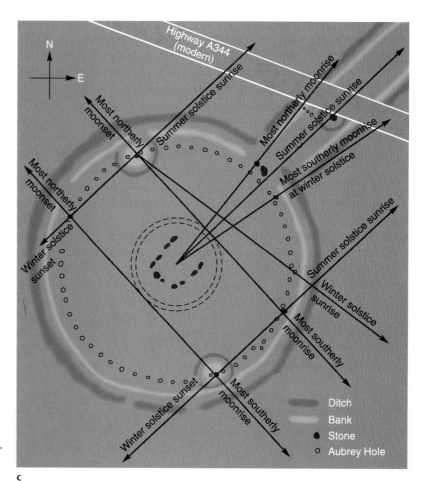

c

Figure 4-1

(a) A stamp issued to mark the return of Comet Halley shows the central horseshoe of upright stones at Stonehenge. (b) The best-known astronomical alignment at Stonehenge is the summer solstice sun rising over the heelstone. (c) Although a number of astronomical alignments have been found at Stonehenge, experts debate their significance.

Figure 4-2

The Big Horn Medicine Wheel in Wyoming was built by Native Americans before the westward spread of European settlers. Its rock cairns appear to be aligned with the rising and setting of the summer solstice sun (red) and with the rising of three bright stars (blue). Such connections with the sky are found in many Native American structures and ceremonies. *(John A. Eddy)*

a

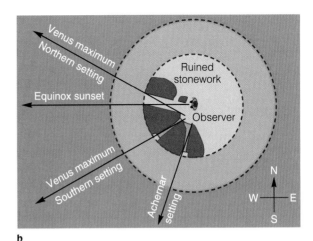

b

Figure 4-3
The Caracol at the Mayan city of Chichén Itzá (a) is a 1000-year-old observatory. The remaining windows in the partially ruined tower contain sight lines that point toward the equinox sunset, the setting places of Venus at its most northerly and most southerly positions, and the setting place of the star Achernar (b). Various evidence suggests that the Caracol was directly associated with the worship of the Venus god, Quetzalcoatl. *(Anthony Aveni)*

ward the rising and setting of the sun at the summer and winter solstices.

Primitive astronomy also flourished in Central and South America. The Mayan and Aztec empires built many temples aligned with the solstice rising of the sun and with the extreme points where Venus rose and set. The Caracol temple in the Yucatán is a good example (Figure 4-3). It is a circular tower containing complicated passageways and a spiral staircase (thus the name *Caracol*—"the snail's shell"). The tubelike windows at the top of the tower point toward the equinox sunset point and the most northerly and most southerly setting points of Venus. Unfortunately, only about one-third of the tower top survives, so the directions of any other windows are forever lost.

Archaeoastronomers are uncovering the remains of ancient astronomical observatories around the world. Some temples in the jungles of southeast Asia, for instance, have astronomical alignments.

Other scholars are looking not at temples but at small artifacts from thousands of years ago. Scratches on certain bone and stone implements seem to follow a pattern and may be an attempt to keep a record of the phases of the moon (Figure 4-4). Some scientists contend that humanity's first attempts at writing were stimulated by a desire to record and predict lunar phases.

Archaeoastronomy is uncovering the earliest roots of astronomy and simultaneously revealing some of the first human efforts at systematic inquiry. The most important lesson of archaeoastronomy is that humans don't have to be technologically sophisticated to admire and study the universe.

One thing about archaeoastronomy is especially sad. Although we are learning how primitive people observed the sky, we may never know what they thought about their universe. Many had no written language. In other cases, the written record has been lost. Dozens, perhaps hundreds, of beautiful Mayan manuscripts, for instance, were burned by Spanish missionaries who believed that the books were the work of Satan. Only four of these books have survived, and all four contain astronomical references. One contains sophisticated tables that allowed the Maya to predict the motion of Venus and eclipses of the moon. We will never know what was burnt.

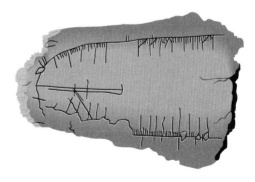

Figure 4-4
A fragment of a 27,000-year-old mammoth tusk found at Gontzi in Ukraine contains scribe marks on its edge, simplified in this drawing. These markings have been interpreted as a record of four cycles of lunar phases. Although controversial, such finds suggest that some of the first human attempts at recording events in written form were stimulated by astronomical phenomena.

The fate of the Mayan books illustrates one reason why histories of astronomy usually begin with the Greeks. Some of their writing has survived, and we can discover what they thought about the shape and motion of the heavens.

The Astronomy of Greece

Greek astronomy was based on the astronomy of Babylonia and Egypt, but these astronomies were heavily influenced by religion and astrology. Those ancient astronomers studied the motions of the heavens as a way of worship and divination. The Greek astronomers studied astronomy in an entirely new way—they tried to understand the universe.

This new attitude toward the heavens, a truly scientific attitude, was made possible by two early Greek philosophers. Thales of Miletus (c. 624–547 BC) lived and worked in what is now Turkey. He taught that the universe is rational and that the human mind can understand why the universe works the way it does. This view contrasts sharply with those of earlier cultures, which believed that the ultimate causes of things are mysteries beyond human understanding. To Thales and his followers, the mysteries of the universe are mysteries because they are unknown, not because they are unknowable.

The second philosopher who made the new scientific attitude possible was Pythagoras (c. 570–500 BC). He and his students noticed that many things in nature seem to be governed by geometrical or mathematical relations. Musical notes, for example, are related in a regular way to the lengths of plucked strings. This led Pythagoras to propose that all nature was underlain by musical principles, by which he meant mathematics. One result of this philosophy was the later belief that the harmony of the celestial movements produced actual music, the music of the spheres. But at a deeper level, the teachings of Pythagoras made Greek astronomers look at the universe in a new way. Thales said that the universe could be understood, and Pythagoras said that the underlying rules were mathematical.

In trying to understand the universe, Greek astronomers did something that Babylonian astronomers had never done—they tried to construct descriptions based on geometrical forms. Anaximander (c. 611–546 BC) described a universe made up of wheels filled with fire: The sun and moon are holes in the wheels through which we see the flames. Philolaus (5th century BC) argued that Earth moves in a circular path around a central fire (not the sun), which is always hidden behind a counterearth located between the fire and Earth. This was the first theory to suppose that Earth is in motion.

Plato (428–347 BC) was not an astronomer, but his teachings influenced astronomy for 2000 years. Plato argued that the reality we see is only a distorted shadow of a perfect, ideal form. If our observations are distorted,

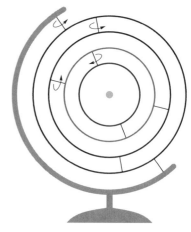

Figure 4-5
The spheres of Eudoxus explain the motions in the heavens by means of nested spheres rotating about various axes at different rates. Earth is located at the center.

then observation can be misleading, and the best path to truth is through pure thought on the ideal forms that underlie nature.

Plato also argued that the most perfect form was the circle and that therefore all motions in the heavens should be made up of combinations of circular motion. Because the most perfect motion is uniform motion, later astronomers tried to describe the motions of the heavens using the principle of **uniform circular motion.**

Pythagoras had taught that Earth is a sphere and that the other heavenly bodies were divine, perfect spheres moving in perfect circles. Eudoxus of Cnidus (409–356 BC), a student of Plato, combined a system of 27 nested spheres rotating at different rates about different axes with the concept of uniform circular motion to produce a mathematical description of the motions of the universe (Figure 4-5).

At the time of the Greek philosophers, it was common to refer to systems such as that of Eudoxus as descriptions of the world, where the word *world* included not only Earth but all of the heavenly spheres. The reality of these spheres was open to debate. Some thought of the spheres as nothing more than mathematical ideas that described motion in the world model, while others began to think of the spheres as real objects made of perfect celestial material. Aristotle, for example, seems to have thought of the spheres as real.

Aristotle (384–322 BC) taught and wrote on philosophy, history, politics, ethics, poetry, drama, and so on (Figure 4-6). Because of his sensitivity and insight, he became the great authority of antiquity, and astronomers for almost 2000 years cited him as their authority in adopting the Greek model of the universe.

Much of what Aristotle wrote about scientific subjects was wrong, but he was not a scientist. He was trying to understand the universe by creating a system of formal philosophy. Modern science depends on evidence and hypothesis, but scientific methods had not

Figure 4-6
Aristotle, honored on this Greek stamp, wrote on such a wide variety of subjects and with such deep insight that he became the great authority on all matters of learning. His opinions on the nature of Earth and the sky were widely accepted for almost two millennia.

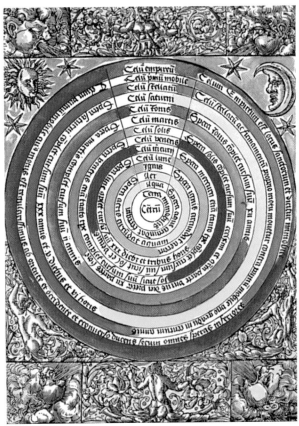

Figure 4-7
According to Aristotle, Earth is motionless (*Terra immobilis*) at the center of the universe. It is surrounded by spheres of water, air, and fire (*ignus*), above which lie spheres carrying the celestial bodies, beginning with the moon (*lune*) in the lowest celestial sphere. This woodcut is from C. Cornipolitanus's book *Chronographia* of 1537. *(Granger Collection, New York)*

been invented at the time of Aristotle. Rather, he attempted to use the most basic observations combined with first principles (ideas that he believed were obviously true) to understand the world. The perfection of the heavens was, for Aristotle, a first principle.

Aristotle believed that the universe was divided into two parts—Earth, corrupt and changeable, and the heavens, perfect and immutable. Like most of his predecessors, he believed that Earth was the center of the universe, so his model was **geocentric** (Earth-centered). The heavens surrounded Earth, and he added more crystalline spheres to bring the total to 56. The lowest sphere, that of the moon, marked the boundary between the imperfect region of Earth and the perfection of the celestial realm above the moon.

Because he believed Earth to be immobile, he had to make these spheres whirl westward around Earth each day and move with respect to one another to produce the motions of the sun, moon, and planets. Like most other Greek philosophers, Aristotle viewed the universe as a perfect heavenly machine that was not many times larger than Earth itself (Figure 4-7).

About a century after Aristotle, the Alexandrian philosopher Aristarchus proposed a theory that Earth rotated on its axis and revolved around the sun. This theory is, of course, correct, but most of the writings of Aristarchus were lost, and his theory was not well known. Later astronomers rejected any suggestion that Earth could move, because it conflicted with Aristotle's theory and because the astronomers saw no parallax.

Parallax (*p*) is the apparent change in the position of an object due to a change in the location of the observer. It is actually an old friend, although you may not have known its name, for you use parallax to judge the distance to things. To see how it works, close your right eye and use your thumb, held at arm's length, to

cover some distant object, a building perhaps. Now look with your right eye. Your thumb seems to move to the left, uncovering the building (Figure 4-8). This

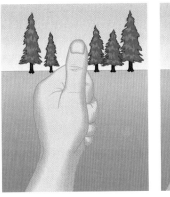

Seen by left eye   Seen by right eye

Figure 4-8
To demonstrate parallax, close one eye and cover a distant object with your thumb held at arm's length. Then look with the other eye; your thumb appears to have shifted position.

apparent shift is parallax, and your brain uses it to estimate distances to objects around you.

Ancient astronomers reasoned that if Earth moved around the sun, we should be viewing the stars from different positions in Earth's orbit at different times. Then we should see parallax shifting the apparent positions of the stars. In the most extreme cases, this should distort the shapes and sizes of the constellations. That is, constellations should look largest when Earth is nearest that part of the sky. Because they did not see this parallax, they concluded that Earth did not move. Actually, they saw no parallax because the stars are much farther away than they supposed, and the parallax is much too small to be visible to the naked eye.

Aristotle had taught that Earth had to be a sphere because it always casts a round shadow during lunar eclipses, but he could only estimate its size. About 200 BC, Eratosthenes, working in the great library in Alexandria, found a way to calculate Earth's radius. He learned from travelers that the city of Syene (Aswan) in southern Egypt contained a well into which sunlight shone vertically on the day of the summer solstice. Thus, the sun was at the zenith at Syene, but on that same day in Alexandria, he noted that the sun was one-fiftieth of the circumference of the sky (about 7°) south of the zenith. Because sunlight comes from such a great distance, its rays arrive at Earth traveling almost parallel. Thus, Eratosthenes could conclude from simple geometry (Figure 4-9) that the distance from Alexandria to Syene was one-fiftieth Earth's circumference.

To find Earth's circumference, Eratosthenes had to know the distance from Alexandria to Syene. Travelers told him it took 50 days to cover the distance, and he knew that a camel can travel about 100 stadia per day. Thus, the total distance was about 5000 stadia. If 5000 stadia is one-fiftieth Earth's circumference, then Earth must be 250,000 stadia around, and dividing by 2π Eratosthenes found Earth's radius to be 40,000 stadia.

We don't know how accurate Eratosthenes was. The stadium (singular of *stadia*) had different lengths in ancient times. If we assume 6 stadia to the kilometer, then Eratosthene's result was too big by only 4 percent. If he used the Olympic stadium, his result was 14 percent too big. In any case, this was a much better measurement of Earth's radius than Aristotle's estimate, which was only about 40 percent of the true radius.

The greatest of the ancient observers was Hipparchus (see Figure 2-5), who lived during the 2nd century BC, about two centuries after Aristotle. He is usually credited with the invention of trigonometry, he compiled the first star catalog, and he discovered precession (Chapter 2). Instead of describing the motion of the sun and moon using nested spheres, as most Greek philosophers did, Hipparchus proposed that the sun and moon traveled around circles with Earth near, but not at, their centers. These off-center circles are now known as **eccentrics.** Hipparchus recognized that he

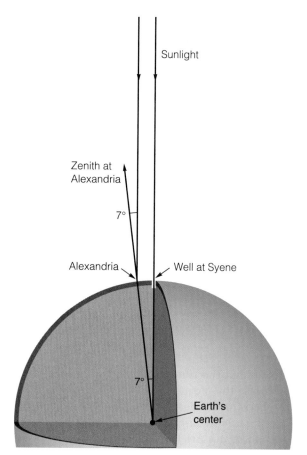

Figure 4-9
On the day of the summer solstice, sunlight fell to the bottom of a well at Syene, but the sun was about one-fiftieth of a circle (7°) south of the zenith at Alexandria. This told Eratosthenes that the distance from Syene to Alexandria was one-fiftieth of the circumference of the earth. Thus, he was able to calculate the radius of the earth.

could produce the same motion by having the sun, for instance, travel around a small circle that followed a larger circle around Earth. The compounded circular motion that he devised became the key element in the masterpiece of the last great astronomer of classical times, Ptolemy.

The Ptolemaic Universe

Claudius Ptolemaeus was one of the great astronomer-mathematicians of antiquity. His nationality and birth date are unknown, but he lived and worked in the Greek settlement at Alexandria about AD 140. He ensured the survival of Aristotle's universe by fitting to it a sophisticated mathematical model.

In agreement with Aristotle, the Ptolemaic model was geocentric (Earth-centered). In addition, it incorporated the Greek belief that the heavenly bodies move perfectly with uniform circular motion. But simple, circular paths centered on Earth do not account for the motions of the planets in the sky. The planets sometimes move faster and sometimes slower, and occasionally they appear to slow to a stop and move backward

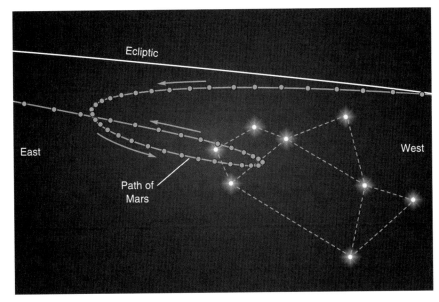

Figure 4-10
The word *planet* comes from the Greek for "wanderer," referring to the motion of the planets against the background of fixed stars. Here the motion of Mars along the ecliptic near the teapot shape of Sagittarius is shown at four-day intervals. Though the planet usually moves eastward, it sometimes slows to a stop and moves westward in retrograde motion.

over a period of months, tracing a **retrograde loop** (Figure 4-10).

To describe the complicated planetary motions and yet preserve a geocentric model with uniform circular motion, Ptolemy adopted a system of wheels within wheels. The planet moves in a small circle called an **epicycle,** and the center of the epicycle moves along a larger circle around the earth called a **deferent** (Figure 4-11). By adjusting the size of the circles and the rate of their motion, Ptolemy could account for most planetary movement. But as a final adjustment, he placed Earth

off center in the deferent circle and specified that the center of an epicycle would appear to move at constant speed only if viewed from a point called the **equant,** located on the other side of the deferent's center. Thus, by using a few dozen circles of various sizes rotating at various rates, Ptolemy's system predicted the positions of the planets (Figure 4-12).

About AD 140, Ptolemy included this work in a book now known as *Almagest.* He never knew that title; he called his work *Mathematical Syntaxis.* With his death, classical astronomy ended forever. Western civilization was slipping into the Middle Ages, and the invading Islamic nations dominated astronomy for almost 1000 years. Islamic astronomers translated, studied, and preserved many classical manuscripts, and in Arabic Ptolemy's book became *Al Magisti* (Greatest). Beginning in the 1200s, Europeans began recovering their classical heritage through translations from Arabic. In Latin, Ptolemy's book became *Almagestum* and thus our modern *Almagest.* For 1000 years, Islamic astronomers studied and preserved Ptolemy's work, but they made no significant improvement in his theory.*

At first the Ptolemaic system predicted the positions of the planets well, but as centuries passed, errors accumulated. A watch that gains only one second a year will keep time well for many years, but the error gradually accumulates. After a century the watch will be 100 seconds fast. So, too, did the errors in the Ptolemaic system gradually accumulate as the centuries

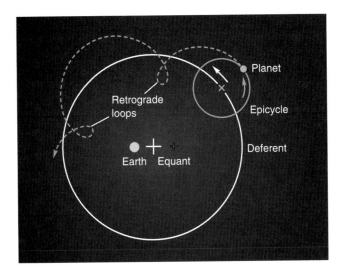

Figure 4-11
To explain the periodic retrograde motion of a planet, Ptolemy proposed that the planet traveled around a small circle called the epicycle, which revolved around a bigger circle called the deferent. To slightly adjust the speed of the planet, he supposed that Earth was slightly off center and that the center of the epicycle moved such that it appeared to move at a constant rate as seen from the equant point. A planet moving in this way would periodically pass through a retrograde loop and appear to move backward (westward) against the background of stars.

*Islamic scholars knew of both Ptolemy's and Eratosthenes's estimates for Earth's size, and they had settled on a final figure that was quite accurate. However, Columbus miscalculated the Arabic mile and thus thought Earth to be much smaller than it really is. That underestimation, combined with his overestimation of the eastward extent of Asia, led him to believe he could sail west to Japan. Had he known Earth's true size, he might never have attempted his voyage.

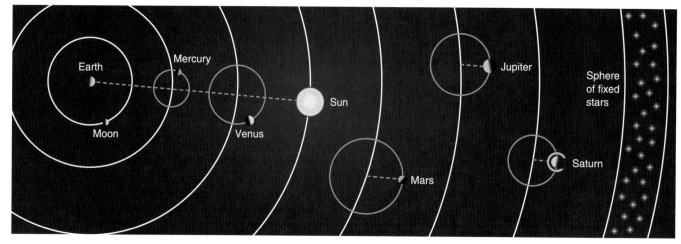

Figure 4-12

The Ptolemaic model of the universe is geocentric and based on uniform circular motion of epicycles moving around deferents. The centers of the epicycles of Mercury and Venus must always lie on the Earth–sun line, but Mars, Jupiter, and Saturn are free of this restriction. Notice that this modern illustration shows rings around Saturn and sunlight illuminating the globes of the planets, features that could not be known before the invention of the telescope.

passed. Islamic and later European astronomers tried to update the system, computing new constants and sometimes adding more epicycles. In the middle of the 13th century, a team of astronomers supported by King Alfonso X of Castile worked for 10 years revising the Ptolemaic system, publishing the result as the *Alfonsine Tables*. It was the last great adjustment of the Ptolemaic system.

Ptolemy was the last of the great classical astronomers, and his work dominated astronomical thought for almost 1500 years. The collapse of the Ptolemaic model and the rise of a new model of the universe make an exciting story, but it is not just the story of an astronomical idea. It includes the invention of science as a new way of knowing and understanding what we are and where we are.

REVIEW Critical Inquiry

How did the astronomy of Hipparchus and Ptolemy violate the principles of the early Greek philosophers Plato and Aristotle?

Hipparchus and Ptolemy lived very late in the history of classical astronomy, and they concentrated more on the mathematical problems and less on philosophical principles. They replaced the perfect spheres of Plato with nested circles in the form of epicycles and deferents. Earth was moved slightly away from the center of the deferent, so their models of the universe were not exactly geocentric, and the epicycles moved uniformly only as seen from the equant. The celestial motions were no longer precisely uniform, and the principles of geocentrism and uniform circular motion were weakened.

The work of Hipparchus and Ptolemy led eventually to a new understanding of the heavens, but first astronomers had to abandon the principle of uniform circular motion. The great philosopher Plato had argued for uniform circular motion from what seemed the most basic of first principles. What were those first principles?

4-2 The Copernican Revolution

Nicolaus Copernicus (Figure 4-13) triggered an Earth-shaking revision in human thought by proposing that the universe is not centered on Earth but is **heliocentric,** centered on the sun. That idea eventually led Galileo before the Inquisition and changed forever how we think of our world and ourselves. This Copernican Rev-

Figure 4-13

Copernicus proposed that the sun and not Earth was the center of the universe. Notice the heliocentric model on this stamp issued in 1973 to commemorate the 500th anniversary of his birth.

olution was much more than an upheaval in astronomy; it marks the birth of modern science.

Copernicus the Revolutionary

When Copernicus proposed that the universe was heliocentric, he risked controversy. According to Aristotle, the most perfect region was the heavens, and the most imperfect was Earth's center. Thus, the classical geocentric universe matched the commonly held Christian geometry of heaven and hell, and anyone who criticized the geometry of the Aristotelian universe challenged Christian belief and thus risked a charge of heresy.

Throughout his life, Copernicus was associated with the church. His uncle, by whom he was raised and educated, was an important bishop in Poland, and after studying canon law and medicine in some of the finest universities in Europe, Copernicus became a canon of the church at the age of 24. He was secretary and personal physician to his powerful uncle for 15 years. When his uncle died, Copernicus moved to quarters adjoining the cathedral in Frauenburg. No doubt his long association with the church added to his reluctance to publish controversial ideas.

He first wrote about a heliocentric universe in 1507, at the age of 34, in a short pamphlet that he distributed anonymously. Over the years, Copernicus worked on his book *De Revolutionibus Orbium Coelestium* (Figure 4-14a), but he hesitated to publish it even though other astronomers knew of his theories.

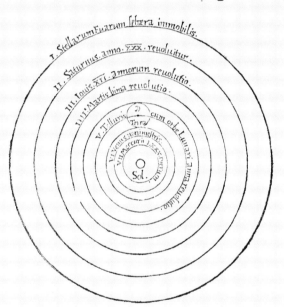

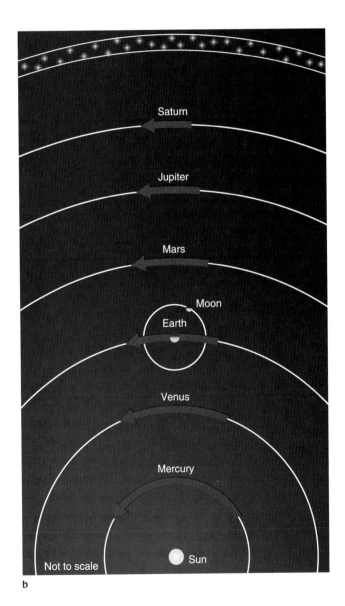

a

b

Figure 4-14

(a) The Copernican universe as reproduced in *De Revolutionibus*. Earth and all the known planets revolve in separate circular orbits about the sun (Sol) at the center. The outermost sphere carries the immobile stars of the celestial sphere. Notice the orbit of the moon around Earth. *(Yerkes Observatory)* (b) The model was elegant not only in its arrangement of the planets but in their motions. Orbital velocities (red arrows) decreased from Mercury, the fastest, to Saturn, the slowest. Compare the elegance of this model with the complexity of the Ptolemaic model in Figure 4-12.

This was a time of rebellion—Martin Luther was speaking harshly about fundamental church teachings, and others, scholars and scoundrels, questioned the authority of the church. Even astronomy could stir argument; moving Earth from its central place was a controversial and perhaps heretical idea. Thus, Copernicus did not permit publication of his book until 1543 when he realized he was dying.

The most important idea in the book was the placement of the sun at the center of the universe. That single innovation had an astonishing consequence—the retrograde motion of the planets was immediately explained in a straightforward way without the large epicycles that Ptolemy used. In the Copernican system, Earth moves faster along its orbit than the planets that lie further from the sun. Consequently, Earth periodically overtakes and passes these planets, and they appear to slow and fall behind (Figure 4-15). Because the planetary orbits do not lie in precisely the same plane, a planet does not resume its eastward motion in precisely the same path it followed earlier. Consequently, it describes a loop whose shape depends on the angle between the orbital planes.

Copernicus could explain retrograde motion without epicycles, and that was impressive. But he could not do away with epicycles completely. Copernicus, a classical astronomer, had tremendous respect for the old concept of uniform circular motion. In fact, he objected strongly to Ptolemy's use of the equant. It seemed arbitrary to Copernicus, a direct violation of the elegance of Aristotle's philosophy of the heavens. Copernicus called equants "monstrous" in that they violated both geocentrism and uniform circular motion. In his model, Copernicus returned to a strong belief in uniform circular motion. Although he did not need epicycles to explain retrograde motion, Copernicus discovered that the sun, moon, and planets suffered small variations in their motions—variations that he could not explain with the concept of uniform circular motion centered on the sun. Today, we recognize those variations as typical of objects following elliptical orbits, but Copernicus held firmly to uniform circular motion, so he had to use small epicycles to produce these minor variations in the motions of the sun, moon, and planets.

Because Copernicus imposed uniform circular motion on his model, it could not accurately predict the motions of the planets. The *Prutenic Tables* (1551) were based on the Copernican model, and they were not significantly more accurate than the *Alfonsine Tables* (1251), which were based on Ptolemy's model. Both could be in error by as much as 2°, four times the angular diameter of the full moon.

The Copernican *model* was inaccurate, but the Copernican *hypothesis* that the universe is heliocentric was correct. There are probably a number of reasons why the hypothesis gradually won acceptance in spite of the inaccuracy of the epicycles and deferents. The

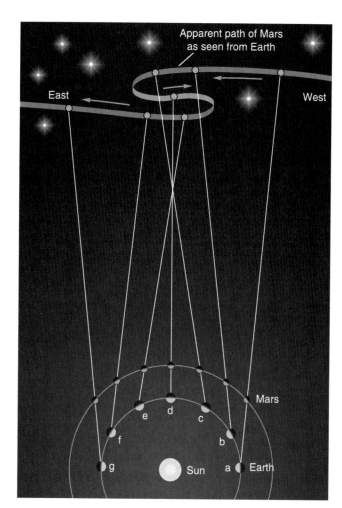

Figure 4-15
The Copernican explanation of retrograde motion. As Earth overtakes Mars (a–c), Mars appears to slow its eastward motion. As Earth passes Mars (d), Mars appears to move westward. As Earth draws ahead of Mars (e–g), Mars resumes its eastward motion against the background stars. Compare with Figure 4-10. The positions of Earth and Mars are shown at equal intervals of one month.

most important factor may be the elegance of the idea. Placing the sun at the center of the universe produced a symmetry among the motions of the planets that was pleasing to the eye and to the intellect (Figure 4-14b). All of the planets moved in the same way at speeds that were simply related to their distance from the sun. Venus and Mercury moved in the same kind of orbits as did all the other planets. Thus, the model may have won support not for its accuracy, but for its elegance.

The most astonishing consequence of the Copernican hypothesis was not what it said about the sun, but what it said about Earth. By placing the sun at the center, Copernicus made Earth move along an orbit just as the other planets did. By making Earth a planet, Copernicus revolutionized humanity's view of its place in the universe and triggered a controversy that would eventually bring Galileo before the Inquisition.

Although astronomers throughout Europe read and admired *De Revolutionibus*, they did not usually ac-

The Copernican Revolution is often cited as the perfect example of a scientific revolution. Over a few decades, astronomers abandoned a way of thinking about the universe that was almost 2000 years old and adopted a new set of ideas and assumptions. The American philosopher of science Thomas Kuhn has referred to a commonly accepted set of scientific ideas and assumptions as a scientific **paradigm.** Thus, the pre-Copernican astronomers had a geocentric paradigm that included uniform circular motion and the perfection of the heavens. That paradigm survived for many centuries until a new generation of astronomers was able to overthrow the old paradigm and establish a new paradigm that included heliocentrism and Earth's motion.

A scientific paradigm is powerful because it shapes our perceptions. It determines what we judge to be important questions and what we judge to be significant evidence. Thus, it is often difficult for us to recognize how our paradigms limit what we can understand. For example, the geocentric paradigm contained problems that seem obvious to us, but because astronomers before Copernicus lived and worked inside that paradigm, they had difficulty seeing the problems. Overthrowing an outdated paradigm is not easy, because we must learn to see nature in an entirely new way. Galileo, Kepler, and Newton saw nature from a new paradigm that would have been almost incomprehensible to astronomers of earlier centuries.

We can find examples of scientific revolutions in many fields, including biology, geology, genetics, and psychology. These scientific revolutions have been difficult and controversial because they have involved the overthrow of accepted paradigms. But that is why scientific revolutions are exciting. They give us not just a new idea or a new theory, but an entirely new insight into how nature works—a new way of seeing the world.

---

cept the Copernican hypothesis. The mathematics was elegant, and the astronomical observations and calculations were of tremendous value. Yet few astronomers believed, at first, that the sun actually was the center of the planetary system and that Earth moved. How the Copernican hypothesis became gradually recognized as correct has been named the Copernican Revolution because it involved not just the adoption of a new idea, but a total revolution in the way astronomers thought about the place of the earth (Window on Science 4-1).

Galileo the Defender

Most people know two facts about Galileo, and both facts are wrong. We should begin our discussion by getting those facts right: Galileo did not invent the telescope, and he was not condemned by the Inquisition for believing that Earth moved around the sun. Then why is Galileo so important that in 1979, almost 400 years after his trial, the Vatican reopened his case? As we discuss Galileo, we will discover that his trial concerned not just the place of the Earth but a new way of finding truth, a method known today as science.

Galileo Galilei (Figure 4-16) was born in Pisa (in what is now Italy) in 1564, and he studied medicine at the university there. His true love, however, was mathematics, and although he had to leave school early because of financial difficulties, he returned only four years later as a professor of mathematics. Three years later, he became professor of mathematics at the university at Padua. He remained there for 18 years.

During this time, Galileo seems to have adopted the Copernican model, although he admitted in a 1597 let-ter to Kepler that he did not support Copernicanism publicly because of the criticism such a declaration would bring. It was the telescope that drove Galileo to publicly defend the heliocentric model.

Galileo did not invent the telescope. It was apparently invented around 1608 by lens makers in Holland. Galileo, hearing descriptions in the fall of 1609, was able to build telescopes in his workshop (Figure 4-17). Galileo was not the first person to look at the sky through a telescope, but he was the first person to observe the sky systematically and apply his observations to the theoretical problem of the day—the place of the Earth (Figure 4-17).

What Galileo saw through his telescopes was so amazing he rushed a small book into print. *Sidereus Nuncius* (*The Sidereal Messenger*) reported three major discoveries. First, the moon was not perfect. It had mountains and valleys on its surface, and Galileo used the shadows to calculate the height of the mountains. The Ptolemaic model held that the moon was perfect, but Galileo showed that it was not only imperfect, it was a world like the Earth.

The second discovery reported in the book was that the Milky Way was made up of a myriad of stars too faint to see with the unaided eye. While intriguing, this discovery could not match the third discovery. Galileo's telescope revealed four new "planets" circling Jupiter, planets known today as the Galilean moons of Jupiter.

The moons of Jupiter supported the Copernican model over the Ptolemaic model. Critics of Copernicus had said Earth could not move because the moon would

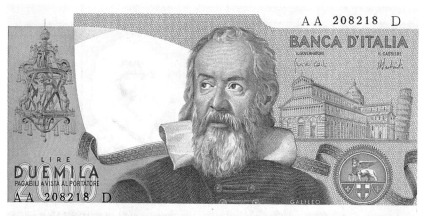

Figure 4-16
Galileo, remembered as the great defender of Copernicanism, also made important discoveries in the physics of motion. He is honored here on an Italian 2000-lira note. The reverse side shows one of his telescopes at lower right and a modern observatory above it.

be left behind. But Jupiter moved yet kept its satellites, so Galileo's discovery proved that Earth, too, could move and keep its moon. Also, the Ptolemaic model included the Aristotelian belief that all heavenly motion was centered on Earth at the center of the universe.

Galileo showed that Jupiter's moons revolve around Jupiter, so there could be other centers of motion.

Soon after *Sidereus Nuncius* was published, Galileo made two additional discoveries. When he observed the sun, he discovered sunspots, raising the suspicion

Figure 4-17
Galileo's telescopic discoveries generated intense interest and controversy. Some critics refused to look through a telescope lest it deceive them. *(Yerkes Observatory)*

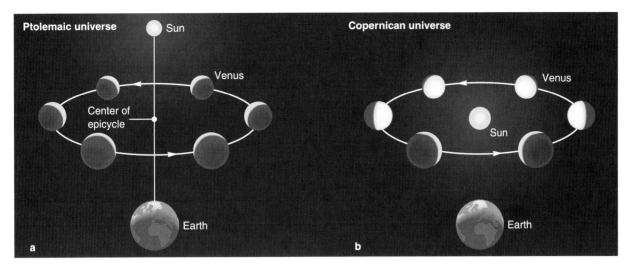

Figure 4-18
(a) If Venus moved in an epicycle centered on the Earth–sun line (see Figure 4-12), it would always appear as a crescent. (b) Galileo's telescope showed that Venus goes through a full set of phases, proving that it must orbit the sun.

that the sun was less than perfect. Further, by noting the movement of the spots, he concluded that the sun was a sphere and that it rotated on its axis. When he observed Venus, Galileo saw that it was going through phases like those of the moon. In the Ptolemaic model, Venus moves around an epicycle centered on a line between Earth and the sun. Thus, it would always be seen as a crescent (Figure 4-18a). But Galileo saw Venus go through a complete set of phases, proving that it did indeed revolve around the sun (Figure 4-18b).

Sidereus Nuncius was very popular and made Galileo famous. He became chief mathematician and philosopher to the Grand Duke of Tuscany in Florence. In 1611, Galileo visited Rome and was treated with great respect. He had long, friendly discussions with the powerful Cardinal Barberini, but he also made enemies. Personally, Galileo was outspoken, forceful, and sometimes tactless. He enjoyed debate, but most of all he enjoyed being right. Thus, in lectures, debates, and letters he offended important people who questioned his telescopic discoveries.

By 1616, Galileo was the center of a storm of controversy. Some critics said he was wrong, and others said he was lying. Some refused to look through a telescope lest it mislead them, and others looked and claimed to see nothing (hardly surprising given the awkwardness of those first telescopes). Pope Paul V decided to end the disruption, so when Galileo visited Rome in 1616 Cardinal Bellarmine interviewed him privately and ordered him to cease debate. There is some controversy today about the nature of Galileo's instructions, but he did not pursue astronomy for some years after the interview. Books relevant to Copernicanism, including *De Revolutionibus,* were banned.

In 1621 Pope Paul V died, and his successor, Pope Gregory XV, died in 1623. The next pope was Galileo's friend Cardinal Barberini, who took the name Urban VIII. Galileo rushed to Rome hoping to have the prohibition of 1616 lifted, and although the new pope did not revoke the orders, he did encourage Galileo. Thus, Galileo began to write his great defense of the Copernican model, finally completing it on December 24, 1629. After some delay, the book was approved by both the local censor in Florence and the head censor of the Vatican in Rome. It was printed in February 1632.

Called *Dialogo Dei Due Massimi Sistemi* (*Dialogue Concerning the Two Chief World Systems*), it confronts the ancient astronomy of Aristotle and Ptolemy with the Copernican model and with telescopic observations as evidence. Galileo wrote the book as a debate among three friends. Salviati, a swift-tongued defender of Copernicus, dominates the book; Sagredo is intelligent but largely uninformed. Simplicio is the dismal defender of Ptolemy. In fact, he does not seem very bright.

The publication of *Dialogo* created a storm of controversy, and it was sold out by August 1632, when the Inquisition ordered sales stopped. The book was a clear defense of Copernicus, and, either intentionally or unintentionally, Galileo exposed the pope's authority to ridicule. Urban VIII was fond of arguing that, as God was omnipotent, He could construct the universe in any form while making it appear to us to have a different form, and thus we could not deduce its true nature by mere observation. Galileo placed the pope's argument in the mouth of Simplicio, and Galileo's enemies showed the passage to the pope as an example of Galileo's disrespect. The pope thereupon ordered Galileo to face the Inquisition.

Galileo was interrogated by the Inquisition four times and was threatened with torture. He must have thought often of Giordano Bruno, tried, condemned, and burned at the stake in Rome in 1600. One of Bruno's offenses had been Copernicanism. But the trial did not center on Galileo's belief in Copernicanism. *Dialogo* had been approved by two censors. Rather, the trial centered on the instructions given Galileo in 1616. From his file in the Vatican, his accusers produced a record of the meeting between Galileo and Cardinal Bellarmine that included the statement that Galileo was "not to hold, teach, or defend in any way" the principles of Copernicus. Many historians believe that this document, which was signed neither by Galileo nor by Bellarmine nor by a legal secretary, was a forgery and that Galileo's true instructions were much less restrictive. But Bellarmine was dead, and Galileo had no defense.

The Inquisition condemned him not for heresy but for disobeying the orders given him in 1616. On June 22, 1633, at the age of 70, kneeling before the Inquisition, Galileo read a recantation admitting his errors. Tradition has it that as he rose he whispered *"E pur si muove"* ("Still it moves"), referring to Earth. Although he was sentenced to life imprisonment, he was actually confined at his villa for the next 10 years, perhaps through the intervention of the pope. He died there on January 8, 1642, 99 years after the death of Copernicus.

Galileo was not condemned for heresy, nor was the Inquisition interested when he tried to defend Copernicanism. He was tried and condemned on a charge we might call a technicality. Then why is his trial so important that historians have studied it for almost four centuries? Why have some of the world's greatest authors, including Bertolt Brecht, written about Galileo's trial? Why in 1979 did Pope John Paul II create a commission to reexamine the case against Galileo?

To understand the trial, we must recognize that it was the result of a conflict between two ways of understanding our universe. Since the Middle Ages, scholars had taught that the only path to true understanding was through religious faith. St. Augustine (AD 354–430) wrote *"Credo ut intelligame,"* which can be translated as "Believe in order to understand." But Galileo and other scientists of the Renaissance used their own observations to try to understand the universe, and when their observations contradicted Scripture, they assumed their observations were correct (Figure 4-19). Galileo paraphrased Cardinal Baronius in saying, "The Bible tells us how to go to heaven, not how the heavens go."

Various passages of Scripture seemed to contradict observation. For example, Joshua is said to have commanded the sun to stand still, not Earth to stop rotating (Joshua 10:12–13). In response to such passages, Galileo argued that we should "read the book of nature"—that is, we should observe the universe with our own eyes.

The trial of Galileo was not about the place of the Earth. It was not about Copernicanism. It wasn't really about the instructions Galileo received in 1616. It was about the birth of modern science as a rational way to understand our universe. The commission appointed

Figure 4-19
Although he did not invent it, Galileo will always be associated with the telescope because it was the source of the observational evidence he used to try to understand the universe. By depending on evidence instead of first principles, Galileo led the way to the invention of modern science as a way to know about the natural world.

by John Paul II in 1979, reporting its conclusions in October 1992, said of Galileo's inquisitors, "This subjective error of judgment, so clear to us today, led them to a disciplinary measure from which Galileo 'had much to suffer.'" Galileo was not found innocent in 1992 so much as the Inquisition was forgiven for having charged him in the first place.

The gradual change from reliance on personal faith to scientific observation came to a climax with the trial of Galileo. Since that time, scientists have increasingly reserved Scripture and faith for personal comfort and have depended on systematic observation to describe the physical world. The final verdict of 1992 was an attempt to bring some balance to this conflict. In his remarks on the decision, Pope John Paul II said, "A tragic mutual incomprehension has been interpreted as the reflection of a fundamental opposition between science and faith. . . . this sad misunderstanding now belongs in the past." Galileo's trial is over, but it continues to echo through history as part of our struggle to understand the place of the Earth in our universe.

How were Galileo's observations of the moons of Jupiter evidence against the Ptolemaic model?

Galileo presented his arguments in the form of evidence and conclusions, and the moons of Jupiter were key evidence. Ptolemaic astronomers argued from first principles given by Aristotle that the heavens were made up of perfect spheres rotating uniformly around Earth at the center. Then Earth could be the only center of motion. Although undermined by Ptolemy's epicycles and equants, this belief was still critical to classical astronomers. When Galileo saw four moons revolving around Jupiter, he knew that Earth was not the only center of motion in the universe, and thus at least that one aspect of classical astronomy had to be wrong.

Furthermore, Ptolemaic astronomers had argued that Earth could not move or it would leave the moon behind. The fact that the moon continued to circle Earth seemed to prove that Earth was immobile. But Galileo saw moons orbit Jupiter, and everyone agreed that Jupiter moved. Thus, if Jupiter could move and keep its moons, then Earth might move and keep its moon.

Of all of Galileo's telescopic observations, the moons of Jupiter caused the most debate. But the craters on the moon and the phases of Venus were also critical evidence. How did they argue against the Ptolemaic model?

Galileo faced the Inquisition because of a conflict between two ways of knowing and understanding the world. The Church taught faith and understanding through revelation, but the scientists of the age were inventing a new way to understand nature that relied on evidence. In astronomy, evidence means observations, so our story turns now from the philosophical problem of the place of the Earth to the observational problem of the motion of the planets.

4-3 The Puzzle of Planetary Motion

While Galileo was defending Copernicanism in Florence and Rome, two other astronomers were working in northern Europe, beyond the sway of the Inquisition. Although they, too, struggled to understand the place of the Earth, they approached the problem differently. They used the most accurate observations of the positions of the planets to try to discover the rules that govern planetary motion. When that puzzle was finally solved, astronomers understood why the Ptolemaic model of the universe does not work well and how to make the Copernican universe a precise predictor of planetary motion.

Tycho the Observer

The great observational astronomer of our story is a Danish nobleman, Tycho Brahe, born December 14, 1546, only three years after the publication of *De Revolutionibus* (Figure 4-20). Great figures in history are often referred to by their last name, but Tycho Brahe is usually called Tycho. Were he alive today, he would no doubt object to such familiarity from his obvious inferiors. He was well known for his vanity and lordly manners.

Tycho's college days were eventful. He was officially studying law with the expectation that he would enter Danish politics, but he made it clear to his family that his real interest was astronomy and mathematics. It was also during his college days that he fought a duel and received a wound that disfigured his nose. For the rest of his life, he wore false noses made of gold and silver and stuck on with wax. The disfigurement probably did little to improve his disposition.

Tycho's first astronomical observations were made while he was a student. In 1563, Jupiter and Saturn passed very near each other in the sky, nearly merging into a single point on the night of August 24. Tycho found that the *Alfonsine Tables* were a full month in error and that the *Prutenic Tables* were in error by a number of days. These discrepancies dismayed Tycho and sparked his interest in the motions of the planets.

In 1572, a brilliant "new star" (now called Tycho's supernova) appeared in the sky. Such changes in the stars puzzled classically trained astronomers. Aristotle had argued that the starry sphere was perfect and unchanging, and therefore such new stars had to lie closer to Earth than the moon. Layers below the moon were thought to be less perfect and thus more changeable.

Tycho observed that the new star moved westward through the night, keeping pace with the stars, so he

Figure 4-20
Tycho Brahe (1546–1601), a Danish nobleman, established an observatory at Hveen and measured planetary positions with high accuracy. His artificial nose is suggested in this engraving.

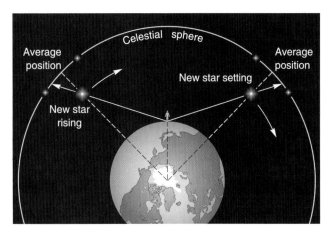

Figure 4-21

If an object lay lower than the celestial sphere, reasoned Tycho Brahe, then it should be seen east of its average position as it was rising and west of its average position when it was setting. Because he did not detect this daily parallax, he concluded the new star of 1572 had to lie on the celestial sphere.

concluded that its position should shift slightly through the night as seen by an observer on Earth. This change in position, called daily parallax, would make the star appear slightly east of its average position when it was in the eastern sky and slightly west of its average position when it was in the western sky (Figure 4-21). Tycho had great confidence in his talents for astronomy and mathematics and was sure he could detect this daily parallax, if it existed, by carefully measuring the position of the new star against background stars.

Tycho saw no daily parallax in the new star and thus concluded that it was farther away than the moon and was probably among the stars of the celestial sphere. The appearance of a new star in the supposedly unchanging heavens beyond the moon forced Tycho to question the Ptolemaic system. He summarized his results in a small book, *De Stella Nova* (*The New Star*), published in 1573.

The book attracted the attention of astronomers throughout Europe, and soon Tycho was summoned to the court of the Danish King Frederik II and offered funds to build an observatory on the island of Hveen just off the Danish coast. Tycho also received a steady source of income as landlord of a coastal district from which he collected rents. (He was not a popular landlord.) On Hveen, Tycho constructed a luxurious home with four towers specially equipped for astronomy and populated it with servants, assistants, and a dwarf to act as jester. Soon Hveen was an international center of astronomical study.

Tycho Brahe's Legacy

Tycho Brahe made no direct contribution to astronomical theory. Because he could measure no parallax for the stars, he concluded that Earth had to be stationary,

thus rejecting the Copernican hypothesis. However, he also rejected the Ptolemaic model because of its inaccurate predictions. Instead, he devised a complex model in which Earth was the immobile center of the universe around which the sun and moon moved. The other planets circled the sun. This model thus incorporated part of the Copernican model, but Earth—rather than the sun—was stationary. In this way, Tycho preserved the central, immobile Earth described in scripture (Figure 4-22a). Although Tycho's model, The Tychonic Universe, was very popular at first, the Copernican model replaced it within a century.

The true value of Tycho's work was observational. Because he was able to devise new and better instruments, he was able to make highly accurate observations of the positions of the stars, sun, moon, and planets. Tycho had no telescopes—they were not invented until the next century—so his observations were made by the naked eye peering along sights. All of his instruments were designed to measure angles in the sky. For example, his quadrant could measure angles up to 90° above the horizon (Figure 4-22b). By designing and building large instruments with great care, he was able to measure angles to high precision. He measured the positions of 777 stars to better than 4 minutes of arc and measured the positions of the sun, moon, and planets almost daily for the 20 years he stayed on Hveen.

Unhappily for Tycho, King Frederik II died in 1588, and his young son took the throne. Suddenly Tycho's temper, vanity, and noble presumptions threw him out of favor. In 1596, taking most of his instruments and books of observations, he went to Prague, the capital of Bohemia, and became imperial mathematician to the Holy Roman Emperor Rudolph II. His goal was to revise the *Alphonsine Tables* and publish the revision as a monument to his new patron. It would be called the *Rudolphine Tables*.

Tycho did not intend to base the *Rudolphine Tables* on the Ptolemaic system but rather on his own Tychonic system, proving once and for all the validity of his hypothesis. To assist him, he hired a few mathematicians and astronomers, including one Johannes Kepler. Then in November 1601, Tycho collapsed at a nobleman's home. Before he died, 11 days later, he asked Rudolph II to make Kepler imperial mathematician. Thus the newcomer, Kepler, became Tycho's replacement (though at one-half Tycho's salary).

Kepler the Analyst

No one could have been more different from Tycho Brahe than Johannes Kepler (Figure 4-23). He was born December 27, 1571, to a poor family in a region now included in southwestern Germany. His father was unreliable and shiftless, principally employed as a mercenary soldier fighting for whoever paid enough. He finally failed to return from a military expedition, either because he was killed or because he found circumstances

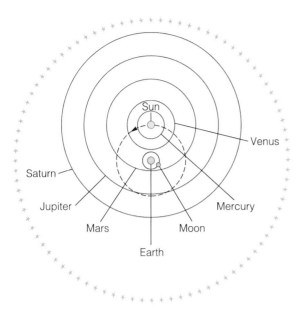

a

Figure 4-22
(a) Tycho Brahe's model of the universe held that Earth was fixed at the center of the starry sphere. The moon and sun circled Earth while the planets circled the sun. (b) Much of Tycho's success was due to his skill in designing large, accurate instruments. In this engraving of his mural quadrant, the figure of Tycho, his dog, and the background scene are painted on the wall within the arc of the quadrant. The observer (Tycho himself) at the extreme right peers through a sight out the loophole in the wall at the upper left to measure an object's angular distance above the horizon. *(Granger Collection, New York)*

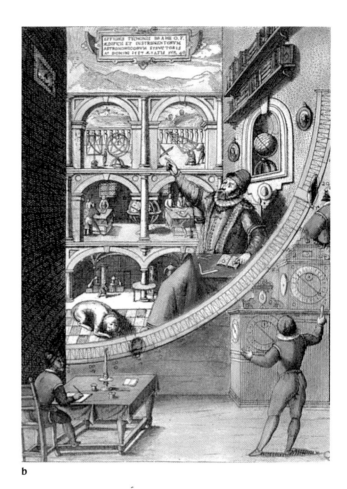

b

more to his liking. Kepler's mother was apparently an unpleasant and unpopular woman. She was accused of witchcraft in her later years, and Kepler had to defend her in a trial that dragged on for three years. She was finally acquitted but died the following year.

Kepler was the oldest of six children, and his childhood was no doubt unhappy. The family was not only poor and often lacking a father, but it was also Protestant in a predominantly Catholic region. In addition, Kepler was never healthy, even as a child, so it is surprising that he did well in the pauper's school he attended, eventually winning a scholarship to the university at Tübingen, where he studied to become a Lutheran pastor.

During his last year of study, Kepler accepted a job in Graz teaching mathematics and astronomy. Evidently, he was not a good teacher. He had few students his first year and none at all his second. His superiors put him to work teaching a few introductory courses and preparing an annual almanac that contained astronomical, astrological, and weather predictions. Through good luck, in 1595 some of his weather predictions were fulfilled and he gained a reputation as an astrologer and seer, and even in later life he earned money from his almanacs.

While still a college student, Kepler had become a believer in the Copernican theory, and at Graz he used his extensive spare time to study astronomy. By 1596,

the same year Tycho left Hveen, Kepler was ready to solve the mystery of the universe. That year he published a book called *The Forerunner of Dissertations on the Universe, Containing the Mystery of the Universe.* Like nearly all scientific works, the book was in Latin, and it is now known as *Mysterium Cosmographicum.*

Figure 4-23
Johannes Kepler (1571–1630) derived three laws of planetary motion from Tycho Brahe's observations of the positions of the planets. This Romanian stamp commemorates the 400th anniversary of Kepler's birth. Ironically, it contains an error—the horns of the crescent moon should be inclined upward from the horizon.

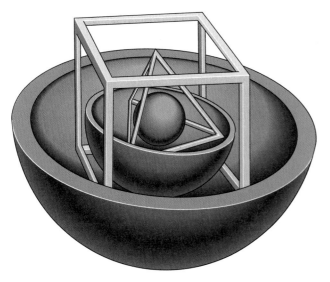

Figure 4-24

In this diagram based on one drawn by Kepler, the sizes of the celestial spheres carrying the outer three planets, Saturn, Jupiter, and Mars, are determined by spacers (blue) consisting of a cube and a tetrahedron. Inside the sphere of Mars (not shown here), the remaining regular solids separate the spheres of the Earth, Venus, and Mercury. The sun lay at the very center of this Copernican universe based on geometrical spacers.

The book begins with a long appreciation of Copernicanism and then goes on to speculate on the spacing of the planetary orbits. Kepler, as a Copernican, knew of six planets circling the sun—Mercury, Venus, Earth, Mars, Jupiter, and Saturn. According to his model, the spheres containing the six planets were separated from one another, and their relative sizes were fixed by the five regular solids*—the cube, tetrahedron, dodecahedron, icosahedron, and octahedron (Figure 4-24). Kepler gave astrological, numerological, and musical arguments for his theory and so followed the tradition of Pythagoras in believing that the order of the universe is underlain by musical (meaning mathematical) principles.

In the second half of the book, Kepler tried to fit the five solids to the planetary orbits and so demonstrated that he was a talented mathematician well versed in astronomy. He sent copies to Tycho and to Galileo, and both recognized his talents.

Life was unsettled for Kepler because of the persecution of Protestants in the region, so when Tycho invited him to Prague in 1600, he went readily, anxious to work with the famous astronomer. Tycho's sudden death in 1601 left Kepler in a position to use the observations from Hveen to analyze the motions of the planets and complete the *Rudolphine Tables.* Tycho's family, recognizing that Kepler was a Copernican and guessing that he would not follow the Tychonic system

*A regular solid is a three-dimensional body each of whose faces is the same. For example, a cube is a regular solid each of whose faces is a square.

in completing the *Rudolphine Tables,* sued to recover the instruments and books of observations. The legal wrangle went on for years. The family did recover the instruments Tycho had brought to Prague, but Kepler had the books, and he kept them.

Whether Kepler had any legal right to Tycho's records is debatable, but he put them to good use. He began by studying the motion of Mars, trying to deduce from the observations how the planet moves. By 1606, he had solved the puzzle of planetary motion. The orbit of Mars is an ellipse and not a circle. Thus he abandoned the 2000-year-old belief in the circular motion of the planets. But the mystery was even more complex. The planets do not move at a uniform speed along their elliptical orbits. Kepler recognized that they move faster when close to the sun and slower when farther away. Kepler therefore abandoned both circular motion and uniform motion.

Kepler published his results in 1609 in a book called *Astronomia Nova* (*New Astronomy*). Like Copernicus's book, *Astronomia Nova* did not become an instant success. It is written in Latin for other scientists and is highly mathematical. In some ways, the book is surprisingly advanced. For instance, Kepler discusses the force that holds the planets in their orbits and comes within a paragraph of discovering the principle of mutual gravitation.

Despite the abdication of Rudolph II in 1611, Kepler continued his astronomical studies. He wrote about a supernova that had appeared in 1604 (now known as Kepler's supernova) and about comets, and he wrote a textbook about Copernican astronomy. In 1619, he published *Harmonice Mundi* (*The Harmony of the World*), in which he returned to the cosmic mysteries of *Mysterium Cosmographicum.* The only thing of note in *Harmonice Mundi* is his discovery that the radii of the planetary orbits are related to the planets' orbital periods. That and his two previous discoveries are now recognized as Kepler's three laws of planetary motion.

Kepler's Three Laws of Planetary Motion

Although Kepler dabbled in the philosophical arguments of the day, he was a mathematician, and his triumph was the solution of the problem of the motion of the planets. The key to his solution was the ellipse.

An **ellipse** is a figure drawn around two points called *foci* in such a way that the distance from one focus to any point on the ellipse and back to the other focus equals a constant. This makes it easy to draw ellipses with two thumbtacks and a loop of string. Press the thumbtacks into a board, loop the string about the tacks, and place a pencil in the loop. If you keep the string taut as you move the pencil, it traces out an ellipse (Figure 4-25a).

The geometry of an ellipse is described by two simple numbers. The **semimajor axis, *a,*** is half of the longest diameter. The **eccentricity** of an ellipse, *e,* is the dis-

Even scientists misuse the words *hypothesis, theory,* and *law.* We must try to distinguish these terms from one another because they are key elements in science.

A **hypothesis** is a single assertion or conjecture that must be tested. It could be true or false. "All Texans love chili" is a hypothesis. To know whether it is true or false, we need to test it against reality by making observations or performing experiments. Copernicus asserted that the universe was heliocentric; his assertion was a hypothesis subject to testing.

In Chapter 2, we saw that a model (Window on Science 2-1) is a description of some natural phenomenon; it can't be right or wrong. A model is not a conjecture of truth but merely a convenient way to think about a natural phenomenon. Thus, a model such as the celestial sphere

cannot be a hypothesis. Copernicus used his hypothesis to build a model of the universe, and, to the extent that the model described the motion of the planets, it was useful.

A **theory** is a system of rules and principles that can be applied to a wide variety of circumstances. A theory may have begun as one or more hypotheses, but it has been tested, expanded, and generalized. Many textbooks refer to the "Copernican theory," but some experts argue that it was not complete in that it lacked a precise description of orbital motion and gravitation, features added by Kepler and Newton. Thus, it is probably more accurate to call it the Copernican hypothesis.

A **natural law** is a theory that almost everyone accepts as true. Such a theory has been tested over and over and applied

to many situations. Thus, natural laws are the most fundamental principles of scientific knowledge. Kepler's laws of planetary motion and Newton's laws of gravity and motion (Chapter 5) are examples of natural laws.

Scientists work by developing a hypothesis and testing it, often by building a model based on the hypothesis. If the model is a successful description of nature, the hypothesis is more likely to be true. As a hypothesis is tested repeatedly, expanded, and applied to many circumstances, we begin to refer to it as a theory. If the theory has been so intensively tested that it is almost universally accepted as true, we refer to it as a natural law.

---

tance between the foci divided by the longest diameter. To draw a circle with the string and tacks as shown in Figure 4-25a, we would move the two thumbtacks together, which shows that a circle is really just an ellipse with eccentricity equal to zero. As we move the thumbtacks farther apart, the ellipse becomes flatter, and the eccentricity moves closer to 1.

Kepler used ellipses to describe the motion of the planets in three fundamental rules that have been tested and confirmed so many times that astronomers now refer to them as natural laws (Window on Science 4-2). They are commonly called Kepler's laws of planetary motion (Table 4-1).

Table 4-1 Kepler's Laws of Planetary Motion

I. The orbits of the planets are ellipses with the sun at one focus.

II. A line from a planet to the sun sweeps over equal areas in equal intervals of time.

III. A planet's orbital period squared is proportional to its average distance from the sun cubed:

$$P_y^2 = a_{AU}^3$$

Figure 4-25
The geometry of elliptical orbits. (a) Drawing an ellipse with two tacks and a loop of string. (b) The semi-major axis, *a*, is half of the longest diameter. (c) Kepler's second law is demonstrated by a planet that moves from *A* to *B* in 1 month and from *A'* to *B'* in the same amount of time. The two blue segments have the same area.

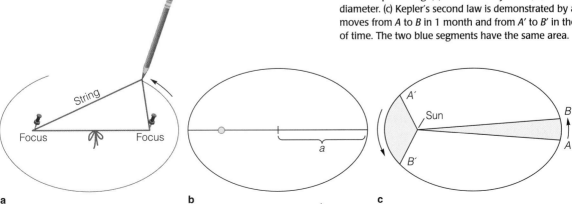

a   b   c

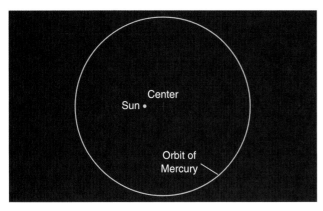

Figure 4-26
The orbits of the planets are nearly circular. In this scale drawing of the orbit of Mercury, the horizontal diameter and the vertical diameter of the orbit differ by less than 3 percent.

Kepler's first law states that the orbits of the planets around the sun are ellipses with the sun at one focus. Thanks to the precision of Tycho's observations and the sophistication of Kepler's mathematics, Kepler was able to recognize the elliptical shape of the orbits, even though they are nearly circular. Of the planets known to Kepler, Mercury has the most elliptical orbit, but even it deviates only slightly from a circle (Figure 4-26).

Kepler's second law states that a line from the planet to the sun sweeps over equal areas in equal intervals of time. This means that when the planet is closer to the sun and the line connecting it to the sun is shorter, the planet moves more rapidly to sweep over the same area that is swept over when the planet is farther from the sun. Thus the planet in Figure 4-25c would move from point A' to point B' in one month, sweeping over the area shown. But when the planet is farther from the sun, one month's motion would be shorter, from A to B.

The time that a planet takes to travel around the sun once is its orbital period, P, and its average distance from the sun equals the semimajor axis of its orbit, a. Kepler's third law tells us that these two quantities are related; the orbital period squared is proportional to the semimajor axis cubed. If we measure P in years and a in astronomical units, we can summarize the third law as:

$$P_y^2 = a_{AU}^3$$

The subscripts remind us that we must measure the period in years and the semimajor axis in astronomical units (AU).

We can use Kepler's third law to make simple calculations. For example, Jupiter's average distance from the sun is 5.20 AU. What is its orbital period? If a equals 5.20, then a^3 equals 140.6. The orbital period must be the square root of 140.6, which equals about 11.8 years.

It is important to notice that Kepler's three laws are empirical. That is, they describe a phenomenon without explaining why it occurs. Kepler derived them from Tycho's extensive observations, not from any fundamental assumption or theory. In fact, Kepler never knew what held the planets in their orbits or why they continued to move around the sun. His books are a fascinating blend of careful observation, mathematical analysis, and mystical theory.

The *Rudolphine Tables*

In spite of Kepler's recurrent involvement with astrology and numerology, he continued to work on the *Rudolphine Tables*. At last, in 1627, they were ready, and he financed their printing himself, dedicating them to the memory of Tycho Brahe. In fact, Tycho's name appears in larger type on the title page than Kepler's own. This is especially surprising when we recall that the tables were based on the heliocentric model of Copernicus and the elliptical orbits of Kepler, and not on the Tychonic system. The reason for Kepler's evident deference was Tycho's family, still powerful and still intent on protecting the memory of Tycho. They even demanded a share of the profits and the right to censor the book before publication, though they changed nothing but a few words on the title page.

The *Rudolphine Tables* was Kepler's masterpiece, the precise model of planetary motion that Copernicus had sought but failed to find. The accuracy of the *Rudolphine Tables* was strong evidence that both Kepler's model for planetary motion and the Copernican hypothesis for the place of Earth were correct. Copernicus would have been pleased.

Kepler solved the problem of planetary motion, and his *Rudolphine Tables* demonstrated his solution. Although he did not understand why the planets moved or why they followed ellipses, insights that had to wait half a century for Isaac Newton, Kepler's three rules worked. In science the only test of a theory is, "Does it describe reality?" Kepler's laws have been used for almost four centuries as a true description of orbital motion.

R E V I E W Critical Inquiry

What were the main differences between the *Alfonsine Tables*, the *Prutenic Tables*, and the *Rudolphine Tables*?

All three of these tables predicted the motions of the heavenly bodies, but only the *Rudolphine Tables* proved accurate. The *Alfonsine Tables,* produced in Toledo around AD 1250, were based on the Ptolemaic model, so they were geocentric and used uniform circular motion; consequently, they were not very accurate. The *Prutenic Tables,* published in 1551, were based on the Copernican model and so were heliocentric. But because the *Prutenic Tables* included the classical principle of uniform circular motion, they were no more accurate than the *Alfonsine Tables*. Kepler's *Rudolphine Tables* were Copernican in that they were heliocentric, but they included Kepler's three laws of planetary motion and thus finally predicted the motions of the planets.

One of the main reasons for the success of the Copernican hypothesis was not that it was accurate but that it was elegant. Compare the motions of the planets and the explanation of retrograde motion in the Copernican model with those in the Ptolemaic and Tychonic models. How was the Copernican model more elegant?

Kepler died November 15, 1630. During his life he had been an astrologer, a mystic, a numerologist, and a seer, but he became one of the world's great astronomers. His work not only overthrew the concept of uniform circular motion that had shackled the minds of astronomers for over 2000 years but also opened the way to the empirical study of planetary motion, a study that lies at the heart of modern astronomy.

4-4 Modern Astronomy

We date the origin of modern astronomy from the 99 years between the deaths of Copernicus and Galileo (1543 to 1642) because it was an age of transition. That period marked the transition between the Ptolemaic model of the universe and the Copernican with the attendant controversy over the place of the Earth. But that same period also marked a transition in the nature of astronomy in particular and science in general, a transition illustrated in the resolution of the puzzle of planetary motion. The puzzle was solved by precise observation and careful computation, techniques that are the foundation of modern science.

The discoveries of Kepler and Galileo found acceptance in the 1600s because the world was in transition. Astronomy was not the only thing changing during this period. The Renaissance is commonly taken to be the period between 1300 and 1600, and thus these 99 years of astronomical history lie at the culmination of the reawakening of learning in all fields (Figure 4-27). The world was open to new ideas and new observations. Martin Luther remade religion, and other philosophers

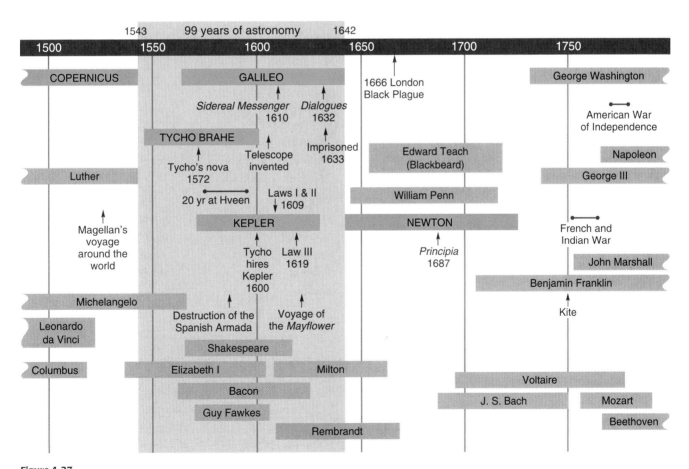

Figure 4-27
The 99 years between the death of Copernicus in 1543 and the birth of Newton in 1642 marked the transition from the ancient astronomy of Ptolemy and Aristotle to the revolutionary theory of Copernicus. This period saw the birth of modern scientific astronomy.

and scholars re-formed their areas of human knowledge. Had Copernicus not published his hypothesis, someone else would have suggested that the universe is heliocentric. History was ready to shed the Ptolemaic system.

In addition, this period marks the beginning of the modern scientific method. Beginning with Copernicus, Tycho, Kepler, and Galileo, scientists depended more and more on evidence, observation, and measurement. This, too, is coupled to the Renaissance and its advances in metalworking and lens making. Before our story began, no astronomer had looked through a telescope, because one could not be made. By 1642, not only telescopes but also other sensitive measuring instruments had transformed science into something new and precise. Also, the growing number of scientific societies increased the exchange of observations and hypotheses among scientists and stimulated more and better work. However, the most important advance was the application of mathematics to scientific questions. Kepler's work demonstrated the power of mathematical analysis, and as the quality of these numerical tools improved, the progress of science accelerated. Our story, therefore, is the story of the birth not just of modern astronomy but of modern science as well.

We will continue our story in the next chapter with the life of Isaac Newton. We will see how his accomplishments had their origins in the work of Galileo and Kepler.

Summary

Archaeoastronomy, the study of the astronomy of ancient cultures, is revealing that primitive astronomers could make sophisticated astronomical observations. The Stone Age people of Britain who built Stonehenge aligned its long avenue and Heelstone to point toward the rising sun at the summer solstice. Other stones mark sight lines toward the rising moon. It seems clear that Stonehenge was a calendrical device, and some believe that its builders used it to predict eclipses.

Other archaeoastronomy sites span the world. A great many stone circles dot England and northern Europe, and related patterns, called medicine wheels, show that North American Indians also practiced astronomy. The Mound Builders of the Midwest, the Indians of the Southwest, and the Aztecs, Maya, and Incas of Central and South America also studied astronomy.

Traditional histories of astronomy usually begin with the Greeks. The Greek philosopher Aristotle held that Earth is fixed at the center of the universe, and Ptolemy based a mathematical model of the moving planets on this geocentric universe. To preserve the concept of uniform circular motion, he used epicycles, deferents, and equants for each of the planets, the sun, and the moon.

Near the end of his life, in 1543, Nicolaus Copernicus published his hypothesis that the sun is the center of the universe. Because the teachings of Aristotle had been adopted as official church doctrine, the heliocentric universe of Copernicus was considered radical.

Galileo Galilei was the first astronomer to use the newly invented telescope for a systematic study of the heavens, and he found that the appearance of the sun, moon, and planets supported the heliocentric model. After he described his findings in a book, the church, in 1616, ordered him to stop teaching Copernicanism. After he published another book on the subject in 1632, he was tried by the Inquisition and ordered to recant. He was imprisoned in his home until his death in 1642.

In northern Europe, beyond the power of the church, Tycho Brahe rejected the Ptolemaic and Copernican systems. In Tycho's own system, Earth is fixed at the center of the universe, and the sun and moon revolve around Earth. The planets orbit the moving sun. Although his hypothesis is wrong, he was a brilliant observer and accumulated 20 years of precise measurements of the positions of the sun, moon, and planets.

At Tycho's death, one of his assistants, Johannes Kepler, became his successor. Analyzing Tycho's observations, Kepler finally discovered how the planets move. His three laws of planetary motion overthrew uniform circular motion. The first law says the orbits of the planets are elliptical rather than circular, and the second law says the planets do not move at a uniform rate but move faster when near the sun and slower when more distant. The third law relates orbital period to orbital size. These studies of the place of the Earth and the nature of planetary motion led to the birth of modern astronomy.

New Terms

archaeoastronomy	heliocentric universe
uniform circular motion	paradigm
geocentric universe	ellipse
parallax (p)	semimajor axis, a
eccentrics	eccentricity, e
retrograde loop	hypothesis
epicycle	theory
deferent	natural law
equant	

Review Questions

1. What evidence do we have that early human cultures observed astronomical phenomena?

2. Why did Plato propose that all heavenly motion was uniform and circular?

3. How do the epicycles of Mercury and Venus differ from those of Mars, Jupiter, and Saturn?

4. Why did Copernicus have to keep epicycles and deferents in his system?

5. Explain how each of Galileo's telescopic discoveries supported the Copernican hypothesis.

6. Galileo was condemned, but Kepler, also a Copernican, was not. Why not?

7. When Tycho observed the new star of 1572, he could detect no parallax. Why did that result undermine belief in the Ptolemaic system?

8. Does Tycho's model of the universe explain the phases of Venus that Galileo observed? Why or why not?

9. How do the first two of Kepler's three laws overthrow one of the basic beliefs of classical astronomy?

Discussion Questions

1. Suggest some reasons why Copernicus may have delayed publication of his book.

2. What parallels and contrasts do you see between the Vatican's recent reconsideration of Galileo's trial and the controversy over teaching evolution in public schools?

3. Why might Tycho have hesitated to hire Kepler? Why did Tycho appoint Kepler, a commoner and a Copernican, to be his scientific heir?

4. Tycho's description of the universe is usually called the Tychonic Universe. What should it be called—a hypothesis, a model, a theory, or a law?

Problems

1. Draw and label a diagram of the eastern horizon from northeast to southeast, and label the rising point of the sun at the solstices and equinoxes. (See Figures 2-18 and 4-18).

2. If you lived on Mars, which planets would describe retrograde loops? Which would never be visible as crescent phases?

3. Galileo's telescope showed him that Venus has a large angular diameter (61 seconds of arc) when it is a crescent and a small angular diameter (10 seconds of arc) when it is nearly full. Use the small-angle formula to find the ratio of its maximum distance to its minimum distance. Is this ratio compatible with the Ptolemaic universe shown in Figure 4-12?

4. Galileo's telescopes were not of high quality by modern standards. He was able to see the rings of Saturn, but he never reported seeing features on Mars. Use the small-angle formula to find the angular diameter of Mars when it is closest to Earth. How does that compare with the maximum angular diameter of Saturn's rings?

5. If a planet has an average distance from the sun of 4 AU, what is its orbital period?

6. If a space probe is sent into an orbit around the sun that brings it as close as 0.5 AU and as far away as 5.5 AU, what will be its orbital period?

7. Pluto orbits the sun with a period of 247.7 years. What is its average distance from the sun?

Critical Inquiries for the Web

1. The trial of Galileo is an important event in the history of science. We now know, and the church now recognizes, that Galileo's view was correct, but what were the arguments on both sides of the issue as it was unfolding? Research the Internet for documents that chronicle the trial, Galileo's observations and publications, and the position of the church. Use this information to outline the case for and against Galileo in the context of the times in which the trial occurred.

2. Take a "virtual tour" of Stonehenge by browsing the Web for information related to this megalithic site and the possibility that it was used for astronomical purposes. (Be careful to use legitimate scientific and historical Web sites in your survey.) Summarize the various astronomical alignments evident in the layout of the site.

Exploring *The Sky*

1. The three laws of Johannes Kepler mathematically describe the motion of the planets around the sun. The following activities help to get a qualitative and visual sense of what these laws mean. In *The Sky,* use a feature called the **3D Solar System Mode,** which gives you a three-dimensional view of the solar system as it might be viewed from a spaceship far from the sun. Set the inner solar system (planets out to Mars) in motion using a time increment of 1 day.

   a. How is Kepler's first law demonstrated in the simulation?

   b. How is Kepler's third law demonstrated in the simulation? Qualitatively, Kepler's third law states that planets further from the sun take longer to make one orbit about the sun.

   c. Use the solar system simulation to estimate the lengths of the years of Mercury, Earth, and Mars. You should be able to find these values to within a day or two. Compare your results to values given in the Appendix table "Properties of Planets."

   How to proceed: To view the 3-D solar system simulation, click on the **View** menu, and select **3-D Solar System Mode.** Or you can simply click on the **3-D Solar System** button on the **View Toolbar.** Use the magnifier button to zoom in on the inner solar system. Then set the time increment to 1 day on the **Time Skip toolbar,** and set the system in motion. It is helpful to have a grid in place to give you spatial reference: Press the **View** menu button, select **Reference Lines,** and under **General Lines,** select **Ecliptic.** Also note that you can change the viewing perspective using the arrow keys on the **Orientation Toolbar.**

 Go to the Brooks/Cole Astronomy Resource Center (www. brookscole.com/astronomy) for critical thinking exercises, articles, and additional readings from InfoTrac College Edition, Brooks/Cole's online student library.

CHAPTER FIVE

Newton, Einstein, and Gravity

Nature and Nature's laws lay hid

in night:

God said, "Let Newton be!" and all

was light.

Alexander Pope

Guidepost

Astronomers are gravity experts. All of the heavenly motions described in the preceding chapters are dominated by gravitation. Isaac Newton gets the credit for discovering gravity, but even Newton couldn't explain what gravity was. Einstein proposed that gravity is a curvature of space, but that only pushes the mystery further away. "What is curvature?" we might ask.

This chapter shows how scientists build theories to explain and unify ob-

servations. Theories can give us entirely new ways to understand nature, but no theory is an end in itself. Astronomers continue to study Einstein's theory, and they wonder if there is an even better way to understand the motions of the heavens.

The principles we discuss in this chapter will be companions through the remaining 22 chapters. Gravity is universal.

a

b

Figure 5-1
Isaac Newton (1642–1727) was the founder of modern physics. He made important discoveries in optics, developed three laws of motion, invented differential calculus, and discovered the law of mutual gravitation. These stamps were issued in 1987 to mark the 300th anniversary of the publication of his book *Principia*.

Figure 5-2
Although Galileo is often associated with the telescope, as on this Italian stamp, he also made systematic studies of the motion of falling bodies and made discoveries that led to the law of inertia.

Isaac Newton was born in Woolsthorpe, England, on December 25, 1642, and on January 4, 1643. This was not a biological anomaly but a calendrical quirk. Most of Europe, following the lead of the Catholic countries, had adopted the Gregorian calendar, but Protestant England continued to use the Julian calendar. Thus, December 25 in England was January 4 in Europe. If we take the English date, then Newton was born in the same year that Galileo Galilei died.

Newton went on to become one of the greatest scientists who ever lived (Figure 5-1), but even he admitted the debt he owed to those who had studied nature before him. He said, "If I have seen farther than other men, it is because I stood on the shoulders of giants."

One of those giants was Galileo (Figure 5-2). Although we remember Galileo as the defender of Copernicanism, he was also a trained scientist who studied the motions of falling bodies.

Newton based his own work on the discoveries of Galileo and others. During his life he studied optics, invented calculus, developed three laws of motion, and discovered the principle of mutual gravitation. Of these, the last two are the most important to us in this chapter because they make it possible to understand the orbital motion of the moon and planets.

Newton's laws of motion and gravity were astonishingly successful in describing the heavens. A friend of Newton, Edmund Halley, used Newton's laws to calculate the orbits of comets and discovered that certain comets that had been seen throughout recorded history were actually a single object returning every 75 years, now known as Comet Halley.

For over two centuries following the publication of Newton's works, astronomers used his laws to describe the universe. Then, early in this century, Albert Einstein proposed a new way to describe gravity. The new theory did not replace Newton's laws but rather showed that they were only approximately correct and could be seriously in error under special circumstances. We will see how Einstein's theories further extend our understanding of the nature of gravity. Just as Newton had stood on the shoulders of Galileo, Einstein stood on the shoulders of Newton.

5-1 Galileo and Newton

Johannes Kepler discovered three laws of planetary motion, but he never understood why the planets move along their orbits. At one place in his writings, he wonders if they are pulled along by magnetic forces emanating from the sun. At another place, he speculates that they are pushed along their orbits by angels beating their wings.

Newton refined Kepler's model of planetary motion but did not perfect it. In science, a model is an intellectual conception of how nature works (Window on Science 2-1). No model is perfect; although Kepler's model was better than Aristotle's, it still included angels to push the planets along their orbits. Newton improved Kepler's model by expanding it into a general theory of motion and gravity. In fact, most scientists now refer to Newton's *law* of gravity. Whether we call it a model, a theory, or a law, we must recognize that it was not perfect. Newton never understood what gravity was. It was as mysterious as an angel pushing the moon inward toward Earth instead of forward along the moon's orbit.

To understand science, we must understand the nature of scientific descriptions. The scientist studies nature by either creating new theories or refining old theories. Yet a theory can never be perfect, because it can never represent the universe in all its intricacies. Instead, a theory must be a limited approximation of a single phenomenon, such as orbital motion. It is fitting that Newton's discoveries all began with Kepler's fellow Copernican, Galileo.

Galileo and Motion

Even before Galileo built his first telescope, he had begun studying the motion of freely moving bodies. After the Inquisition condemned and imprisoned him in 1633, he continued his study of motion. He seems to have realized that he would have to understand motion before he could truly understand the Copernican system. That he was eventually able to formulate principles that later led Newton to the laws of motion and the theory of gravity is a tribute to Galileo's ability to set aside authority and think for himself.

The authority of the age was Aristotle, whose ideas on motion were hopelessly confused. Aristotle said that the world is made up of four classical elements: earth, water, air, and fire. According to this idea, all earthly things—wood, rock, flesh, bone, metals, and so on—are made up of mixtures of earth and water. The motions of bodies are determined by their natural tendencies to move toward their proper places in the cosmos. Earth and water, and all things composed of them, move toward the center of the cosmos, which, in Aristotle's geocentric universe, is Earth's center (Figure 5-3). Thus, objects fall downward because they are moving toward their proper place.* On the other hand, fire and air move upward toward the heavens, their proper place in the cosmos.

Aristotle called these motions **natural motions** to distinguish them from **violent motions** produced, for instance, when we push on an object and make it move other than toward its proper place. According to Aristotle, such motions stop as soon as the force is removed. To explain how an arrow could continue to move upward even after it had left the bowstring, he said currents in the air around the arrow carried it forward even though the bowstring was no longer pushing it.

These ideas about the proper places of objects, natural and violent motion, and the necessity of a force to preserve motion were still accepted theory in Galileo's time. In fact, in 1590, when Galileo was 26, he wrote a short work called *De Motu* (*On Motion*) that deals with the proper places of objects and their natural motions.

*This is one reason why Aristotle had to have a geocentric universe. If Earth's center had not also been the center of the cosmos, his explanation of gravity would not have worked.

Figure 5-3
In this drawing of Aristotle's universe, earth and water are located at the center, surrounded by layers of the remaining two elements: air and fire. *(From Peter Apian,* Cosmographia, *1539)*

In Galileo's time and for the two preceding millennia, scholars had commonly tried to resolve problems of science by referring to authority. To analyze the flight of a cannonball, for instance, they would turn to the writings of Aristotle and other classical philosophers and try to deduce what those philosophers would have said on the subject. This generated a great deal of discussion but little real progress. Galileo broke with this tradition and conducted his own experiments.

He began by studying the motions of falling bodies, but he quickly discovered that the velocities were so great and the times so short that he could not measure them accurately. Consequently, he began using polished bronze balls rolling down gently sloping inclines. In that instance, the velocity is lower and the time longer. Using an ingenious water clock, he was able to measure the time the balls took to roll given distances down the incline, and he correctly recognized that these times are proportional to the times taken by falling bodies.

He found that falling bodies do not fall at constant rates, as Aristotle had said, but are accelerated. That is, they move faster with each passing second. Near Earth's surface, a falling object will have a velocity of 9.8 m/s (32 ft/s) at the end of 1 second, 19.6 m/s (64 ft/s) after 2 seconds, 29.4 m/s (96 ft/s) after 3 seconds, and so on. Each passing second adds 9.8 m/s (32 ft/s) to the object's velocity (Figure 5-4). In modern terms, this is called the **acceleration of gravity** at Earth's surface.

Galileo also discovered that the acceleration does not depend on the weight of the object. This, too, is contrary to the teachings of Aristotle, who believed that

Figure 5-4
Galileo found that a falling object is accelerated downward. Each second, its velocity increases by 9.8 m/s (32 ft/s).

heavy objects, containing more earth and water, fall with higher velocity. Galileo found that the acceleration of a falling body is the same whether it is heavy or light. According to some accounts, he demonstrated this by dropping balls of iron and wood from the top of the Leaning Tower of Pisa to show that they would fall to-gether and hit the ground at the same time (Figure 5-5a).

a

b

Figure 5-5
(a) According to tradition, Galileo demonstrated that the acceleration of a falling body is independent of its weight by dropping balls of iron and wood from the Leaning Tower of Pisa. In fact, air resistance would have confused the result. (b) In a historic television broadcast from the moon on August 2, 1971, David Scott dropped a hammer and a feather at the same instant. In the vacuum of the lunar surface, they fell together. *(NASA)*

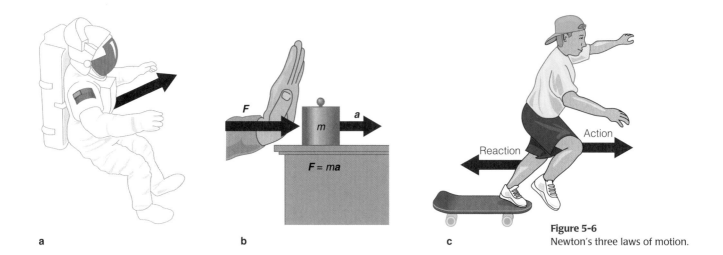

a   b   c

Figure 5-6
Newton's three laws of motion.

In fact, he probably didn't perform this experiment. It would not have been conclusive anyway because of air resistance. More than 300 years later, Apollo 15 astronaut David Scott, standing on the airless moon, demonstrated Galileo's discovery by dropping a feather and a steel geologist's hammer. They hit the lunar surface at the same time (Figure 5-5b).

Having described natural motion, Galileo turned his attention to violent motion—that is, motion directed other than toward an object's proper place in the cosmos. He pointed out that an object rolling down an incline is accelerated toward Earth and that an object rolling up the same incline is decelerated. If the surface were perfectly horizontal and frictionless, he reasoned, there could be no acceleration or deceleration to change the object's velocity, and, in the absence of friction, the object would continue to move forever. In his own words, "any velocity once imparted to a moving body will be rigidly maintained as long as the external causes of acceleration or retardation are removed."

This is contrary to Aristotle's belief that motion can continue only if a force is present to maintain it. In fact, Galileo's statement is a perfectly valid summary of the law of inertia, which became Newton's first law of motion.

Galileo published his work on motion in 1638, two years after he had become entirely blind and only four years before his death. The book was called *Mathemat-*

ical Discourses and Demonstrations Concerning Two New Sciences, Relating to Mechanics and to Local Motion. It is known today as *Two New Sciences.*

The book is a brilliant achievement for a number of reasons. To understand motion, Galileo had to abandon the authority of the ancients, devise his own experiments, and draw his own conclusions. In a sense, this was the first example of experimental science. But Galileo also had to generalize his experiments to discover how nature worked. Though his apparatus was finite and plagued by friction, he was able to imagine an infinite, totally frictionless plane on which a body moves at constant velocity. In his workshop, the law of inertia was obscure, but, in his imagination, it was clear and precise.

Newton and the Laws of Motion

Newton's three laws of motion (Table 5-1) are critical to our understanding of orbital motion. They apply to any moving object, from an automobile driving along a highway to galaxies colliding with each other.

The first law is really a restatement of Galileo's law of inertia. An object continues at rest or in uniform motion in a straight line unless acted upon by some force. An astronaut drifting in space will travel at a constant rate in a straight line forever if no forces act on him or her (Figure 5-6a).

Newton's first law also explains why a projectile continues to move after all forces have been removed—for instance, why an arrow continues to move after leaving the bowstring. The object continues to move because it has momentum. We can think of an object's **momentum** as a measure of the amount of motion.

Momentum depends on velocity and mass.* A low-velocity object such as a paper clip tossed across a room has little momentum, and we could easily catch it in our hand. But the same paper clip fired at the speed

---

Table 5-1 Newton's Three Laws of Motion

I. A body continues at rest or in uniform motion in a straight line unless acted upon by some net force.

II. The acceleration of a body is inversely proportional to its mass, directly proportional to the net force, and in the same direction as the net force.

III. To every action, there is an equal and opposite reaction.

---

*Mathematically, momentum is the product of mass and velocity.

Mass: A Fundamental Parameter of Science

Five fundamental parameters—mass, energy, thermal energy, density, and pressure—are so critical in science that we will discuss them individually in separate Windows on Science. Here we will discuss **mass**, the measure of the amount of matter in an object. A bowling ball, for example, contains a large amount of mass; but a child's rubber ball contains less matter than the bowling ball, and we say it is less massive.

Mass is not the same as weight. Our weight is the force that Earth's gravity exerts on the mass of our bodies. Because gravity pulls us downward, we press against the bathroom scale and we can measure our weight. Floating in space, we would have no weight at all; a bathroom scale would be useless. But our bodies would still contain the same amount of matter, so we would still have mass.

Sports analogies illustrate the importance of mass in dramatic ways. A bowling ball, for example, must be massive in order to have a large effect on the pins it strikes. Imagine trying to knock down all the pins with a bowling ball no more massive than a balloon. Even in space, where the bowling ball would be weightless, a low-mass bowling ball would have little effect on the pins. On the other hand, runners want track shoes that have low mass and thus are easy to move. Imagine trying to run a 100-meter dash wearing track shoes that were as massive as bowling balls. They would be very hard to move, and it would be difficult to accelerate away from the starting block. The shot put takes muscle because the shot is massive, not because it is heavy. Imagine throwing the shot in space where it would have no weight. It would still be massive, and it would take great effort to start it moving.

Mass is a unique measure of the amount of material in an object. Using the metric system (Appendix), we will measure mass in kilograms.

---

of a rifle bullet would have tremendous momentum, and we would not dare try to catch it.

Momentum also depends on the **mass** of an object (Window on Science 5-1). To see how, imagine that, instead of tossing us a paper clip, someone tosses us a bowling ball. A bowling ball contains much more mass than a paper clip and therefore has much greater momentum at the same velocity.

Newton's first law explains the consequences of the conservation of momentum. When we say that momentum is conserved, we mean that it remains constant until something acts to change it. A moving object has a given amount of momentum. To change that momentum, we must exert some force on the object to change either the speed or the direction. Newton's first law and the concept of momentum came from the work of Galileo.

Newton's second law of motion discusses forces, and Galileo did not talk about forces. He spoke instead of accelerations. Newton saw that an acceleration is the result of a force acting on a mass (Figure 5-6b). Newton's second law is commonly written as:

$$F = ma$$

Once again, we must carefully define terms. An **acceleration** is a change in velocity, and a **velocity** is a directed rate of motion. By rate of motion, we mean, of course, a speed, but the word *directed* has a special meaning. Speed itself does not have any direction associated with it, but velocity does. If you drive a car in a circle at 55 mph, your speed is constant, but your velocity is changing. Thus, an object experiences an ac-

celeration if its speed changes or if its direction of motion changes. Every automobile has three accelerators—the gas pedal, the brake pedal, and the steering wheel. All three change the auto's velocity.

The acceleration of a body is proportional to the force applied. This is reasonable. If we push gently against a grocery cart, we do not expect a large acceleration. The second law of motion also says that the acceleration is inversely proportional to the mass of the body. This, too, is reasonable. If the cart were filled with bricks and we pushed it gently, we would expect very little result. If it were full of inflated balloons, however, it would move easily in response to a gentle push. Finally, the second law says that the resulting acceleration is in the direction of the force. This is also what we would expect. If we push on a cart that is not moving, we expect it to begin moving in the direction we push.

The second law of motion is important because it establishes a precise relationship between cause and effect (Window on Science 5-2). Objects do not just move. They accelerate due to the action of a force. Moving objects do not just stop. They decelerate due to a force. Also, moving objects don't just change direction for no reason. Any change in direction is a change in velocity and requires the presence of a force. Aristotle said that objects move because they have a tendency to move. Newton said that objects move due to a specific cause, a force.

Newton's third law of motion specifies that for every action there is an equal and opposite reaction. In other

One of the most often used and least often stated principles of science is cause and effect, and we could argue that Newton's second law of motion was the first clear statement of the principle. Ancient philosophers such as Aristotle argued that objects moved because of tendencies. They said that earth and water, and objects made of earth and water, had a natural tendency to move toward the center of the universe. This natural motion had no cause but was inherent in the nature of the objects. But Newton's second law says $F = ma$. If an object (of mass m) changes its motion (a in the equation), then it must be acted on by a force (F in the equation). Any effect (a) must be the result of a cause (F).

The principle of cause and effect goes far beyond motion. The principle of cause and effect gives scientists confidence that every effect has a cause. Hearing loss in certain laboratory rats, color changes in certain chemical dyes, and explosions on certain stars are all effects that must have causes. All of science is focused on understanding the causes of the effects we see. If the universe were not rational, then we could never expect to discover causes. Newton's second law of motion was arguably the first statement that the behavior of the universe depends rationally on causes.

---

words, forces must occur in pairs directed in opposite directions. For example, if you stand on a skateboard and jump forward, the skateboard will shoot away backward. As you jump, your feet exert a force against the skateboard, which accelerates it toward the rear. But forces must occur in pairs, so the skateboard must exert an equal but opposite force on your feet that accelerates your body forward (Figure 5-6c).

Mutual Gravitation

The three laws of motion led Newton to consider the force that causes objects to fall. The first and second laws tell us that falling bodies accelerate downward because some force must be pulling downward on them. Newton wondered what that force could be.

Newton was also aware that some force has to act on the moon. The moon follows a curved path around Earth, and motion along a curved path is accelerated motion. The second law says that a force is required in order to make the moon follow that curved path.

Newton wondered if the force that holds the moon in its orbit could be the same force that causes rocks to roll downhill—gravity. He was aware that gravity extends at least as high as the tops of mountains, but he did not know if it could extend all the way to the moon. He believed that it could, but he thought it would be weaker at greater distances. He also guessed that its strength would decrease as the square of the distance increased.

This relationship, the **inverse square law,** was familiar to Newton from his work on optics, where it applied to the intensity of light. A screen set up 1 unit from a candle flame receives a certain amount of energy on each square meter. However, if that screen is moved to a distance of 2 units, the light that originally illuminated 1 m² must cover 4 m² (Figure 5-7). Thus, the intensity of the light is inversely proportional to the square of the distance to the screen.

Newton made two assumptions that enabled him to predict the strength of Earth's gravity at the distance of the moon. He assumed that the strength of gravity follows the inverse square law and that the critical distance is not the distance from Earth's surface but the distance from Earth's center. Because the moon is about 60 Earth radii away, Earth's gravity at the distance of the moon should be about 60² times less than at Earth's

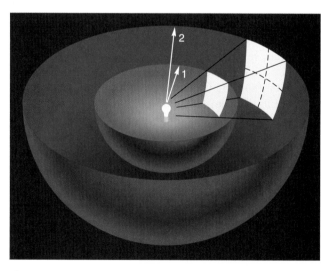

Figure 5-7
The inverse square law. A light source is surrounded by spheres with radii of 1 unit and 2 units. The light falling on an area of 1 m² on the inner sphere spreads to illuminate an area of 4 m² on the outer sphere. Thus, the brightness of the light source is inversely proportional to the square of the distance.

surface. Instead of being 9.8 m/s² at Earth's surface, it should be about 0.0027 m/s² at the distance of the moon.

Now, Newton wondered, could this acceleration keep the moon in orbit? He knew the moon's distance and its orbital period, so he could calculate the actual acceleration needed to keep it in its curved path. The answer is 0.0027 m/s². To the accuracy of Newton's data for Earth's radius, it was exactly what his assumptions predicted. The moon is held in its orbit by gravity, and gravity obeys the inverse square law.

Newton's third law says that forces always occur in pairs, and this quickly led him to realize that gravity is mutual. If Earth pulls on the moon, then the moon must pull on Earth. Gravitation is a general property of the universe. The sun, the planets, and all their moons must also attract each other by mutual gravitation. In fact, every particle in the universe must attract every other particle, and Newtonian gravity is often called universal mutual gravitation.

Because we do not find ourselves attracting particles of mass—books, rocks, passing birds, and so on—drawn to us by our personal gravity, Newton concluded that the gravitational force depends on the mass of the object. Large masses, like Earth, have strong gravity, but smaller masses, like people, have weaker gravity. Combining this dependence on mass with the inverse square law led to the famous formula for the gravitational force between masses M and m:

$$F = -G\frac{Mm}{r^2}$$

The constant G is the gravitational constant, and r is the distance between the masses. The negative sign tells us the force is attractive, pulling the masses together and making r decrease. In plain English, Newton's law of gravitation states: The force of gravity between two masses M and m is proportional to the product of the masses and inversely proportional to the square of the distance between them.

Newton's description of gravity was a difficult idea for physicists of his time to accept because it is an example of action at a distance. Earth and moon exert forces on each other although there is no physical connection between them. Modern scientists resolve this problem by referring to gravity as a **field**. Earth's presence produces a gravitational field directed toward Earth's center. The strength of the field decreases according to the inverse square law. Any particle of mass in that field experiences a force that depends on the mass of the particle and the strength of the field at the particle's location. The resulting force is directed toward the center of the field.

The field is an elegant way to describe gravity, but it does not tell us what gravity is. For that we must wait until later in this chapter, when we discuss Einstein's theory of curved space-time.

What do the words *universal* and *mutual* mea...
term "universal mutual gravitation"?

Newton realized that gravity was a force that drew ṭ moon toward Earth. But his third law of motion said that forces always occur in pairs, so if Earth attracted the moon, then the moon had to attract Earth. That is, gravitation had to be *mutual* between any two objects.

When Newton looked beyond the Earth–moon system, he realized that Earth's gravity must attract the sun as well as the moon, and if Earth attracted the sun, then the sun had to attract Earth. Furthermore, if the sun had gravity that acted on Earth, then the same gravity would act on all the planets and even on the distant stars. Step by step, Newton's third law of motion led him to conclude that gravitation had to apply to all masses in the universe. That is, it had to be *universal*.

Aristotle explained gravity in a totally different way. How did Aristotle account for a falling apple? Could that explanation account for a hammer falling on the surface of the moon?

Newton first came to understand gravity by thinking about the orbital motion of the moon. Gravitation is tremendously important in astronomy, and its most important manifestation is orbital motion.

5-2 Orbital Motion

Newton's laws of motion and gravitation make it possible to understand why the planets move along their orbits. We can understand how they are held in their curved paths, and we can even discover why Kepler's laws work.

Orbiting Earth

To illustrate the principle of orbital motion, imagine that we position a large cannon at the top of a mountain, point it horizontally, and fire it (Figure 5-8). The cannonball falls to Earth some distance from the foot of the mountain. The more gunpowder we use, the faster the ball travels and the farther from the foot of the mountain it falls. If we use enough powder, the ball travels so fast it never strikes the ground. Earth's gravity pulls it toward Earth's center, but Earth's surface curves away from it at the same rate at which it falls. We say it is in orbit. If the cannonball is high above the atmosphere, where there is no friction, it will continue to fall around Earth forever. Real Earth satellites do fall back to Earth sometimes, but that is caused by friction with the tenuous upper layers of Earth's atmosphere.

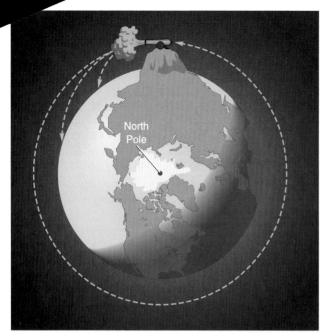

Figure 5-8
A cannon on a high mountain could put its projectile into orbit if it could achieve a high enough velocity. Newton published a similar figure in *Principia*.

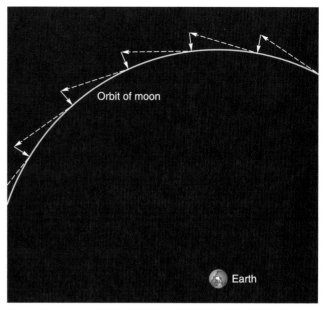

Figure 5-9
If no forces acted on the moon it would follow a straight line (dashed arrows) and disappear into space. Gravity accelerates the moon toward Earth, as shown by the inward pointing arrows. Each moment, the moon's forward motion and its inward acceleration combine to produce the curved path of the moon's orbit.

The first law of motion says that an object in motion tends to stay in motion in a straight line unless acted upon by some force. Thus, the cannonball in our example travels in a curve around Earth only because Earth's gravity acts to pull it away from its straight-line motion.

Similarly, the moon would move in a straight line at constant speed (dashed arrows in Figure 5-9) were it not for gravity accelerating it toward Earth. Each second, the moon moves 1020 m (3350 ft) eastward along its orbit and falls about 1.6 mm (about $\frac{1}{16}$ inch) toward Earth. The combination of these two motions results in a closed orbit around Earth.

A space capsule orbiting Earth is not "beyond Earth's gravity," to use a phrase common in old science fiction movies. Like the moon, the space capsule is accelerated toward Earth. That acceleration, combined with its lateral motion, places it in a closed orbit. Astronauts inside are weightless not because they are beyond Earth's gravity, but because they are falling at the same rate as their capsule. Rather than saying the astronauts are weightless, we should more accurately say they are in free fall.

To be precise, we should avoid saying that the moon orbits Earth. In fact, the moon and Earth orbit each other. Gravitation is mutual, and if Earth pulls on the moon, then the moon pulls on Earth. The two bodies revolve around their common **center of mass,** the balance point of the system. If Earth and the moon could be connected by a massless rod and placed in a uniform gravitational field such as that near Earth's surface, the system would balance at its center of mass like a child's seesaw (Figure 5-10). As the system revolves, the moon and Earth each describe an orbit around the center of mass.

As we might expect from experience on a seesaw, the balance point, or center of mass, is closest to the more massive body. The center of mass of the Earth–moon system lies only 4708 km (2926 miles) from Earth's center. This places the center of mass inside Earth. As the moon revolves around its orbit, Earth swings about center of mass.

Orbital Velocity

If we were about to ride a rocket into orbit, we would have a critical question. How fast do we have to travel to stay in orbit? An object's **circular velocity** is the lateral velocity the object must have to remain in a circular orbit. If we assume the mass of our spaceship is small compared with the mass of the object we expect to orbit, Earth in this case, then the circular velocity is:

$$V_c = \sqrt{\frac{GM}{r}}$$

In this formula, M is the mass of the central body in kilograms, r is the radius of the orbit in meters, and G is the gravitational constant, 6.67×10^{-11} N m²/kg², where N stands for a newton, a measure of force. This formula is all we need to calculate how fast an object must travel to stay in a circular orbit.

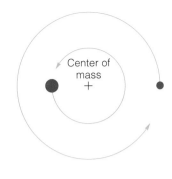

Figure 5-10
The center of mass is the balance point between two objects. (a) If two bodies were connected by a massless rod and placed in a uniform gravitational field like that at Earth's surface, they would balance at their center of mass. (b) If two bodies orbit each other, their center of mass remains fixed, and the objects move around it.

a   b

For example, how fast does the moon travel in its orbit? Earth's mass is 5.98×10^{24} kg, and the radius of the moon's orbit is 3.84×10^8 m. Then the moon's velocity is:

$$V_c = \sqrt{\frac{6.67 \times 10^{-11} \times 5.98 \times 10^{24}}{3.84 \times 10^8}} = \sqrt{\frac{39.9 \times 10^{13}}{3.84 \times 10^8}}$$
$$= \sqrt{1.04 \times 10^6} = 1020 \text{ m/s}$$

This calculation shows that the moon travels 1.02 km along its orbit each second. That is the circular velocity of the moon.

A satellite just above Earth's atmosphere is only about 200 km above Earth's surface, or 6578 km from Earth's center, so Earth's gravity is much stronger and the satellite must travel much faster to stay in a circular orbit. We can use the formula above to find that the circular velocity just above Earth's atmosphere is about 7790 m/s, or 7.79 km/s. This is about 17,400 miles per hour and shows why putting satellites into Earth orbit takes such large rockets. Not only must the rocket lift the satellite above the earth's atmosphere, but the rocket must tip over and accelerate the satellite to circular velocity.

Geosynchronous Orbits

One orbit that is particularly useful is called a **geosynchronous orbit.** An object in such an orbit remains over the same spot on Earth (Figure 5-11). Communications satellites, for example, are often placed in a geosynchronous orbit.

The orbital velocity of an Earth satellite depends on its distance from Earth's center. If an object orbits near Earth, it has a high velocity and a small orbit, and as a result its orbital period is short, about 90 minutes. At a greater distance from Earth, the orbital velocity is lower and the orbit is larger. Thus, the satellite has a longer orbital period. At a distance of about 42,000 km (26,000 miles), the orbital period is 24 hours.

If we launch an Earth satellite eastward around Earth in a geosynchronous orbit above Earth's equator, it will remain above a fixed point on Earth. Earth ro-

tates eastward in 24 hours, and the satellite circles its orbit in 24 hours. If we aim a communications antenna at the satellite, we do not have to move the antenna in order to track the satellite. The many television dishes that dot the landscape are aimed at communications satellites orbiting 42,000 km above Earth's equator in

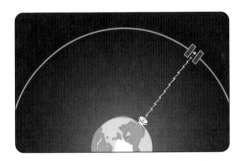

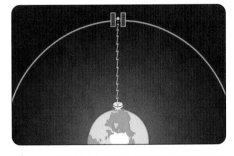

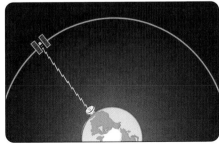

Figure 5-11
An Earth satellite in a geosynchronous orbit revolves eastward around Earth in an orbit with a period of 24 hours. Thus, it remains fixed above a given spot on Earth, and a dish antenna can remain pointed at the satellite 24 hours a day. Communications satellites are often put into geosynchronous orbit.

geosynchronous orbit. Communications satellites may be one of the most commonly used applications of the orbital physics worked out by Kepler and Newton.

Open and Closed Orbits

Kepler's laws refer to elliptical orbits, which include circular orbits because a circle is a special kind of ellipse. These orbits are called **closed orbits** because they return back on themselves. If you were in a spacecraft orbiting Earth in a closed orbit, you would return to the same place in the orbit time after time.

If your spaceship were in an elliptical orbit, your distance from Earth and your orbital velocity would vary. Recall from Chapter 3 that the point on the orbit of closest approach to Earth is called perigee, and the point of greatest distance is called apogee. Notice that apogee and perigee occur on opposite sides of an elliptical orbit (Figure 5-12). Kepler's second law tells us that the velocity of your spaceship would be greatest at perigee and lowest at apogee. To remain in a safe orbit, you would want to be certain that your orbit remained above Earth's atmosphere even at perigee. If your perigee was too low, friction with the atmosphere would slow your spaceship, your orbit would decay, and you would eventually reenter the atmosphere.

Newton's laws reveal the existence of another kind of orbit. An **open orbit** (or trajectory) leads away from the central body, never to return. If you were in a spacecraft following an open orbit around Earth, you would be leaving Earth along a path that did not return. Open orbits are also called **escape orbits.**

We can analyze these open and closed orbits by thinking again of the cannon in Figure 5-8. If the cannonball travels at the circular velocity, it will fall in a circular orbit around Earth. Of course, if the velocity is much too small, the ball will hit Earth. However, if the velocity of the ball is only slightly less than the circular velocity, the ball will follow an elliptical orbit with its highest point, apogee, at the cannon (Figure 5-13). If the velocity of the ball is slightly greater than the circular velocity, it will again follow an elliptical path, but with the lowest part of its orbit, perigee, at the cannon. All of these orbits are closed—the cannonball returns to its starting point. If we really performed this experiment, the cannonball would circle Earth in about 90 minutes, returning to its starting point and smashing the cannon to fragments.

If the cannonball's velocity equals or exceeds **escape velocity,** the velocity needed to escape from the surface of a body, the cannonball will follow an open orbit. If the velocity equals the escape velocity, the orbit is a parabola, and if the velocity exceeds the escape velocity, the orbit is a hyperbola. The mathematical definitions of the parabola and hyperbola are not important here, but in both cases the cannonball leaves Earth, never to return.

Calculating Escape Velocity

If we launch a rocket upward, it will consume its fuel in a few moments and reach its maximum speed. From that point on, it will coast upward. How fast must a rocket travel to coast away from Earth and escape? We know, of course, that no matter how far it travels, it can never escape from Earth's gravity. The effects of Earth's gravity extend to infinity. It is possible, however, for a rocket to travel so fast initially that gravity can never slow it to a stop. Thus, it could leave Earth.

The escape velocity is the velocity required to escape from the surface of an astronomical body. Here we are interested in escaping from Earth or a planet; in later chapters, we will consider the escape velocity from stars, galaxies, and even a black hole.

The escape velocity, V_e, is given by a simple formula:

$$V_e = \sqrt{\frac{2GM}{R}}$$

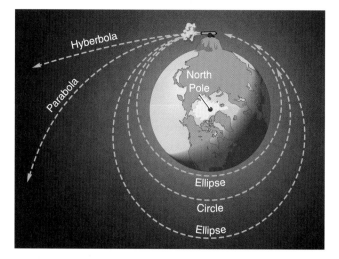

Figure 5-13
If an object's orbital velocity is less than escape velocity, it will follow an elliptical or circular orbit. If its velocity equals or exceeds escape velocity, it will follow a parabolic or hyperbolic orbit and escape from Earth.

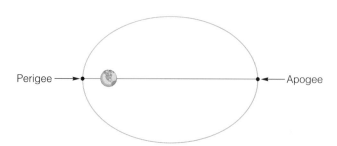

Figure 5-12
In an elliptical orbit around Earth, the point of closest approach is called perigee. The point on the orbit most distant from Earth is called apogee. Note that apogee and perigee lie at the ends of the major axis of the ellipse.

Here G is the gravitational constant 6.67×10^{-11} N m^2/kg^2, M is the mass of the astronomical body in kilograms, and r is its radius in meters. (This formula is very similar to the formula for circular velocity.)

We can find the escape velocity from Earth by looking up its mass, 5.98×10^{24} kg, and its radius, 6.38×10^6 m. Then the escape velocity is:

$$V_e = \sqrt{\frac{2 \times 6.67 \times 10^{-11} \times 5.98 \times 10^{24}}{6.38 \times 10^6}} = \sqrt{\frac{7.98 \times 10^{14}}{6.38 \times 10^6}}$$
$$= \sqrt{1.25 \times 10^8} = 11{,}200 \text{ m/s} = 11.2 \text{ km/s}$$

This is equal to about 25,000 miles per hour.

Notice that the formula tells us that escape velocity depends on both mass and radius. A massive body might have a low escape velocity if it has a very large radius. We will meet such objects when we discuss giant stars. On the other hand, a rather low-mass body can have a very large escape velocity if it has a very small radius, a condition we will discuss when we meet black holes.

Circular velocity and escape velocity are two aspects of Newton's laws of gravity and motion. Once Newton understood gravity and motion, he could do what Kepler had failed to do—he could explain why the planets obey Kepler's laws of planetary motion.

Kepler's Laws Reexamined

Now that we understand Newton's laws, gravity, and orbital motion, we can understand Kepler's laws of planetary motion in a new way.

Kepler's first law says that the orbits of the planets are ellipses with the sun at one focus. Kepler wondered why the planets keep moving along these orbits, and now we know the answer. They move because there is nothing to slow them down. Newton's first law says that a body in motion stays in motion unless acted on by some force. The gravity of the sun accelerates the planets inward toward the sun and holds them in their orbits, but it doesn't pull backward on the planets, so they don't slow to a stop. With no friction, they must continue to move.

The orbits of the planets are ellipses because gravity follows the inverse square law. In one of his most famous problems, Newton proved that if a planet moves in a closed orbit under the influence of an attractive force that follows the inverse square law, then the planet must follow an elliptical path.

Kepler's second law says that a planet moves faster when it is near the sun and slower when it is farther away. Once again, Newton's discoveries explain why. Earlier we saw that a body moving on a frictionless surface will continue to move in a straight line until it is acted on by some force; that is, the object has momentum. But an object set rotating on a frictionless surface

will continue rotating until something acts to speed it up or slow it down. Such an object has **angular momentum,** a measure of the rotation of the body about some point. A planet circling the sun has a given amount of angular momentum, and, with no outside influences to speed it up or slow it down, it must conserve its angular momentum. That is, its angular momentum must remain constant.

Mathematically, a planet's angular momentum is the product of its mass, velocity, and distance from the sun. This explains why a planet must speed up as it comes closer to the sun along an elliptical orbit. Its angular momentum is conserved, so as its distance from the sun decreases, its velocity must increase. In the same way, the planet's velocity must decrease as its distance from the sun increases.

This conservation of angular momentum is actually a common human experience. Skaters spinning slowly can draw their arms and legs closer to their axis of rotation and, through conservation of angular momentum, spin faster (Figure 5-14). To slow their rotation, they again extend their arms. Similarly, divers can spin rapidly in the tuck position and slow their rotation by stretching into the extended position.

Kepler's third law is also explained by a conservation law, but in this case it is the law of conservation of energy (Window on Science 5-3). A planet orbiting the sun has a specific amount of energy that depends only on its average distance from the sun. That energy can be divided between energy of motion and energy stored in the gravitational attraction between the planet and the sun. The energy of motion depends on how fast the planet moves, and the stored energy depends on the size of its orbit. The relation between these two kinds of energy is fixed by Newton's laws. That means there has to be a fixed relationship between the rate at which

Figure 5-14
Skaters demonstrate conservation of angular momentum when they spin faster by drawing their arms and legs closer to their axis of rotation.

Five fundamental parameters—mass, energy, thermal energy, density, and pressure—are so critical in science that we will discuss them individually in separate Windows on Science. Here we will discuss **energy.** Physicists define energy as the ability to do work, but we might paraphrase that definition as the ability to produce a change. Certainly a moving body has energy. A planet moving along its orbit, a cement truck rolling down the highway, and a golf ball sailing down the fairway all have the ability to produce a change. Imagine colliding with any of these objects! But energy need not be represented by motion. Sunlight falling on a green plant, on photographic film, or on unprotected skin can produce chemical changes, and thus light is a form of energy.

Energy can take many forms. Batteries and gasoline are examples of chemical energy, and uranium fuel rods contain nuclear energy. A tank of hot water contains thermal energy. Even a weight on a high shelf can represent stored energy. Imagine a bowling ball falling off a high shelf onto your desk. That would produce significant change.

Much of science is the study of how energy flows from one place to another place, producing changes. A biologist might study the way a nerve cell transmits energy along its length to a muscle, while a geologist might study how energy flows as heat from Earth's interior and deforms Earth's surface. In such processes, we see energy being transformed from one state to another. Sunlight (energy) is absorbed by ocean plants and stored as sugars and

starches (energy). When the plant dies, it and other ocean life are buried and become oil (energy), which we pump to the surface and burn in automobile engines to produce motion (energy).

Aristotle believed that all change originated in the motion of the starry sphere and flowed down to Earth. Modern science has found a more sophisticated description of the continual change we see around us, but we are still interested in the way energy flows through the world and produces change. Energy is the pulse of the natural world.

Using the metric system (Appendix), we will measure energy in **joules** (abbreviated J). One joule is about as much energy as that given up when an apple falls from a table to the floor.

---

a planet moves around its orbit and the size of the orbit—between its orbital period P and the orbit's semimajor axis a. This is just Kepler's third law.

Newton's Version of Kepler's Third Law

The equation for circular velocity is actually a version of Kepler's third law, as we can prove with three lines of simple algebra. The result is one of the most useful formulas in astronomy.

The equation for circular velocity, as we have seen, is:

$$V_c = \sqrt{\frac{GM}{r}}$$

The orbital velocity of a planet is simply the circumference of its orbit divided by the orbital period:

$$V = \frac{2\pi r}{P}$$

If we substitute this for V in the equation for circular velocity and solve for P^2, we get:

$$P^2 = \frac{4\pi^2}{GM} r^3$$

Here M is just the total mass of the system in kilograms. For a planet orbiting the sun, we can use the mass of the sun for M, because the mass of the planet is negligible compared to the mass of the sun. In a later chapter, we will apply this formula to two stars orbiting each

other, and then the mass M will be the sum of the two masses. For a circular orbit, r equals the semimajor axis a, so this formula is a general version of Kepler's third law, $P^2 = a^3$. In Kepler's version, we use the units AU and years, but in this formula, known as Newton's version of Kepler's third law, we use units of meters, seconds, and kilograms. G, of course, is the gravitational constant.

This is a very powerful formula. Astronomers use it to find the masses of bodies by observing orbital motion. If, for example, we observe a moon orbiting a planet and we can determine the size of its orbit, r, and the orbital period, P, we can use this formula to solve for M, the total mass of the planet plus the moon. There is no other way to find masses in astronomy, and we will use this formula in later chapters to find the masses of stars, galaxies, and planets.

This discussion is a good illustration of the power of Newton's work. By carefully defining motion and gravity and by giving them mathematical expression, Newton was able to derive new truths, among them Newton's version of Kepler's third law. His work changed science and made it into something new, precise, and exciting.

Astronomy after Newton

Newton published his work in July 1687 in a book called *Philosophiae Naturalis Principia Mathematica* (*Mathematical Principles of Natural Philosophy*), now

When you read about any science, you should notice that scientific theories face in two directions. They look back into the past and explain phenomena we have previously observed. For example, Newton's laws of motion and gravity explained how the planets moved. But theories also face forward in that they enable us to make predictions about what we should find as we explore further. Thus, Newton's laws allowed astronomers to calculate the orbits of comets, predict their return, and eventually understand their origin.

Scientific predictions are important in two ways. First, if a theory leads to a prediction and scientists later discover the prediction was true, the theory is confirmed, and scientists gain confidence that it is a true description of nature. But predictions are important in science in a second way. Using an existing theory to make a prediction may lead us into an unexplored avenue of knowledge. Thus, the first theories of genetics made predictions that confirmed the genetic theory of inheritance, but those predictions also created a new understanding of how living creatures evolve.

As you read about any scientific theory, think about both what it can explain and what it can predict.

known simply as *Principia* (Figure 5-15). It is one of the most important books ever written. The principles changed astronomy, science, and the way we think about nature.

Principia changed astronomy and ushered in the age of gravitational astronomy. No longer did astronomers appeal to the whim of the gods to explain things in the heavens. No longer did they speculate on why the planets move. They now knew that the motions of the heavenly bodies are governed by simple, universal rules that describe the motions of everything from planets to falling apples. Suddenly the universe was understandable in simple terms.

Newton's laws of motion and gravity made it possible for astronomers to calculate the orbits of planets and moons. Not only could they explain how the heavenly bodies move, they could predict future motions (Window on Science 5-4). This subject, known as gravitational astronomy, dominated astronomy for almost 200 years and is still important. It included the calculation of the orbits of comets and asteroids and the theoretical prediction of the existence of two planets, Neptune and Pluto.

Principia also changed science in general. The works of Copernicus and Kepler had been mathematical, but no book before had so clearly demonstrated the power of mathematics as a language of precision. Newton's arguments were couched in geometrical terms instead of the new analytical methods developed by European mathematicians, but *Principia* was so powerful an illustration of the quantitative study of nature that scientists around the world adopted mathematics as their most powerful tool.

Also, *Principia* changed the way we think about nature. Newton showed that the rules that govern the universe are simple. Particles move according to three rules of motion and attract each other with a force called gravity. These motions are predictable, and that makes

Figure 5-15
Newton, working from the discoveries of Galileo and Kepler, derived three laws of motion and the principle of mutual gravitation. He and some of his discoveries are honored on this English pound note. Notice the diagram of orbital motion in the background and the open copy of *Principia* in Newton's hands.

the universe a vast machine based on a few simple rules. It is complex only in that it contains a vast number of particles. In Newton's view, if he knew the location and motion of every particle in the universe, he could, in principle, derive the past and future of the universe in every detail. This mechanical determinism has been undermined by modern quantum mechanics, but it dominated science for more than two centuries during which scientists thought of nature as a beautiful clockwork that would be perfectly predictable if we knew how all the gears meshed.

Most of all, Newton's work broke the last bonds between science and formal philosophy. Newton did not speculate on the good or evil of gravity. He did not debate its meaning. Not more than a hundred years before, scientists would have argued over the "reality" of gravity. Newton didn't care for these debates. He wrote, "It is enough that gravity exists and suffices to explain the phenomena of the heavens."

Figure 5-16
Einstein's greatest accomplishments, the special theory of relativity and the general theory of relativity, were developed as he analyzed the nature of motion and time.

REVIEW Critical Inquiry

How do Newton's laws of motion explain the orbital motion of the moon?

If Earth and the moon did not attract each other, the moon would move in a straight line in accord with Newton's first law of motion and vanish into space. Instead, gravity pulls the moon toward Earth's center, and the moon accelerates toward Earth. This acceleration is just enough to pull the moon away from its straight-line motion and cause it to follow a curve around Earth. In fact, it is correct to say that the moon is falling, but because of its lateral motion it continuously misses Earth.

Every orbiting object is falling toward the center of its orbit but is moving laterally fast enough to compensate for the inward motion, and it follows a curved orbit. If this is true, then how can astronauts float inside spacecraft in a "weightless" state? Why might "free fall" be a more accurate term?

—

Newton's laws of motion and gravitation are critical in astronomy not only because they describe orbital motion but also because they describe the interaction of astronomical bodies. Newton made other discoveries, but we reserve these for later chapters. Now we move forward two centuries to see how Einstein described gravity in a new and powerful way.

5-3 Einstein and Relativity

In the early years of the last century, Albert Einstein (1879–1955) (Figure 5-16) began thinking about how motion and gravity interact. He soon gained international fame by showing that Newton's laws of motion and gravity were only partially correct. The revised theory became known as the theory of relativity. As we will see, there are really two theories of relativity.

Special Relativity

Einstein began by thinking about how moving observers see events around them. His analysis led him to the first postulate of relativity, also known as the principle of relativity:

> **First postulate** (the principle of relativity)
> Observers can never detect their *uniform* motion except relative to other objects.

You may have experienced the first postulate while sitting on a train in a station. You suddenly notice that the train on the next track has begun to creep out of the station. However, after several moments you realize that it is your own train that is moving and that the other train is still motionless on its track.

Consider another example. Suppose you are floating in a spaceship in interstellar space and another spaceship comes coasting by (Figure 5-17a). You might conclude that it is moving and you are not, but someone in the other ship might be equally sure that you are moving and it is not. The principle of relativity says that there is no experiment you can perform to decide which ship is moving and which is not. This means that there is no such thing as absolute rest—all motion is relative.

Because neither you nor the people in the other spaceship could perform any experiment to detect your absolute motion through space, the laws of physics must have the same form in both spaceships. Otherwise, experiments would give different results in the two ships, and you could decide who was moving. Thus, a more general way of stating the first postulate refers to these laws of physics:

First postulate (alternate version) The laws of physics are the same for all observers, no matter what their motion, so long as they are not *accelerated.*

The words *uniform* and *accelerated* are important. If either spaceship were to fire its rockets, then its velocity would change. The crew of that ship would know it because they would feel the acceleration pressing them into their couches. Accelerated motion, therefore, is different—we can always tell which ship is accelerating and which is not. The postulates of relativity discussed here apply only to observers in uniform motion. That is why the theory is called **special relativity.**

The first postulate fit with Einstein's conclusion that the speed of light must be constant for all observers. No matter how you move, your measurement of the speed of light has to give the same result (Figure 5-17b). This became the second postulate of special relativity:

Second postulate The velocity of light is constant and will be the same for all observers independent of their motion relative to the light source.

Once Einstein had accepted the basic postulates of relativity, he was led to some startling discoveries. Newton's laws of motion and gravity worked well as long as distances were small and velocities were low. But when we begin to think of very large

distances or very high velocities, Newton's laws are no longer adequate to describe what happens. Instead, we must use relativistic physics. For example, special relativity shows that the observed mass of a moving particle depends on its velocity. The higher the velocity, the

a

b

Figure 5-17

(a) The principle of relativity says that observers can never detect their uniform motion, except relative to other observers. Thus, neither of these travelers can decide who is moving and who is not. (b) If the principle of relativity is correct, the velocity of light must be a constant for all observers. If the velocity of light depended on the motion of the observer through space, then these travelers could decide who was moving and who was not.

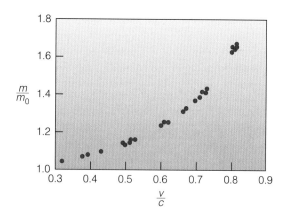

Figure 5-18
The observed mass of moving electrons depends on their velocity. As the ratio of their velocity to the velocity of light, *v/c*, gets larger, the mass of the electrons in terms of their mass at rest, m/m_0, increases. Such relativistic effects are quite evident in particle accelerators, which accelerate atomic particles to very high velocities.

greater the mass of the particle. This is not significant at low velocities, but it becomes very important as the velocity approaches the velocity of light. Such increases in mass are observed whenever physicists accelerate atomic particles to high velocities (Figure 5-18).

This discovery led to yet another insight. The relativistic equations that describe the energy of a moving particle predict that the energy of a motionless particle is not zero. Rather, its energy at rest is m_0c^2. This is of course the famous equation:

$$E = m_0c^2$$

The *c* is the speed of light and the m_0 is the mass of the particle when it is at rest. (We must specify mass this way because one of the consequences of relativity is that a particle's mass depends on its velocity.) This simple formula suggests that mass and energy are related, and we will see in later chapters how nature can convert one into the other inside stars.

For example, suppose that we convert 1 kg of matter into energy. We must express the velocity of light as 3×10^8 m/s, and our result is 9×10^{16} joules (J) (approximately equal to a 20-megaton nuclear bomb). (Recall that a joule is a unit of energy roughly equivalent to the energy given up when an apple falls from a table to the floor.) Our simple calculation shows that the energy equivalent of even a small mass is very large.

Other relativistic effects include the slowing of moving clocks and the shrinkage of lengths measured in the direction of motion. A detailed discussion of the

major consequences of the special theory of relativity is beyond the scope of this book. Instead, we must consider Einstein's second advance, the general theory.

The General Theory of Relativity

In 1916, Einstein published a more general version of the theory of relativity that dealt with accelerated as well as uniform motion. This **general theory of relativity** contained a new description of gravity.

Einstein began by thinking about observers in accelerated motion. Imagine an observer sitting in a spaceship. Such an observer cannot distinguish between the force of gravity and the inertial forces produced by the acceleration of the spaceship (Figure 5-19). This led Einstein to conclude that gravity and motion through space-time are related, a conclusion now known as the equivalence principle:

Equivalence principle Observers cannot distinguish locally between inertial forces due to

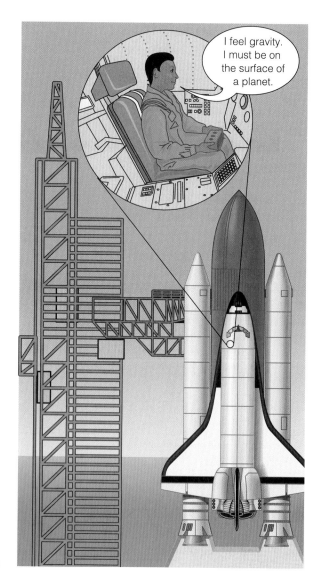

a

Figure 5-19
(a) An observer in a closed spaceship on the surface of a planet feels gravity. (b) In space, with the rockets smoothly firing and accelerating the spaceship, the observer feels inertial forces that are equivalent to gravitational forces.

acceleration and uniform gravitational forces due to the presence of a massive body.

The importance of the general theory of relativity lies in its description of gravity. Einstein concluded that gravity, inertia, and acceleration are all associated with the way space and time are related. This relation is often referred to as curvature, and a one-line description of general relativity explains a gravitational field as a curved region of space-time:

Gravity according to general relativity
Mass tells space-time how to curve, and the curvature of space-time (gravity) tells mass how to accelerate.

Thus, we feel gravity because Earth's mass causes a curvature of space-time. The mass of our bodies responds to that curvature by accelerating toward Earth's center. According to general relativity, all masses cause curvature, and the larger the mass, the more severe the curvature.

Confirmation of the Curvature of Space-Time

Einstein's general theory of relativity has been confirmed by a number of experiments, but two are worth mentioning here because they were among the first tests of the theory. One involves Mercury's orbit, and the other involves eclipses of the sun.

Johannes Kepler understood that the orbit of Mercury is elliptical, but only since 1859 have astronomers known that the long axis of the orbit sweeps around the sun in a motion called precession (Figure 5-20). The total observed precession is 5600.73 seconds of arc per century (as seen from Earth), or about 1.5° per century. This precession is produced by the gravitation of Venus, Earth, and the other planets. However, when

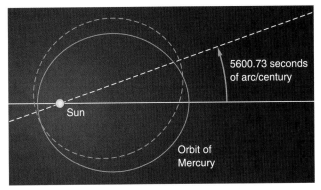

a

b

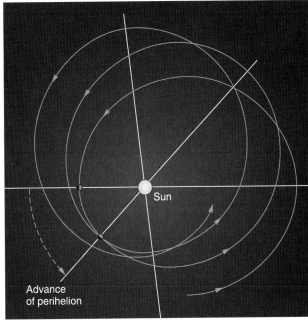

b

Figure 5-20
(a) Mercury's orbit precesses 5600.73 seconds of arc per century—43.11 seconds of arc per century faster than predicted by Newton's laws. (b) Even when we ignore the influences of the other planets, Mercury's orbit is not a perfect ellipse. Curved space-time near the sun distorts the orbit from an ellipse into a rosette. The advance of Mercury's perihelion is exaggerated about a million times in this figure.

astronomers used Newton's description of gravity, they calculated that the precession should amount to only 5557.62 seconds of arc per century. Thus, Mercury's orbit is advancing 43.11 seconds of arc per century faster than Newton's law predicted.

This is a tiny effect. Each time Mercury returns to perihelion, its closest point to the sun, it is about 29 km (18 miles) past the position predicted by Newton's laws. This is such a small distance compared with the planet's diameter of 4850 km that it could never have been detected had it not been cumulative. Each orbit, Mercury gains 29 km, and in a century it gains over 12,000 km— more than twice its own diameter. Thus, this tiny effect, called the advance of perihelion of Mercury's orbit, accumulated into a serious discrepancy in the Newtonian description of the universe.

The advance of perihelion of Mercury's orbit was one of the first problems to which Einstein applied the principles of general relativity. First he calculated how much the sun's mass curves space-time in the region of Mercury's orbit, and then he calculated how Mercury moves through the space-time. The theory predicted that the curved space-time should cause Mercury's orbit to advance by 43.03 seconds of arc per century, well within the observational accuracy of the excess (Figure 5-20b).

Einstein was elated with this result, and he would be even happier with modern studies that have shown that Mercury, Venus, Earth, and even Icarus, an asteroid that comes close to the sun, have orbits observed to be slipping forward due to the curvature of space-time near the sun (Table 5-2).

This same effect has been detected in pairs of stars that orbit each other. In some cases, the advance of perihelion agrees with general relativity; in many cases, the sizes and masses of the stars are not well enough known for us to be certain of the theoretical rate of advance we should expect. But in a few cases, the stars' orbits appear to be changing faster than predicted. This may be a critical test for Einstein's theory. This shows how science continues to test theories over and over even after they are widely accepted.

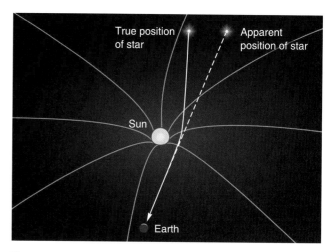

Figure 5-21
Like a depression in a putting green, the curved space-time near the sun deflects light from distant stars and makes them appear to lie slightly farther from the sun than their true positions.

A second test of the curvature of space-time was directly related to the motion of light through the curved space-time near the sun. The equations of general relativity predicted that light would be deflected by curved space-time just as a rolling golf ball is deflected by undulations in a putting green. Einstein predicted that starlight grazing the sun's surface would be deflected by 1.75 seconds of arc (Figure 5-21). Starlight passing near the sun is normally lost in the sun's glare, but during a total solar eclipse stars beyond the sun could be seen. As soon as Einstein published his theory, astronomers rushed to observe such stars and thus test the curvature of space-time.

The first solar eclipse following Einstein's announcement in 1916 was June 8, 1918. It was cloudy. The next occurred on May 29, 1919, only months after the end of World War I, and was visible from Africa and South America. British teams went to both Brazil and Príncipe, an island off the coast of Africa. First, they photographed that part of the sky where the sun would be located during the eclipse and measured the positions of the stars on the plates. Then during the eclipse they photographed the same star field with the eclipsed sun located in the middle. After measuring the plates, they found slight changes in the positions of the stars. During the eclipse, the positions of the stars on the plates were shifted outward, away from the sun (Figure 5-22). If a star had been located at the edge of the solar disk, it would have been shifted outward by about 1.8 seconds of arc. This represents good agreement with the theory's prediction.

This test has been repeated at many total solar eclipses since 1919, with similar results. The most accurate results were obtained in 1973 when a Texas–Princeton team measured a deflection of 1.66 ± 0.18 seconds of arc—good agreement with Einstein's theory.

Table 5-2 Precession in Excess of Newtonian Physics

Planet	Observed Excess Precession (seconds of arc per century)	Relativistic Prediction (seconds of arc per century)
Mercury	43.11 ± 0.45	43.03
Venus	8.4 ± 0.48	8.6
Earth	5.0 ± 1.2	3.8
Icarus	9.8 ± 0.8	10.3

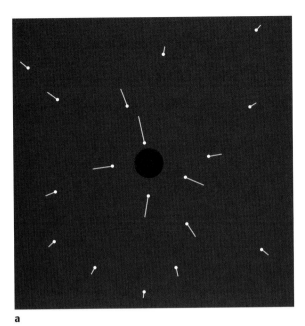 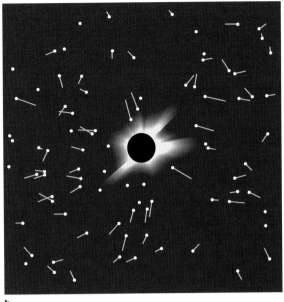

a   b

Figure 5-22

(a) Schematic drawing of the deflection of starlight by the sun's gravity. Dots show the true positions of the stars as photographed months before. Lines point toward the positions of the stars during the eclipse. (b) Actual data from the eclipse of 1922. Random errors of observation cause some scatter in the data, but in general the stars appear to move away from the sun by 1.77 seconds of arc at the edge of the sun's disk. The deflection of stars is magnified by a factor of 2300 in both (a) and (b).

The general theory of relativity is critically important in modern astronomy. We will discuss it again when we meet black holes, distant galaxies, and the big bang universe. The theory revolutionized modern physics by providing a theory of gravity based on the geometry of curved space-time. Thus, Galileo's inertia and Newton's mutual gravitation are shown to be fundamental properties of space and time.

REVIEW Critical Inquiry

What does the equivalence principle tell us?

The equivalence principle says that there is no observation we can make inside a closed spaceship to distinguish between uniform acceleration and gravitation. Of course, we could open a window and look outside, but then we would no longer be in a closed spaceship. As long as we make no outside observations, we can't tell whether our spaceship is firing its rockets and accelerating through space or resting on the surface of a planet where gravity gives us weight.

Einstein took the equivalence principle to mean that gravity and acceleration through space-time are somehow related. The general theory of relativity gives that relationship mathematical form and shows that gravity is really a disturbance in space-time that physicists refer to as curvature. Thus, we say "mass tells space-time how to curve, and space-time tells mass how to move."

The equivalence principle led Einstein to an explanation for gravity.

But what about the second postulate of special relativity? Why does it have to be true if the first postulate is true? And what does the second postulate tell us about the nature of uniform motion?

Our discussion of the origin of astronomy began with the builders of Stonehenge and reaches the modern day with Einstein's general theory of relativity. Now that we have seen where astronomy came from, we are ready to see how it helps us understand the nature of the universe. Our first question should be "How do astronomers get information?" The answer involves the astronomer's most basic tool, the telescope; that is the subject of the next chapter.

Summary

Galileo took the first step toward understanding motion and gravity when he began to study falling bodies. He found that a falling object is accelerated; that is, it falls faster and faster with each passing second. The rate at which it accelerates, termed the acceleration of gravity, is 9.8 m/s^2 (32 ft/s^2) at

Earth's surface and does not depend on the weight of the object. According to tradition, Galileo demonstrated this by dropping balls of iron and wood from the Leaning Tower of Pisa to show that they would fall together. Finally, Galileo stated the law of inertia. In the absence of friction, a moving body on a horizontal plane will continue moving forever.

Newton adopted Galileo's law of inertia as his first law of motion. The second law of motion establishes the relationship between the force acting on a body, its mass, and the resulting acceleration. The third law says that forces occur in pairs acting in opposite directions.

Newton also developed an explanation for the accelerations that Galileo had discovered—gravity. By considering the motion of the moon, Newton was able to show that objects attract each other with a gravitational force that is proportional to the product of their masses and inversely proportional to the square of the distance between them.

If we understand Newton's laws of motion and gravity, we can better understand orbital motion. An object in space near Earth would move along a straight line and quickly leave Earth were it not for Earth's gravity accelerating the object toward Earth's center and forcing it to follow a curved path, an orbit. If there is no friction, the object will fall around its orbit forever.

Newton's laws also illuminate the meaning of Kepler's three laws of planetary motion. The planets follow elliptical orbits because gravity follows the inverse square law. The planets move faster when closer to the sun and slower when farther away because they conserve angular momentum. The same law makes ice skaters spin faster when they draw their arms and legs nearer their bodies. A planet's orbital period squared is proportional to its orbital radius cubed because the moving planet conserves energy.

In fact, Newtonian gravity and motion show that an ellipse is only one of a number of orbits that a body can follow. The circle and ellipse are closed orbits that return to their starting points. If an object moves at circular velocity, V_c, it will follow a circular orbit. If its velocity equals or exceeds the escape velocity, V_e, it will follow a parabola or hyperbola. These orbits are termed "open" because the object never returns to its starting place.

Newton's laws changed astronomy and our view of nature. They made it possible for astronomers to predict the motions of the heavenly bodies using the analytic power of mathematics. Newton's laws also show that the apparent complexity of the universe is based on a few simple principles, or natural laws.

Einstein published two theories that extended Newton's laws of motion and gravity. The special theory of relativity, published in 1905, applies to observers in uniform motion. The theory holds that the speed of light is a constant for all observers and that mass and energy are related by the expression $E = m_0 c^2$.

The general theory of relativity, published in 1916, holds that a gravitational field is a curvature of space-time caused by the presence of a mass. Thus, Earth's mass curves space-time, and the mass of our bodies responds to that curvature by accelerating toward Earth's center. This curvature of space-time was confirmed by the slow advance in perihelion of the orbit of Mercury and by the deflection of starlight observed during a 1919 total solar eclipse.

New Terms

natural motion	circular velocity
violent motion	geosynchronous orbit
acceleration of gravity	closed orbit
momentum	open (escape) orbit
mass	escape velocity
acceleration	angular momentum
velocity	energy
inverse square law	joule (J)
field	special relativity
center of mass	general theory of relativity

Review Questions

1. Why wouldn't Aristotle's explanation of gravity work if Earth was not the center of the universe?

2. According to the principles of Aristotle, what part of the motion of a baseball pitched across the home plate is natural motion? What part is violent motion?

3. If we drop a feather and a steel hammer at the same moment, they should hit the ground at the same instant. Why doesn't this work on Earth, and why does it work on the moon?

4. What is the difference between mass and weight? between speed and velocity?

5. Why did Newton conclude that some force had to pull the moon toward Earth?

6. Why did Newton conclude that gravity had to be mutual and universal?

7. How does the concept of a field explain action at a distance? Name another kind of field also associated with action at a distance.

8. Why can't a spacecraft go "beyond Earth's gravity"?

9. What is the center of mass of the Earth–moon system? Where is it?

10. How do planets orbiting the sun and skaters conserve angular momentum?

11. Why is the period of an open orbit undefined?

12. How does the first postulate of special relativity imply the second?

13. When we ride a fast elevator upward, we feel slightly heavier as the trip begins and slightly lighter as the trip ends. How is this phenomenon related to the equivalence principle?

14. From your knowledge of general relativity, would you expect radio waves from distant galaxies to be deflected as they pass near the sun? Why or why not?

Discussion Questions

1. How did Galileo idealize his inclines to conclude that an object in motion stays in motion until it is acted on by some force?

2. Give an example from everyday life to illustrate each of Newton's laws.

Problems

1. Compared with the strength of Earth's gravity at its surface, how much weaker is gravity at a distance of 10 Earth radii from Earth's center? at 20 Earth radii?

2. Compare the force of gravity on the surface of the moon with the force of gravity at Earth's surface.

3. If a small lead ball falls from a high tower on Earth, what will be its velocity after 2 seconds? after 4 seconds?

4. What is the circular velocity of an Earth satellite 1000 km above Earth's surface? (*Hint:* Earth's radius is 6380 km.)

5. What is the circular velocity of an Earth satellite 36,000 km above Earth's surface? What is its orbital period? (*Hint:* Earth's radius is 6380 km.)

6. What is the orbital period of an imaginary satellite orbiting just above Earth's surface? Ignore friction with the atmosphere.

7. Repeat the previous problem for Mercury, Venus, the moon, and Mars.

8. Describe the orbit followed by the slowest cannonball in Figure 5-8 on the assumption that the cannonball could pass freely through Earth. (Newton got this problem wrong the first time he tried to solve it.)

9. If you visited an asteroid 30 km in radius with a mass of 4×10^{17} kg, what would be the circular velocity at its surface? A major league fastball travels 90 mph. Could a good pitcher throw a baseball into orbit around the asteroid?

10. What is the orbital period of a satellite orbiting just above the surface of the asteroid in Problem 9?

11. What would be the escape velocity at the surface of the asteroid in Problem 9? Could a major league pitcher throw a baseball off of the asteroid?

Critical Inquiries for the Web

1. Einstein's general theory of relativity predicts the curvature of space-time, but here on Earth we have little opportunity to observe such effects. Find an astronomical situation in which space-time curvature is evident from our observations, and describe the effect of the curvature on what we see when we view these objects.

2. Communications satellites are obvious uses of the geosynchronous orbit, but can you think of other uses for such orbits? Find an Internet site that uses or displays information gleaned from geosynchronous orbit that provides a useful service.

Go to the Brooks/Cole Astronomy Resource Center (www. brookscole.com/astronomy) for critical thinking exercises, articles, and additional readings from InfoTrac College Edition, Brooks/Cole's online student library.

CHAPTER SIX

Light and Telescopes

He burned his house down for the

fire insurance

And spent the proceeds on a telescope.

Robert Frost
The Star-Splitter

Guidepost

Previous chapters have described the sky as it appears to our unaided eyes, but modern astronomers turn powerful telescopes on the sky. Chapter 6 introduces us to the modern astronomical telescope and its delicate instruments.

The study of the universe is so challenging, astronomers cannot ignore any source of information; that is why they use the entire spectrum, from gamma rays to radio waves. This chapter shows how critical it is for astronomers to understand the nature of light.

In each of the 22 chapters that follow, we will study the universe using information gathered by the telescopes and instruments described in this chapter.

What do fleas living on rats have to do with modern astronomy? That may sound like the beginning of a bad joke, but it is actually related to the subject of this chapter. We will examine the tools that modern astronomers use, and those tools are connected by an interesting sequence of events to rats and their fleas.

The most horrible disease in history—the black plague—was spread by fleabites, and the fleas lived on the rats that infested the cities. Plague broke out in London in 1665, and, although no one knew how the disease spread, people who lived in the country were less likely to get the plague than city dwellers. All who could left the cities. When the plague reached Cambridge, the colleges were closed, and both students and faculty fled to the English countryside. One who fled was the young Isaac Newton. From the summer of 1665 to 1667, he spent most of his time in his mother's cottage in the small village of Woolsthorpe. While he was there, he conducted an experiment that changed the history of science.

Boring a hole in a shutter, he admitted a thin beam of sunlight into his darkened room. A glass prism placed in the beam threw a rainbow of color—a spectrum—across the wall. When he used a second prism to recombine the colors, they produced white light. From this and other experiments conducted in his bedroom, he concluded that white light was made up of a mixture of all the colors of the rainbow.

When the plague abated, Newton returned to the university, where he began experimenting with telescopes. He discovered that telescopes made of lenses produced colored fringes around bright objects in the field of view because the glass lenses broke the light into colors, just as his prism broke up the sunlight. To solve the problem, Newton designed and built a telescope containing a mirror instead of a lens. Although his first model hardly exceeded 1 inch in diameter, when Newton presented it to the Royal Society in 1671, it established his reputation as a scientist.

For a century after Newton's first telescope, astronomers did little with such devices, but as instrument makers grew more skilled, large telescopes became the principal tool of the astronomer. Telescopes are important in astronomy because they gather light and concentrate it for study. The larger the telescope, the more light it gathers. Thus astronomers are still striving to build bigger telescopes to gather more light from the objects in the sky. Like Newton's original telescope, almost all modern telescopes use mirrors rather than lenses to avoid spreading the light into its component colors.

6-1 Radiation: Information from Space

Just as a book on bread baking might begin with a discussion of flour, our chapter on telescopes begins with a discussion of light—not just visible light, but the entire range of radiation from the sky.

Light as a Wave and a Particle

Light is merely one form of radiation called **electromagnetic radiation** because it is associated with changing electric and magnetic fields that travel through space and transfer energy from one place to another. When light enters our eye, the fluctuating electric and magnetic fields carry energy that stimulates nerve endings, and we see what we call light.

The oscillating electric and magnetic fields that constitute electromagnetic radiation move through space at about 300,000 km/s (186,000 mi/s). This speed is commonly referred to as the speed of light, c, but it is in fact the speed of all such radiation in a vacuum.

Electromagnetic radiation is a wave phenomenon; that is, it is associated with a periodically repeating disturbance, or wave. We are familiar with waves in water. If we disturb a quiet pool of water, waves spread across the surface. Imagine that we use a meter stick to measure the distance between the successive peaks of a wave. This distance is the **wavelength,** usually represented by the Greek letter lambda (λ). If we were measuring ripples in a pond, we might find that the wavelength is a few centimeters, whereas the wavelength of ocean waves might be a hundred meters or more. There is no restriction on the wavelength of electromagnetic radiation. Wavelengths can range from smaller than the diameter of an atom to larger than that of Earth.

Because all electromagnetic radiation travels at the speed of light, wavelength is related to **frequency,** the number of cycles that pass in one second. Short-wavelength radiation has a high frequency; long-wavelength radiation has a low frequency. To understand this, imagine watching an electromagnetic wave race past us while we count its peaks (Figure 6-1). If the wavelength is short,

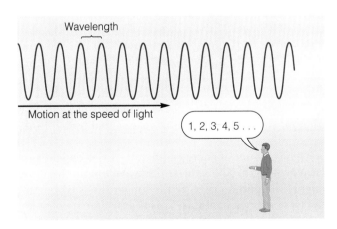

Figure 6-1
All electromagnetic waves travel at the speed of light. The wavelength is the distance between successive peaks. The frequency of the wave is the number of peaks that pass us in one second.

we will count many peaks in one second; if the wavelength is long, we will count few peaks per second. The dials on radios are marked in frequency, but they could just as easily be marked in wavelength. The relation between wavelength and frequency is a simple one:

$$\lambda = \frac{c}{f}$$

That is, the wavelength equals the speed of light c divided by the frequency f. Notice that the larger (higher) the frequency, the smaller (shorter) the wavelength. In most cases, astronomers use wavelength rather than frequency.

Radio waves can have wavelengths from a few millimeters for microwaves to kilometers. In contrast, the wavelength of light is so short that we must use more convenient units. In this book, we will use **nanometers (nm)** because this unit is consistent with the International System of units. One nanometer is 10^{-9} meter, and visible light has wavelengths that range from about 400 nm to about 700 nm. Another unit that astronomers commonly use, and a unit that you will see in many references on astronomy, is the **Angstrom (Å).** One Angstrom is 10^{-10} meter, and visible light has wavelengths between 4000 Å and 7000 Å.

You may find radio astronomers describing wavelengths in centimeters or millimeters, and infrared astronomers often refer to wavelengths in micrometers (or microns). One micrometer (μm) is 10^{-6} meter. Whatever unit is used to describe the wavelength, we must keep in mind that all electromagnetic radiation is the same phenomenon.

What exactly is electromagnetic radiation? Although we have been discussing its wavelength, it is incorrect, or at least incomplete, to say that electromagnetic radiation is a wave. It sometimes has the properties of a wave and sometimes has the properties of a particle. For instance, the beautiful colors in a soap bubble arise from the wave nature of light. On the other hand, when light strikes the photoelectric cell in a camera's light meter, it behaves like a stream of particles carrying specific amounts of energy. Throughout his life, Newton believed that light was made up of particles, but we now recognize that light can behave as both particle and wave. Our model of light is thus more complete than Newton's. We will refer to "a particle of light" as a **photon,** and we can recognize its dual nature by thinking of it as a bundle of waves.

The amount of energy a photon carries depends on its wavelength. The shorter the wavelength, the more energy the photon carries; the longer the wavelength, the less energy it contains. This is easy to remember because short wavelengths have high frequencies, and we expect rapid fluctuations to be more energetic. We can express this relationship in a simple formula:

$$E = \frac{hc}{\lambda}$$

Here h is Planck's constant (6.6262×10^{-34} joule s), c is the speed of light (3×10^{8} m/s), and λ is the wavelength in meters. A photon of visible light carries a very small amount of energy, but a photon with a very short wavelength can carry much more.

The Electromagnetic Spectrum

A spectrum is an array of electromagnetic radiation in order of wavelength. We are most familiar with the spectrum of visible light, which we see in rainbows. The colors of the spectrum differ in wavelength, with red having the longest wavelength and violet the shortest. The visible spectrum is shown at the top of Figure 6-2.

The average wavelength of visible light is about 0.00005 cm. We could put 50 light waves end to end across the thickness of a sheet of household plastic wrap. Measured in nanometers, the wavelength of visible light ranges from about 400 to 700 nm. Just as we sense the wavelength of sound as pitch, we sense the wavelength of light as color. Light near the short-wavelength end of the visible spectrum (400 nm) looks violet to our eyes, and light near the long-wavelength end (700 nm) looks red.

Figure 6-2 shows how the visible spectrum makes up only a small part of the entire electromagnetic spectrum. Beyond the red end of the visible spectrum lies **infrared radiation,** where wavelengths range from 700 nm to about 0.1 cm. Our eyes are not sensitive to this radiation, but our skin senses it as heat. A "heat lamp" is just a bulb that gives off principally infrared radiation.

Beyond the infrared part of the electromagnetic spectrum lie radio waves. Microwaves have wavelengths of a millimeter to a few centimeters and are used for radar and long-distance telephone communication. Longer wavelengths are used for UHF and VHF television transmissions. FM, military, governmental, and ham radio signals have wavelengths up to a few meters, and AM radio waves can have wavelengths of kilometers.

The distinction between the wavelength ranges is not sharp. Long-wavelength infrared radiation and the shortest microwave radio waves are the same. Similarly, there is no clear division between the short-wavelength infrared and the long-wavelength part of the visible spectrum. It is all electromagnetic radiation.

Look once again at the electromagnetic spectrum in Figure 6-2, and notice that electromagnetic waves shorter than violet are called **ultraviolet.** Electromagnetic waves even shorter are called X rays, and the shortest are gamma rays. Again, the boundaries between these wavelength ranges are not clearly defined.

X rays and gamma rays can be dangerous, and even ultraviolet photons have enough energy to do us harm.

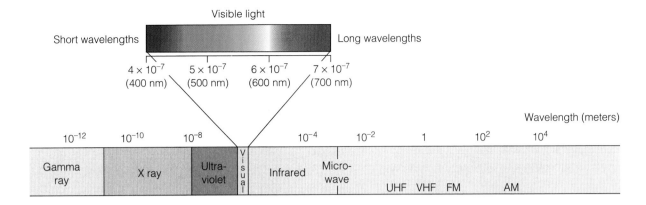

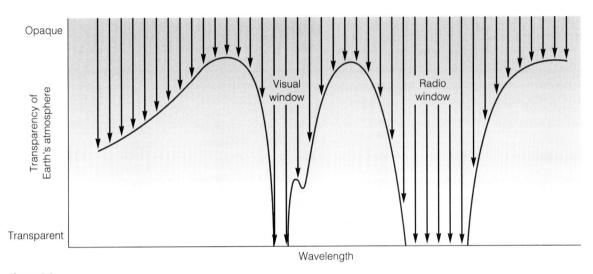

Figure 6-2
The spectrum of visible light, extending from red to violet, is only part of the electromagnetic spectrum. Most radiation is absorbed in Earth's atmosphere, and only radiation in the visual window and the radio window can reach Earth's surface.

Small doses produce a suntan, sunburn, and skin cancers. Contrast this to the lower-energy infrared photons. Individually they have too little energy to affect skin pigment, a fact that explains why you can't get a tan from a heat lamp. Only by concentrating many low-energy photons in a small area, as in a microwave oven, can we transfer significant amounts of energy.

We are interested in electromagnetic radiation because it brings us clues to the nature of stars, planets, and other celestial objects. Earth's atmosphere is opaque to most electromagnetic radiation, as shown by the graph at the bottom of Figure 6-2. Gamma rays, X rays, and some radio waves are absorbed high in Earth's atmosphere, and a layer of ozone (O_3) at an altitude of about 30 km absorbs ultraviolet radiation. Water vapor in the lower atmosphere absorbs the longer wavelength infrared radiation. Only visible light, some shorter wavelength infrared, and some radio waves reach Earth's surface through two wavelength regions called **atmospheric windows.** Obviously, if we wish to study the sky from Earth's surface, we must look out through one of these windows.

REVIEW Critical Inquiry

What could we see if our eyes were sensitive only to X rays?

Sometimes the critical analysis of an idea is easier if we try to imagine a totally new situation. In this case, we might at first expect to be able to see through walls, but remember that our eyes detect only light that already exists. There are almost no X rays bouncing around at Earth's surface, so if we had X-ray eyes, we would be in the dark and would be unable to see anything. Even when we look up at the sky, we would see nothing, because Earth's atmosphere is not transparent to X rays. If Superman can see through walls, it is not because his eyes can detect X rays.

But suppose our eyes were sensitive only to infrared waves or to radio waves. Would we be in the dark?

Now that we know something about electromagnetic radiation, we can study the tools astronomers use to gather and analyze that radiation.

6-2 Astronomical Telescopes

The telescope is the symbol of the astronomer. Later in this chapter, we will see how astronomers are using special telescopes to observe radio waves and X rays and how some telescopes are venturing into space. Here, however, we consider the traditional Earth-based astronomical telescope that gathers visible light.

Astronomers have used two kinds of telescopes. Most early astronomical telescopes, beginning with those made by Galileo, were **refracting telescopes** using a lens to bend the light to a focus. The newer **reflecting telescopes** use a concave mirror to reflect the light to a focus (Figure 6-3). Reflecting telescopes have some strong advantages.

Refracting Telescopes

The main element in a refracting telescope is a lens, a piece of glass carefully shaped so that light striking it is refracted, or bent, into an image. The **focal length** of a lens is the distance from the lens to the point where it focuses parallel rays of light. If the surface of the lens is strongly curved, the light is brought to a focus close to the lens, and we say it has a short focal length. If the

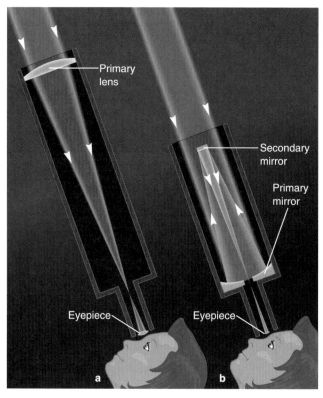

Figure 6-3

(a) A refracting telescope uses a primary lens to focus starlight into an image that is magnified by a lens called an eyepiece. The primary lens has a long focal length, and the eyepiece has a short focal length. (b) A reflecting telescope uses a primary mirror to focus the light by reflection. A small secondary mirror reflects the starlight back down through a hole in the middle of the primary lens to the eyepiece.

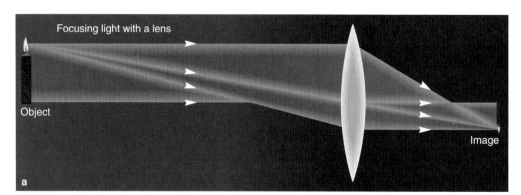

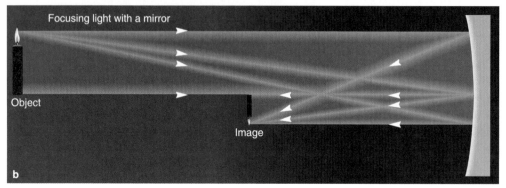

Figure 6-4

(a) To see how a lens can focus light, we trace four light rays from the flame and base of a candle through a lens, where they are refracted to form an inverted image. (b) A mirror forms an image by reflection from a concave surface. Notice that the light is reflected from the aluminized front surface of the mirror and does not enter the glass.

glass is less strongly curved, the lens has a longer focal length, and the image is formed farther from the lens.

The image in an astronomical telescope is inverted. This is true for all such optical systems, whether they use a lens or a mirror (Figure 6-4). Even your eye forms an inverted image. The image could be reinverted by an extra lens, but this is an unnecessary expense and a possible source of distortion. Consequently, most astronomical telescopes, like microscopes, produce inverted images.

To build a refracting telescope, we need a lens of relatively long focal length to form an image of the object we wish to view. This lens is often called the **objective lens** because it is closest to the object. To view the image, we add a lens of short focal length called an **eyepiece** to enlarge the image and make it easy to see (Figure 6-3). Thus, the eyepiece acts as a magnifier. By changing eyepieces, we can easily change the magnification of the telescope.

Refracting telescopes suffer from a serious optical distortion (aberration) that limits what we can see through them. When light is refracted through glass, shorter wavelengths bend more than longer wavelengths, and blue light comes to a focus closer to the lens than does red light (Figure 6-5a). If we focus the eyepiece on the blue image, the red light is out of focus, producing a red blur around the image. If we focus on the red image, the blue light blurs. This color separation is called **chromatic aberration.**

A telescope designer can partially correct for this aberration by replacing the single objective lens with one made of two lenses ground from different kinds of glass. Such lenses, called **achromatic lenses,** can be designed to bring any two colors to the same focus (Figure 6-5b). Because our eyes are most sensitive to red and yellow light, we might bring these two colors to the same focus, but blue and violet would still be out of focus, producing a hazy blue fringe around bright objects.

Refracting telescopes were popular through the 19th century, but they are no longer economical for professional astronomy. A large achromatic lens is very expensive because it contains four matched optical surfaces and must be made of high-quality glass. Refractors can't be made larger than about 1 m in diameter because such large lenses sag under their own weight. Also, large refractors have very long telescope tubes that require large observatory domes. Modern telescopes, like Newton's first telescope, focus light with mirrors.

Reflecting Telescopes

In a reflecting telescope, a concave mirror, the **objective mirror,** focuses the starlight into an image that can be viewed with an eyepiece (Figure 6-3b). Objective mirrors are usually made of special kinds of glass or quartz covered with a thin layer of aluminum to act as a reflecting surface.

Reflecting telescopes may have different optical arrangements (Figure 6-6). The objective mirror forms an image at the location called the **prime focus** at the upper end of the telescope tube. In larger telescopes, astronomers can ride inside a small compartment to photograph objects using the prime focus, or they can place cameras and instruments there which are controlled from a remote control room. To make observing more convenient, a small **secondary mirror** can be used to reflect light out the side of the telescope in an arrangement called a **Newtonian telescope** or back down the telescope tube through a hole in the objective mirror. This kind of telescope is called a **Cassegrain telescope.** Larger telescopes are sometimes designed so they can be used at either prime focus or Cassegrain focus (Figure 6-7).

Many smaller telescopes used by amateurs use a thin correcting plate at the top of the telescope tube to produce sharp images. Because the plate is very thin and is not strongly curved, it does not produce significant chromatic aberration. These telescopes are called **Schmidt-Cassegrain telescopes,** and a typical example is shown in Figure 6-6.

Nearly all recently built telescopes are reflectors. Because the light does not enter the glass, there is no chromatic aberration, the glass need not be of perfect optical quality, and the mirror can be supported over

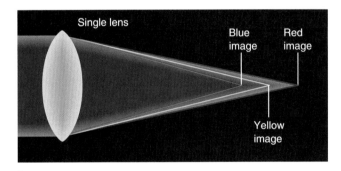

Figure 6-5
(a) A normal lens suffers from chromatic aberration because short wavelengths bend more than long wavelengths. (b) An achromatic lens, made in two parts, can bring any two colors to the same focus, but other colors remain slightly out of focus.

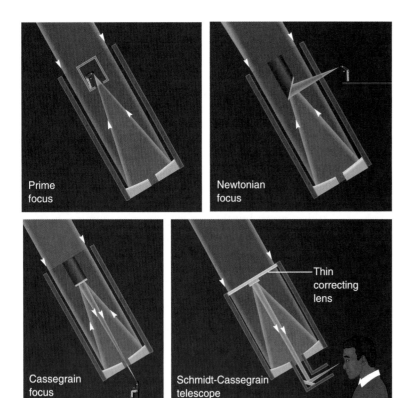

Figure 6-6
The mirror of a reflecting telescope forms the image in the middle of the telescope tube, so astronomers have devised various focal arrangements to bring the light to a convenient place for observing. Observing in a prime focus cage is possible only in the largest telescopes. Most large telescopes use the Cassegrain focus, but many amateur astronomers use the smaller Schmidt-Cassegrain telescopes.

its back to reduce sagging. Also, reflecting telescopes tend to be shorter and thus require smaller mountings and smaller observatory buildings.

Telescope Mountings

A telescope mounting can be more expensive than the telescope itself. The mounting must support the optics, protect against vibration, move accurately to designated objects, and then compensate for Earth's rotation. Because Earth rotates eastward, the telescope mounting must contain a **sidereal drive** (literally, a star drive) that can move the telescope smoothly westward to follow the object being studied. To simplify this motion, most telescope mountings are **equatorial mountings** with one of the two axes of rotation, the **polar axis**, parallel to Earth's axis. Rotation around the polar axis

moves the telescope parallel to the celestial equator (Figure 6-8).

Improvements in computer technology are making telescope mountings and observatories much less expensive. Instead of building a large, awkward equatorial mounting inside a large, heavy dome, astronomers now take advantage of the short focal lengths typical of reflecting telescopes. Such telescopes are so short they can be mounted like a cannon; that is, the mounting can move the telescope both in altitude (perpendicular to the horizon) and in azimuth (parallel to the horizon). Such **alt-azimuth mountings** are compact, but they must move simultaneously about both axes at varying rates in order to track stars in different parts of the sky. This complicated motion is made possible by computers that guide the telescope.

Such computer-controlled alt-azimuth mountings are stronger and smaller and do not require buildings as large as those required by equatorial mountings. In fact, some of the newest telescopes are designed so that the telescope is part of the observa-

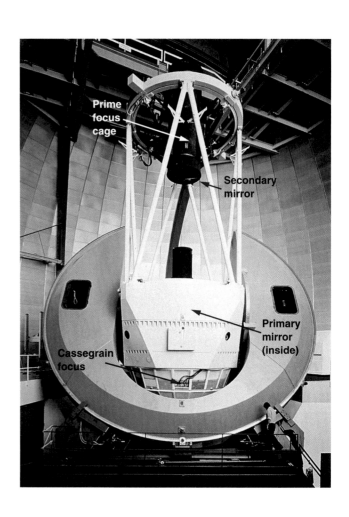

Figure 6-7
The 4-meter Mayall telescope at Kitt Peak National Observatory can be used as a prime focus telescope or, with the secondary mirror in place, as a Cassegrain telescope. Note the human figure at the lower right for scale. *(© Association of Universities for Research in Astronomy, Inc., Kitt Peak National Observatory)*

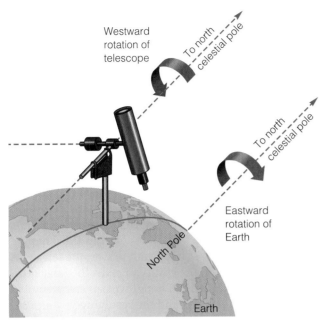

Figure 6-8
Westward motion around the polar axis of an equatorial mounting counters the Earth's eastward rotation and keeps the telescope pointed at a given star.

tory building. Like the turret of a tank, the entire building rotates as the telescope moves.

Nearly all large astronomical telescopes and many smaller ones are operated from control rooms. Thinking of the old days of observing all through the night in the unheated dome beside their telescopes, astronomers often refer to these control rooms as "warm rooms." Comfort, however, is not the main goal. Astronomical telescopes have become so large and so sophisticated that computers are needed to control their motions and record data; and computers, unlike astronomers, must be protected from extremes of temperature and humidity. From a control room, astronomers can work through the night controlling the computers that control the telescope.

Some telescopes can be operated from control rooms in remote locations. The two giant Keck telescopes sit at 13,600 feet (4100 m) atop an extinct volcano in Hawaii. Astronomers can observe with the telescopes from sea-level control rooms at the foot of the mountain. Other telescopes can be operated by astronomers from computers in their offices on the other side of the world. A few telescopes are designed so that computers can operate them every night with no human present, and the astronomer merely sends instructions to the computer before dark falls and retrieves the data the next day.

Astronomical telescopes are not only large and heavy, but they require intricate computer control systems. We might wonder what advantages they give us.

The Powers of a Telescope

A telescope can aid our eyes in only three ways—the three powers of a telescope. They make images brighter, more detailed, and larger.

Most interesting celestial objects are faint sources of light, so we need a telescope that can gather large amounts of light to produce a bright image. **Light-gathering power** refers to the ability of a telescope to collect light. Catching light in a telescope is like catching rain in a bucket—the bigger the bucket, the more rain it catches (Figure 6-9). Light-gathering power is proportional to the area of the telescope objective. A lens or mirror with a large area gathers a large amount of light. The area of a circular lens or mirror of diameter D is $\pi(\frac{D}{2})^2$. To compare the relative light-gathering powers (LGP) of two telescopes A and B, we can calculate the ratio of the areas of their objectives, which reduces to the ratio of their diameters (D) squared.

$$\frac{LGP_A}{LGP_B} = \left(\frac{D_A}{D_B}\right)^2$$

For example, suppose we compare a telescope 24 cm in diameter with a telescope 4 cm in diameter. The ratio of the diameters is 24/4, or 6, but the larger telescope does not gather 6 times as much light. Light-gathering power increases as the ratio of diameters squared, so it gathers 36 times more light than the smaller telescope. This example shows the importance of diameter in astronomical telescopes. Even a small increase in diameter produces a large increase in light-gathering

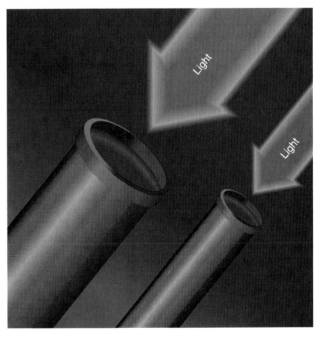

Figure 6-9
Gathering light is like catching rain in a bucket. A large-diameter telescope gathers more light and has a brighter image than a smaller telescope of the same focal length.

power and allows astronomers to study much fainter objects.

The second power, **resolving power,** refers to the ability of the telescope to reveal fine detail. Because light acts as a wave, it produces a small **diffraction fringe** around every point of light in the image, and we cannot see any detail smaller than the fringe (Figure 6-10). Astronomers can't eliminate diffraction fringes, but the larger a telescope is in diameter, the smaller the diffraction fringes are. Thus the larger the telescope, the better its resolving power.

For optical telescopes, we estimate the resolving power by calculating the angular distance between two stars that are just barely visible through the telescope as two separate images. The resolving power, α, in seconds of arc, equals 11.6 divided by the diameter of the telescope in centimeters:

$$\alpha = \frac{11.6}{D}$$

For example, the resolving power of a 25-cm telescope is 11.6 divided by 25, or 0.46 second of arc. No matter how perfect the telescope optics, this is the smallest detail we can see through that telescope.

In addition to resolving power, two other factors—lens quality and atmospheric conditions—limit the detail we can see through a telescope. A telescope must contain high-quality optics to achieve its full potential resolving power. Even a large telescope shows us little detail if its optics are marred with imperfections. Also, when we look through a telescope, we are looking through miles of turbulent air in Earth's atmosphere, which makes the image dance and blur, a condition called **seeing.** On a night when the atmosphere is unsteady and the images are blurred, the seeing is bad. Even under good seeing conditions, the detail visible through a large telescope is limited, not by its diffraction fringes, but by the air through which the telescope must look. A telescope performs better on a high moun-taintop where the air is thin and steady, but even there Earth's atmosphere limits the detail the best telescopes can reveal to about 0.5 second of arc.

This limitation on the amount of information in an image is related to the limitation on the accuracy of a measurement. All measurements have some built-in uncertainty (Window on Science 6-1), and scientists must learn to work within those limitations.

A technique called **adaptive optics** uses a high-speed computer to monitor atmospheric distortion and adjust the optics to partially compensate for seeing. This can produce dramatically sharper images. Of course, the ultimate solution to the problem of atmospheric seeing is to send the telescope above Earth's atmosphere. We will discuss space telescopes later in this chapter.

The third and least important power of a telescope is **magnifying power,** the ability to make the image bigger. Because the amount of detail we can see is limited by the seeing conditions and the resolving power, very high magnification does not necessarily show us more detail. Also, we can change the magnification by changing the eyepiece, but we cannot alter the telescope's light-gathering power or resolving power.

We calculate the magnification of a telescope by dividing the focal length of the objective by the focal length of the eyepiece:

$$M = \frac{F_o}{F_e}$$

For example, if a telescope has an objective with a focal length of 80 cm and we use an eyepiece whose focal length is 0.5 cm, the magnification is 80/0.5, or 160 times.

The search for light-gathering power and high resolution explains why nearly all major observatories are located far from major cities and usually on high mountains. Astronomers avoid cities because **light pollution,** the brightening of the night sky by light scattered from artificial outdoor lighting, can make it impossible to see

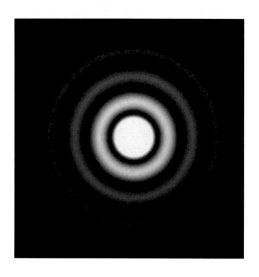
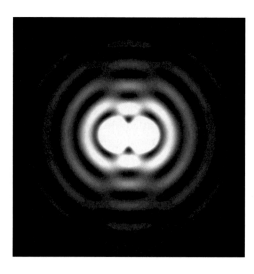

Figure 6-10
(a) Stars are so far away that their images are points, but the wave nature of light surrounds each star image with diffraction fringes (much magnified in this computer model). (b) Two stars close to each other have overlapping diffraction fringes and become impossible to detect separately. *(Computer model by M. A. Seeds)*

Have you ever seen a movie in which the hero magnifies a newspaper photo and reads some tiny detail? It isn't really possible, because newspaper photos are made up of tiny dots of ink, and no detail smaller than a single dot will be visible no matter how much you magnify the photo. In fact, all images are made up of elements of some sort, and that means there is a limit to the amount of detail you can see in an image. In an astronomical image, the resolution is often set by seeing. It is foolish to attempt to see a detail in the image that is smaller than the resolution.

This limitation is true of all measurements in science. A zoologist might be trying to measure the length of a live snake, or a sociologist might be trying to measure the attitudes of people toward drunk driving, but both face limits to the resolution of their measurements. The zoologist might specify that the snake was 43.28932 cm long, and the sociologist might say that 98.2491 percent of people oppose drunk driving, but a critic might point out that it isn't possible to make these measurements that accurately. The resolution of the techniques does not justify the accuracy implied.

Science is based on measurement, and whenever we make a measurement we should ask ourselves how accurate that measurement can be. The accuracy of the measurement is limited by the resolution of the measurement technique, just as the amount of detail in a photograph is limited by the resolution of the photo.

faint objects (Figure 6-11). In fact, many residents of cities are unfamiliar with the beauty of the night sky because they can see only the brightest stars. Astronomers prefer to place their telescopes on carefully selected high mountains. The air there is thin and more transparent, but, most important, astronomers select

Figure 6-11
This satellite view of the continental United States at night shows the light pollution produced by outdoor lighting. Not only does the glare drown out the fainter stars and interfere with astronomy, but it wastes electrical power. Many astronomers work with city governments to enact laws that improve lighting on the ground and reduce light scattered into the night sky. *(National Geophysical Data Center)*

Figure 6-12
The two 8-meter Gemini telescopes are located atop an 8895-ft-high mountain in the Chilean Andes and atop the 13,600-ft-high volcano Mauna Kea in Hawaii. Here the northern Gemini dome is seen against star trails circling the north celestial pole. *(Gemini Observatory/ AURA/NOAO/NSF)*

mountains where the air flows smoothly and is not turbulent. This produces the best seeing. Building an observatory on top of a high mountain far from civilization is difficult and expensive, but the dark sky and steady seeing make it worth the effort (Figure 6-12).

Buying a Telescope

Thinking about how we should shop for a new telescope will not only help us if we decide to buy one but will also illustrate some important points about astronomical telescopes.

Assuming we have a fixed budget, we should buy the highest-quality optics and the largest-diameter telescope we can afford. Of the two things that limit what we see, optical quality is under our control. We can't make the atmosphere less turbulent, but we should buy good optics. If we buy a telescope from a toy store and it has plastic lenses, we shouldn't expect to see very much. Also, we want to maximize the light-gathering power of our telescope, so we want to purchase the largest-diameter telescope we can afford. Given a fixed budget, that means we should buy a reflecting telescope (which includes Schmidt-Cassegrains) rather than a refracting telescope. Not only will we get more diameter per dollar, but our telescope will not suffer from chromatic aberration.

We can safely ignore magnification. Department stores and camera stores may advertise telescopes by quoting their magnification, but it is not an important number. What we can see is fixed by light-gathering power, optical quality, and Earth's atmosphere. Besides, we can change the magnification by changing eyepieces.

Other things being equal, we should choose a telescope with a solid mounting that will hold the telescope steady and allow us to point at objects easily. A sidereal drive would be very useful even on a small telescope, and computer-controlled pointing systems are available for a price on many small telescopes. A good telescope on a poor mounting is almost useless.

We might be buying a telescope to put in our backyard, but we must think about the same issues astronomers consider when they design giant telescopes to go on mountaintops. In fact, some of the newest telescopes solve these traditional problems in new ways.

New-Generation Telescopes

For most of this century, astronomers faced a serious limitation in the size of telescopes. A large telescope mirror sags under its own weight and cannot focus light accurately. To avoid this sag, astronomers made telescope mirrors very thick, but that produced two further problems—weight and cost.

Supporting such a heavy mirror required a massive telescope. The two largest conventional telescopes in the world are the 6-m (236-in.) reflector in the former Soviet Union (Figure 6-13) and the 5-m (200-in.) Hale Telescope on Mount Palomar. The 5-m mirror alone weighs 14.5 tons, and its mounting weighs about 530 tons.

Grinding a conventional telescope mirror to shape is expensive because it requires the slow removal of large amounts of glass. The 5-m mirror, for instance, was begun in 1934 and finished in 1948, and 5 tons of glass were ground away.

Astronomers have been developing new techniques to make large mirrors that weigh and cost less. For example, the Steward Observatory Mirror Laboratory has built a revolving oven under the football stands at the University of Arizona that can produce preshaped mirrors. The oven turns like a merry-go-round, and the molten glass flows outward in the mold to form a concave upper surface. After they slowly cool, the preshaped mirrors can be quickly ground to final shape. The oven is currently able to cast mirrors as large as 8.4 m in diameter.

A mirror of conventional thickness with a diameter of 8 m would be too heavy to support in a telescope, so astronomers have devised ways to make the mirrors thinner (Figure 6-14). One technique is to make the mirror in segments. Small segments are less expensive and sag less under their own weight. The largest general-purpose optical telescopes in the world are the twin 10-m Keck I and Keck II Telescopes in Hawaii. Each

a

b

Figure 6-13
Until recent decades, telescope mirrors were single, thick pieces of glass that required massive telescope mountings and large observatory domes. (a) The world's largest traditional telescope is the 6-m reflector built by the former U.S.S.R. (b) The 5-m (200-in.) Hale Telescope on Mount Palomar contains a mirror weighing 14.5 tons. The telescope dome weighs 1000 tons.

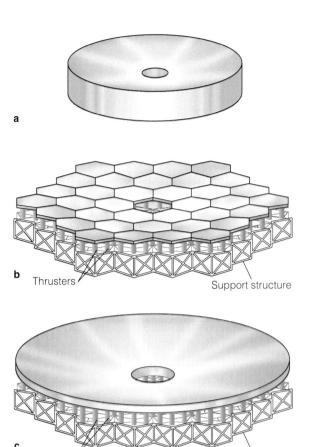

a

b   Thrusters   Support structure

c   Thrusters   Support structure

Figure 6-14
The problem and two solutions: Conventional telescope mirrors (a) had to be made very thick so they could not sag out of shape. To reduce the weight and increase the diameter, astronomers have begun making mirrors of segments (b) or as relatively thin disks (c). To hold these mirrors in shape, computer-controlled thrusters (red) are positioned under the mirrors.

of these giant telescopes uses 36 hexagonal mirror segments held in alignment by computer-controlled thrusters to form a single mirror (Figure 6-15). The special purpose Hobby-Eberly Telescope in Texas uses 91 hexagonal segments to make up a mirror 11 m in diameter; it is especially designed to focus starlight into a sophisticated spectrograph.

Another way to reduce the weight of telescope mirrors is to make them thin. Thin mirrors sag so easily under their own weight that they are called floppy mirrors, but a computer can control their shape in what astronomers call **active optics.** The New Technology Telescope at the European Southern Observatory in Chile contains a 3.58-m (141-in.) mirror that is only 24 cm (10 in.) thick.

The Multiple Mirror Telescope (MMT) was originally built using six round mirrors, each 1.8 m (72 in.) in diameter, on one large mounting. The mirrors could combine their light to produce the equivalent of a 4.5-m (176-in.) telescope. The success of active optics and floppy mirrors has led astronomers to replace the six mirrors with a single thin mirror 6.5 m (256 in.) in diameter.

Segmented or floppy, thin telescope mirrors have yet another advantage—they cool quickly. During the day, a telescope mirror warms to air temperature, but after sunset the air cools quickly. A massive telescope mirror cannot cool rapidly and uniformly, and because some parts of the mirror cool faster than other parts, the material expands and contracts, producing strains in the mirror that distort the image. Thin telescope mirrors cool more quickly and more uniformly as night falls and thus produce better images.

New technology is allowing astronomers to build new, giant telescopes. The Very Large Telescope (VLT), built by the European Southern Observatory, is located high in the remote Andes mountains in northern Chile. The VLT consists of four telescopes (Figure 6-16a) with computer-controlled main mirrors 8.2 m in diameter and only 17.5 cm (6.9 inches) thick. The four telescopes will be able to work together as a single giant telescope

a   b

Figure 6-15
Segmented mirrors can be assembled to create very-large-diameter telescopes. (a) The hexagonal mirror segments in the Keck I Telescope surround a worker crouching in the Cassegrain opening. Each segment is 1.8 m (70 inches) in diameter and is supported and aligned by computer-controlled actuators. *(© Russ Underwood/ W. M. Keck Observatory)* (b) The 91 hexagonal mirrors of the Hobby-Eberly Telescope in Texas are visible here reflecting the supporting frame on the interior of the dome. Computer controlled, the mirror segments make up a mirror 11 m in diameter. *(Photo courtesy Dr. Thomas G. Barnes III, McDonald Observatory, University of Texas at Austin)*

or separately. The Gemini project is completing a pair of 8-m thin-mirror telescopes (Figure 6-16b), which are located in Chile and in Hawaii so they can cover the entire sky. Italian and American astronomers are building the Large Binocular Telescope, which will carry a pair of 8.4-m mirrors on a single mounting. Astronomers are now drawing plans for truly gigantic telescopes with segmented mirrors 30 m and even 100 m in diameter (larger than a football field). Computer control of the optics makes such huge telescopes worth considering.

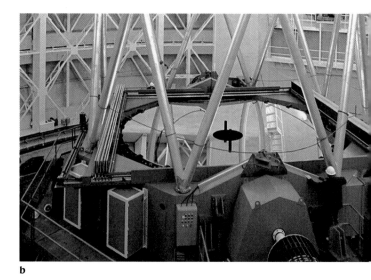

a   b

Figure 6-16
(a) The four telescopes of the VLT are housed in separate domes, but they can combine their light to work with the light-gathering power of a telescope over 16 m in diameter. *(ESO)* (b) The two Gemini Telescopes use a thin 8-m mirror carried in an alt-azimuth mount under computer control. *(Gemini Observatory/AURA/NOAO/NSF)*

Why do astronomers build observatories at the tops of mountains?

Astronomers have joked that the hardest part of building a new observatory is constructing the road to the top of the mountain. It certainly isn't easy to build a large, delicate telescope at the top of a high mountain, but it is worth the effort. A telescope on top of a high mountain is above the thickest part of Earth's atmosphere. There is less air to dim the light, and there is less water vapor to absorb infrared radiation. Even more important, the thin air on a mountaintop causes less disturbance to the image, and thus the seeing is better. A large telescope on Earth's surface has a resolving power much better than the distortion caused by Earth's atmosphere. Thus, it is limited by seeing, not by its own diffraction. It really is worth the trouble to build telescopes atop high mountains.

Astronomers not only build telescopes on mountaintops; they also build gigantic telescopes many meters in diameter. What are the problems and advantages in building such giant telescopes?

Astronomers sometimes refer to a telescope that produces distorted images as a "light bucket." In a sense, all astronomical telescopes are light buckets, because the light they focus into images tells us very little until it is recorded and analyzed by special instruments attached to the telescopes.

6-3 Special Instruments

Looking through a telescope doesn't tell us much. To use an astronomical telescope to learn about stars, we must be able to analyze the light the telescope gathers. Special instruments attached to the telescope make that possible.

Imaging Systems

The original imaging device in astronomy was the photographic plate. It could record faint objects in long exposures and could be stored for later analysis. But photographic plates have been almost entirely replaced in astronomy by electronic imaging systems.

Most modern astronomers use **charge-coupled devices (CCDs)** to record images. A CCD is a specialized computer chip containing roughly a million microscopic light detectors arranged in an array about the size of a postage stamp. These devices can be used like a small photographic plate, but they have dramatic advantages. They can detect both bright and faint objects in a single exposure, are much more sensitive than a photographic plate, and can be read directly into computer memory for later analysis. Although CCDs for astronomy are extremely sensitive and therefore expensive, less sophisticated CCDs are used in most video cameras and digital electronic cameras.

The image from a CCD is stored in computer memory as numbers, so it is easy to manipulate the image to bring out details that would not otherwise be visible. For example, astronomical images are often reproduced as negatives with the sky white and the stars dark . This makes the faint parts of the image easier to see (Figure 6-17a). Astronomers also manipulate images to produce **false-color images** in which the colors represent different levels of intensity and are not related to the true colors of the object (Figure 6-17b).

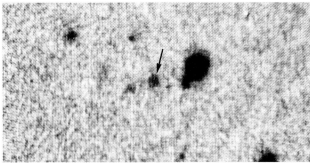

a

b

Figure 6-17
(a) The arrow points to an electronic image of a faint galaxy so distant that light took 10 billion years to reach us. This image is reproduced as a negative. The sky is light, and the stars and galaxy are black. *(Hyron Spinrad)* (b) This false-color image of the galaxy NGC1232 displays different levels of brightness as different colors. *(CARA)*

Measurements of intensity and color were made in the past using a photometer, a highly sensitive light meter attached to a telescope. Today, however, most such measurements are made on CCD images. Because the CCD image is easily digitized, brightness and color can be measured to high precision.

The Spectrograph

To analyze light in detail, we need to spread the light out according to wavelength into a spectrum, a task performed by a **spectrograph.** We can understand how this works if we reproduce the experiment performed by Isaac Newton in 1666. Placing a prism in a beam of sunlight spreads it into a beautiful spectrum. From this

Newton concluded that white light was made of a mixture of all the colors.

Newton didn't think in terms of wavelength, but we can use that modern concept to see that the light passing through the prism is bent at an angle that depends on the wavelength. Violet (short wavelength) bends most, and red (long wavelength) least. Thus, the white light entering the prism is spread into a spectrum (Figure 6-18a). A typical prism spectrograph contains more than one prism to spread the light farther and lenses to guide the light into the prism and to focus the light onto a photographic plate.

Nearly all modern spectrographs use a grating in place of a prism. A **grating** is a piece of glass with thousands of microscopic parallel lines scribed onto its sur-

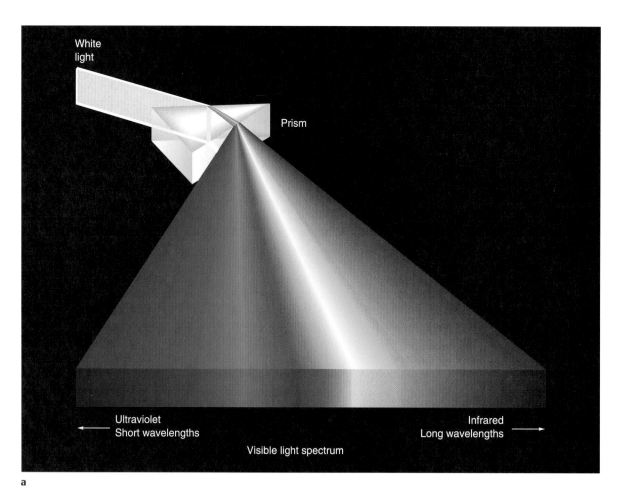

a

b

Figure 6-18
(a) A prism bends light by an angle that depends on the wavelength of the light. Short wavelengths bend most, and long wavelengths least. Thus, white light passing through a prism is spread into a spectrum. (b) In this negative image, the bright spectrum of the star is the dark streak running from left to right in the middle of the image. The gaps in the star's spectrum are produced by different gases in the star. Above and below the stellar spectrum are the lines of a comparison spectrum, which astronomers use to calibrate the spectrum of the star and measure wavelengths to high precision. *(Caltech)*

face. Different wavelengths of light reflect from the grating at slightly different angles, so white light is spread into a spectrum and can be recorded, often by a CCD camera.

Because astronomers need to measure the precise location of features in a spectrum, most spectrographs can add a **comparison spectrum** above and below the spectrum of the observed object (Figure 6-18b). The features in this comparison spectrum have known wavelengths carefully measured in the laboratory, and thus the astronomer can use the comparison spectrum as a precise calibration of the wavelengths in the spectrum being studied.

Because astronomers understand how light interacts with matter, a spectrum carries a tremendous amount of information (as we will see in the next chapter), and that makes a spectrograph the astronomer's most powerful instrument. An astronomer recently remarked, "We don't know anything about an object till we get a spectrum," and that is only a slight exaggeration.

REVIEW Critical Inquiry

What is the difference between light going through a lens and light passing through a prism?

A refracting telescope producing chromatic aberration and a prism dispersing light into a spectrum are two examples of the same thing, but one is bad and one is good. When light passes through the curved surfaces of a lens, different wavelengths are bent by slightly different amounts, and the different colors of light come to focus at different focal lengths. This produces the color fringes in an image called chromatic aberration, and that's bad. But the surfaces of a prism are made to be precisely flat, so all of the light enters the prism at the same angle, and any given wavelength is bent by the same amount wherever it meets the prism. Thus, white light is dispersed into a spectrum. We could call the dispersion of light by a prism "controlled chromatic aberration," and that's good.

CCDs have been very good for astronomy, and they are now widely used. Explain why they are more useful than photographic plates.

So far, our discussion has been limited to visual wavelengths. Now it is time to consider the rest of the electromagnetic spectrum.

6-4 Radio Telescopes

All the telescopes and instruments we have discussed look out through the visible light window in Earth's atmosphere, but there is another window running from a wavelength of 1 cm to about 1 m (see Figure 6-2). By

building the proper kinds of instruments, we can study the universe through this radio window.

Operation of a Radio Telescope

A radio telescope usually consists of four parts: a dish reflector, an antenna, an amplifier, and a recorder (Figure 6-19). The components, working together, make it possible for astronomers to detect radio radiation from celestial objects.

The dish reflector of a radio telescope, like the mirror of a reflecting telescope, collects and focuses radiation. Because radio waves are much longer than light waves, the dish need not be as smooth as a mirror. In some radio telescopes, the reflector may not even be dish-shaped, or the telescope may contain no reflector at all.

Though a radio telescope's dish may be many meters in diameter, the antenna may be as small as your hand. Like the antenna on a TV set, its only function is to absorb the radio energy and direct it along a cable to an amplifier. After amplification, the signal goes to some kind of recording instrument. Most radio observatories record data on magnetic tape or feed it directly to a computer. However it is recorded, an observation with a radio telescope measures the amount of radio energy coming from a specific point on the sky.

Because humans can't see radio waves, astronomers must convert them into something perceptible. One way is to measure the strength of the radio signal at various places in the sky and draw a map in which contours mark areas of uniform radio intensity. We might compare such a map to a seating diagram for a baseball stadium in which the contours mark areas in which the seats have the same price (Figure 6-20a). Contour maps are very common in radio astronomy and are often reproduced using false colors (Figure 6-20b)

Limitations of the Radio Telescope

A radio astronomer works under three handicaps: poor resolution, low intensity, and interference. We saw that the resolving power of an optical telescope depends on the diameter of the objective lens or mirror. It also depends on the wavelength of the radiation. At very long wavelengths, like those of radio waves, images become fuzzy because of the large diffraction fringes. As with an optical telescope, the only way to improve the resolving power is to build a bigger telescope. Consequently, radio telescopes must be quite large.

Even so, the resolving power of a radio telescope is not good. A dish 30 m in diameter receiving radiation with a wavelength of 21 cm has a resolving power of about 0.5°. Such a radio telescope would be unable to show us any details in the sky smaller than the moon. Fortunately, radio astronomers can combine two or more radio telescopes to improve the resolving power. Such

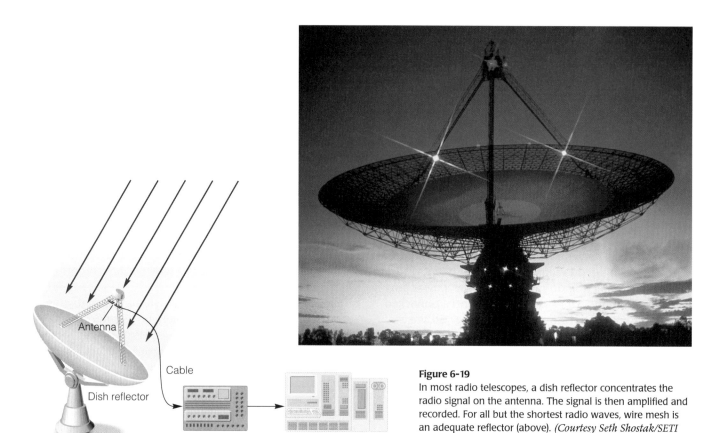

Figure 6-19
In most radio telescopes, a dish reflector concentrates the radio signal on the antenna. The signal is then amplified and recorded. For all but the shortest radio waves, wire mesh is an adequate reflector (above). *(Courtesy Seth Shostak/SETI Institute)*

a linkup of radio telescopes is called a **radio interferometer** and has the resolving power of a radio telescope whose diameter equals the separation of the radio telescopes (Figure 6-21a). For example, the Very Large Array (VLA) radio interferometer shown in Figure 6-21b uses multiple radio dishes spread across the New Mexico–Arizona desert to simulate a single radio telescope with

a diameter of 40 km (25 mi). It can produce radio maps with a resolution better than 1 second of arc. The Very Long Baseline Array (VLBA) consists of matching radio dishes spread from Hawaii to the Virgin Islands and has an effective diameter of about 8000 km. HALCA, the Highly Advanced Laboratory for Communications and Astronomy, consists of an 8-m radio dish in orbit

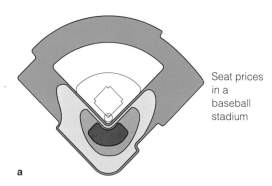

Seat prices in a baseball stadium

a

Figure 6-20
(a) A contour map of a baseball stadium shows regions of similar admission prices. The most expensive seats are those behind home plate. (b) A false-color-image radio map of Tycho's supernova remnant, the expanding shell of gas produced by the explosion of a star in 1572. The radio contour map has been color-coded to show intensity. Red is the strongest radio intensity, and violet the weakest. *(Courtesy NRAO)*

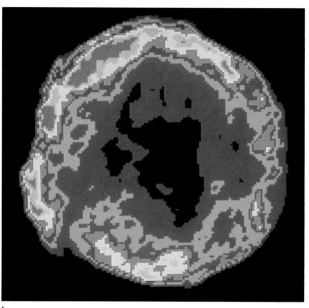

b

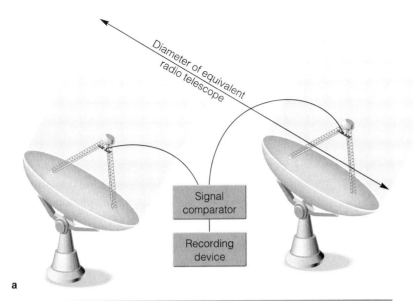

Figure 6-21

(a) A radio interferometer consists of two or more radio telescopes whose signals are combined to give the resolving power of a much larger instrument. (b) The Very Large Array (VLA) radio interferometer uses up to 27 dishes along the 20-km arms of a Y to simulate a radio telescope 36 km in diameter. *(National Radio Astronomy Observatory, operated by Associated Universities, Inc., under a contract with the National Science Foundation)*

around Earth. It can be used with ground-based radio telescopes to produce maps as good as a radio telescope 32,000 km in diameter. In these ways, radio astronomers use interferometers to compensate for the low resolving power of a single radio telescope.

The second handicap radio astronomers face is the low intensity of the radio signals. We saw earlier that the energy of a photon depends on its wavelength. Photons of radio energy have such long wavelengths that their individual energies are quite low. In order to get strong signals focused on the antenna, the radio astron-

omer must build large collecting dishes.

The largest radio dish in the world is 300 m (1000 ft) in diameter. So large a dish can't be supported in the usual way, so it is built into a mountain valley in Arecibo, Puerto Rico. The reflecting dish is a thin metallic surface supported above the valley floor by cables attached near the rim, and the antenna hangs above the dish on cables from three towers built on three mountain peaks that surround the valley (Figure 6-22). Although this telescope can look only overhead, the operators can change its aim slightly by moving the antenna and by waiting for Earth's rotation to point the telescope in the proper direction. This may sound clumsy, but the telescope's ability to detect weak radio sources, together with its good resolution, makes it one of the most important radio observatories in the world. Chinese astronomers are now designing a similar 500-m radio telescope they hope to build in a mountain valley in China.

The third handicap the radio astronomer faces is interference. A radio telescope is an extremely sensitive radio receiver listening to radio signals thousands of times weaker than artificial radio and TV transmissions. Such weak signals are easily drowned out by interference. Sources of such interference include everything from poorly designed transmitters in Earth satellites to automobiles with faulty ignition systems. To avoid this kind of interference, radio astronomers locate their telescopes as far from civilization as possible. Hidden deep in mountain valleys, they are able to listen to the sky protected from human-made radio noise.

Advantages of Radio Telescopes

Building large radio telescopes in isolated locations is expensive, but three factors make it all worthwhile. First, and most important, a radio telescope can show us where clouds of cool hydrogen are located between the stars. Because 90 percent of the atoms in the universe are hydrogen, that is important information. Large clouds of cool hydrogen are completely invisible to normal telescopes, because they produce no visible light of their own and reflect too little to be detected on photographs. However, cool hydrogen emits a radio signal at the specific wavelength of 21 cm. (We will see how the hydrogen produces this radiation when we discuss the gas clouds in space in Chapter 12.) The only way

Figure 6-22
The largest radio telescope in the world is the 300-m (1000-ft) dish suspended in a valley in Arecibo, Puerto Rico. The antenna hangs above the dish on cables stretching from towers. The Arecibo Observatory is part of the National Astronomy and Ionosphere Center, which is operated by Cornell University under contract with the National Science Foundation. *(David Parker/Science Photo Library)*

Why do optical astronomers build big telescopes, while radio astronomers build groups of widely separated smaller telescopes?

Optical astronomers build large telescopes to maximize light-gathering power, but the problem for radio telescopes is resolving power. Because radio waves are so much longer than light waves, a single radio telescope can't see details in the sky much smaller than the moon. By linking radio telescopes miles apart, radio astronomers build a radio interferometer that can simulate a radio telescope miles in diameter and thus increase the resolving power.

The difference between the wavelengths of light and radio waves makes a big difference in building the best telescopes. But why don't radio astronomers want to build their telescopes on mountaintops as optical astronomers do?

Our atmosphere causes trouble for Earth's astronomers in two ways. It distorts images, and it absorbs many wavelengths. The only way to avoid these limitations completely is to send telescopes above the atmosphere, into space.

6-5 Space Astronomy

Ground-based telescopes can operate only at wavelengths in the visual and radio windows of the atmosphere. Most of the rest of the electromagnetic radiation—infrared, ultraviolet, X ray, and gamma ray—never reaches Earth's surface. To observe at these wavelengths, telescopes must fly above the atmosphere in high-flying aircraft, rockets, balloons, and satellites. The only exception is some observations of the shorter infrared wavelengths that can be made from high mountains.

Infrared Astronomy

Some infrared radiation does leak through our atmosphere. This radiation enters narrow, partially open atmospheric windows scattered from 1200 nm to about 40,000 nm. Infrared astronomers usually measure wavelength in micrometers (10^{-6} meters), so they refer to this wavelength range as 1.2 to 40 micrometers. In this range, called the near infrared, much of the radiation is absorbed by water vapor, carbon dioxide, and oxygen molecules in Earth's atmosphere, so it is an advantage to place telescopes on mountains where the air is thin and dry. A number of important infrared telescopes, for example, observe from the 4150-m (13,600-ft) summit of Mauna Kea in Hawaii (Figure 6-23). At this altitude,

we can detect these clouds of gas is with a radio telescope that receives the 21-cm radiation. These hydrogen clouds are the places where stars are born, and that is one reason that radio telescopes are important.

Nevertheless, there is a second reason. Because radio signals have relatively long wavelengths, they can penetrate the vast clouds of dust that obscure our view at visual wavelengths. Light waves are short, and they interact with tiny dust grains floating in space; thus, the light is scattered and never penetrates the dust to reach optical telescopes on Earth. However, radio signals from far across the galaxy pass unhindered through the dust, giving us an unobscured view.

Finally, a radio telescope can detect objects that are more luminous at radio wavelengths than at visible wavelengths. This includes everything from the coldest clouds of gas to the hottest stars. Some of the most distant objects in the universe, for instance, are detectable only at radio wavelengths.

they are above much of the water vapor, which is the main absorber of infrared.

The far-infrared range, which includes wavelengths longer than 40 micrometers, can tell us about planets, comets, forming stars, and other cool objects, but these wavelengths are absorbed high in the atmosphere. To observe in the far infrared, telescopes must venture to high altitudes. Remotely operated infrared telescopes suspended under balloons have reached altitudes as high as 41 km (25 mi). For many years, a NASA jet transport carried a 91-cm infrared telescope and a crew of astronomers to altitudes of 12,000 m (40,000 ft) to get above 99 percent of the water vapor in Earth's atmosphere. Now retired from service, that airborne observatory will soon be replaced with the Stratospheric Observatory for Infrared Astronomy (SOFIA), a Boeing 747 that will carry a 2.5-m telescope to the fringes of the atmosphere.

The ultimate solution is to place infrared telescopes in space above the atmosphere. In the early 1980s, the Infrared Astronomy Satellite (IRAS) mapped the sky at infrared wavelengths (Figure 6-24). In the middle 1990s, European astronomers launched the Infrared Space

Figure 6-23
Comet Hale-Bopp hangs in the sky over the 3-meter NASA Infrared Telescope atop Mauna Kea. The air at high altitudes is steady and so dry that it is transparent to shorter infrared photons. Infrared astronomers can observe with the lights on in the telescope dome. Their instruments are usually insensitive to visible light. *(Courtesy William Keel)*

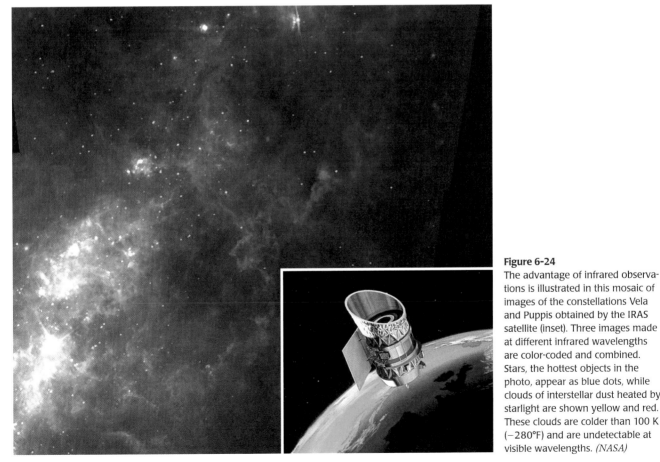

Figure 6-24
The advantage of infrared observations is illustrated in this mosaic of images of the constellations Vela and Puppis obtained by the IRAS satellite (inset). Three images made at different infrared wavelengths are color-coded and combined. Stars, the hottest objects in the photo, appear as blue dots, while clouds of interstellar dust heated by starlight are shown yellow and red. These clouds are colder than 100 K (−280°F) and are undetectable at visible wavelengths. *(NASA)*

Observatory, which carried detectors more sensitive than IRAS. NASA now plans for the Space Infrared Telescope Facility (SIRTF), a major infrared observatory in space, to begin observing in 2002.

If a telescope observes at far-infrared wavelengths, then it must be cooled. Infrared radiation is heat, and if the telescope is warm it will emit many times more infrared radiation than that coming from a distant object. Imagine trying to look at a dim, moonlit scene through binoculars that are glowing brightly. In a telescope observing near-infrared wavelengths, only the detector, the element on which the infrared radiation is focused, must be cooled. To observe in the far infrared, however, as IRAS did, the entire telescope must be cooled.

Ultraviolet Astronomy

Ultraviolet radiation with wavelengths shorter than about 290 nm, the far ultraviolet, is completely absorbed by the ozone layer in our atmosphere. The ozone layer extends from 20 km to about 40 km above Earth's surface. Telescopes to observe in the far ultraviolet must get above the ozone layer, and that means they must go into space.

One of the most successful ultraviolet observatories was the International Ultraviolet Explorer (IUE). Launched in 1978, it carried a 45-cm (18-in.) telescope and used TV systems to record spectra from 320 nm to 115 nm. Although IUE was expected to last only a year or two, it survived and was widely used by astronomers from all around the world until, partially

shut down because of budget cutbacks, it finally failed in 1996.

The Extreme Ultraviolet Explorer (EUVE) telescope was launched in 1992 to observe at wavelengths from 100 nm to 10 nm. With four telescopes and more modern detectors than IUE, EUVE surveyed the entire sky and studied specific targets at its shorter wavelength range.

Observations in the infrared reveal cool material, but observations in the ultraviolet tend to show hot, excited regions. Hot stars and hot gas are mapped by such telescopes.

X-Ray Astronomy

Beyond the UV, at wavelengths from 10 nm to 0.01 nm, lie the X rays. These photons can be produced only by high-energy processes, so we see them coming from very hot regions in stars and from violent events such as matter smashing onto a neutron star—a subject we will discuss in Chapter 15. X-ray images can give us information about the heavens that we can get in no other way.

Although early X-ray observations of the sky were made from balloons and small rockets in the 1960s, the age of X-ray astronomy did not really begin until 1970, when an X-ray telescope named Uhuru (Swahili for "freedom") was put into orbit. Uhuru detected nearly 170 separate sources of celestial X rays. In the late 1970s, the three High Energy Astronomy Observatories (HEAO) satellites, carrying more sensitive and more sophisticated equipment, pushed the total to many hundreds.

a

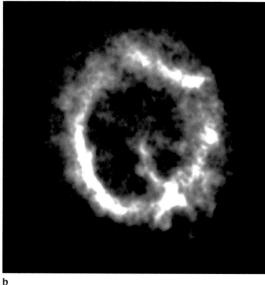

b

Figure 6-25
(a) From Earth orbit, Chandra can photograph the sky at X-ray wavelengths. Its images will help astronomers study a wide range of astrophysical problems ranging from the birth of stars out of the gas in space to eruptions in the most distant galaxies. (b) This X-ray image of the remains of an exploded star reveals an expanding shell of hot gas about 1000 years old. *(NASA)*

The second HEAO satellite, named the Einstein Observatory, used special optics to produce X-ray images.

For many years, astronomers have worked to create a major X-ray observatory in space. In 1999, the Advanced X-ray Astrophysics Facility (AXAF) was carried into orbit by the Space Shuttle and renamed Chandra in honor of the late Indian-American Nobel Laureate Subrahmanyan Chandrasekhar, who worked on theoretical astrophysics in many areas of astronomy which the X-ray observatory Chandra will study (Figure 6-25).

Focusing X rays is difficult because the high-energy X-ray photons do not reflect off conventional mirrors but rather penetrate into the mirrors. The optics in Chandra are especially designed mirrors in which the X-ray photons graze the surface of the mirrors at very small angles. Under these circumstances, the photons are reflected to form an image. Just as a CCD chip records visible light photons, the detector in Chandra has been designed to absorb and record X-ray photons to produce an X-ray image. Chandra can detect X-ray emitting objects 50 times fainter than any previous X-ray telescope and can resolve details 10 times smaller.

Gamma-Ray Telescopes

Gamma rays have wavelengths even shorter than X rays, and that means they have even higher energies. They are not understood very well, but they appear to be produced by the hottest and most violent objects in the universe: exploding stars, erupting galaxies, neutron stars, and black holes.

We don't understand gamma rays well because they are so difficult to detect. First, they cannot be focused or detected as X rays can. Thus, gamma-ray telescopes are more complex and can see less detail than similar X-ray telescopes. Also, gamma rays are such high-energy photons that natural processes produce only a few. Gamma-ray telescopes must count gamma rays one at a time. For example, a gamma-ray telescope observed an intense source for a month and counted only 3000 gamma rays—one every 15 minutes.

In addition, gamma rays are almost totally absorbed by our atmosphere, so gamma-ray telescopes must observe from orbit. NASA has operated three important gamma-ray telescopes aboard satellites, including one on the Small Astronomy Satellite 2 (SAS 2). The Soviet Union orbited two, the more recent of which was aboard the satellite COS B, launched in 1975 by the European Space Agency. It surveyed the sky and detected gamma rays from clouds of gas in space, from the center of our galaxy, and from the remains of exploded stars.

NASA launched the 17-ton Gamma Ray Observatory in April 1991. At 10 to 50 times the sensitivity of any previous gamma-ray telescope, it first mapped the entire sky and then observed selected targets.

Cosmic Rays

All of the radiation we have discussed in this chapter has been electromagnetic radiation. **Cosmic rays,** however, are not really rays; they are subatomic particles traveling at tremendous velocities that strike our atmosphere from space. Almost no cosmic rays reach the ground, but they do smash into gas atoms in the upper atmosphere, and fragments of these collisions shower down on us day and night over our entire lives. These secondary cosmic rays are passing through you as you read this sentence.

Some cosmic-ray research can be done from high mountains or high-flying aircraft, but to study cosmic rays in detail, detectors must go into space. A number of cosmic-ray detectors have been carried into orbit, but this area of astronomical research is just beginning to bear fruit.

We can't be sure where cosmic rays come from. Because they are atomic particles with electric charges, they are deflected by the magnetic fields spread through our galaxy, and that means we can't tell where they are coming from. The space between the stars is a glowing fog of cosmic rays. Some lower-energy cosmic rays come from the sun, but many cosmic rays are probably produced by the violent explosions of dying stars. At present, cosmic rays largely remain a mystery. We will discuss them again in future chapters.

The Hubble Space Telescope

Astronomers have dreamed of having a giant telescope in orbit since the 1920s. Such a telescope would not only allow observation at nonvisible wavelengths but would also avoid the turbulence in Earth's atmosphere. That dream was fulfilled on April 24, 1990, when the Hubble Space Telescope (HST) was released into orbit (Figure 6-26a). Named after the man who discovered the expansion of the universe (Chapter 15), it is the largest orbiting telescope ever built. As big as a large bus, the HST carries a 2.4-m (96-in.) mirror. The light can be directed to any of its spectrographs and cameras.

The telescope has exceeded its original design criteria. It can detect objects so faint it can look 50 times farther into space than can Earth-based telescopes. With its precise optics and its location above Earth's distorting atmosphere, it can resolve details 10 times smaller than can be resolved by any conventional telescope on Earth. Every few years, astronauts visit the telescope in orbit and install new instruments.

HST is controlled from the Space Telescope Science Institute located at Johns Hopkins University in Baltimore, Maryland. Astronomers there plan the telescope's observing schedule to maximize the scientific return while engineers study the operation of the telescope. HST transmits its data to the Institute, where it

a   b

Figure 6-26

(a) The Hubble Space Telescope is the largest orbiting telescope ever launched. It was carried into orbit by the space shuttle in 1990. (b) This image of Mars recorded by the Hubble Space Telescope reveals thin clouds drifting around high volcanoes at the left and details of the polar cap at the top. *(NASA)*

is analyzed. Because the telescope is a national facility, anyone may propose research projects for the telescope. Competition is fierce, and only the most worthy projects win approval. Nevertheless, even some amateur astronomers have had projects approved for the Space Telescope.

The Hubble Space Telescope has been a phenomenal success, exploring such mysteries as the age of the universe, the birth and death of stars, and weather on nearby planets (Figure 6-26b). Visits by astronauts will keep it in operation for years to come, but astronomers are already planning an even larger space telescope that will someday orbit above Earth's atmosphere.

REVIEW Critical Inquiry

Why can infrared astronomers observe from high mountaintops, while X-ray astronomers must observe from space?

Infrared radiation is absorbed by water vapor in Earth's atmosphere. If we built our infrared telescope on top of a high mountain, we would be above most of the water vapor in the atmosphere, and we could collect some infrared radiation from the stars. The longer-wavelength infrared radiation is absorbed much higher in the atmosphere, so we couldn't observe it from our mountaintop. Similarly, X rays are absorbed in the uppermost layers of the atmosphere, and we would not be able to find any mountain high enough to get an X-ray telescope above those absorbing layers. To observe the stars at X-ray wavelengths, we would need to put our telescope in space, above Earth's atmosphere.

X-ray and far-infrared telescopes must observe from space, but the Hubble Space Telescope observes in the visual wavelength range. Then why must the Hubble Space Telescope observe from orbit?

The tools of the astronomer are designed to gather radiation from the sky and extract information. Perhaps no tool is as important as the spectrograph, because no form of observation is as loaded with information as a spectrum. In the next chapter, we will see how we can harvest the information in a star's spectrum.

Summary

Electromagnetic radiation is an electric and magnetic disturbance that transports energy at the speed of light. The electromagnetic spectrum includes radio waves, infrared radiation, visible light, ultraviolet radiation, X rays, and gamma rays.

We can think of "a particle of light," a photon, as a bundle of waves that sometimes acts as a particle and sometimes as a wave. The energy a photon carries depends on its wavelength. The wavelength of visible light, usually measured in nanometers (10^{-9} m), ranges from 400 nm to 700 nm. Infrared and radio photons have longer wavelengths and carry less energy. Ultraviolet, X-ray, and gamma-ray photons have shorter wavelengths and carry more energy.

Astronomical telescopes are of two types, refractor and reflector. A refractor uses a lens to bend the light and focus it into an image. Because of chromatic aberration, refracting telescopes cannot bring all colors to the same focus, resulting in color fringes around the images. An achromatic lens partially corrects for this, but such lenses are expensive and cannot be made larger than about 1 m in diameter.

Reflecting telescopes use a mirror to focus the light and are less expensive than refracting telescopes of the same diameter. In addition, reflecting telescopes do not suffer from chromatic aberration. Thus, most recently built telescopes are reflectors.

The largest traditional telescopes in the world are the 6-m telescope in the former Soviet Union and the 5-m telescope in California. They have thick, solid mirrors of great weight. New telescopes use thin mirrors controlled by computers or mirrors made up of segments. The two Keck Telescopes in Hawaii use segments to make up mirrors 10 m in diameter.

The powers of a telescope are light-gathering power, resolving power, and magnifying power. The first two of these depend on the telescope's diameter; thus, astronomical telescopes often have large diameters.

Special instruments attached to a telescope analyze the light it gathers. The photographic plate records vast amounts of detail for later analysis, but electronic imaging systems such as charge-coupled devices have largely replaced photographic plates. CCDs are more sensitive, record both faint and bright objects, and can be read directly into computer memory. Spectrographs, using prisms or gratings, break the starlight into a spectrum, which then can be photographed or electronically recorded.

To observe radio signals from celestial objects, we need a radio telescope, which usually consists of a dish reflector, an antenna, an amplifier, and a recorder. Such an instrument measures the intensity of radio signals over the sky and constructs radio maps. The poor resolution of the radio telescope can be improved by combining it with another radio telescope to make a radio interferometer. The three principal advantages of radio telescopes are that they can detect the very cold gas clouds in space, they can detect regions of very hot gas produced by exploding stars or erupting galaxies, and they can look through the dust clouds that block our view at optical wavelengths.

Observations at some wavelengths in the near infrared are possible from high mountaintops or from high-flying aircraft. At these altitudes, the air is thin and dry, and the infrared radiation can reach the telescope. At longer infrared wavelengths, the telescope must observe from above Earth's atmosphere—from orbit. To observe at the short ultraviolet wavelengths, a telescope must be above the ozone layers in Earth's atmosphere. That means it must be in orbit. X-ray and gamma-ray telescopes must also be placed in orbit above Earth's absorbing atmosphere.

The largest orbiting telescope is the Hubble Space Telescope, launched in the spring of 1990. The Chandra X-ray telescope was put into orbit in 1999, and SIRTF, a major infrared observatory in space, will begin observing in 2002.

New Terms

electromagnetic radiation

wavelength

frequency

nanometer (nm)

Angstrom (Å)

photon

infrared radiation

ultraviolet radiation

atmospheric window

refracting telescope

reflecting telescope

focal length

objective lens

eyepiece

chromatic aberration

achromatic lens

objective mirror

prime focus

secondary mirror

Newtonian focus

Cassegrain telescope

Schmidt-Cassegrain telescope

sidereal drive

equatorial mounting

polar axis

alt-azimuth mounting

light-gathering power

resolving power

diffraction fringe

seeing

adaptive optics

magnifying power

light pollution

active optics

charge-coupled device (CCD)

false-color image

spectrograph

grating

comparison spectrum

radio interferometer

cosmic ray

Review Questions

1. Why do nocturnal animals usually have large pupils in their eyes? How is that related to astronomical telescopes?

2. Astronomers have not constructed a single major refracting telescope since about 1900. Why not?

3. How have computers and new technology changed the design of astronomical telescopes and their mountings?

4. Small telescopes are often advertised as "200 power" or "magnifies 200 times." As someone knowledgeable about astronomical telescopes, how would you improve such advertisements?

5. An astronomer recently said, "Some people think I should give up photographic plates." Why might she change to something else?

6. What purpose do the colors in a false-color image or false-color radio map serve?

7. Why can we observe in the infrared from high mountains and aircraft but must go into space to observe in the far ultraviolet?

8. The moon has no atmosphere at all. At what wavelengths could we observe if we had an observatory on the lunar surface?

Discussion Questions

1. Why does the wavelength response of the human eye match so well the visual window of Earth's atmosphere?

2. Most people like beautiful sunsets with brightly glowing clouds, bright moonlit nights, and twinkling stars. Most astronomers don't. Why?

Problems

1. The thickness of the plastic in plastic bags is about 0.001 mm. How many wavelengths of red light is this?

2. Measure the actual wavelength of the wave in Figure 6-1. In what portion of the electromagnetic spectrum would it belong?

3. Compare the light-gathering powers of the 5-m telescope and a 0.5-m telescope.

4. How does the light-gathering power of the largest telescope in the world compare with that of the human eye? (*Hint:* Assume that the pupil of your eye can open to about 0.8 cm.)

5. What is the resolving power of a 25-cm telescope? What do two stars 1.5 seconds of arc apart look like through this telescope?

6. Most of Galileo's telescopes were only about 2 cm in diameter. Should he have been able to resolve the two stars mentioned in Problem 5?

7. How does the resolving power of the 5-m telescope compare with that of the Hubble Space Telescope? Why will the HST outperform the 5-m telescope?

8. If we build a telescope with a focal length of 1.3 m, what focal length should the eyepiece have to give a magnification of 100 times?

9. Astronauts observing from a space station need a telescope with a light-gathering power 15,000 times that of the human eye, capable of resolving detail as small as 0.1 second of arc, and having a magnifying power of 250. Design a telescope to meet their needs. Could you test your design by observing stars from Earth?

10. A spy satellite orbiting 400 km above Earth is supposedly capable of counting individual people in a crowd. What minimum-diameter telescope must the satellite carry? (*Hint:* Use the small-angle formula.)

Critical Inquiries for the Web

1. Research in chemistry, physics, and biology is supported in part by industry. Because astronomy has few indus- trial applications, it is not well supported by industry. Visit Web sites for major observatories, and find out who pays their bills.

2. Visit Web sites for the major observatories in space, such as Hubble Space Telescope and Chandra, and check on their latest status. What kinds of observations are they making, and what kinds of discoveries are they making?

3. Astronomers are leaders in efforts to reduce electric power wasted on outdoor lighting because it produces light pol- lution. Find Web sites on light pollution, and see if your town is making any effort to preserve the beauty of the night sky and simultaneously save electrical power.

Exploring *The Sky*

1. Although the largest of modern telescopes are usually mounted with computer-controlled **alt-azimuth mount- ings,** many of the smaller telescopes in existence, includ- ing amateur telescopes, use **equatorial mountings.** The polar axis of such mountings must be aligned parallel to Earth's axis of rotation and must be rotated at the side- real rate to compensate for Earth's rotation. To set up such a telescope, the polar axis of the mounting is set to point to the celestial north pole, the northern point in the sky about which all stars seem to rotate. In practice, however, it is difficult to determine the exact location of the celes- tial north pole, and it is often sufficient to point the polar axis to Polaris, the North Star. Use *The Sky* to determine the error in using Polaris instead of the celestial north pole. That is, find the angle between Polaris and the ce- lestial north pole.

How to proceed: Set up *The Sky* to show you a northern view of the sky. To locate Polaris, first find Ursa Minor by clicking on the **Constellation Lines** and **Common Names** buttons on the **View** toolbar. Polaris is the last star in the handle of what is called the "Little Dipper." Click on Polaris, and the **Object Information** window will ap- pear to give a variety of information about Polaris, in- cluding the **declination.** The declination of any object in the sky is one of the coordinates used to define its loca- tion; it corresponds to latitude on Earth. The angle be- tween Polaris and the celestial north pole is the differ- ence between the declination of the celestial north pole, which is 90 degrees, and that of Polaris.

 Go to the Brooks/Cole Astronomy Resource Center (www. brookscole.com/astronomy) for critical thinking exercises, articles, and additional readings from InfoTrac College Edi- tion, Brooks/Cole's online student library.

CHAPTER SEVEN

Starlight and Atoms

Awake! for Morning in the Bowl

of Night

Has flung the Stone that puts the

Stars to Flight:

And Lo! the Hunter of the East has

caught

The Sultan's Turret in a Noose

of Light.

The Rubáiyát of Omar Khayyám
Trans. Edward FitzGerald

Guidepost

Some chapters in textbooks do little more than present facts. The chapters in this book attempt to present astronomy as organized understanding. But this chapter is special. It presents us with a tool. The interaction of light with matter gives astronomers clues about the nature of the heavens, but the clues are meaningless unless astronomers understand how atoms leave their traces on starlight. Thus, we dedicate an entire chapter to understanding how atoms interact with light.

This chapter marks a transition in the way we look at nature. Earlier chapters described what we see with our eyes and explained those observations using models and theories. With this chapter, we turn to modern astrophysics, the application of physics to the study of the sky. Now we can search out secrets of the stars that lie beyond the grasp of our eyes.

If this chapter presents us with a tool, then we should use it immediately. The next chapter will apply our new tool to understanding the sun.

Figure 7-1
So vast that light takes 20 years to cross its diameter, the cloud of gas called the Eagle Nebula glows deep red. Nothing in the appearance of the swirling gas and the cluster of bright stars just above and to the right of the nebula's center helps us make sense of what we see. But by analyzing the light from the gas and stars, astronomers have learned that the nebula is mostly hydrogen gas and that the star cluster is a young group of stars that have formed recently from the nebula. *(AURA/NOAO/NSF)*

No laboratory jar on Earth holds a sample labeled "star stuff," and no instrument has ever probed inside a star. The stars are far beyond our reach, and the only information we can obtain about them comes to us hidden in light (Figure 7-1). Whatever we want to know about the stars we must catch in a noose of light.

We Earthbound humans knew almost nothing about stars until the early 19th century, when the Munich optician Joseph von Fraunhofer studied the solar spectrum and found it interrupted by some 600 dark lines. As scientists realized that the lines were related to the various atoms in the sun and found that stellar spectra had similar patterns of lines, the door to an understanding of stars finally opened.

In this chapter, we will go through that door by considering how atoms interact with light to produce spectral lines. We begin with the hydrogen atom because it is the most common atom in the universe as well as the simplest. Other atoms are larger and more complicated, but in many ways their properties resemble those of hydrogen. Another reason for studying hydrogen is to consider how the structure of its atoms can give rise to the 21-cm-wavelength radiation that is so important to radio astronomers (see Chapter 6).

Once we understand how an atom's structure can interact with light to produce spectral lines, we will recognize certain patterns in stellar spectra. By classifying the spectra according to these patterns, we can arrange the stars in a sequence according to temperature. One of the most important pieces of information revealed in a star's spectrum is its temperature.

But, properly analyzed, a stellar spectrum can tell us much more. The spectrum gives us information about the chemical composition of the star and the star's motion relative to Earth.

Five fundamental parameters—mass, energy, thermal energy, density, and pressure—are so critical in science that we will discuss them individually in separate Windows on Science. Here we will discuss thermal energy, the measure of the amount of agitation in the particles in an object. Because temperature and thermal energy are often confused, we must be careful to distinguish between them. Temperature is an intensity, and thermal energy is an amount.

Temperature is a measure of the average motion of the atoms and molecules that make up an object. In a hot object, the particles vibrate at higher speeds than in a cool object. If you have your temperature taken, it will probably be 98.6°F, an indication that the atoms and molecules

in your body are vibrating at a normal pace. If you measure the temperature of a month-old baby, the thermometer should register the same temperature, showing that the atoms and molecules in the baby's body are moving at the same average velocity as the atoms and molecules in your body. Thus, temperature is a measure of the intensity of the motion.

Thermal energy is a measure of the total energy stored in a body as motion among its particles. Although you and the infant have the same temperature, you contain much more mass than the infant, so you must contain much more thermal energy. Thus, the thermal energy in your body and in the baby's body have the same intensity (temperature) but different amounts.

Many people say "heat" when they should say thermal energy. Heat is the energy that moves from a hot object to a cool object. If two objects have the same temperature, you and the infant for example, there is no transfer of thermal energy and no heat. This is a fine distinction, but when you hear someone say "heat," check to see if they don't really mean thermal energy.

You may have burned your tongue on cheese pizza, but you probably haven't burnt yourself on green beans. At the same temperature, cheese holds more thermal energy than green beans. It isn't the temperature that burns us, but the thermal energy.

7-1 Starlight

If you look at the stars in the constellation Orion, you will notice that they are not all the same color. Betelgeuse, in the upper left corner, is quite red; Rigel, in the lower right corner, is blue. These differences in color arise from the way the stars produce light and give us our first clue to the temperatures of stars.

Temperature and Heat

Temperature is one of the defining characteristics of a star. That is, if we know a star's temperature and a few other properties, such as size, we can understand the star. But if we don't know a star's temperature, we can know almost nothing about it.

A gas is made up of particles—atoms and molecules—that are in constant motion, colliding with one another millions of times a second. The **temperature** of a gas is a measure of the average velocity of the particles. If a gas is hot, the particles are moving very rapidly; if it is cool, the particles are moving more slowly.

We must take care when we discuss heat. When most people say "heat" they are really thinking of the **thermal energy** stored in a material as motion among its particles. The atoms in a gas are free to move, of course, but the atoms in a liquid or a solid can also move. If the particles in a material are agitated, we say the material is hot, meaning that it contains a lot of thermal energy. Notice the distinction between temperature and thermal energy. A cup of coffee and a giant

vat of coffee can have the same temperature, but the vat of coffee obviously contains more thermal energy (Window on Science 7-1).

We will measure the temperature of stars on the **Kelvin temperature scale.** The Kelvin scale sets its zero point at **absolute zero,** the temperature at which the particles of a gas have no remaining motion that we can extract as heat. Absolute zero is −273.2°C, which is −459.7°F. (See Appendix A). The temperatures of the stars range from 2000K to 40,000K or more and can be deduced from an analysis of starlight.

The Origin of Starlight

The starlight we see comes from gases in the outer surface of the star—the photosphere. The gases deep inside the star also emit light, but it is absorbed before it can escape, and the low-density gas above the photosphere is too thin to emit significant amounts of light. Thus, the photosphere of a star is that layer of gases dense enough to emit significant amounts of light but thin enough to allow the photons to escape. (Recall that we met the photosphere of the sun in Chapter 3.)

To see how the photosphere of a star can produce light, we must consider two things: how photons are produced and how the temperature of a material is related to motion among its atoms.

First, a photon can be produced by a changing electric field. An **electron** is a negatively charged subatomic particle, and if we disturb the motion of an electron, the sudden change in the electric field around it can produce an electromagnetic wave. For example, if you run a comb through your hair while standing near an AM radio, you can produce popping noises on the radio. The moving comb disturbs the electrons in both the comb and your hair, building static electricity. The sudden sparks of static electricity produce electromagnetic waves that the radio picks up as noise. This illustrates an important principle: When we change the motion of an electron, we generate an electromagnetic wave.

Second, recall that temperature is a measure of the average motion among the particles in a material. In the gases of a hot star, the atoms move faster, on average, than the atoms in a cool star.

Now we can understand why a hot object glows. The hotter an object is, the more violent the motion among its particles. The agitated particles collide with electrons, and when electrons are accelerated, part of the energy is carried away as a photon. The radiation emitted by a heated object is called **black body radiation,** a name that refers to the way a perfect emitter of radiation would behave. A perfect emitter would also be a perfect absorber and at room temperature would look black. Thus we often use the term black body radiation to refer to objects that glow brightly.

Black body radiation is quite common. In fact, it is responsible for the light emitted by an incandescent lightbulb. Electricity flowing through the filament of the lightbulb heats it to high temperature, and it glows. We also recognize the light emitted by a heated horseshoe in the blacksmith's forge as black body radiation. Many objects in astronomy, including stars, emit radiation as if they were nearly perfect black bodies.

Hot objects emit black body radiation, but so do cold objects. Ice cubes are cold, but their temperature is higher than absolute zero, so they contain some thermal energy and must emit some black body radiation. The coldest gas drifting in space has a temperature only a few degrees above absolute zero, but it too emits black body radiation.

We must discuss two important features of black body radiation. First, the hotter an object is, the more black body radiation it emits. Hot objects emit more radiation because their agitated particles travel faster and collide more often. Thus we expect a glowing coal from a fire to emit more total energy than an ice cube of the same size.

The second feature we must discuss is the relationship between the temperature of the object and the wavelengths of the photons it emits. The wavelength of a photon emitted when a particle collides with an electron depends on the violence of the collision. Only a violent collision can produce a short wavelength (high-energy) photon. Because extremely violent collisions don't occur very often, short wavelength photons are rare. Similarly, most collisions are not extremely gentle, so long wavelength (low-energy) photons are also

rare. Consequently, black body radiation is made up of photons with a distribution of wavelengths, and very short and very long wavelengths are rare. The **wavelength of maximum intensity (λ_{max}),** the wavelength at which the object emits the most radiation, occurs at some intermediate wavelength.

Figure 7-2 shows the intensity of radiation versus wavelength for three objects of different temperatures. As we expect, the objects emit most of their radiation at intermediate wavelengths, and the hotter object (top) emits more total radiation than the cooler object (bottom). We can also see that the wavelength of maximum intensity depends on temperature. The hotter the ob-

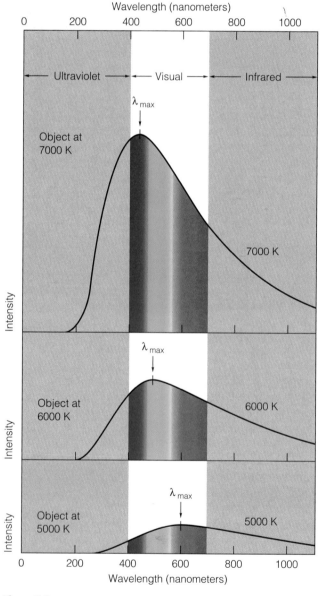

Figure 7-2

Black body radiation from three bodies at different temperatures demonstrates that a hot body radiates more total energy and that the wavelength of maximum intensity is shorter for hotter objects. The hotter object here will look blue to our eyes, while the cooler object will look red.

ject, the shorter the wavelength of maximum. Notice in Figure 7-2 how the temperature determines the color of a glowing black body. The hotter object emits more blue light than red and thus looks blue, and the cooler object emits more red than blue and consequently looks red. We will see this among stars. Hot stars look blue and cool stars red.

Notice that cool objects may emit little visible radiation but are still producing black body radiation. For example, the human body has a temperature of 310 K and emits black body radiation mostly in the infrared part of the spectrum. Infrared security cameras can detect burglars by the radiation they emit, and mosquitoes can track you down in total darkness by homing in on your infrared radiation. Although we humans emit lots of infrared radiation, we rarely emit higher-energy photons; we almost never emit X-ray or gamma-ray photons. Our wavelength of maximum intensity lies in the infrared part of the spectrum.

Two Radiation Laws

Black body radiation can be described by two simple laws that will help us understand the nature of starlight. One law is related to energy and one to color.

As we saw in the previous section, a hot surface emits more black body radiation than a cool surface. That is, it emits more energy. Recall from Chapter 5 that we measure energy in units called joules (J); 1 joule is about the energy of an apple falling from a table to the floor. The total radiation given off by 1 square meter of the object in joules per second equals a constant number, represented by σ, times the temperature raised to the fourth power.* This relationship is often called the Stefan-Boltzmann law:

$$E = \sigma\, T^4 \ (\text{J/s/m}^2)$$

How does this help us understand stars? Suppose a star the same size as the sun had a surface temperature that was twice as hot as the sun's surface. Then each square meter of that star would radiate not twice as much energy but 2^4 or 16 times as much energy. From this law we see that a small difference in temperature can produce a very large difference in the amount of energy emitted.

The second radiation law is related to the color of stars. In the previous section, we saw that hot stars look blue and cool stars look red. Wien's law tells us that the wavelength at which a star radiates the most energy, the wavelength of maximum (λ_{max}), depends only on the star's temperature:

$$\lambda_{max} = \frac{3{,}000{,}000}{T}$$

That is, the wavelength of maximum radiation in nanometers equals 3 million divided by the temperature on the Kelvin scale.

This is a powerful tool in astronomy, because it means we can relate the temperature of a star and its wavelength of maximum. For example, we might find a star that has a surface temperature of 3000 K. Then its wavelength of maximum would be 3,000,000/3000, or 1000 nm—in the near infrared. Later we will meet objects much hotter than most stars; such objects radiate most of their energy at very short wavelengths. The hottest stars, for instance, radiate most of their energy in the ultraviolet.

The Color Index

Hot stars look blue, but if we want to know the temperature of the star's surface, we must have a way of measuring color. In astronomy, the **color index** is a measure of color.

To measure the color index of a star, we need to measure brightness through a set of standard filters. The most commonly used filters are the blue (B) and visual (V). Each filter isolates a specific part of the spectrum. The B filter, for example, will not allow any light through except those photons with wavelengths between about 400 nm and 480 nm. The V filter is transparent to photons with wavelengths between about 500 nm and 600 nm, roughly approximating the sensitivity of the human eye.

If we measure the magnitude of a star through each of these filters, the difference between the two magnitudes is a number related to the color of the star. This is commonly written B-V and is referred to as the B-V color index.

For example, a hot star radiates much more blue light than red and will be brighter through the B filter (Figure 7-2). That means that the B magnitude will be smaller, and the B-V color index will be negative. The bluest stars have a B-V color index of about -0.4 and have surface temperatures of about 50,000 K. Red stars radiate much more red light than blue, so they will be fainter through the blue filter. Then B-V for a red star will be positive. The reddest stars have a B-V color index of about 2 and surface temperatures of about 2000 K.

R E V I E W   Critical Inquiry

Why are hot stars bluer than cool stars?

Temperature is a measure of the average velocity of the randomly moving particles of a material. If a star is hot, then its atoms must be bouncing around at high velocities. When they interact with electrons and cause the emission of photons, most of the interactions are quite

---

*For the sake of completeness, we should note that the constant σ equals 5.67×10^{-8} J/m²s degree⁴.

violent, and most of the photons that are emitted are very energetic. If photons have high energies, they have short wavelengths, so the gas in a hot star tends to emit larger numbers of blue photons than red photons, and the star looks blue.

In a cool star, the atoms are not moving as rapidly, collisions are less violent, and the average photon emitted is not as energetic. Lower-energy photons have longer wavelengths and look red, so cooler stars emit a larger proportion of red photons.

This interesting bit of physics explains why hot stars look blue and cool stars look red. But the surface of a hot star emits more energy than that of a cool star. How is this related to the agitation of the atoms?

The colors of the stars are beautiful, but now we know that their color reveals their temperature. To get even more information from starlight, we must understand the structure of atoms and how atoms interact with light.

7-2 Atoms

The atoms in the surface layers of stars leave their marks on the light the stars emit. By understanding what atoms are and how they interact with light, we can decode the spectra of the stars. We begin this section by constructing a working model of an atom.

A Model Atom

To think about atoms and how they can interact with light, we must create a working model of an atom. In Chapter 2, we created a working model of the sky, the celestial sphere. We identified and named the important parts and described how they were located and how they interacted. We begin our study of atoms by creating a model of an atom.

In our model atom, we identify a positively charged **nucleus** at the center; this nucleus consists of two kinds of particles. **Protons** carry a positive electrical charge, and **neutrons** have no charge. Thus, the nucleus has a net positive charge.

The nucleus in our model atom is surrounded by a whirling cloud of orbiting electrons, low-mass particles with a negative charge. In a normal atom, the number of electrons equals the number of protons, and the positive and negative charges balance to produce a neutral atom. Because protons and neutrons each have a mass 1836 times greater than that of an electron, most of the mass of the atom lies in the nucleus. The hydrogen atom is the simplest of all atoms. The nucleus is a

single proton orbited by a single electron, with a total mass of only 1.67×10^{-27} kg, about a trillionth of a trillionth of a gram.

An atom is mostly empty space. To see this we can construct a simple scale model. The nucleus of a hydrogen atom is a proton with a diameter of about 0.0000016 nm, or 1.6×10^{-15} m. If we multiply this by one trillion (10^{12}) we can represent the nucleus of our model atom with a grape seed, which is about 0.16 cm in diameter. The region of a hydrogen atom containing the whirling electron has a diameter of about 0.4 nm or 4×10^{-10} m. Multiplying by a trillion magnifies the diameter to about 400 m, or about 4.5 football fields laid end to end (Figure 7-3). When you imagine a grape seed in the midst of a sphere 4.5 football fields in diameter, you can see that an atom is mostly empty space.

Different Kinds of Atoms

There are over a hundred kinds of atoms called chemical elements. Which element an atom is depends only on the number of protons in the nucleus. For example, carbon has six protons in its nucleus. An atom with one more proton than this is nitrogen, and an atom with one fewer proton is boron.

Although the number of protons in an atom of an element is fixed, we can change the number of neutrons in an atom's nucleus without changing the atom significantly. For instance, if we add a neutron to the carbon nucleus, we still have carbon, but it is slightly heavier than normal carbon. Atoms that have the same number of protons but a different number of neutrons are **iso-**

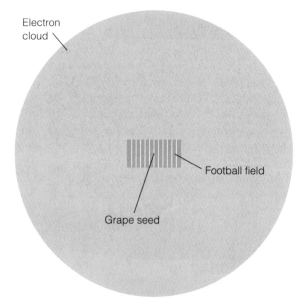

Figure 7-3
Magnifying a hydrogen atom by 10^{12} makes the nucleus the size of a grape seed and the diameter of the electron cloud about 4.5 times longer than a football field. The electron itself is still too small to see.

topes. Carbon has two stable isotopes. One form contains six protons and six neutrons for a total of 12 particles and is thus called carbon-12. Carbon-13 has six protons and seven neutrons in its nucleus (Figure 7-4).

Protons and neutrons are bound tightly into the nucleus, but the electrons are held loosely in the electron cloud. Running a comb through your hair creates a static charge by removing a few electrons from their atoms. This process is called **ionization,** and the atom that has lost one or more electrons is an **ion.** A carbon atom is neutral if it has six electrons to balance the positive charge of the six protons in its nucleus. If we ionize the atom by removing one or more electrons, the atom is left with a net positive charge. Under some circumstances, an atom may capture one or more extra electrons, giving it more negative charges than positive. Such a negatively charged atom is also considered an ion.

Atoms that collide may form bonds with each other by exchanging or sharing electrons. Two or more atoms bonded together form a **molecule.** Atoms do collide in stars, but the high temperatures cause violent collisions that are unfavorable for chemical bonding. Only in the coolest stars are the collisions gentle enough to permit the formation of chemical bonds. We will see later that the presence of molecules such as titanium oxide (TiO) in a star is a clue that the star is cool. In later chapters, we will see that molecules can form in cool gas clouds in space and in the atmospheres of planets.

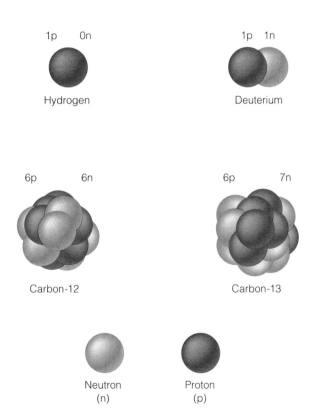

Figure 7-4
Some common isotopes. A rare isotope of hydrogen, deuterium, contains a proton and a neutron in its nucleus. Two isotopes of carbon are carbon-12 and carbon-13.

Electron Shells

We mentioned the whirling cloud of orbiting electrons in a general way, but the specific way electrons behave within the cloud is very important in astronomy.

The electrons are bound to the atom by the attraction between their negative charge and the positive charge on the nucleus. This attraction is known as the **Coulomb force,** after the French physicist Charles-Augustin de Coulomb (1736–1806). If we wish to ionize an atom, we need a certain amount of energy to pull an electron away from its nucleus. This energy is the electron's **binding energy,** the energy that holds it to the atom.

An electron may orbit the nucleus at various distances. If the orbit is small, the electron is close to the nucleus, and a large amount of energy is needed to pull it away. Therefore, its binding energy is large. An electron orbiting farther from the nucleus is held more loosely, and less energy will pull it away. It therefore has less binding energy. The size of an electron's orbit is related to the energy that binds it to the atom.

Nature permits atoms only certain amounts (quanta) of binding energy, and the laws that describe how atoms behave are called the laws of **quantum mechanics** (Window on Science 7-2). Much of our discussion of atoms is based on the laws of quantum mechanics.

Because atoms can have only certain amounts of binding energy, our model atoms can have orbits of only certain sizes, called **permitted orbits.** These are like steps in a staircase: You can stand on the number-one step or the number-two step, but not on the number-one-and-one-quarter step. The electron can occupy any permitted orbit but not orbits in between.

The arrangement of permitted orbits depends primarily on the charge of the nucleus, which in turn depends on the number of protons. Thus, each kind of element has its own pattern of permitted orbits (Figure 7-5). Isotopes of the same elements have nearly the same pattern because they have the same number of protons. However, ionized atoms have orbital patterns that differ from their un-ionized forms. Thus the arrangement of permitted orbits differs for every kind of atom and ion.

REVIEW Critical Inquiry

How many hydrogen atoms would it take to cross the head of a pin?

This is not a frivolous question. In answering it, we will discover how small atoms really are, and we will see how powerful physics can be as a way to understand nature. First, we will assume that the head of a pin is about 1 mm in diameter. That is 0.001 m. The size of a hydrogen atom is represented by the diameter of the electron cloud, roughly 0.4 nm. Because 1 nm equals 10^{-9} m, we

Quantum mechanics is the set of rules that describe how atoms and subatomic particles behave. When we think about large objects such as stars, planets, aircraft carriers, and hummingbirds, we don't have to think about quantum mechanics, but on the atomic scale, particles behave in ways that seem unfamiliar to us.

One of the principles of quantum mechanics specifies that we can not know simultaneously the exact location and motion of a particle. This is why physicists refer to the electrons in an atom as if they were a cloud of negative charge surrounding the nucleus. Because we can't know the position and motion of the electron, we can't really describe it as a small particle following an orbit. We can use that image as a model to help our imaginations, but the reality is much more interesting, and describing the electrons as a charge cloud gives us a better and more sophisticated model of an atom.

This raises some serious questions about reality. Is an electron really a particle at all? Quantum mechanics describes particles as waves, and waves as particles. If we can't know simultaneously the position and motion of a specific particle, how can we know how it will react to a collision with a photon or another particle? The answer is that we can't know, and that seems to violate the principle of cause and effect (Window on Science 5-2).

Needless to say, we can't explore these discrepancies here. Scientists and philosophers of science continue to struggle with the meaning of reality on the quantum-mechanical level. Here we should note that the reality we see on the scale of stars and hummingbirds is only part of nature. We have constructed some models to help us think about nature on the scale of atoms, but the truth is much more interesting and much more exciting than anything we see on larger scales. Although we use models of atoms to study stars, there is still much to learn about the atoms themselves.

multiply and discover that 0.4 nm equals 4×10^{-10} m. To find out how many atoms would stretch 0.001 m, we divide the diameter of the pinhead by the diameter of an atom. That is, we divide 0.001 m by 4×10^{-10} m, and we get 2.5×10^6. It would thus take 2.5 million hydrogen atoms lined up side by side to cross the head of a pin.

This shows how tiny an atom is and also how powerful basic physics is. A bit of arithmetic gives us a view of nature beyond the capability of our eyes. Now use another bit of arithmetic to calculate how many hydrogen atoms you would need to add up to the mass of a paper clip (1 g).

The electron orbits in atoms are familiar territory to astronomers because the electrons in those orbits can interact with light. Such interactions fill starlight with clues to the nature of the stars.

7-3 The Interaction of Light and Matter

If light did not interact with matter, you would not be able to see these words. In fact, you would not exist, because, among other problems, photosynthesis would

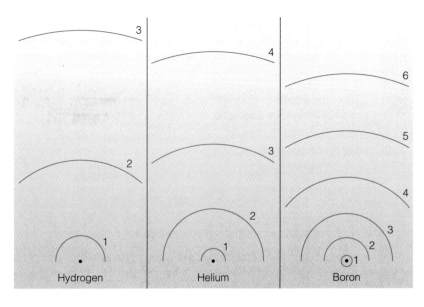

Figure 7-5
The electron in atoms may occupy only certain, permitted orbits. Because different elements have different charges on their nuclei, the elements have different patterns of permitted orbits.

be impossible and there would be no grass, wheat, bread, beef, cheeseburgers, or any other kind of food. The interaction of light and matter makes our lives possible, and it also makes it possible for us to understand our universe.

We begin our study of light and matter by considering the hydrogen atom. As we noted earlier, hydrogen is both simple and common. Roughly 90 percent of all atoms in the universe are hydrogen.

The Excitation of Atoms

Each orbit in an atom represents a specific amount of binding energy, so physicists commonly refer to the orbits as **energy levels.** Using this terminology, we can say that an electron in its smallest and most tightly bound orbit is in its lowest permitted energy level. We can move the electron from one energy level to another by supplying enough energy to make up the difference between the two energy levels. It is like moving a flowerpot from a low shelf to a high shelf; the greater the distance between the shelves, the more energy we need to raise the pot. The amount of energy needed to move the electron is the energy difference between the two energy levels.

If we move the electron from a low energy level to a higher energy level, we say the atom is an **excited atom.** That is, we have added energy to the atom in moving its electron. If the electron falls back to the lower energy level, that energy is released.

An atom can become excited by collision. If two atoms collide, one or both may have electrons knocked into a higher energy level. This happens very commonly in hot gas, where the atoms move rapidly and collide often.

Another way an atom can get the energy that moves an electron to a higher energy level is to absorb a photon. Only a photon with exactly the right amount of energy can move the electron from one level to another. If the photon has too much or too little energy, the atom cannot absorb it. Because the energy of a photon depends on its wavelength, only photons of certain wavelengths can be absorbed by a given kind of atom. Figure 7-6 shows the lowest four energy levels of the hydrogen atom along with three photons the atom could absorb. The longest-wavelength photon only has enough energy to excite the electron to the second en-

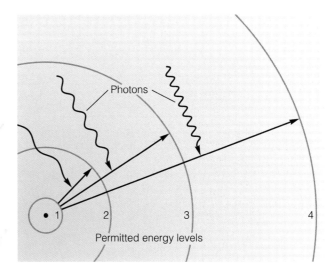

Figure 7-6
A hydrogen atom can absorb only those photons that move the atom's electron to one of the higher-energy orbits. Here three different photons are shown along with the change they would produce if they were absorbed.

ergy level, but the shorter-wavelength photons can excite the electron to higher levels. A photon with too much or too little energy cannot be absorbed. Because the hydrogen atom has many more energy levels than shown in Figure 7-6, it can absorb photons of many different wavelengths.

Atoms, like humans, cannot exist in an excited state forever. The excited atom is unstable and must eventually (usually within 10^{-6} to 10^{-9} seconds) give up the energy it has absorbed and return its electron to the lowest energy level. Because the electrons eventually tumble down to this bottom level, physicists call it the **ground state.**

When the electron drops from a higher to a lower energy level, it moves from a loosely bound level to one more tightly bound. The atom then has a surplus of energy—the energy difference between the levels—that it can emit as a photon. Study the sequence of events in Figure 7-7 to see how an atom can absorb and emit photons. Because each type of atom or ion has its unique set of energy levels, each type absorbs and emits photons with a unique set of wavelengths. Thus we can identify the elements in a gas by studying the characteristic wavelengths of light absorbed or emitted.

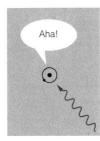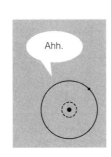

Figure 7-7
An atom can absorb a photon only if the photon has the correct amount of energy. The excited atom is unstable and within a fraction of a second returns to a lower energy level, reradiating the photon in a random direction.

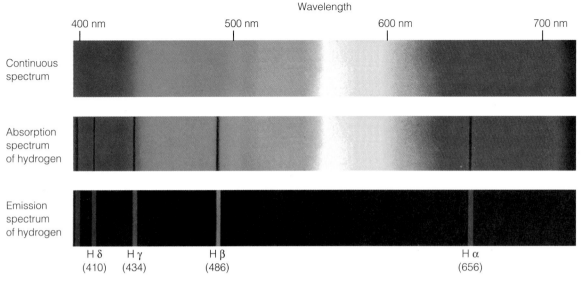

Wavelength

400 nm   500 nm   600 nm   700 nm

Continuous spectrum

Absorption spectrum of hydrogen

Emission spectrum of hydrogen

H δ (410)   H γ (434)   H β (486)   H α (656)

Figure 7-8
The three types of spectra. A continuous spectrum (top) contains no bright or dark lines, but an absorption spectrum (middle) is interrupted by dark absorption lines. An emission spectrum (bottom) is dark except at certain wavelengths where emission lines occur. Note that the lines in the absorption spectrum of hydrogen have the same wavelengths as the lines in the emission spectrum of hydrogen.

The process of excitation and emission is a common sight in urban areas at night. A neon sign glows when atoms of the neon gas in the tube are excited by electricity flowing through the tube. As the electrons in the electric current flow through the gas, they collide with the neon atoms and excite them. As we have seen, immediately after an atom is excited, its electron drops back to a lower energy level, emitting the surplus energy as a photon of a certain wavelength. The photons emitted by excited neon produce a reddish-orange glow. Signs of other colors, erroneously called "neon," contain other gases or mixtures of gases instead of pure neon.

The Formation of a Spectrum

To see how an astronomical object like a star can produce a spectrum, we can construct an imaginary experiment with an incandescent lightbulb and a cloud of hydrogen gas in space. We will discover there are three kinds of spectra and that our single experiment is capable of producing all three.

Our incandescent lightbulb contains a hot filament and emits black body radiation. As we saw in the previous section, a black body radiates at all wavelengths, and if we spread the light from the bulb out to form a spectrum we will find all wavelengths represented. This is called a **continuous spectrum,** as illustrated in the top spectrum in Figure 7-8.

In our imaginary experiment, the light from the bulb must pass through the hydrogen gas before it can reach our telescope (Figure 7-9). Most of the photons will pass through the gas unaffected because they have wavelengths the hydrogen atoms cannot absorb, but a

few photons will have the right wavelengths. These photons cannot pass through the gas because they are absorbed by the first atom they meet. The atom is excited for a fraction of a second, and the electron then drops back to a lower energy level and a new photon is emitted. The original photon was traveling through the gas toward our telescope, but the new photon is emitted in some random direction. Very few of these new photons leave the cloud in the direction of our telescope, so the light that finally enters the telescope has very few photons at the wavelengths the atoms can absorb. When we form a spectrum from this light, pho-

Table 7-1 Kirchhoff's Laws

Law I: The Continuous Spectrum

A solid, liquid, or dense gas excited to emit light will radiate at all wavelengths and thus produce a continuous spectrum.

Law II: The Emission Spectrum

A low-density gas excited to emit light will do so at specific wavelengths and thus produce an emission spectrum.

Law III: The Absorption Spectrum

If light comprising a continuous spectrum is allowed to pass through a cool, low-density gas, the resulting spectrum will have dark lines at certain wavelengths. That is, it will be an absorption spectrum.

tons of these wavelengths are missing, and the spectrum has dark lines at the positions these photons would have occupied. These dark lines, like those the Munich optician Fraunhofer saw in the solar spectrum in 1814 and 1815, are called **absorption lines** because the atoms absorbed the photons. A spectrum containing absorption lines is an **absorption spectrum** (also called a **dark-line spectrum**). The middle spectrum in Figure 7-8 illustrates the absorption spectrum of hydrogen.

What happens to the photons that were absorbed? They bounce from atom to atom, being absorbed and emitted over and over until they escape from the cloud. If, instead of aiming our telescope at the bulb, we swing it to one side so that no light from the bulb enters the telescope, we can photograph a spectrum of the light emitted by the gas atoms (Figure 7-10). In this case, the only photons entering the telescope are photons that were absorbed and re-emitted. A spectrum of this light is almost entirely dark except for the wavelengths corresponding to the photons the gas can absorb and re-emit. Thus we will see a spectrum containing only bright lines on a dark background. These bright lines are called **emission lines,** and a spectrum with emission lines is an **emission spectrum** (also called a **bright-line spectrum**). The lower spectrum in Figure 7-8 illustrates the emission spectrum of hydrogen. The spectrum of a neon sign is an emission spectrum. Also, the bluish-purple color of mercury-vapor streetlights and the pink-orange color of sodium-vapor streetlights are produced by the emission lines of these elements.

The properties of the three spectra—continuous, absorption, and emission—can be described by the three rules known as **Kirchhoff's laws** (Table 7-1). Law I states that a heated solid, a liquid, or a dense gas will produce a continuous spectrum. We see this in the filament of a lightbulb. Law II states that a low-density gas excited to emit light will produce an emission spectrum. We see this in neon signs. And Law III states that if light comprising a continuous spectrum passes through a cool, low-density gas, the result will be an absorption spectrum. We see this in stars.

Stellar spectra are absorption spectra formed as light from the bright surface, the photosphere, travels outward through the stellar atmosphere (Figure 7-11). The

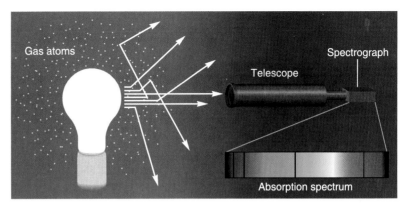

Figure 7-9
Photons of the proper wavelengths can be absorbed by the gas atoms and re-emitted in random directions. Because most of these particular photons do not reach the telescope, the spectrum is dark at the wavelengths of the missing photons.

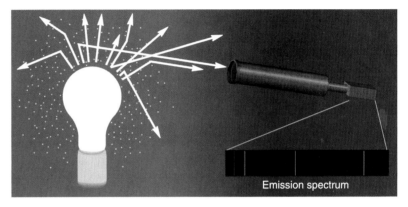

Figure 7-10
Pointing the telescope away from the bulb, we can receive only those photons the atoms can absorb and re-emit, producing emission lines in the spectrum.

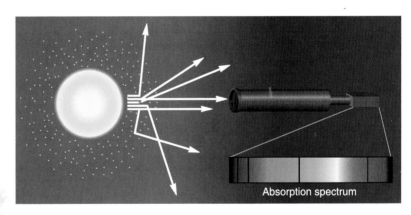

Figure 7-11
A star produces an absorption spectrum because its atmosphere absorbs certain wavelength photons in the spectrum.

bright surface of the star is sufficiently dense and hot that collisions between gas atoms and electrons produce a continuous spectrum just as does the filament of the lightbulb in Figure 7-9. The lower-density gases in the stellar atmosphere absorb certain wavelength photons, and thus the light that finally reaches our telescope is missing those specific wavelengths, and we see an absorption spectrum.

The absorption lines in stellar spectra provide a windfall of data about the star's atmosphere. By studying the spectral lines we can identify the elements in the stellar atmosphere and find the temperature of the atoms. To see how to get all this information, we need to look carefully at the way the hydrogen atom produces lines in a star's spectrum.

The Hydrogen Spectrum

As you must have gathered by now, each element has its own spectrum, unique as a human fingerprint; each element can be recognized by its spectrum across trillions of miles. To see how hydrogen produces its unique spectrum, we need a detailed diagram of its permitted orbits drawn so that the size of each orbit is proportional to its energy. That is, we need an accurate diagram of its energy levels such as that in Figure 7-12. An energy-level diagram will allow us to study the way hydrogen atoms interact with light.

A **transition** occurs in an atom when an electron changes energy levels. In our diagram of a hydrogen atom, arrows pointing from one level to another represent transitions. If the transition results in the absorption or emission of a photon, the length of the arrow tells us its energy. Long arrows represent large amounts of energy and thus short-wavelength photons. Short arrows represent smaller amounts of energy and longer-wavelength photons.

We can divide the possible transitions in a hydrogen atom into groups called series, according to their lowest energy level. Those arrows whose lower ends rest on the ground state represent the **Lyman series;** those resting on the second energy level, the **Balmer series;** and those resting on the third, the **Paschen series.** In principle, each series contains an infinite number of transitions, and there are an infinite number of series. Figure 7-12 shows the first few transitions in the first few series.

The Lyman series transitions are large changes in energy, as shown by the long arrows in Figure 7-12. These energetic transitions produce lines in the ultraviolet part of the spectrum, where they are invisible to the human eye. The Paschen series also lies outside the visible part of the spectrum. These transitions are small changes in energy and thus produce spectral lines in the infrared.

Balmer series transitions produce the only spectral lines of hydrogen in the visible part of the spectrum. Figure 7-12 shows that the first few Balmer series transitions are intermediate between the energetic Lyman

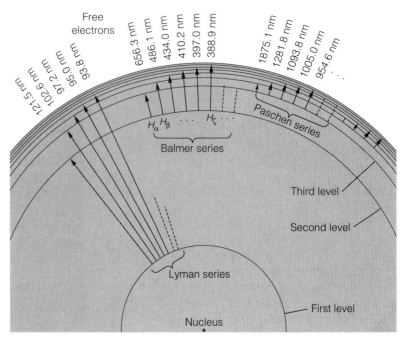

Figure 7-12
The levels in this diagram are spaced to represent the energy the hydrogen atom's electron can have. The transitions, drawn as arrows, can be grouped into series according to the lowest level. This drawing shows only a few of the infinity of transitions and series possible. Note that the arrows point upward and thus represent transitions that would absorb energy.

Figure 7-13
The Lagoon Nebula in Sagittarius is a cloud of gas and dust about 60 ly in diameter. Its gases are excited by the ultraviolet radiation of the hot, young stars within, and it glows in the pink color produced by the mixture of the red, blue, and violet Balmer lines. *(AURA/NOAO/NSF)*

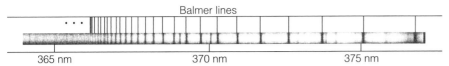

Balmer lines

365 nm   370 nm   375 nm

Figure 7-14
The Balmer lines photographed in the near ultraviolet. *(The Observatories of the Carnegie Institution of Washington)*

transitions and the low-energy Paschen transitions. These Balmer lines are labeled by Greek letters for easy identification. H_α is a red line, H_β is blue, and H_γ is violet. These three lines create the purple-pink color characteristic of glowing clouds of hydrogen in space (Figure 7-13). The remaining Balmer lines have wavelengths too short to see, though they can be photographed easily (Figure 7-14).

R E V I E W   Critical Inquiry

What spectrum would we see if we observed molten iron?

Molten iron is a dense liquid, and the atoms and molecules collide so often they emit all wavelengths and we would see a continuous spectrum. That is Kirchhoff's first law. But in order to see the molten iron, we would have to look through the hot vapors rising from it. Photons on their way to our spectrograph would pass through these gases, and atoms in the gases would absorb certain wavelengths. So what we would really see would be the continuous spectrum of the molten iron with weak absorption lines caused by the gases above the iron. This is what Kirchhoff's third law describes.

But suppose we had a very sensitive spectrograph that could look at the hot gases above the molten iron from the side so as to avoid looking directly at the molten iron. What kind of spectrum would the hot gases emit?

Whatever kind of spectrum astronomers look at, the most common spectral lines are the Balmer lines of hydrogen, the only hydrogen lines we can study from Earth's surface. In the next section, we will see how Balmer lines can tell us a star's temperature.

7-4 Stellar Spectra

In later chapters, we use spectra to study galaxies and planets, but we begin here by studying the spectra of stars. Such spectra are the easiest to understand, and the nature of stars is central to our study of all celestial objects.

The Balmer Thermometer

We can use the Balmer absorption lines as a thermometer to find the temperatures of stars. Earlier we saw how to estimate temperature from color, but the strengths of the Balmer lines in a star's spectrum give a much more accurate estimate of the star's temperature.

The Balmer thermometer works because the Balmer absorption lines are produced only by atoms whose electrons are in the second energy level (Figure 7-12). If the star is cool, there are few violent collisons between atoms to excite the electrons, and most atoms have their electron in the ground state. If most electrons are in the ground state, they can't absorb photons in the Balmer series. As a result, we should expect to find weak Balmer absorption lines in the spectra of cool stars.

In hot stars, on the other hand, there are many violent collisions between atoms, exciting electrons to high energy levels or knocking the electron clear out of some atoms; that is, some atoms are ionized. Thus, few atoms have their electron in the second orbit to form Balmer absorption lines, and we should expect hot stars, like cool stars, to have weak Balmer absorption lines.

At some intermediate temperature, the collisions are just right to excite large numbers of electrons into the second energy level. With many atoms excited to the second level, the gas absorbs Balmer-wavelength photons strongly and thus produces strong Balmer lines.

To summarize, the strength of the Balmer lines depends on the temperature of the star's surface layers. Both hot and cool stars have weak Balmer lines, but medium-temperature stars have strong Balmer lines.

Theoretical calculations can predict just how strong the Balmer lines should be for stars of various temperatures. The details of these calculations are not important to us, but the results are. The curve in Figure 7-15a shows the strength of the Balmer lines for various stellar temperatures. We could use this as a temperature indicator except that the curve gives us two answers. A star with Balmer lines of a certain strength might have either of two temperatures, one high and one low. We must examine other spectral lines to choose the correct temperature.

We have seen how the strength of the Balmer lines depends on temperature. Temperature has a similar effect on the spectral lines of other elements, but the temperature at which they reach maximum strength differs for each element (Figure 7-15b). If we add these elements to our graph, we get a handy tool for finding the stars' temperatures (Figure 7-15c).

How do we use this tool? We determine a star's temperature by comparing the strengths of its spectral lines with our graph. For instance, if we photographed a spectrum of a star and found medium-strength Balmer lines and strong helium lines, we could conclude it had a temperature of about 20,000 K. But if the star had

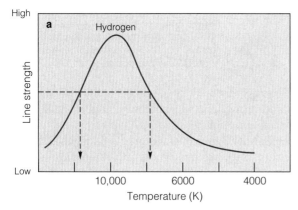

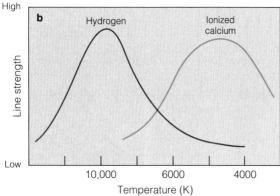

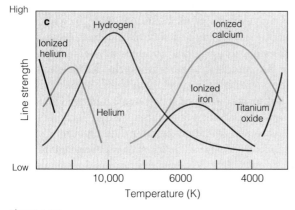

Figure 7-15
(a) The strength of the Balmer lines in a stellar spectrum depends on the temperature of the star. The dashed lines indicate that a star with medium-strength Balmer lines could have one of two possible temperatures. (b) The curves for lines of hydrogen and ionized calcium reach maximum strength at different temperatures. (c) The strengths of the spectral lines produced by each atom, ion, and molecule depend on the temperature of the star. Astronomers can compare the strengths of lines in a stellar spectrum with a diagram such as this to find the temperature of the star.

weak hydrogen lines and strong lines of ionized iron, we would assign it a temperature of about 5800 K, similar to that of the sun.

The spectra of stars cooler than about 3000 K contain dark bands produced by molecules such as titanium oxide (TiO). Because of their structure, molecules can absorb photons at many wavelengths, producing numerous, closely spaced spectral lines that blend to-

gether to form bands. These molecular bands appear only in the spectra of the coolest stars because, as mentioned before, molecules in cool stars are not subject to the violent collisions that would break up molecules in hotter stars. Thus the presence of dark bands in a star's spectrum indicates that the star is very cool.

From stellar spectra, astronomers have found that the hottest stars have surface temperatures above 40,000 K and the coolest about 2000 K. Compare these with the surface temperature of the sun, which is about 5800 K.

Spectral Classification

We have seen that the strengths of spectral lines depend on the surface temperature of the star. From this we can predict that all stars of a given temperature should have similar spectra. If we learn to recognize the pattern of spectral lines produced by a 6000-K star, for instance, we need not use Figure 7-15c every time we see that kind of spectrum. We can save time by classifying stellar spectra rather than analyzing each one individually.

The first widely used classification system was devised by astronomers at Harvard during the 1890s and 1900s. One of them, Annie J. Cannon, personally inspected and classified the spectra of over 400,000 stars. The spectra were first classified in groups labeled A through Q, but some groups were later dropped, merged with others, or reordered. The final classification includes the seven **spectral classes,** or **types,** still used today: O, B, A, F, G, K, M.*

This sequence of spectral types, called the **spectral sequence,** is important because it is a temperature sequence. The O stars are the hottest, and the temperature continues to decrease down to the M stars, the coolest. For maximum precision, astronomers divide each spectral class into 10 subclasses. For example, spectral class A consists of the subclasses A0, A1, A2, . . . A8, A9. Next comes F0, F1, F2, and so on. This finer division gives us a star's temperature to an accuracy within about 5 percent. The sun, for example, is not just a G star, but a G2 star, with a temperature of 5800 K.

We classify a star by the lines and bands in its spectrum as shown in Table 7-2. For example, if it has weak Balmer lines and lines of ionized helium, it must be an O star. This table is based on the same information we used in Figure 7-15c.

Figure 7-16 shows spectra for stars of different spectral types. Notice that the spectra extend from the visible into the near ultraviolet to reveal the calcium H and K lines; the human eye cannot see to wavelengths shorter than about 420 nm.

With the rise of computers, astronomers began displaying spectra as graphs of intensity versus wavelength.

---

*Generations of astronomy students have remembered the spectral sequence using the mnemonic "Oh, Be A Fine Girl (Guy), Kiss Me."

Table 7-2 Spectral Classes

Spectral Class	Approximate Temperature (K)	Hydrogen Balmer Lines	Other Spectral Features	Naked-Eye Example
O	40,000	Weak	Ionized helium	Meissa (O8)
B	20,000	Medium	Neutral helium	Achernar (B3)
A	10,000	Strong	Ionized calcium weak	Sirius (A1)
F	7,500	Medium	Ionized calcium weak	Canopus (F0)
G	5,500	Weak	Ionized calcium medium	Sun (G2)
K	4,500	Very weak	Ionized calcium strong	Arcturus (K2)
M	3,000	Very weak	TiO strong	Betelgeuse (M2)

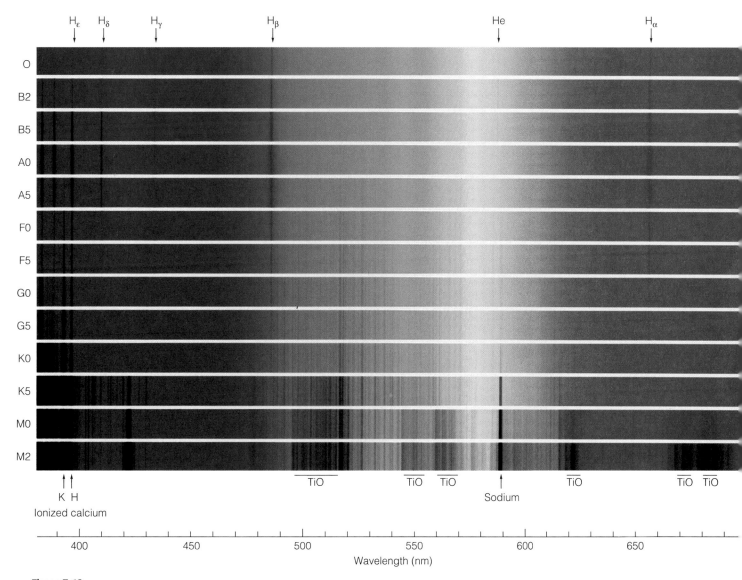

Figure 7-16
These stellar spectra were produced by computer simulation of real spectra. The spectra run from the hot O star at the top to the cool M2 star at the bottom. In the spectra of the hot stars we see weak Balmer lines and barely visible helium lines. The Balmer lines are strongest about A0 but are very weak in cool stars. The two ultraviolet lines of ionized calcium are strong in cooler stars, while sodium lines and titanium oxide (TiO) bands are strong in the spectra of the coolest stars. These spectra have been extended past the short-wavelength limit for the human eye, about 420 nm, to show the H and K lines of calcium. *(Courtesy Roger Bell and Michael Briley)*

Such graphs show more detail than photographic plates, and they can be measured directly. Figure 7-17 shows graphical spectra for stars of different spectral types. In these spectra, absorption lines are sharp dips in the curve. Compare Figures 7-16 and 7-17 to find similar lines and bands in the spectra, and you will see how precise the graphical spectra are.

When we look at spectra for stars from O to M we see how spectral lines change from class to class. Note that the Balmer lines are strongest in A stars, where the temperature is moderate but still high enough to excite the electrons in hydrogen atoms to the second energy level, where they can absorb Balmer wavelength photons. In the hotter stars (O and B), the Balmer lines are weak because the higher temperature excites the electrons to energy levels above the second or ionizes the atoms. The Balmer lines in cooler stars (F through M) are also weak, but for a different reason. The lower tem-

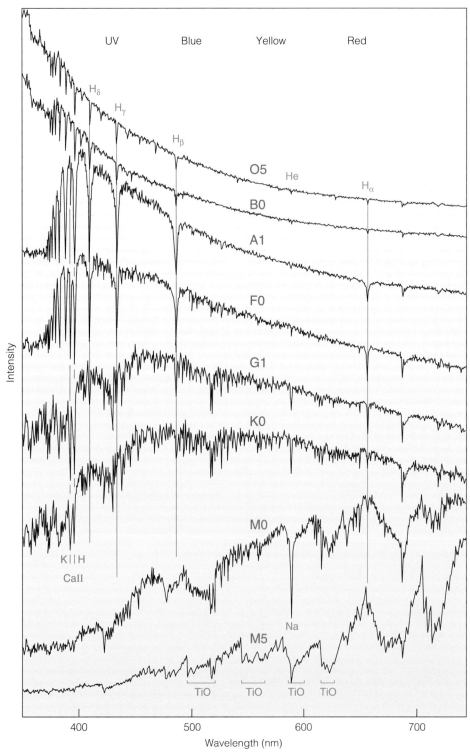

Figure 7-17
Modern digital spectra are often represented by graphs of intensity versus wavelength. Dark absorption lines are dips in intensity. The hottest stars are at the top and the coolest at the bottom. Hydrogen Balmer lines are strongest at about A0, while lines of ionized calcium (CaII) are strong in K stars. Titanium oxide (TiO) bands are strongest in the coolest stars. Compare these spectra with Figure 7-15c and 7-16. *(Courtesy NOAO, G. Jacoby, D. Hunter, and C. Christian)*

perature cannot excite many electrons to the second energy level, so few hydrogen atoms are capable of absorbing Balmer wavelength photons.

The spectral lines of other atoms also change from class to class. Helium is visible only in the spectra of the hottest classes, and titanium oxide bands only in the coolest. Two lines of ionized calcium, labeled H and K, increase in strength from A to K and then decrease from K to M. Because the strength of these spectral lines depends on temperature, it requires only a few moments to study a star's spectrum and determine its temperature.

The century-old study of spectral types might seem a closed book, but in 1998 astronomers surveying the sky in infrared wavelengths found a new class of star. The extremely cool, faint stars have been added to the spectral sequence after M and are known as spectral type L stars or **L dwarfs.** They are clearly a new class of star and not an extension of the M stars. The spectra of M stars contain bands produced by metal oxides such as titanium oxide, but spectra of the cooler L stars contain bands produced by chromium hydride and iron hydride (Figure 7-18). Their temperatures must be less than 2000 K for these molecules to form. The L dwarfs are faint, difficult to find, and difficult to study at visual wavelengths. Nevertheless, they are helping astronomers understand the coolest stars.

The Composition of the Stars

It seems as though it should be easy to find the composition of the sun and stars just by looking at their spectra, but it turns out to be a difficult task that wasn't well understood until the 1920s. The story of how astronomers first discovered the composition of the stars is worth telling, not only because it is the story of an important American astronomer who never got proper credit, but also because the story illustrates the temperature dependence of spectral features. The story begins in England.

As a child in England, Cecilia Payne (1900–1979) excelled in classics, languages, mathematics, and literature, but her first love was astronomy. After finishing Newnham College in Cambridge, she left England, sensing that there were no opportunities in England for a woman of science. In 1922, Payne arrived at Harvard, where she eventually earned her Ph.D., although the degree was awarded by Radcliffe because Harvard did not then admit women.

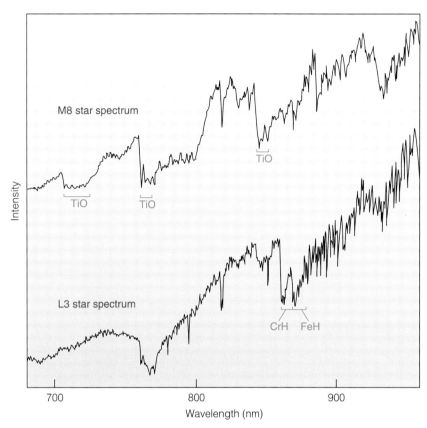

Figure 7-18

The spectrum of an M star and an L star compared: M-star spectra contain strong bands produced by the titanium oxide molecule (TiO), but L-star spectra contain weak or absent TiO bands. Notice the missing TiO band near 710 nm in the L-star spectrum above. Instead, the spectra of L stars contain bands produced by chromium hydride (CrH) and iron hydride (FeH). *(Adapted from data by J. Davy Kirkpatrick)*

In her thesis, Payne attempted to relate the strength of the absorption lines in stellar spectra to the physical conditions in the atmospheres of the stars. This was not easy because, as we have seen in this chapter, a given spectral line can be weak because the atom is rare or because the temperature is too high or too low for that atom to be able to absorb efficiently. If we see sodium lines in a star's spectrum, we can be sure that the star contains sodium atoms, but if we see no sodium lines, we must consider the possibility that the star is too hot or too cool for sodium to produce spectral lines.

Payne's problem was to untangle these two factors and find the true temperatures of the stars and the true abundance of the atoms in their atmospheres. Recent advances in atomic physics gave her the theoretical tools she needed. About the time Payne left Newnham College, Indian physicist Meghnad Saha published his work on the ionization of atoms. Drawing from such theoretical work, Payne was able to show that over 90 percent of the atoms in stars (including the sun) were hydrogen and most of the rest helium (Table 7-3). The heavier atoms seemed more abundant only because

Table 7-3 The Most Abundant Elements in the Sun

Element	Percentage by Number of Atoms	Percentage by Mass
Hydrogen	91.0	70.9
Helium	8.9	27.4
Carbon	0.03	0.3
Nitrogen	0.008	0.1
Oxygen	0.07	0.8
Neon	0.01	0.2
Magnesium	0.003	0.06
Silicon	0.003	0.07
Sulfur	0.002	0.04
Iron	0.003	0.1

they are better at absorbing photons at the temperatures of stars.

At the time, astronomers found it hard to believe Payne's abundances of hydrogen and helium. They especially found the abundance of helium unacceptable. After all, hydrogen lines are at least visible in most stellar spectra, but helium lines are almost invisible in the spectra of all but the hottest stars. Rather, nearly all astronomers assumed that the stars had roughly the same composition as Earth's surface; that is, they believed that the stars were composed mainly of heavier atoms such as carbon, silicon, iron, aluminum, and so on. Even the most eminent astronomers dismissed Payne's result as illusory. Faced with this pressure and realizing the limited opportunities available to women in science in the 1920s, Payne could not press her discovery.

It was 1929 before astronomers generally understood the importance of temperature on measurements of composition derived from stellar spectra. At that point, astronomers recognized that stars are mostly hydrogen and helium, but Payne received no credit.

Payne worked for many years as a staff astronomer at the Harvard College Observatory with no formal position on the faculty. She married Russian astronomer Sergei Gaposchkin in 1934 and was afterward known as Cecilia Payne-Gaposchkin. In 1956, when Harvard accepted women to its faculty, she was appointed a full professor and chair of the Harvard astronomy department.

Cecilia Payne-Gaposchkin's work on the chemical composition of the stars illustrates the importance of fully understanding the interaction between light and matter. Only a detailed understanding of the physics could lead her to the correct composition. As we turn our attention to other information that can be derived from stellar spectra, we again discover the importance of understanding light.

The Doppler Effect

Surprisingly, one of the pieces of information hidden in a spectrum is the velocity of the light source. Astronomers can measure the wavelengths of lines in a star's spectrum and find the velocity of the star. The **Doppler effect** is the apparent change in the wavelength of radiation caused by the motion of the source.

If a star is moving toward Earth, the lines in its spectrum will be shifted slightly toward shorter wavelengths. That is, they are shifted toward the blue end of the spectrum—a **blue shift.** If a star is moving away from Earth, the lines are shifted slightly toward the red end of the spectrum—a **red shift.** These Doppler shifts are small and don't change the color of the star. But it is easy to detect these changes in wavelength in a star's spectrum.

Figure 7-19a shows the Doppler effect in two spectra of the star Arcturus. The lines in the top spectrum are slightly blue shifted because the spectrum was recorded when Earth, following its orbit, was moving toward Arcturus. The lines in the bottom spectrum are red shifted because it was recorded six months later, when Earth was moving away from Arcturus.

The Doppler effect tells us how rapidly the distance between us and the source of light is increasing or decreasing. It does not matter whether we are moving or the star is moving. Only the relative velocity is important. Also, the Doppler shift is only sensitive to the part of the velocity directed away from us or toward us. This part of the velocity is called the **radial velocity (V_r).** The **transverse velocity** is the part of the star's velocity perpendicular to our line of sight (Figure 7-20). The Doppler effect cannot be used to measure the transverse velocity.

The Doppler effect occurs for any kind of radiation, not just light. Radio astronomers, for instance, observe the Doppler effect when they measure the wavelength of radio waves coming from objects moving toward or away from Earth. Whatever wavelength range we consider—radio, X rays, gamma rays—we will still refer to a lengthening of the observed wavelength as a red shift and a shortening of the observed wavelength as a blue shift.

We can even detect the Doppler shift in sound. Sounds with long wavelengths have a lower pitch, and sounds with short wavelengths have a higher pitch. When we hear a truck roar past us on the highway, we hear the sound of the truck drop in pitch as it passes. Its sound is blue shifted while it is approaching and red shifted after it passes.

We can understand how the Doppler shift works by thinking about a fire truck approaching us with a bell clanging once a second. When the bell clangs, the sound travels ahead of the truck to reach our ears. One second later, the bell clangs again, but not at the same place. During that one second, the fire truck moved

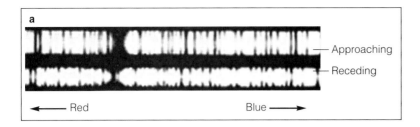

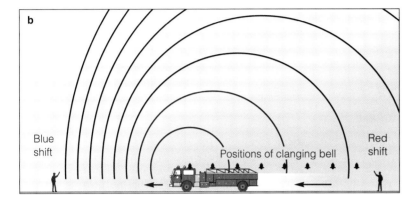

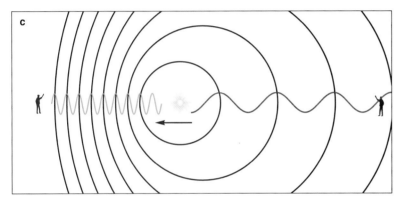

Figure 7-19

The Doppler effect. (a) A blue shift appears in the spectrum of a star approaching Earth (top spectrum). A red shift appears in the spectrum of a star moving away from Earth (bottom spectrum). *(The Observatories of the Carnegie Institution of Washington)* (b) The clanging bell on a moving fire truck produces sounds that move outward (black circles). An observer ahead of the truck hears the clangs closer together, while an observer behind the truck hears them farther apart. (c) A moving source of light emits waves that move outward (black circles). An observer in front of the light source observes a shorter wavelength (a blue shift), and an observer behind the light source observes a longer wavelength (a red shift).

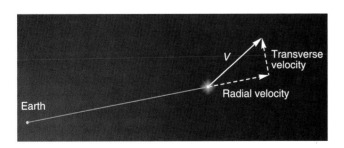

Figure 7-20

The radial velocity of a star is the part of the star's velocity *V* that is directed away from or toward Earth. The Doppler effect can tell us a star's radial velocity but cannot tell us the transverse velocity—the part of the velocity perpendicular to the radial direction.

closer to us, so the bell is closer at its second clang. Now the sound has a shorter distance to travel and reaches our ears a little sooner than it would have if the fire truck were not approaching. The third time the bell clangs, it is even closer. By timing the bell, we could observe that the clangs are slightly less than 1 second apart, all because the fire truck is approaching. If the fire truck were moving away from us, we would hear the clangs sounding more than one second apart, because each successive clang of the bell occurs farther from us.

Figure 7-19b shows a fire truck moving toward one observer and away from another observer. The position of the bell at each clang is shown by the small black bells with the sound spreading outward as black circles. We can see that the clangs are squeezed together ahead of the fire truck and stretched apart behind.

Now we can substitute a source of light waves for the clanging of the bell. If the source is approaching, then each time the source emits the peak of a wave it will be slightly closer to us, and we will observe a shorter wavelength. If it is moving away, we will observe a longer wavelength. This is shown in Figure 7-19c, where the peaks of the waves appear compressed in front of the moving light source and stretched out behind the source.

Of course, how great the change in wavelength is depends on the velocity. Just as a slowly moving car has a smaller Doppler effect in sound than does a high-speed airplane, a slowly moving star has a smaller Doppler effect in light than does a high-speed star. Thus, we can measure the velocity by measuring the amount by which the spectral lines are shifted in wavelength. A large Doppler shift means a high radial velocity.

This can be expressed as a simple ratio relating the radial velocity V_r divided by the speed of light c to the change in wavelength $\Delta\lambda$ divided by the unshifted wavelength λ_0:

$$\frac{V_r}{c} = \frac{\Delta\lambda}{\lambda_0}$$

This expression is quite accurate for the low radial velocities of stars, but we will need a better version later when we discuss objects moving with very high velocities.

For example, we might observe a line in a star's spectrum with a wavelength of 600.1 nm. Laboratory measurements show that the line should have a wavelength of 600 nm. That is, its unshifted wavelength is 600 nm. What is the star's radial velocity? First we note that the change in wavelength is 0.1 nm:

$$\frac{V_r}{c} = \frac{0.1}{600} = 0.000167$$

Multiplying by the speed of light, 3×10^5 km/s, gives the radial velocity, 50 km/s. Because the wavelength is shifted to the red (lengthened), the star must be receding from us.

You may be quite familiar with this method of speed measurement. Police radar uses the Doppler shift in reflected radio waves to determine the velocity of approaching cars.

Understanding the Doppler shift leads us to a final illustration of the information hidden in stellar spectra. Even the shapes of the spectral lines can tell us secrets about the stars.

The Shapes of Spectral Lines

When astronomers refer to the shape of a spectral line, they mean the variation of intensity across the line. An absorption line, for instance, is darkest in the center and grows brighter to each side. This shape is represented by a **line profile,** a graph of brightness as a function of wavelength across a spectral line (Figure 7-21).

The exact shape of a line profile can tell us a great deal about a star, but the most important characteristic is the width of the line. Spectral lines are not perfectly narrow; if they were, we could not see them. They have a natural width because nature allows an atom some leeway in the energy it may absorb or emit. The quantum mechanics behind this effect is beyond the scope of our discussion, but the result is very simple. In the absence of all other effects, spectral lines have a natural width of about 0.001 to 0.00001 nm—very narrow indeed.

The natural widths of spectral lines are not important in most branches of astronomy because other effects smear out the lines and make them much broader. For example, if a star spins rapidly, the Doppler effect will broaden the spectral lines. As the star rotates, one side will recede from us, and the other side will approach. Light from the receding side will be red shifted, and light from the approaching side will be blue shifted, so any spectral lines will be broadened.

Another important process is called **Doppler broadening.** To consider this process, let us imagine that we photograph the spectrum of a jar full of hydrogen atoms (Figure 7-22). Because the gas has some thermal energy (it is not at absolute zero), the gas atoms are in motion. Some will be coming toward our spectrograph, and some will be receding. Most, of course, will not be traveling very fast, but some will be moving very quickly. The photons emitted by the atoms approaching us will have slightly shorter wavelengths because of the Dop-

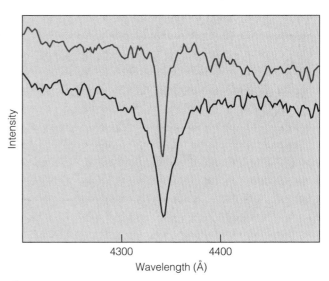

Figure 7-21
A line profile is a graph that shows the intensity of starlight as a function of wavelength across a spectral line. In these two examples, the spectra show the same spectral line from the spectra of two A1 stars. Thus, both stars have the same temperature. Differences in the density of the gas in the atmospheres of the stars cause the differences in the widths of the spectral lines. Such differences in the width of the spectral lines are easily measured from such line profiles. *(Courtesy NOAO, G. Jacoby, D. Hunter, and C. Christian)*

pler effect, and photons emitted by atoms receding from us will have slightly longer wavelengths. Thus, the Doppler shifts due to the motions of the individual atoms will smear the spectral line out and make it broader. We have described the Doppler broadening of an emission line, but the effect is the same for absorption lines.

The extent of Doppler broadening depends on the temperature of the gas. If the gas is cold, the atoms travel at low velocities, and the Doppler shifts are small (Figure 7-22a). If the gas is hot, however, the atoms travel faster, Doppler shifts are larger, and the lines will be wider (Figure 7-22b).

Another form of broadening, **collisional broadening,** is caused by collisions between atoms, and thus it depends on the density of the gas (Window on Science 7-3). Densities in astronomy cover an enormous range, from one atom per cubic centimeter in space to millions of tons of atoms per cubic centimeter inside dead stars. Clearly, we should expect such densities to affect the way atoms collide with one another and thus how they absorb and emit photons.

Collisional broadening spreads out spectral lines when the atoms absorb or emit photons while they are colliding with other atoms, ions, or electrons. The collisions disturb the energy levels in the atoms, making it possible for the atoms to absorb a slightly wider range of wavelengths. Thus, the spectral lines are wider. Because atoms in a dense gas collide more often than atoms in a low-density gas, collisional broadening depends on the density of the gas, but temperature is also a fac-

Five fundamental parameters—mass, energy, heat, density, and pressure—are so critical in science that we have been discussing them individually in separate Windows on Science. Here we will discuss **density,** the measure of the amount of matter in a given volume. Density is expressed as mass per volume, such as grams per cubic centimeter. The density of water, for example, is about 1 g/cm³.

To get a feel for density, imagine holding a brick in one hand and a similar-sized block of styrofoam in the other hand. We can easily tell that the brick contains more matter than the styrofoam block even though both are the same size. The brick weighs more than the styrofoam, but it isn't really the weight that we should consider. Rather, we should think about the

mass of the two objects. In space, where they have no weight, the brick and the styrofoam would still have mass, and we could tell by moving them about that the brick contains more mass than the styrofoam. For example, imagine tapping each object gently against your ear. The massive brick would be easy to distinguish from the low-mass styrofoam block even in weightlessness.

When we think of density, we divide mass by volume, and our minds make that comparison at an instinctive level. We can sense the density of an object just by handling it. Gift shops sometimes sell imitation rocks made of styrofoam as humorous gifts. "Rocks" made of styrofoam seem odd when we handle them because our brain expects rocks to be dense.

Density is a common parameter in science because it is a general property of materials. Lead, for example, has a density of about 7 g/cm³, and rock has a density of 3 to 4 g/cm³. Water and ice have densities of about 1 g/cm³. If we knew that a small moon had a density of 1.5 g/cm³, we could immediately draw some conclusions about what kinds of materials it might be made of—ice and a little rock, but not much lead. The density of an object is a basic clue to its composition.

Density also determines how materials behave. The low-density gases in a nebula behave in one way, but the same gases under high density inside a star behave in a different way. Thus, astronomers must consider the density of materials when they describe astronomical objects.

tor. Atoms in a hot gas travel faster and collide more often and more violently than atoms in a cool gas. Once again, the physics of the interaction of light and matter gives us a tool to understand starlight. In later chapters, we will see how collisional broadening affects the widths of spectral lines that are formed in normal stars like the sun, in giant stars, and in gas in interstellar space.

Figure 7-22

Doppler broadening. The atoms of a gas are in constant motion. Photons emitted by atoms moving toward the observer will have slightly shorter wavelengths, and those emitted by atoms moving away will have slightly longer wavelengths. This broadens the spectral line. If the gas is cool (a), the atoms do not move very fast, the Doppler shifts are small, and the line is narrow. If the gas is hot (b), the atoms move faster, the Doppler shifts are larger, and the line is broader.

R E V I E W Critical Inquiry

Why are helium lines weak and calcium lines strong in the visible spectrum of the sun?

To analyze this problem, we must recall that the ability of an atom or ion to absorb light depends on temperature. Helium is quite abundant in the sun, but the surface of the sun is too cool to excite helium atoms and enable them to easily absorb visible-wavelength photons. On the other hand, calcium is easily ionized, and calcium atoms that have lost an electron are very good absorbers of photons at temperatures like those at the

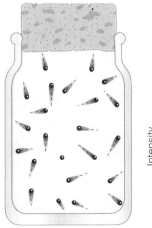
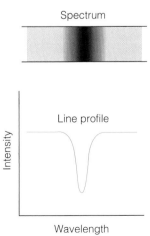

a

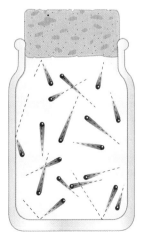
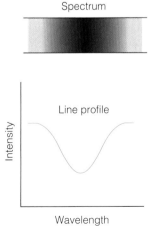

b

sun's surface. Thus, calcium lines are strong in the solar spectrum even though calcium ions are rare, and helium lines are weak even though helium atoms are common.

If we are to analyze a star's spectrum and consider the strength of spectral lines, we must take into account the temperature of the star. Why don't we need to take temperature into account when we use the Doppler effect to measure a star's radial velocity?

The spectra of the stars are filled with clues about the gas emitting the light. Much of astronomy is based on unraveling these clues. We will begin in the following chapter by studying the nearest star—the sun.

Summary

Stars emit black body radiation from the dense gases of their photospheres, and as that radiation passes through the less dense gas of the star's atmosphere, the atoms absorb photons of certain wavelengths to produce dark lines in the spectrum.

A heated solid, liquid, or dense gas emits black body radiation that contains all wavelengths and thus produces a continuous spectrum. Black body radiation is most intense at the wavelength of maximum, λ_{max}, which depends on the temperature of the radiating body. Hot objects emit mostly short-wavelength radiation, whereas cool objects emit mostly long-wavelength radiation. This effect gives us clues to the temperatures of stars—hot stars appear blue and cool stars red.

A low-density gas excited to emit radiation will produce an emission (or bright-line) spectrum. Light from a source of a continuous spectrum passing through a low-density gas will produce an absorption (or dark-line) spectrum. These three kinds of spectra are described by Kirchhoff's laws.

The lines in spectra are produced by the electrons that surround the nucleus of the atom. The electrons may occupy only certain permitted energy levels, and photons may be absorbed or emitted when electrons move from one energy level to another. Because each kind of atom and ion has a different set of energy levels, each kind of atom and ion can absorb or emit only photons of a certain wavelength. The hydrogen atom can produce the Lyman series lines in the ultraviolet, the Balmer series in the visual and near ultraviolet, the Paschen series in the infrared, and many more.

In cool stars, the Balmer lines are weak because most atoms are not excited out of the ground state. In hot stars, the Balmer lines are weak because most atoms are excited to higher orbits or ionized. Only at medium temperatures are the Balmer lines strong. We can use this effect as a thermometer for determining the temperature of a star. In its simplest form, this amounts to classifying the stars' spectra in the spectral sequence O, B, A, F, G, K, M.

Stellar spectra can tell us the chemical compositions of the stars, but we must be careful to consider the temperature of the star in our analysis. In general, 90 percent of the atoms in a star are hydrogen.

When a source of radiation is approaching us, we observe shorter wavelengths, a blue shift, and when it is receding, we observe longer wavelengths, a red shift. This Doppler effect makes it possible for the astronomer to measure a star's radial velocity, that part of its velocity directed toward or away from the earth.

The widths of spectral lines are affected by a number of processes. With no outside influences, a spectral line has a natural width that is very small. Doppler broadening, due to the thermal motion of the gas atoms, depends on the temperature of the gas. Collisional broadening occurs when the atoms emitting or absorbing photons collide with other atoms, ions, or electrons. The collisions disturb the atomic energy levels and slightly alter the wavelengths the atoms can absorb or emit. In a dense, hot gas, where the gas atoms collide often, collisional broadening is important.

New Terms

temperature	absorption line
thermal energy	absorption spectrum (dark-line spectrum)
Kelvin temperature scale	
absolute zero	emission line
electron	emission spectrum (bright-line spectrum)
black body radiation	
wavelength of maximum (λ_{max})	Kirchhoff's laws
	transition
color index	Lyman series
nucleus	Balmer series
proton	Paschen series
neutron	spectral class or type
isotope	spectral sequence
ionization	L dwarf
ion	Doppler effect
molecule	blue shift
Coulomb force	red shift
binding energy	radial velocity (V_r)
quantum mechanics	transverse velocity
permitted orbit	line profile
energy level	Doppler broadening
excited atom	collisional broadening
ground state	density
continuous spectrum	

Review Questions

1. Why can a good blacksmith judge the temperature of a piece of heated iron by its color?

2. Why do hot stars look bluer than cool stars?

3. Use black body radiation to explain why you could hold a cigarette between your lips and light it with a match, but you couldn't hold the cigarette in your lips while you

light it by sticking the tip into a bonfire. The match and bonfire have about the same temperature. (This is just one of the hazards of smoking.)

4. What is the difference between a neutral atom, an ion, and an excited atom?

5. How do the energy levels in an atom determine which wavelength photons it can absorb or emit?

6. What kind of spectrum would you expect to record if you observed molten lava? In practice, to view molten lava you must look through gases boiling out of the lava. What kind of spectrum might you see in that case?

7. Why do the strengths of the Balmer lines depend on the temperature of the star?

8. Why does a stellar spectrum tell us about the surface layers but not the deeper layers of the star?

9. Explain the similarities between Figure 7-15, Figure 7-16, Figure 7-17, and Table 7-2.

10. Why do we not see TiO bands in the spectra of hot stars?

11. In observing a star's spectrum, we note that the lines of a certain element are not present. Are we safe in concluding that the star does not contain this element? Why or why not?

12. If a star moves exactly perpendicular to a line connecting it to Earth, will the Doppler effect change its spectrum? Why or why not?

13. Why would we expect a star that rotates very rapidly to have broad spectral lines? What else could broaden the lines?

Discussion Questions

1. In what ways is our model of an atom a scientific model? How can we use it when it is not a completely correct description of an atom?

2. Can you think of classification systems we commonly use to simplify what would otherwise be very complex measurements? Consider foods, movies, cars, grades, clothes, and so on.

Problems

1. Human body temperature is about 310 K (98.6°F). At what wavelength do humans radiate the most energy? What kind of radiation do we emit?

2. If a star has a surface temperature of 20,000 K, at what wavelength will it radiate the most energy?

3. Infrared observations of a star show that it is most intense at a wavelength of 2000 nm. What is the temperature of the star's surface?

4. If we double the temperature of a black body, by what factor will the total energy radiated per second per square meter increase?

5. If one star has a temperature of 6000 K and another star has a temperature of 7000 K, how much more energy per

second will the hotter star radiate from each square meter of its surface?

6. Transition A produces light with a wavelength of 500 nm. Transition B involves twice as much energy as A. What wavelength light does it produce?

7. Determine the temperatures of the following stars based on their spectra. Use Figure 7-15.
   a. medium-strength Balmer lines, strong helium lines
   b. medium-strength Balmer lines, weak ionized calcium lines
   c. TiO bands very strong
   d. very weak Balmer lines, strong ionized calcium lines

8. To which spectral classes do the stars in Problem 7 belong?

9. In a laboratory, the Balmer beta line has a wavelength of 486.1 nm. If the line appears in a star's spectrum at 486.3 nm, what is the star's radial velocity? Is it approaching or receding?

10. The highest-velocity stars an astronomer might observe have velocities of about 400 km/s. What change in wavelength would this cause in the Balmer gamma line? (*Hint:* Wavelengths are given in Figure 7-12).

Critical Inquiries for the Web

1. To what extent is it possible to tell the spectral type of a star based on observation of its color? Search for color index data on the stars in Table A-9 (Appendix A) to see how B-V indices correlate with spectral type. Observational follow-up: Is it possible to rate star colors in such a way as to determine spectral type through visual observation? (Search the Web for indications that others have tried this.)

2. Is it possible to tell the direction of motion of the sun in space by observing radial velocities of other stars? Search the Internet for stellar radial velocity data for stars in a variety of directions on the celestial sphere, and look for trends that might indicate our motion in space.

Exploring *The Sky*

1. Classification of stars into spectral classes provides invaluable information about the surface temperature of a star. Use *The Sky* to determine the spectral class (or type) and the approximate temperature of the following easily observed bright stars: Antares in Scorpius, Betelgeuse in Orion, Aldebaran in Taurus, Sirius in Canis Major, and Rigel in Orion.

   How to proceed: use **Find** to obtain the **Object Information** window. Once you have found the spectral class, use Table 7-2 to find the approximate temperature of the star.

 Go to the Brooks/Cole Astronomy Resource Center (www.brookscole.com/astronomy) for critical thinking exercises, articles, and additional readings from InfoTrac College Edition, Brooks/Cole's online student library.

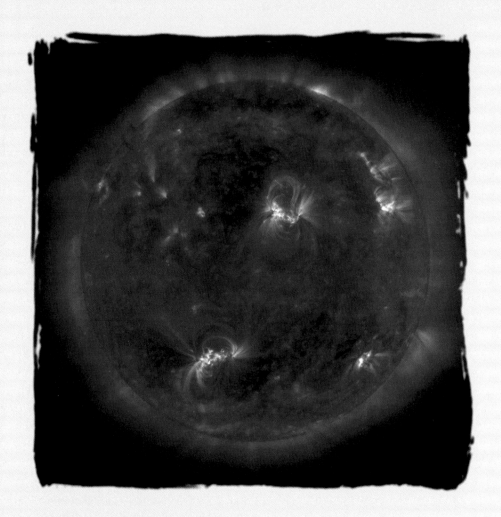

CHAPTER EIGHT

The Sun—
Our Star

All cannot live on the piazza, but

everyone may enjoy the sun.

Italian Proverb

Guidepost

The preceding chapter described how we can get information from a spectrum. In this chapter, we apply these techniques to the sun, to learn about its complexities.

This chapter gives us our first close look at how scientists work, how they use evidence and hypothesis to understand nature. Here we will follow carefully developed logical arguments to understand our sun.

Most important, this chapter gives us our first detailed look at a star. The chapters that follow will discuss the many kinds of stars that fill the heavens, but this chapter shows us that each of them is both complex and beautiful; each is a sun.

A wit once remarked that solar astronomers would know a lot more about the sun if it were farther away. This comment contains a grain of truth; the sun is only a humdrum star, and there are billions like it in the sky, but the sun is the only one close enough to show surface detail. Solar astronomers can see so much detail in the swirling currents of gas and arching bridges of magnetic force that present theories seem inadequate to describe it. Yet the sun is not a complicated object. It is just a star.

In their general properties, stars are very simple. They are great balls of hot gas held together by their own gravity. Their gravity would make them collapse into small, dense bodies were they not so hot. The tremendously hot gas inside stars has such a high pressure that the stars would surely explode were it not for their own confirming gravity. Like soap bubbles, stars are simple structures balanced between opposing forces that individually would destroy them. Thus, we study the sun as a close-up example of a star.

Another reason to study the sun is that life on Earth depends critically on the sun. Should the sun's energy output vary by even a small amount, life on Earth might vanish. In addition, we get nearly all our energy from the sun—oil and coal are merely stored sunlight—and our pleasant climate is maintained by energy from the sun. In fact, the sun's atmosphere of very thin gas reaches out past Earth's orbit, and thus any change in the sun, such as an eruption or a magnetic storm, can have a direct effect on Earth.

Finally, we study the sun because it is beautiful. Our analysis of sunlight will reveal that the sun is both powerful and delicate. Thus, we study the sun not only because it is *a star,* not only because it is *our star,* but because it is the sun.

8-1 The Solar Atmosphere

With a radius 109.1 times Earth's (109.1 $R_\oplus$),* the sun is gaseous throughout (Data File One). When we look at the sun, we see only the surface layers, what astronomers call the atmosphere. These layers extend from the visible surface out to about 5×10^6 km (7 $R_\odot$). If we include the low-density gas flowing away from the sun, the atmosphere extends to envelop Earth.

Heat Flow in the Sun

Simple logic tells us that energy in the form of heat is flowing outward from the sun's interior. The solar spectrum reveals that the temperature of the sun's surface is about 5800 K. At that temperature, every square

*In astronomy the symbols $\odot$ and $\oplus$ represent the sun and Earth, respectively.

The Sun Data File One

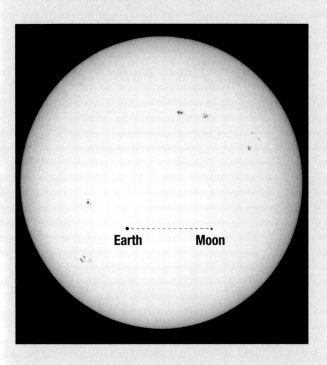

An image of the sun in visible light shows a few sunspots. The Earth–moon system is added for scale. *(Daniel Good)*

Average distance from Earth	1.00 AU (1.495979×10^8 km)
Maximum distance from Earth	1.0167 AU (1.5210×10^8 km)
Minimum distance from Earth	0.9833 AU (1.4710×10^8 km)
Average angular diameter seen from Earth	0.53° (32 minutes of arc)
Period of rotation	25 days at equator
Radius	6.9599×10^5 km
Mass	1.989×10^{30} kg
Average density	1.409 g/cm³
Escape velocity at surface	617.7 km/s
Luminosity	3.826×10^{26} J/s
Surface temperature	5800 K
Central temperature	15×10^6 K
Spectral type	G2 V
Apparent visual magnitude	−26.74
Absolute visual magnitude	4.83

centimeter of the sun's surface must be radiating more energy than a 6000-watt lightbulb. With all that energy radiating into space, the sun's surface would cool rapidly if energy did not flow up from the interior to keep the surface hot.

Not until the 1930s did astronomers understand how the sun makes its energy. Nuclear reactions occur in the core of the sun and generate tremendous heat, and that heat flowing upward keeps the surface hot. We will discuss these nuclear reactions in detail later in this chapter, but here we must recognize the importance of the energy flowing outward through the sun's surface.

As we study the atmosphere of the sun, we will find many phenomena that are driven by the energy flowing outward. Like a pot of boiling soup on a hot stove, the surface of the sun is in constant activity as the heat flows up from below.

The Photosphere

The visible surface of the sun, the photosphere, is not a solid surface. In fact, the sun is gaseous from its outer atmosphere right down to its center. The photosphere is the thin layer of gas from which we receive most of the sun's light. It is less than 500 km deep and has an average temperature of about 5800 K. If the sun magically shrank to the size of a bowling ball, the photosphere would be no thicker than a layer of tissue paper wrapped around the ball (Figure 8-1a). For comparison, the chromosphere lies above the photosphere and is only a few times deeper, but the corona, beginning above the chromosphere, extends far from the sun (Figure 8-1b).

Below the photosphere, the gas is denser and hotter and therefore radiates plenty of light, but that light cannot escape from the sun because of the outer layers of gas. Thus, we cannot detect light from these deeper layers. Above the photosphere, the gas is less dense and so is unable to radiate much light. The photosphere is the layer in the sun's atmosphere that is dense enough to emit plenty of light but not so dense that the light can't escape.

One reason the photosphere is so shallow is related to the hydrogen atom. Because the temperature of the photosphere is sufficient to ionize some atoms, there are a large number of free electrons in the gas. Neutral hydrogen atoms can add an extra electron and become an H^- (H-minus) ion, but this extra electron is held so loosely that almost any photon has energy enough to free it. In the process, of course, the photon is absorbed. Thus, the H^- ions are very good absorbers of photons and make the gas of the photosphere very opaque. Light from below cannot escape easily, and we see a well-defined surface—the thin photosphere.

Although the photosphere appears to be substantial, it is really a very-low-density gas. Even in the

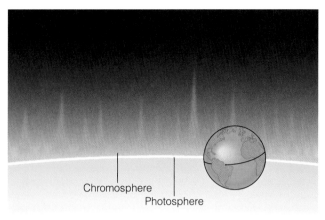

Figure 8-1
(a) A cross section at the edge of the sun shows the relative thickness of the photosphere and chromosphere. Earth is shown for scale. On this scale, the disk of the sun would be more than 1.5 m (5 feet) in diameter. (b) The corona extends from the top of the chromosphere to great height above the photosphere. This photograph, made during a total solar eclipse, shows only the inner part of the corona. *(Daniel Good)*

deepest and densest layers visible, the photosphere is 3400 times less dense than the air we breathe. To find gases as dense as the air we breathe, we would have to descend about 7×10^4 km below the photosphere, about 10 percent of the way to the sun's center. With a fantastically efficient insulation system, we could fly a spaceship right through the photosphere.

The spectrum of the sun is an absorption spectrum, and that can tell us a great deal about the photosphere. We know from Kirchhoff's third law that an absorption spectrum is produced when a source of a continuous spectrum is viewed through a gas. In the case of the photosphere, the deeper layers are dense enough to produce a continuous spectrum, but atoms in the photo-

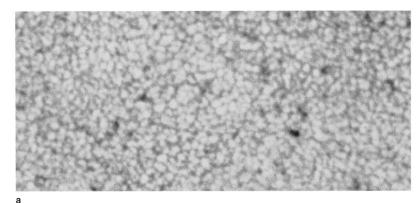

a

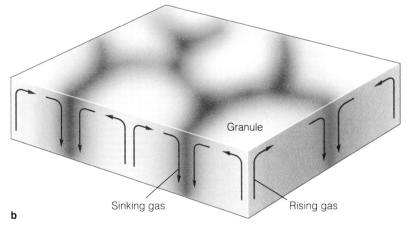

b

Figure 8-2
(a) A visible-light photo of the sun's surface shows granulation. *(AURA/NOAO/NSF)* (b) This model explains granulation as the tops of rising convection currents just below the photosphere. Heat flows upward as rising currents of hot gas and downward as sinking currents of cool gas. The rising currents heat the solar surface in small regions that we see as granules.

Granule

Sinking gas

Rising gas

sphere absorb photons of specific wavelengths, producing the absorption lines we see.

In good photographs, the photosphere has a mottled appearance because it is made up of dark-edged regions. The regions are called granules, and the visual pattern is called **granulation** (Figure 8-2a). Each granule is about the size of Texas and lasts for only 10 to 20 minutes before fading away. Faded granules are continuously replaced by new granules. Spectra of these granules show that the centers are a few hundred degrees hotter than the edges, and Doppler shifts reveal that the centers are rising and the edges are sinking at speeds of about 1 km/second.

From this evidence, astronomers recognize granulation as the surface effects of convection just below the photosphere. **Convection** occurs when hot fluid rises and cool fluid sinks, as when, for example, a convection current of hot gas rises above a candle flame. You can create convection in a liquid by adding a bit of cool nondairy creamer to an unstirred cup of hot coffee. The cool creamer sinks, warms, rises, cools, sinks again, and so on, creating small regions on the surface of the coffee that mark the tops of convection currents. Viewed from above, these regions look much like solar granules. Rising currents of hot gas heat small regions of the photosphere, which, being slightly hotter, emit more black body radiation and look brighter. The cool sinking gas of the edges emits less light and thus looks darker (Figure 8-2b). Thus the granulation reveals that energy flows upward from below by convection and heats the photosphere.

Recent spectroscopic studies of the solar surface have revealed another kind of granulation. **Supergranules** are regions about 30,000 km in diameter (about 2.3 times Earth's diameter) and include about 300 granules. These supergranules are regions of very slowly rising currents that last a day or two. They may be the

surface traces of larger currents of gas deeper under the photosphere.

The edge, or **limb,** of the solar disk is dimmer than the center (see the figure in Data File One). This **limb darkening** is caused by the absorption of light in the photosphere. When we look at the center of the solar disk, we are looking directly down into the sun, and we see deep, hot, bright layers in the photosphere. But when we look near the limb of the solar disk, we are looking at a steep angle and cannot see as deeply. The photons we see come from shallower, cooler, dimmer layers in the photosphere. Limb darkening proves that the temperature in the photosphere decreases with height, as we would expect if energy is flowing up from below.

The Chromosphere

Above the photosphere lies the chromosphere. The depths of the layers in the sun's atmosphere are not precisely defined, but we can think of the chromosphere as being an irregular layer with a depth on average less than Earth's diameter (see Figure 8-1). Because the chromosphere is roughly 1000 times fainter than the photosphere, we can see it with our unaided eyes only during a total solar eclipse when the moon covers the brilliant photosphere. Then, the chromosphere flashes into view as a thin line of pink just above the photosphere. The word *chromosphere* comes from the Greek word *chroma,* meaning "color." The pink color is produced by the combined light of three bright emission lines—the red, blue, and violet Balmer lines.

Astronomers know a great deal about the chromosphere from its spectrum. The chromosphere produces an emission spectrum, and Kirchhoff's second law tells us the chromosphere must be an excited, low-density gas. The density is about 10^8 times less dense that the air we breathe.

Atoms in the lower chromosphere are ionized, and atoms in the higher layers of the chromosphere are even more highly ionized. That is, they have lost more electrons. From this, astronomers can find the temperature in different parts of the chromosphere. Just above the photosphere the temperature falls to a minimum of about 4500 K, and then rises rapidly (Figure 8-3). The region where the temperature increases fastest is called the **transition region** because it makes the transition from the lower temperatures of the photosphere and chromosphere to the extremely high temperatures of the corona.

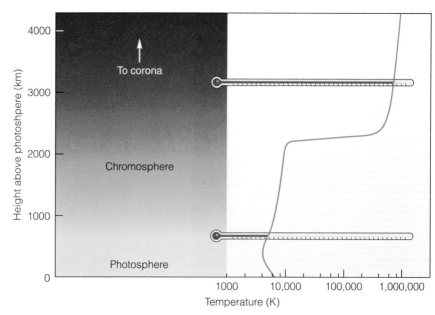

Figure 8-3
If we could place thermometers in the sun's atmosphere, we could construct a graph of temperature versus height as shown at the right. The curve shows that the temperature at the photosphere is about 5800 K, falls slightly in the lower chromosphere, and then rises rapidly.

Solar astronomers can take advantage of some simple physics to study the chromosphere. The gases of the chromosphere are transparent to nearly all visible light, but atoms in the gas are very good at absorbing photos of specific wavelengths. This produces certain

a

b

Figure 8-4
(a) An H-alpha filtergram reveals that the chromosphere contains complex structure that is not visible in normal photographs of the sun. *(AURA/NOAO/NSF)* (b) In this H-alpha filtergram spicules appear, looking like black grass blades, at the edges of supergranules. The longest spicules here are as long as two Earth diameters. *(© 1971 AURA/NOAO/NSF)*

dark absorption lines in the absorption spectrum of the photosphere. A photon at one of those wavelengths is very unlikely to escape from deeper layers. A **filtergram** is a photograph made using light in one of these dark absorption lines. Those photons can only have escaped from higher in the atmosphere. In this way filtergrams reveal detail in the upper layers of the chromosphere.

Figure 8-4 shows filtergrams made at the wavelength of the hydrogen alpha Balmer line. These images reveal complex structure in the chromosphere. **Spicules,** flamelike jets of gas, rise upward into the chromosphere and last 5 to 15 minutes. The spicules appear to be cooler gas from the lower chromosphere extending upward into hotter regions. (See Figure 8-1.) Seen at the limb (edge) of the sun's disk, these spicules blend together and look like flames covering a burning prairie, but filtergrams of spicules located at the center of the solar disk show that they spring up around the edge of supergranules like weeds around flagstones.

Spectroscopic analysis of the chromosphere alerts us that it is a low-density gas in constant motion where the temperature increases rapidly with height. Just above the chromosphere lies even hotter gas.

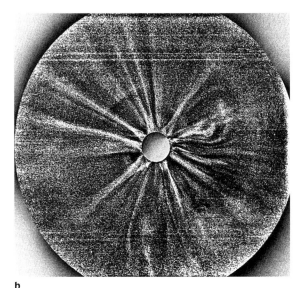

a

Figure 8-5
(a) This image from the Solar Maximum Mission satellite (inset) has been enhanced to produce a false-color image showing details of streamers in the corona. *(NASA)* (b) This coronagraph image of the sun was made from a jet 11 km (36,000 ft) above the earth. Coronal streamers extend beyond 20 solar radii. *(C. Keller, Los Alamos National Laboratory)*

b

The Solar Corona

The outermost part of the sun's atmosphere is called the corona, after the Greek word for "crown." The corona is so dim that it is not visible in the daytime sky because of the glare of the brilliant photosphere. During a total solar eclipse, however, the moon covers the brilliant photosphere, and the corona shines with a pearly glow not even as bright as the full moon. (See Figure 8-1b.)

Especially designed telescopes called **coronagraphs** can block the light of the photosphere and photograph the inner corona of the uneclipsed sun. Such telescopes work best from high mountaintops where the air is thin or from orbit where there is no air at all to scatter light. Using such images, astronomers can study the changing structure of streamers in the corona (Figure 8-5a).

The corona has been imaged out to 12 solar radii or more, well over 10 percent of the way to Earth's orbit (Figure 8-5b).

The spectrum of the corona can tell us a great deal about the coronal gases and simultaneously illustrate how astronomers can analyze a spectrum. Some of the light from the outer corona is just reflected sunlight, and that must mean the outer corona contains dust particles that reflect sunlight in all directions. Some of the coronal light consists of a continuous spectrum, and that appears to form when sunlight from the photosphere is scattered off of free electrons in the ionized coronal gas. We might expect this light to reproduce the absorption spectrum of the photosphere, but it does not because the coronal gas has a temperature over 1 million K, and the electrons travel very fast. When photons scatter off of these high-speed electrons, the photons suffer random Doppler shifts that smear out any absorption lines to produce a continuous spectrum.

Superimposed on the corona's continuous spectrum are emission lines of highly ionized gases. In the lower corona, the atoms are not as highly ionized as they are at higher altitudes, and this tells us that the temperature of the corona rises with altitude. Just above the chromosphere, the temperature is about 500,000 K; but in the outer corona the temperature can be as high as 2 million K or more.

The spectrum of the corona tells us that it is exceedingly hot gas, but it is not very bright. Its density is very low, only 1 to 10 atoms/cm^3, better than the best vacuum on Earth. Because of this low density, the hot gas does not emit much radiation.

Astronomers have wondered for years how the corona and chromosphere can be so hot. Heat flows from hot regions to cool regions, so how can the heat from the photosphere, with a temperature of only 5800 K, flow out into the much hotter chromosphere and corona? The answer appears to be the magnetic field that emerges from the sun's photosphere. Stirred by turbulence below the photosphere, the magnetic field extending outward whips about, and that churns the chromosphere and corona and heats them. Thus energy flows outward in this instance as the agitation of the magnetic fields.

Gas from the solar atmosphere follows the magnetic fields pointing outward and flows away from the sun in a breeze called the **solar wind.** Like an extension of the corona, the low-density gases of the solar wind blow past Earth at 300 to 800 km/s with gusts as high as 1000 km/s. Thus Earth is bathed in the corona's hot breath.

Because of the gas carried away by the solar wind, the sun is slowly losing mass. This is a minor loss for the star, amounting to only about 10^7 tons per year, only about 10^{-14} of a solar mass per year. In a few billion years, the sun, like many other stars, will lose mass rapidly. We will see how this affects stars in later chapters.

Do other stars have chromospheres, coronae, and stellar winds like the sun? Ultraviolet and X-ray observations suggest the answer is yes. The spectra of many stars contain emission lines in the far ultraviolet that could only have formed in the low-density, high-temperature gases of a chromosphere and corona. Also, many stars are sources of X rays, which appear to have been produced by chromospheres. Finally, the spectra of some stars have slightly blue-shifted spectral lines showing that gas is flowing away from those stars. Thus, observational evidence gives us good reason to believe that the sun, for all its complexity, is a typical star.

Helioseismology

It seems that the light we receive from the sun can tell us nothing about the layers below the photosphere, but solar astronomers have found a way to explore the sun's interior by **helioseismology,** the study of the modes of vibration of the sun. Just as geologists can study Earth's interior by observing how sound waves produced by earthquakes are reflected and transmitted by the layers of Earth's interior, so too can solar astronomers explore the sun's interior by studying how vibrations (sound waves) travel through the sun.

The waves are caused by random motions in the sun's atmosphere, and as these waves propagate down into the sun, they are bent back up to the surface by the increasing density. Short-wavelength waves penetrate less deeply and travel shorter distances than longer-wavelength waves, as shown in Figure 8-6a. When the waves reach the surface, they cause sections of the photosphere to move up and down by very small amounts, and these motions can be mapped as Doppler shifts. One of the strongest (loudest) waves in the sun causes the surface to oscillate with a period of about 5 minutes. Although 10 million different wavelengths are possible, the waves penetrate to different depths where conditions can weaken or strengthen a wave. By discovering which wavelength waves are actually present, solar astronomers can determine the temperature, density, pressure, composition, and motion at different depths inside the sun.

Of course, with 10 million possible wavelengths, the observations and analysis are difficult. A single wave produces a complicated pattern of motion on the solar surface (Figure 8-6b). Large amounts of data are necessary, so helioseismologists use a network of telescopes around the world operated by the Global Oscillation Network Group (GONG). The sun never sets on GONG. The network can observe the sun continuously for weeks at a time. The SOHO satellite in space has also observed solar oscillations and can detect motions as slow as 1 mm/s (0.002 mph). Solar astronomers can then use high-speed computers to separate the different patterns on the solar surface and measure the strength of the waves at many different wavelengths.

Helioseismology sounds almost magical, but we can understand it better if we think of a duck pond. If we stood at the shore of a duck pond and looked down at the water, we would see ripples arriving from all parts of the pond. Because every duck on the pond contributes to the ripples, we could, in principle, study the ripples near the shore and draw a map showing the position and velocity of every duck on the pond. Of course, it would be difficult to untangle the different ripples, but all of the information would be there, lapping at the rocks at our feet.

Helioseismology is giving astronomers new information about the sun's interior. For example, measurements of temperature and density can help us under-

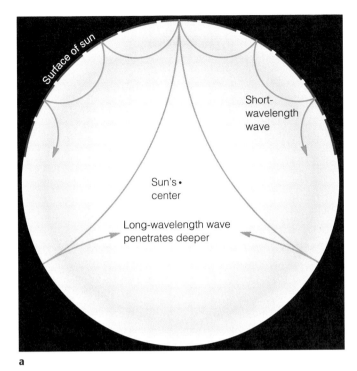

a

b

Figure 8-6

Helioseismology: (a) This cross section through the sun's equator shows how waves with short wavelengths do not penetrate as deeply as longer-wavelength waves. Where waves reflect off of the underside of the solar surface they cause a pattern of rising and falling regions (red and blue in this diagram) that can be observed by the Doppler effect. (b) This model shows only one of the 10 million possible patterns that are superimposed on the sun's surface. *(AURA/NOAO/NSF)*

stand the sun's energy generation, and measurements of the sun's internal motions can help explain the sun's magnetic fields. As we will see in the next section, the sun's magnetic field is highly complex.

REVIEW Critical Inquiry

How deeply into the sun can we see?

This is a simple question, but it has a very interesting answer. When we look into the layers of the sun, our sight does not really penetrate into the sun. Rather, our eyes record photons that have escaped from the sun and traveled outward through the layers of the sun's atmosphere. If we observe at a wavelength at the center of a dark absorption line, then the photosphere and lower chromosphere are opaque, photons can't escape to our eyes, and the only photons we can see come from the upper chromosphere. What we see are the details of the upper chromosphere—a filtergram. On the other hand, if we observe at a wavelength that is not easily absorbed (a wavelength between spectral lines), the atmosphere is more transparent, and photons from deep inside the photosphere can escape to our eyes. There is a limit, however, set by the H$^-$ ion, a hydrogen atom with an extra electron. At a certain depth, there is so much of this ion that the sun's atmosphere is opaque for almost all wavelengths, few photons can escape, and we can't see deeper.

By choosing the proper wavelength, solar astronomers can observe to different depths in the sun's photo-sphere. But the corona is so thin and the gas below the photosphere so dense that this method doesn't work in these regions. How can we observe the corona and the deeper layers of the sun?

So far we have thought of the sun as a static, unchanging ball of gas with energy flowing outward from the interior and through the atmospheric layers. In fact, the sun is a highly variable body whose appearance is constantly changing. It is now time to think of the active sun.

8-2 Solar Activity

The sun is unquiet. It is home to slowly changing spots larger than Earth and vast eruptions that dwarf our imaginations. All of these seemingly different forms of solar activity have one thing in common—magnetic fields. The weather on the sun is magnetic.

Sunspots and Active Regions

Solar activity is often visible with even a small telescope, but we should exercise great caution in observing the sun. Not only is sunlight intense, but it also includes infrared radiation (heat), which can burn our eyes. Looking at the sun with the unaided eye can cause permanent damage, and it is even more dangerous to look at the sun through a telescope, which concentrates the sunlight. Figure 8-7a shows a safe way to observe the sun with a small telescope.

When we observe the sun at visual wavelengths, we see the gases of the photosphere. Dark spots on the photosphere are called **sunspots,** and observations over a period of days will show the sun rotating and the sunspots growing larger or shrinking away (Figure 8-7b).

If we look carefully at sunspots, we see that they consist of a dark center called the umbra and an outer border called the penumbra (Figure 8-8a). Sunspots come in a wide range of sizes, but we might observe that the average spot is about 40 seconds of arc in diameter in our telescope. The small-angle formula (Chapter 3) tells us that such a sunspot is about twice the diameter of Earth. Such spots last a week or so and tend to occur in pairs or in groups of two major spots accompanied by smaller spots. A large group of sunspots is called an **active region,** may contain up to 100 spots, and could last as long as two months (Figure 8-8b).

Sunspots look dark because they are cooler than the photosphere. By noting which atoms are excited in the spectra of sunspots, astronomers can tell that the umbra has a temperature of about 4240 K compared with 5800 K for the photosphere. Because the amount of black body radiation emitted depends on the temperature to the fourth power (the Stefan-Boltzmann law in Chapter 7), this small difference in temperature makes a big difference in brightness, and the spot looks quite dark compared to the photosphere. In fact, a sunspot umbra emits quite a bit of light. If the sun were removed and only an average-size sunspot were left behind, it would glow a brilliant orange-red and would be brighter than the full moon.

A clue to the origin of sunspots appeared in 1908 when American astronomer George Ellery Hale discovered magnetic fields in sunspots. A magnetic field affects the permitted energy levels in an atom. With no magnetic field present, a particular atom might absorb photons of a certain wavelength, producing a spectral line. If the atom is located in a magnetic field, however, the energy levels are split into multiple levels, and the single spectral line could appear as three or more spectral lines at slightly different wavelengths. This is known as the **Zeeman effect** (Figure 8-9). The separation of the spectral lines depends on the strength of the magnetic field, so Hale could measure the strength of magnetic fields on the sun by looking for this effect. He found that the field in a sunspot is about 1000 times stronger than the sun's average field. Modern solar telescopes can use the Zeeman effect to map the magnetic field across the sun's surface. The magnetic map in Figure 8-9b shows that the magnetic field is strongest at the location of active regions.

This suggests that the powerful magnetic field causes a sunspot by inhibiting circulation. Ionized gas is made up of electrically charged particles, and such

Figure 8-7
Looking through a telescope at the sun is dangerous, but you can always view the sun safely with a small telescope by projecting its image on a white screen (a). If you sketch the location and structure of sunspots on successive days (b), you will see the rotation of the sun and gradual changes in the size and structure of sunspots.

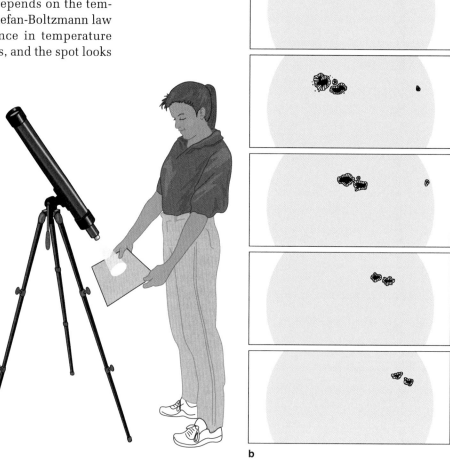

a

b

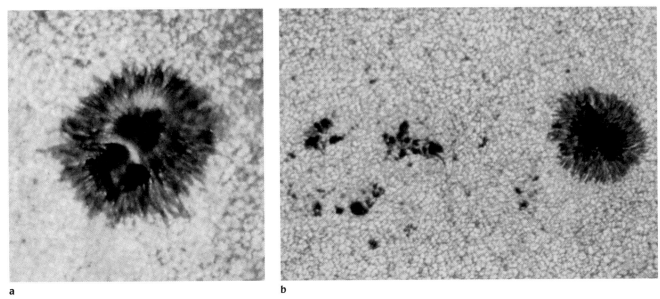

a   b

Figure 8-8
(a) This visible-light photograph of a sunspot is shown digitally enhanced to reveal radial structure in the penumbra and in the umbra. These markings are caused by the magnetic field in the sunspot. *(Courtesy William Livingston)* (b) A dark sunspot slightly larger than Earth (right) is part of a group of spots. Notice the granulation of the solar surface throughout the photo. *(NOAO)*

particles cannot move freely in a magnetic field. Astronomers often say the magnetic field is "frozen into" the ionized gas. This simply means that the gas and magnetic field are locked together. Rising currents of hot gas just under the photosphere might be slowed by the magnetic fields in sunspots, causing a decrease in the temperature and producing a dark spot. The energy unable to emerge through the sunspot is evidently deflected and emerges in the photosphere around the sunspot (Figure 8-9c).

The magnetic field that emerges through sunspots extends up into the chromosphere and corona above active regions, and high-temperature gas becomes trapped in these magnetic fields. At far ultraviolet wavelengths, this trapped gas is much brighter than the cool photosphere and outlines the arched magnetic fields, giving us further evidence that active regions are dominated by magnetic fields (Figure 8-10).

Do other stars have sunspots, or rather "starspots," on their surfaces? This is a difficult question, because, except for the sun, the stars are so far away that no surface detail is visible. Some stars, however, vary in brightness in ways that suggest they are

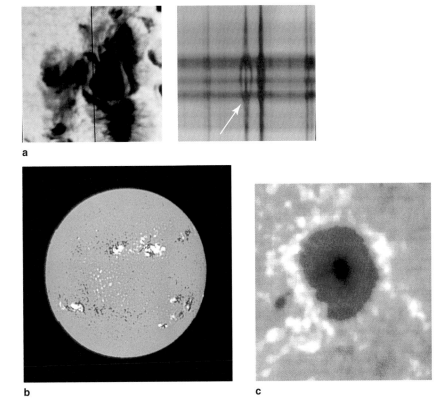

a

b   c

Figure 8-9
(a) The slit of the spectrograph has been placed across a complex sunspot allowing light from the sunspot to enter the spectrograph. The short segment of the resulting spectrum, shown at right, contains a typical spectral line split by the Zeeman effect (arrow). *(AURA/NOAO/NSF)* (b) This Zeeman map of the sun shows regions of strong magnetic field (white) at the location of active regions. *(AURA/NOAO/NSF)* (c) In this infrared image, brightening around a sunspot is apparently caused by energy unable to emerge though the magnetic field of the sunspot. *(Dan Gezari, NASA/Goddard)*

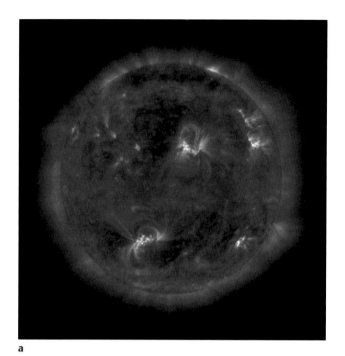

a

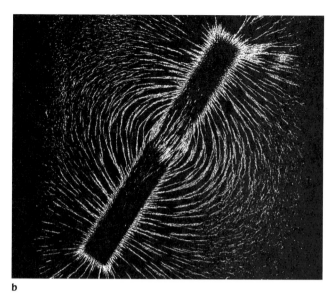

b

Figure 8-10

(a) This far-ultraviolet image of the sun shows high-temperature gas trapped in the arched magnetic fields above active regions. At visual wavelengths we would see a pair of sunspots at the location of each of these active regions. *(SOHO/EIT, ESA, and NASA)* (b) Compare the arched shapes of the trapped gas above active regions to the shape of the magnetic field around a bar magnet as revealed by iron filings. *(Grundy Observatory)*

mottled by randomly placed, dark spots. As the star rotates, its total brightness changes slightly, depending on the number of spots facing in our direction. This has been suggested as an explanation for the variation of the RS Canum Venaticorum stars, whose spots may cover as much as 25 percent of the surface.

Also, some stars show spectral features that suggest the presence of magnetic fields and starspots. Ultraviolet observations reveal stars whose spectra contain emission lines commonly produced by the regions around spots on the sun. This suggests that these stars, too, have spots. One team of astronomers has used the Doppler shifts in the spectrum of the rotating star HR-1099 to construct a map showing the distribution of dark spots on its surface (Figure 8-11). Such results suggest that the sunspots we see on our sun are not unusual.

The Sunspot Cycle

The total number of sunspots visible on our sun is not constant. In 1843, the German amateur astronomer Heinrich Schwabe noticed that the number of sunspots varies in a period of about 11 years. This is now known as the sunspot cycle (Figure 8-12). At sunspot maximum, there are often as many as 100 spots visible at any one time, but at sunspot minimum there are only a few small spots. A sunspot minimum occurred in the mid-1990s with a maximum in the very first years of the 21st century.

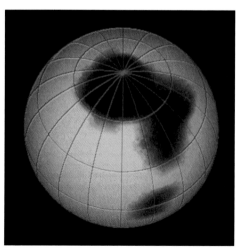

a

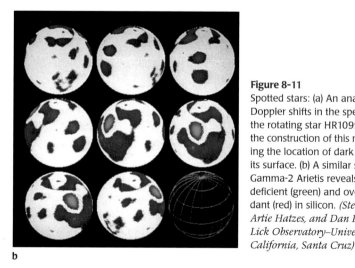

b

Figure 8-11

Spotted stars: (a) An analysis of Doppler shifts in the spectrum of the rotating star HR1099 allowed the construction of this map showing the location of dark spots on its surface. (b) A similar study of Gamma-2 Arietis reveals spots deficient (green) and overabundant (red) in silicon. *(Steven Vogt, Artie Hatzes, and Dan Penrod, Lick Observatory–University of California, Santa Cruz)*

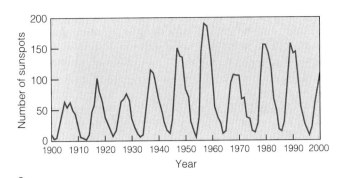

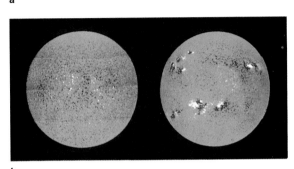

b

Figure 8-12
(a) The number of sunspots varies in an 11-year period. (b) Magnetograms of the sun show the magnetic field over the sun's disk at sunspot minimum (left) and at sunspot maximum (right). *(AURA/ NOAO/NSF)*

At the beginning of each sunspot cycle, the spots begin to appear in the sun's middle latitudes about 35° above and below the sun's equator. As the cycle proceeds, the spots appear at lower latitudes until, near the end of the cycle, they are appearing within 5° of the sun's equator. If we plot the latitude of the appearance of sunspots versus time, the diagram takes on the shape of butterfly wings (Figure 8-13). Such diagrams are known as **Maunder butterfly diagrams,** named after E. Walter Maunder of the Royal Greenwich Observatory, who first published the diagram in 1922.

The Sun's Magnetic Cycle

Sunspots are magnetic phenomena, so the 11-year cycle of sunspots must be caused by cyclical changes in the sun's magnetic field. To explore that idea, we must begin with the sun's rotation.

The sun does not rotate as a rigid body. It is a gas from its outermost layers down to its center, so some parts of the sun rotate faster than other parts. The equatorial region of the photosphere rotates faster than do regions at higher latitudes (Figure 8-14a). At the equator, the photosphere rotates once every 25 days, but at a latitude of 45° one rotation takes 27.8 days. Helioseismology shows that deeper layers of gas rotate slower than the surface and that gases near the poles rotate the slowest of all (Figure 8-14b). This phenomenon is called **differential rotation,** and it is clearly linked with the magnetic cycle.

Astronomers believe that the sun's overall magnetic field is produced by its rotation and the outward flow of energy. The gases of the sun are highly ionized, so they are very good conductors of electricity. When an electrical conductor rotates rapidly and is stirred by convection, it can convert some of the energy flowing outward as convection into a magnetic field. This process is called the **dynamo effect,** and it is believed to produce Earth's magnetic field as well. Helioseismologists have found evidence that the sun's magnetic field is generated in convection currents deep under the photosphere. Once the magnetic field is created, the convection and differential rotation stretch, twist, and tangle the field to produce a magnetic cycle.

The magnetic behavior of sunspots gives us an insight into how the magnetic cycle works. Sunspots tend to occur in pairs, and the magnetic field around the pair resembles that around a bar magnet with one end magnetic north and the other end magnetic south. At any one time, sunspot pairs south of the sun's equator have reversed polarity compared to those north of the sun's equator. Figure 8-14c illustrates this by showing sunspot

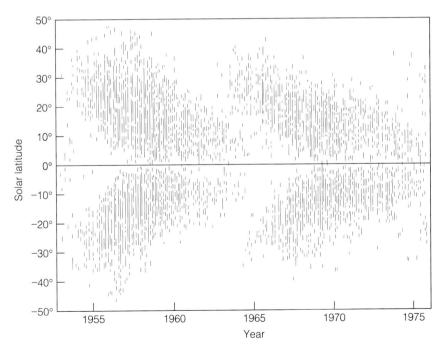

Figure 8-13
The Maunder butterfly diagram is a plot of the latitude on the sun where sunspots first appear. Early in a sunspot cycle, the spots appear at higher latitudes. Later in the cycle, they appear nearer the sun's equator.

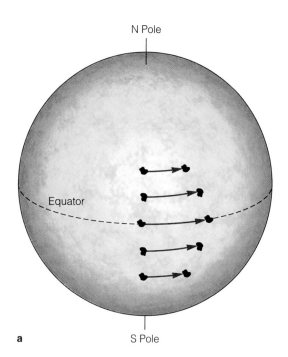

N Pole

Equator

S Pole

a

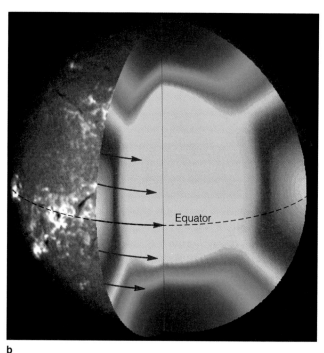

Equator

b

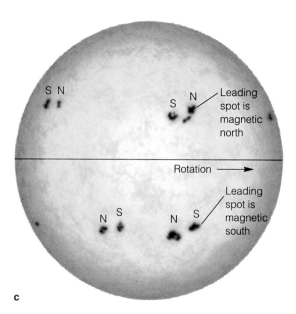

S N

S N — Leading spot is magnetic north

Rotation ⟶

N S   N S — Leading spot is magnetic south

c

Figure 8-14

(a) At the photosphere, the sun rotates faster at the equator than at higher latitudes. If we started five sunspots in a row, they could not stay aligned as the sun rotates. (b) Helioseismology shows how the interior of the sun rotates relative to the surface. In this cutaway diagram, the photosphere at the equator rotates the fastest and the poles the slowest. *(Courtesy Kenneth Libbrecht, Caltech)* (c) The magnetic polarity of sunspot groups in the sun's southern hemisphere is reverse of that in the northern hemisphere.

pairs south of the sun's equator with magnetic south poles leading and sunspots north of the sun's equator with magnetic north poles leading. At the end of an 11-year sunspot cycle, the new spots appear with reversed magnetic polarity.

This magnetic cycle is not fully understood, but the **Babcock model** (named for its inventor) explains the magnetic cycle as a progressive tangling of the solar magnetic field. Because the electrons in an ionized gas are free to move, the gas is a very good conductor of electricity, and any magnetic field in the gas is "frozen" into the gas. If the gas moves, the magnetic field must move with it. Thus the sun's magnetic field is frozen into its gases, and the differential rotation wraps this

field around the sun like a long string caught on a hubcap. Rising and sinking gas currents twist the field into ropelike tubes, which tend to float upward. Where these magnetic tubes burst through the sun's surface, sunspot pairs occur (Figure 8-15).

The Babcock model explains the reversal of the sun's magnetic field from cycle to cycle. As the magnetic field becomes tangled, adjacent regions of the sun's surface are dominated by magnetic fields that point in different directions. After about 11 years of tangling, the field becomes so complex that adjacent regions of the solar surface begin changing their magnetic field to agree with neighboring regions. Quickly the entire field rearranges itself into a simpler pattern, and differential rotation begins winding it up to start a new cycle. But the newly organized field is reversed, and the next sunspot cycle begins with magnetic north replaced by magnetic south. Thus the complete magnetic cycle is 22 years long, and the sunspot cycle is 11 years long.

This magnetic cycle may even explain the Maunder butterfly diagrams. As a sunspot cycle begins, the twisted tubes of magnetic force first begin to float upward and produce sunspot pairs at higher latitude.

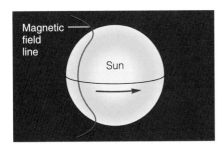

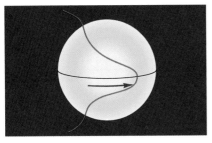

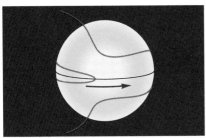

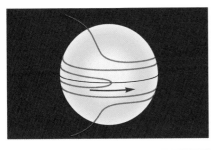

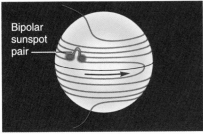

Figure 8-15
The Babcock model suggests that the differential rotation of the sun winds up the magnetic field. When the field becomes tangled and bursts through the surface, it forms sunspot pairs. For the sake of clarity, only one line in the sun's magnetic field is shown here.

Thus the first spots in a cycle appear further north and south of the equator. Later in the cycle, when the field is more tightly wound, the tubes of magnetic force arch up through the surface closer to the equator. Thus the later sunspot pairs in a cycle appear closer to the equator.

Notice the power of a scientific model. The Babcock model may in fact be incorrect in some or all de-

tails, but it gives us a framework on which to organize all of the complex solar activity. Even though our models of the sky (see Chapter 2) and the atom (see Chapter 7) were only partially correct, they served as organizing themes to guide our thinking. Similarly, although the precise details of the solar magnetic cycle are not yet understood, the Babcock model gives us a general picture of the behavior of the sun's magnetic field (Window on Science 8-1).

Magnetic Cycles on Other Stars

Because we believe that the sun is a representative star, we should expect other stars to have similar cycles of starspots. We can't see individual spots from Earth, of course, but certain features in stellar spectra are associated with magnetic fields. Regions of strong magnetic fields on the solar surface emit strongly at the central wavelengths of the H and K lines of ionized calcium. This calcium emission appears in the spectra of other sunlike stars and tells us that these stars, too, have strong magnetic fields on their surfaces. These stars presumably have starspots as well.

In 1966, astronomers began measuring the strength of this H and K emission in the spectra of stars. The stars to be observed were selected to be similar to the sun. With temperatures ranging from 1000 K hotter than the sun to 3000 K cooler, these stars were considered most likely to have sunlike magnetic activity on their surfaces.

The observations show that the strength of the emissions in the spectra of these stars varies from year to year. The H and K emission averaged over the sun's disk varies with the sunspot cycle, and similar periodic variations can be seen in the spectra of the stars studied (Figure 8-16). The star 107 Piscium, for instance, appears to have a starspot cycle lasting 9 years. Thus, we can be sure that stars like the sun do have magnetic fields and are subject to magnetic cycles.

The rotation periods of these stars are also apparent from the observations. If a star has a major region of magnetic activity and rotates every 20 days, we see the emission appear for 10 days and then disappear for 10 days as the rotation of the star carries the region to the far side of the star.

These observations confirm our belief that the sun is an average sort of star, that is, not peculiar. Most other stars like our sun have magnetic fields and starspots and go through magnetic cycles. More important, these studies are helping us understand our sun better.

Prominences and Flares

A **prominence** is composed of hot gas trapped in magnetic fields extending above active regions into the lower corona. We can tell that the gas in a prominence is hot because it is ionized and produces an emission

While many textbooks describe science as the process of testing hypotheses by observation and experiment, you should not think that every astronomer approaches the telescope with the expectation of making an observation that will disprove long-held beliefs and trigger a revolution in science. Then what is the daily grind of science really about?

First, many observations and experiments merely confirm already tested hypotheses. The biologist knows that all worker bees in a hive are sisters, but a careful study of the DNA from different workers further confirms that hypothesis. By repeatedly confirming a hypothesis,

scientists build confidence in the hypothesis and may be able to extend it to a wider application. Of course, there is always the chance that a new observation or experiment will disprove the hypothesis, but that is usually very unlikely. Much of the daily grind of science is confirmation.

Another aspect of routine science is consolidation, the linking of a hypothesis to other well-studied phenomena. Chemists may understand certain kinds of carbon molecules shaped like rings, but by repeated study they find a carbon molecule shaped like a hollow sphere. To consolidate their findings, they must show that the chemical bonding in the two mole-

cules follows the same rules and that the molecules have certain properties in common. No hypothesis is overthrown, but the chemists consolidate their knowledge and understand carbon molecules better.

The Babcock model of the solar magnetic cycle is an astronomical example of the scientific process. Solar astronomers know that the model explains some solar features but has shortcomings. Although most astronomers don't expect to discard the entire model, they work through confirmation and consolidation to better understand how the solar magnetic cycle works and how it is related to cycles in other stars.

---

spectrum. At visible wavelengths, the Balmer lines give prominences a bright pink color, and they are visible during a total solar eclipse protruding beyond the edge of the moon. (See Figure 3-17a.) We can tell that prominences are dominated by magnetic fields because of their arched shapes (Figure 8-17).

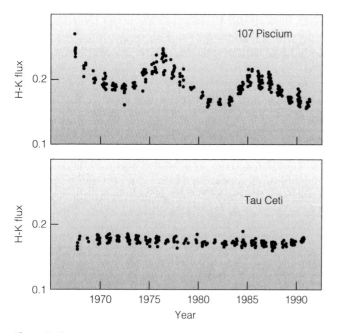

Figure 8-16
The average amount of emission in the H and K lines of calcium (mean H-K flux) is related to magnetic activity. In the sun, this emission is stronger when sunspot activity is higher. The sunlike star 107 Piscium appears to have a magnetic cycle, while the star Tau Ceti does not. *(Adapted from data by Baliunas and Saar)*

We can identify two kinds of prominences. Eruptive prominences burst out of active regions and in a few hours extend into the lower corona only to fall back. Quiescent prominences sometimes hang in the lower corona for many days, sometimes with gas streaming downward into the active region along magnetic fields.

Although we think of the gas trapped in a prominence as hot and excited, it is not as hot as the gas in the lower corona. Thus prominences are complex objects that originate in the hot gas of an active region and extend up into the even hotter regions of the corona.

A solar **flare** is a violent outburst on the solar surface that grows to a maximum for a few minutes and decays in an hour or less. During that time it emits vast amounts of X-ray, ultraviolet, and visible radiation, plus streams of high-energy protons and electrons. Figure 8-18 shows a **shock wave** (a naturally occurring sonic boom) rushing away from a solar flare. A large flare can release energy equivalent to 10 billion megatons of TNT. (Large hydrogen bombs on Earth typically release about 10 megatons of energy.)

Solar flares are linked to the sun's magnetic field. Flares almost always occur near active regions, where the magnetic field is strong and twisted into great arches. Small flares can recur over and over at the same place. A large active region may experience 100 small flares a day. Flares are believed to occur when arches of magnetic force encounter each other and reconnect in such a way that they cancel each other out. These **reconnections** release the tremendous energy stored in the magnetic fields as X rays and as high-speed electrons that heat the gas nearby (Figure 8-19a). During a major solar

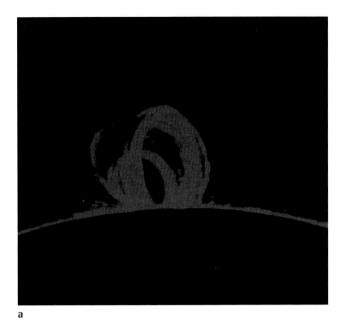
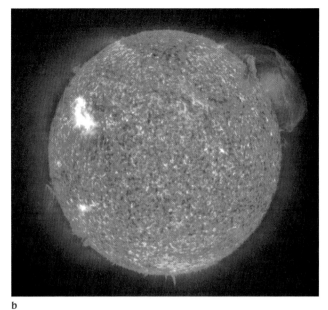

a   b

Figure 8-17

(a) A loop prominence near the edge of the sun shows by its shape the magnetic fields that control the gas. *(Sacramento Peak Observatory)* (b) This far-ultraviolet photograph of the sun shows a very large arched prominence extending out into the corona. Note the complexity of the magnetic field. *(SOHO/EIT, ESA, and NASA)*

flare, the gas can be heated to over 5,000,000 K. X-ray observations detect photons of exactly the energies produced in certain nuclear reactions, and that assures us that the gas in a solar flare is very hot indeed. Although flares are common on the sun, most are detected only in filtergrams. So-called white-light flares, which are detectable in visible-light photographs, occur only about once a year.

Solar flares can have important effects on Earth. The X-ray and ultraviolet radiation reaches Earth in only 8 minutes and increases the ionization in Earth's upper atmosphere. This alters the reflection of short-wave radio signals and can absorb them completely, thereby interfering with communications. Flares can eject high-energy particles at a third the speed of light, but most particles ejected from flares have lower velocities and reach Earth hours or days after the flare as gusts in the solar wind. Such gusts interact with Earth's magnetic field and generate tremendous electrical currents (as much as a million megawatts), which flow down into Earth's atmosphere near the magnetic poles. There they can excite the atoms of the upper atmosphere at altitudes of 100 to 400 km to glow in displays called **auroras** (Figure 8-19b). These currents can also disturb Earth's magnetic field and cause magnetic storms in which compasses behave erratically.

Other effects of solar flares include surges in high-voltage power lines and radiation hazards to passengers in supersonic transports and in spacecraft. The U.S. Air Force watches for flares from observatories

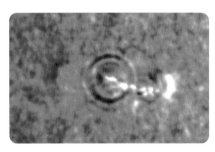

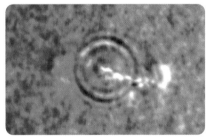

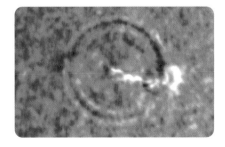

Figure 8-18

This sequence of images recorded about 5 minutes apart shows the result of a solar flare. A shock wave spreads outward across the solar disk as a ripple up to 3 km high, traveling 50 km/s, and equivalent to an earthquake 40,000 times stronger than the quake that destroyed San Francisco in 1906. *(SOHO/MDI, ESA, and NASA)*

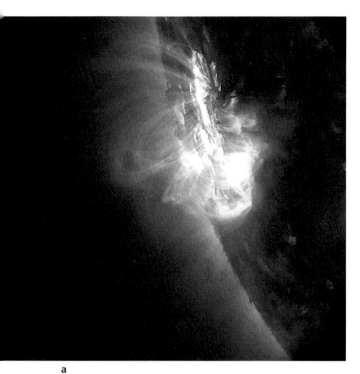

a   b

Figure 8-19

(a) This ultraviolet image of the edge of the sun shows an outburst of hot gas extending up into the corona produced when two magnetic fields just above the photosphere recombined, canceled each other out, and released their energy to the surrounding gas. *(NASA)* (b) Eruptions on the sun can bathe Earth in high-energy particles that produce auroras commonly seen at high latitudes. Notice the two vertical curtains of light within the glowing cloud in this photograph of an aurora seen in southern Australia. *(David Miller)*

around the world, and solar astronomers have developed ways of predicting which flares will affect Earth.

Coronal Activity

Even the corona of the sun takes part in the magnetic cycle. At sunspot minimum, when there are few active regions and the magnetic field is not strongly tangled, eclipse observers see a small, slightly flattened corona. But at sunspot maximum, when the solar magnetic field produces many active regions, eclipse observers are treated to a blazing corona that is nearly circular. Figure 8-1b shows the corona near sunspot maximum.

We don't need to depend on total solar eclipses to study the corona. Specialized telescopes on Earth and ultraviolet and X-ray telescopes in space can image the corona every day. The results show that the corona is composed of streamers shaped by the solar magnetic field. Some parts of the corona are filled with hot gas trapped by magnetic fields that loop back down to the solar surface. In other regions the magnetic field does not loop back, and the hot gas streams away from the sun. These regions are called **coronal holes** because they appear as dark regions in ultraviolet and X-ray images of the sun (Figure 8-20). There are permanent

coronal holes at the sun's north and south poles, but coronal holes in other regions come and go as they are shaped by the changing magnetic field.

The corona is strongly affected by matter ejected from active regions on the sun's surface. When arcs of magnetic field reconnect, tremendous energy can be released, and great bursts of hot gas are expelled into the corona, as shown in Figure 8-21. If one of these eruptions strikes Earth, it can have dramatic effects on communications and electrical power networks.

It should not surprise us that Earth is affected by disturbances in the corona. The solar wind is derived from the coronal holes, so it is not an exaggeration to say that Earth orbits within the expanding corona.

The Solar Constant

If the sun's energy output varied by even a small amount, life on Earth might end. The continued existence of our civilization and our species depends on the constancy of our sun, but we know very little about the variation of the sun's energy output.

The energy production of the sun can be measured by adding up all of the energy falling on 1 square meter

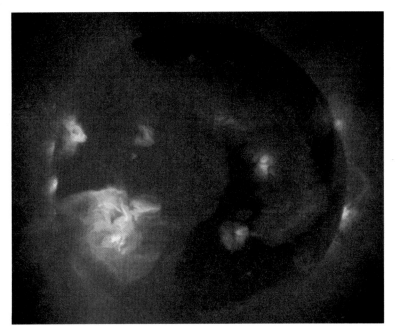

Figure 8-20
An X-ray image of the sun reveals intensely hot regions in the solar atmosphere, linked by magnetic fields. Bright regions lie above active regions on the solar surface and reveal magnetic fields that loop back to the solar surface and confine hot gas. Dark regions are coronal holes, where the magnetic field does not loop back to the surface and thus does not restrict the motion of hot gas. *(NASA and ISAS)*

the solar constant may have varied in the past. The "little ice age" was a period of unusually cool weather in Europe and America that lasted from about 1430 to 1850. The average temperature worldwide was about 1 K cooler than it is now. Although many things affect climate, this cooling could have been caused by a decrease in the solar constant of only 1 percent.

This period of cool weather included an era of few sunspots now known as the **Maunder Minimum*** (Figure 8-22). Between 1645 and 1715, there is almost no record of sunspots, even though the telescope had been perfected and was being actively used by astronomers. Reports of total solar eclipses during this period make no mention of the corona or chromosphere, and there is almost no record of auroral displays. The Maunder Minimum seems to have been a period of reduced solar activity. Solar astronomer John Eddy believes that the historical record shows traces of at least 11 similar periods extending back to 3000 BC. If that is true, our sun is not a perfectly stable star, and the survival of humanity may

of Earth's surface during 1 second. Of course, some correction for the absorption of Earth's atmosphere is necessary, and we must count all wavelengths from X rays to radio waves. The result, which is called the **solar constant,** amounts to about 1360 joules per square meter per second. A change in the solar constant of only 1 percent could change Earth's average temperature by 1 to 2°C (about 1.8 to 3.6°F). For comparison, during the last ice age Earth's average temperature was about 5°C cooler than it is now.

Some of the best measurements of the solar constant were made by instruments aboard the Solar Maximum Mission satellite. These have shown variations in the energy received from the sun of about 0.01 percent that lasted for days or weeks. Superimposed on that random variation is a long-term decrease of about 0.018 percent per year that has been confirmed by observations made by sounding rockets and balloons and by the NIMBUS 7 satellite. This long-term decrease may be related to a cycle of activity on the sun with a period longer than the 22-year magnetic cycle.

Small, random fluctuations will not affect our climate, but a long-term decrease over a decade or more could cause worldwide cooling. History contains some evidence that

*Ironically, the Maunder Minimum coincides with the reign of Louis XIV of France, the "Sun King."

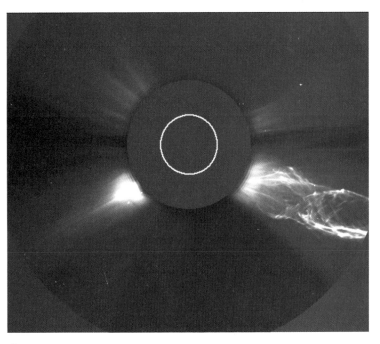

Figure 8-21
In this ultraviolet image of the corona, the disk of the sun (white circle) is hidden behind a circular screen. A violent eruption near the surface of the sun has ejected hot gases out into the corona toward the lower right. Such matter usually escapes from the sun completely. *(SOHO/LASCO, ESA, and NASA.)*

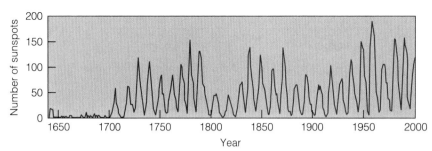

Figure 8-22
Counts of sunspots gathered from historical sources show that there were very few sunspots seen between 1645 and 1715. Called the Maunder Minimum, this period of low solar activity suggests that the sun's cycle of magnetic activity may not be continuous.

depend on our ability to adapt to changes in the solar "constant."

R E V I E W   Critical Inquiry

What kind of activity would the sun have if it didn't rotate differentially?

This is a really difficult question because we can see only one star close up and thus have no other examples. Nevertheless, we can make an educated guess by thinking about the Babcock model. If the sun didn't rotate differentially, with its equator traveling faster than higher latitudes, then the magnetic field would not get wound up, and it would not get twisted and rise through the surface to form sunspot pairs. Of course, the magnetic field would not get progressively more wound up, so there would also be no magnetic cycle. Loops of magnetic field might not rise above the surface to trap ionized gases and form prominences. Furthermore, twists and kinks might never form in the field, and thus there would be no flares. In fact, with no differential rotation to stir up the magnetic field, the chromosphere and corona might not be heated to such high temperatures.

This is very speculative, but it suggests some interesting ideas about other stars. Most stars probably rotate differentially, so we should expect many stars to have activity on their surfaces much as the sun does. What observational evidence do we have for this expectation?

The sun is beautiful and complex, with great eruptions and spots sweeping across its surface like magnetic weather. All of this activity is driven by the steady flow of energy outward from the sun's center through its photosphere and into space. How does the sun make that energy?

8-3 Nuclear Fusion in the Sun

Astronomers often use the wrong words to describe energy generation in the sun and stars. Astronomers will say, "The star ignites hydrogen burning." We use the word *ignite* to mean *catch on fire,* and we use *burn* to

mean *on fire.* What goes on inside stars isn't really burning in the usual sense.

We must also use the word *atom* with care. The interior of the sun is so hot the gas is totally ionized. That is, the electrons are not attached to the atomic nuclei, and the gas is an atomic soup of rapidly moving particles colliding with each other at high velocity. When we discuss nuclear reactions inside stars, we refer to atomic nuclei and not to atoms.

How exactly can the nucleus of an atom yield energy? The answer lies in the forces that hold the nuclei together.

Nuclear Binding Energy

The sun generates its energy by breaking and reconnecting the bonds between the particles *inside* atomic nuclei. This is quite different from the way we generate energy by burning wood in a fireplace. The process of burning wood extracts energy by breaking and reconnecting chemical bonds between atoms in the wood. Chemical bonds are formed by the electrons in atoms, and we saw in Chapter 7 that the electrons are bound to the atoms by the electromagnetic force. Thus chemical energy originates in the force that binds electrons to atoms.

There are only four known forces in nature: the force of gravity, the electromagnetic force, the **weak force,** and the **strong force.** The weak force is involved in the radioactive decay of certain kinds of nuclear particles, and the strong force binds together atomic nuclei. Thus nuclear energy comes from the strong force.

Nuclear power plants on Earth generate energy through **nuclear fission** reactions that split uranium nuclei into less massive fragments. A uranium nucleus contains a total of 235 protons and neutrons, and it splits into a range of fragments containing roughly half as many particles. Because the fragments produced are more tightly bound than the uranium nuclei, binding energy is released during uranium fission.

Stars make energy in **nuclear fusion** reactions that combine light nuclei into heavier nuclei. The most common reaction including that in the sun fuses hydrogen nuclei (single protons) into helium nuclei (two protons and two neutrons). Because the nuclei produced are more tightly bound than the original nuclei, energy is released. Notice in Figure 8-23 that both fu-

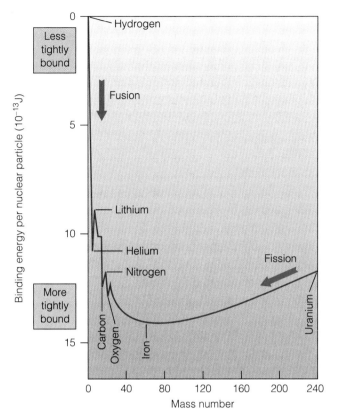

Figure 8-23
The red line in this graph shows the binding energy, the energy that holds an atomic nucleus together, for all the different atoms plotted by their atomic mass number, the number of protons and neutrons in their nucleus. Both fission and fusion nuclear reactions move downward in the diagram (arrows) toward more tightly bound nuclei. Iron has the most lightly bound nucleus, so no nuclear reactions can begin with iron and release energy.

sion and fission reactions move downward in the diagram toward more tightly bound nuclei. Thus both produce energy by releasing the binding energy of atomic nuclei.

Hydrogen Fusion

Not until the 1930s did astronomers realize how the sun generates energy. The sun fuses together four hydrogen nuclei to make one helium nucleus. Because one helium nucleus has 0.7 percent less mass than four hydrogen nuclei, it seems that some mass vanishes in the process:

$$\begin{aligned} 4 \text{ hydrogen nuclei} &= 6.693 \times 10^{-27} \text{ kg} \\ \underline{1 \text{ helium nucleus}} &= \underline{6.645 \times 10^{-27} \text{ kg}} \\ \text{difference in mass} &= 0.048 \times 10^{-27} \text{ kg} \end{aligned}$$

However, this mass does not actually vanish; it merely changes form. The equation $E = m_0 c^2$ reminds us that mass and energy are related, and under certain circumstances mass may become energy, and vice versa. Thus, the 0.048×10^{-27} kg does not vanish but merely

becomes energy. To see how much, we use Einstein's equation:

$$\begin{aligned} E &= m_0 c^2 \\ &= (0.048 \times 10^{-27} \text{ kg})(3 \times 10^8 \text{ m/s})^2 \\ &= 0.43 \times 10^{-11} \text{ J} \end{aligned}$$

This is a very small amount of energy, hardly enough to raise a housefly one-thousandth of an inch. Because one reaction produces such a small amount of energy, it is obvious that many reactions are necessary to supply the energy needs of a star. The sun, for example, needs 10^{38} reactions per second, transforming 5 million tons of mass into energy every second, just to stay hot enough to resist its own gravity.

Two nuclei can fuse only if they come close together, but atomic nuclei have positive charges and repel each other with an electrostatic force. This repulsion between the positively charged nuclei has been called the **Coulomb barrier.** To overcome this barrier, the atomic nuclei must collide at high velocity. If the particles in a material are moving at high velocities, we say that the material is hot. Thus, atomic fusion reactions can only occur if the gas is very hot—at least 10^7 K.

Fusion reactions in the sun also require that the gas be very dense—denser than solid lead. We know that the sun requires 10^{38} reactions per second to manufacture sufficient energy. But fusion occurs in only a small percentage of all collisions, so the sun requires many collisions between nuclei each second. Only where the gas is very dense are there enough collisions to meet the sun's energy needs.

Nuclear fusion requires high temperature and high density. Only in the central regions of the sun is the gas hot enough and dense enough for hyrodgen to fuse. Thus all of the energy that flows outward through the sun's surface is created in the sun's core. Sunlight is nuclear power.

We can symbolize the fusion reactions in the sun with a simple nuclear reaction:

$$4 \text{ }^1\text{H} \rightarrow {}^4\text{He} + \text{energy}$$

In this equation, ^{1}H represents a proton, the nucleus of the hydrogen atom, and ^{4}He represents the nucleus of a helium atom. The superscripts indicate the approximate weight of the nuclei (the number of protons plus the number of neutrons). The actual steps in the process are more complicated than this convenient summary suggests. Instead of waiting for four hydrogen nuclei to collide simultaneously, a highly unlikely event, the process can proceed step by step in a chain of reactions—the proton–proton chain.

The **proton–proton chain** is a series of three nuclear reactions that builds a helium nucleus by adding together protons. This process is efficient at temperatures above 10,000,000 K. The sun, for example, manufactures over 90 percent of its energy in this way.

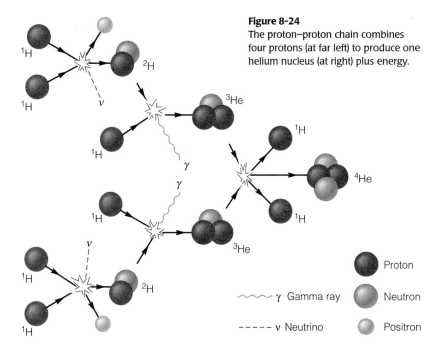

Figure 8-24
The proton–proton chain combines four protons (at far left) to produce one helium nucleus (at right) plus energy.

Proton

Neutron

Positron

$\sim\sim\sim$ γ Gamma ray

$-----$ ν Neutrino

The three steps in the proton–proton chain entail these reactions:

$$^1H + {}^1H \rightarrow {}^2H + e^+ + \nu$$
$$^2H + {}^1H \rightarrow {}^3He + \gamma$$
$$^3He + {}^3He \rightarrow {}^4He + {}^1H + {}^1H$$

In the first step, two hydrogen nuclei (two protons) combine to form a heavy hydrogen nucleus, emitting a particle called a positron, e^+ (a positively charged electron), and a **neutrino,** ν (a subatomic particle originally thought to have zero mass and to travel at the velocity of light). In the second reaction, the heavy hydrogen nucleus absorbs another proton and, with the emission of a gamma ray, γ, becomes a lightweight helium nucleus. Finally, two light helium nuclei combine to form a common helium nucleus and two hydrogen nuclei. Because the last reaction needs two ^3He nuclei, the first and second reactions must occur twice (Figure 8-24). The net result of this chain reaction is the transformation of four hydrogen nuclei into one helium nucleus plus energy.

The energy appears in the form of gamma rays, positrons, the energy of motion of the nuclei, and neutrinos. The gamma rays are photons that are absorbed by the surrounding gas before they can travel more than a fraction of a millimeter. This heats the gas and helps maintain the pressure. The positrons produced in the first reaction combine with free electrons, and both particles vanish, converting their mass into gamma rays. Thus, the positrons also help keep the center of the star hot. In addition, when fusion produces new nuclei, they fly apart at high velocity. This energy of motion helps raise the temperature of the gas. The neutrinos, however, resemble photons except that they almost never interact with other particles. The average neu-

trino could pass unhindered through a lead wall 1 ly thick. Thus, the neutrinos do not help heat the gas but race out of the star at the speed of light, carrying away roughly 2 percent of the energy produced.

It is time to ask the critical question that lies at the heart of science. What is the evidence to support our theorietical explanation of how the sun makes its energy? The search for that evidence introduces us to one of the great problems of modern astronomy.

The Solar Neutrino Problem

The center of a star seems forever hidden from us, but the sun is transparent to neutrinos because these subatomic particles almost never interact with normal matter. Nuclear reactions in the sun's core produce floods of neutrinos that rush out of the sun and off into space. If we could detect these neutrinos, we could probe the sun's interior.

Because neutrinos almost never interact with atoms, we never feel the flood of over 10^{12} solar neutrinos that flow through our bodies every second. Even at night, neutrinos from the sun rush through Earth as if it weren't there, up through our beds, through us, and onward into space. Obviously we are lucky to be transparent to neutrinos, but it means that neutrinos are extremely hard to detect. Certain nuclear reactions, however, can be triggered by a neutrino of the right energy; and in the late 1960s, chemist Raymond Davis, Jr., began using such a reaction to detect solar neutrinos.

Davis filled a 100,000-gallon tank with the cleaning fluid perchloroethylene (C_2Cl_4). Theory predicts that about once a day, a solar neutrino will convert a chlorine atom in the tank into radioactive argon, which can be detected later by its radioactive decay. To protect the detector from cosmic rays from space, the tank was buried nearly a mile deep in a South Dakota gold mine (Figure 8-25). Of course, the mile of rock overhead has no effect on the neutrinos.

The result of the Davis experiment startled astronomers. The cleaning fluid detects too few neutrinos—not one neutrino per day as predicted by models of the sun, but about one every three days. The experiment has been refined, tested, and calibrated for three decades; but it has not found the missing neutrinos. In 1988, the Japanese detector, Kamiokande II, buried deep in a lead mine, confirmed the deficiency in solar neutrinos. The detectors count roughly one-third of the expected number of solar neutrinos.

The missing neutrinos left astronomers wondering if they misunderstood the nuclear reactions that power the sun. Over the years since the Davis experiment first

Scientists like to claim that every scientific belief is based on evidence, that every theory has been tested, and that the moment a theory fails a test, it is discarded or revised. The truth is much more complicated than that, and the solar neutrino problem is a good illustration. If the detection of solar neutrinos contradicts the theory of stellar structure, why hasn't the theory been abandoned?

While scientists do indeed have tremendous respect for evidence, they also have a faith in theories that have been tested successfully many times. If a theory has been tested and confirmed over and over, they may even begin to call it a natural law, and that means scientists have great faith in its truth. That is, they have confidence that the theory or law is a good description of how nature works.

Nevertheless, it is not unusual for an experiment or an observation to contradict well-established theories. In many cases, the experiments and observations are simple mistakes, or they have not been interpreted correctly. Scientists resist abandoning a well-tested theory even when an observation continues to contradict it. If confidence in the theory is stronger than the evidence, scientists begin by testing the evidence. Can it be right? Do we understand it correctly? Of course, if the evidence cannot be impeached and it continues to contradict the theory, scientists must eventually abandon or modify the theory no matter how many times it has previously been tested and confirmed. This ultimate reliance on evidence is the distinguishing characteristic of science.

It is only human nature to hang on to the principles you have come to trust, and that confidence in well-tested scientific principles helps scientists avoid rushing to faulty judgments. For example, claims for perpetual motion machines occasionally crop up in the news, but the world's scientists don't instantly abandon the laws of energy and motion pending an analysis of the latest claim. Of course, if such a claim did prove true, the entire structure of scientific knowledge would come crashing down, but because the known laws of energy and motion have been well tested and no perpetual motion machine has ever been successful, scientists know which way to bet. Like the keel on a ship, confidence in well-tested theories and laws keeps the scientific boat from rocking before every little breeze.

Some forms of faith must be absolute and unshakable—religious faith, for example. But scientific faith may be better described as scientific confidence because it must be open to change. If, ultimately, a single experiment or observation conclusively contradicts our most cherished law of nature, we as scientists must abandon that law and find a new way to understand nature. We can see this scientific confidence at work in many controversies, from the origin of the human race, to the meaning of IQ measurements; but one of the best examples is the solar neutrino problem. While we struggle to understand the origin of solar neutrinos, we continue to have confidence that we do indeed understand how stars work.

revealed a deficiency in solar neutrinos, other neutrino detectors have been built, and they confirm the basic observation. We count fewer solar neutrinos than theory predicts.

The solar neutrino problem has been one of the great scientific mysteries of the late twentieth century (Window on Science 8-2), and two kinds of solutions were proposed. Some scientists wondered if there are errors in the models of the sun, while others suggested that we don't yet understand neutrinos well enough to count

Figure 8-25

The Davis solar neutrino experiment counts neutrinos from the sun using 100,000 gallons of cleaning fluid held in a tank nearly a mile underground. A miniscule fraction of the solar neutrinos that pass through the cleaning fluid convert chlorine atoms into argon atoms that can be counted because they are radioactive. *(Brookhaven National Laboratory)*

them with certainty. After decades of research, the second suggestion seems more likely.

Physicists know that neutrinos come in three types, called "flavors," named electron, muon, and tau neutrinos. The Davis experiment can detect only electron neutrinos, so a group of scientists suggested a way that neutrinos might change back and forth—oscillate—among the three flavors while the neutrinos traveled through matter. If the electron neutrinos produced inside the sun could oscillate among the three flavors as they sped out of the sun, we would only detect a third as many as we expected. The other two-thirds of the electron neutrinos would have become tau or muon neutrinos, which the Davis experiment could not count.

To catch neutrinos in the act of oscillating, scientists used the Super Kamiokande neutrino detector, which contains 12.5 million gallons of hyper-pure water and is buried under a mountain in Japan. High-energy particles striking Earth's atmosphere from space should break up atoms and shower Earth with twice as many muon neutrinos as electron neutrinos. From under the mountain, the detector found the predicted number of muon and electron neutrinos coming from above, but equal numbers of muon and electron neutrinos coming up through Earth. This suggests that the neutrinos oscillated as they traveled through Earth's dense matter, just as the theory suggests solar neutrinos oscillate as they travel through the dense matter in the sun. Other experiments are beginning to produce data that will help physicists test the results from Super Kamiokande and better understand the behavior of neutrinos.

Neutrino oscillation would solve the solar neutrino mystery, but it is also exciting because it requires that neutrinos have mass. Neutrinos are so numerous in the universe that even a very small mass for each one could exert enough gravitational attraction to affect the evolution of the universe as a whole. We will discuss this problem in Chapter 19.

If neutrino oscillation is confirmed by further experiments, the solar neutrino problem may be solved, and that will give astronomers even added confidence that they understand how the sun makes its energy.

REVIEW Critical Inquiry

Why does nuclear fusion require that the gas be very hot?

Only under certain conditions do the nuclei of atoms fuse together to form a new nucleus. Inside a star, the gas is ionized, which means the electrons have been stripped off the atoms, and the nuclei are bare and have a positive charge. For hydrogen fusion, the nuclei are single protons. These atomic nuclei repel each other because of their positive charges, so they must collide with each other violently to overcome that repulsion and get close enough together to fuse. If the atoms in a gas are moving rapidly, we say it has a high temperature, and so nuclear fusion requires that the gas have a very high temperature. If the gas is cooler than about 10 million K, hydrogen can't fuse because the protons don't collide violently enough to overcome the repulsion of their positive charges.

It is easy to see why nuclear fusion in the sun requires high temperature, but why does it require high density?

The sun is beautiful and complex, with great eruptions, prominences, active regions, and dark spots sweeping across its surface like magnetic weather. All of this activity is driven by the outward flow of energy generated by hydrogen fusion in the core. Presumably other stars are like the sun, but other stars are so far away we can't see their surfaces. Through the largest telescopes, they look like nothing more than points of light. How can we learn about the stars? We begin to answer that question in the next chapter.

Summary

The atmosphere of the sun consists of three layers: photosphere, chromosphere, and corona. The photosphere, or visible surface, is a thin layer of low-density gas from which visible photons most easily escape. It is marked by granulation, a pattern produced by circulation below the photosphere.

The chromosphere is most easily visible during total solar eclipses when it flashes into view for a few seconds and produces the flash spectrum. This spectrum shows that the chromosphere is hot, ionized gas. Filtergrams taken in the light emitted by specific atoms show that the chromosphere is filled with large jets called spicules.

The corona is the sun's outermost atmospheric layer. It is composed of very-low-density, very hot gas extending many solar radii from the sun. Although it can be studied with a special telescope called a coronagraph, its outer layers are visible only during total solar eclipses. Its high temperature—millions of Kelvins—is believed to be maintained by interaction with the solar magnetic field. Parts of the corona give rise to the solar wind, a breeze of low-density ionized gas streaming away from the sun. Thus, Earth, which is bathed in the solar wind, is orbiting within the sun's outer atmosphere.

Sunspots are the most prominent example of solar activity. A sunspot appears to have a dark center, called the umbra, and a slightly lighter border, called the penumbra. A sunspot seems dark because it is slightly cooler than the rest of the photosphere. The average sunspot is about twice the size of Earth and contains magnetic fields about 1000 times stronger than the sun's average field. Sunspots are thought to form because the magnetic field inhibits convection. The average number of sunspots visible varies with a period of about 11 years and appears to be related to the solar magnetic cycle.

Alternate sunspot cycles have reversed magnetic polarity, and this has been explained by the Babcock model of the magnetic cycle. In this model, the differential rotation of the sun winds up the magnetic field. Tangles in the field rise to the surface and cause active regions containing sunspots. When the field becomes strongly tangled, it reorders itself into a simpler but reversed field, and the cycle starts over.

Prominences and flares are other examples of solar activity. Prominences occur in the chromosphere; their arched shape shows that they are formed of ionized gas trapped in the magnetic field. Flares, too, seem to be related to the magnetic field. They are sudden eruptions of X-ray, ultraviolet, and visible radiation and high-energy particles that occur among the twisted magnetic fields around sunspot groups.

Activity in the corona is also guided by the magnetic field. The corona seems to be composed of streamers of thin, hot gas escaping from the magnetic field. In some regions of the corona, the magnetic field does not loop back to the sun, and the gas escapes unimpeded. These regions are called coronal holes and are believed to be the source of the solar wind.

The sun generates its energy near its center where the temperature and density are high enough for nuclear fusion reactions that combine hydrogen to make helium. Observations of too few neutrinos coming from the sun's core have been difficult to understand, but neutrino oscillation may explain the missing neutrinos.

New Terms

granulation	Babcock model
convection	prominence
supergranule	flare
limb	shock wave
limb darkening	reconnection
transition region	aurora
filtergram	coronal hole
spicule	solar constant
coronagraph	Maunder Minimum
solar wind	weak force
helioseismology	strong force
sunspot	nuclear fission
active region	nuclear fusion
Zeeman effect	Coulomb barrier
Maunder butterfly diagram	proton–proton chain
differential rotation	neutrino
dynamo effect	

Review Questions

1. Why can't we see deeper than the photosphere?
2. What evidence do we have that granulation is caused by convection?
3. How are granules and supergranules related? How do they differ?
4. How can a filtergram reveal structure in the chromosphere?
5. What evidence do we have that the corona has a very high temperature?
6. What heats the chromosphere and corona to high temperature?
7. How are astronomers able to explore the layers of the sun below the photosphere?
8. What evidence do we have that sunspots are magnetic?
9. How does the Babcock model explain the sunspot cycle?
10. What does the spectrum of a prominence tell us? What does its shape tell us?
11. How can solar flares affect Earth?
12. Why does nuclear fusion require high temperatures?
13. Why does nuclear fusion in the sun occur only near the center?
14. How can astronomers detect neutrinos from the sun?
15. How can neutrino oscillation explain the solar neutrino problem?

Discussion Questions

1. What energy sources on Earth cannot be thought of as stored sunlight?
2. What would the spectrum of an auroral display look like? Why?
3. What observations would you make if you were ordered to set up a system that could warn astronauts in orbit of dangerous solar flares? Such a warning system exists.

Problems

1. The radius of the sun is 0.7 million km. What percentage of the radius is taken up by the chromosphere?
2. The smallest detail visible with ground-based solar telescopes is about 1 second of arc. How large a region does this represent on the sun? (*Hint:* Use the small-angle formula.)
3. What is the angular diameter of a star like the sun located 5 ly from Earth? Is the Hubble Space telescope able to detect detail on the surface of such a star?
4. If a sunspot has a temperature of 4240 K and the solar surface has a temperature of 5800 K, how many times brighter is the surface compared to the sunspot? (*Hint:* Use the Stefan-Boltzmann law, Chapter 7.)
5. A solar flare can release 10^{25} J. How many megatons of TNT would be equivalent? (*Hint:* A 1-megaton bomb produces about 4×10^{15} J.)
6. The United States consumes about 2.5×10^{19} J of energy in all forms in a year. How many years could we run the

United States on the energy released by the solar flare in Problem 5?

7. Neglecting energy absorbed or reflected by our atmosphere, the solar energy hitting 1 square meter of Earth's surface is 1360 J/s (the solar constant). How long does it take a baseball diamond (90 ft on a side) to receive 1 megaton of solar energy? (*Hint:* See Problem 5.)

8. How much energy is produced when the sun converts 1 kg of mass into energy?

9. How much energy is produced when the sun converts 1 kg of hydrogen into helium? (*Hint:* How does this problem differ from Problem 8?)

10. A 1-megaton nuclear weapon produces about 4×10^{15} J of energy. How much mass must vanish when a 5-megaton weapon explodes?

Critical Inquiries for the Web

1. Do disturbances in one layer of the solar atmosphere produce effects in other layers? We have seen that filtergrams are useful in identifying the layers of the solar atmosphere and the structures within them. Visit a Web site that provides daily solar images. Then, choose today's date (or one near it) and examine the sun in several wavelengths to explore the relationships among disturbances in various layers.

2. How can images like those discussed in the previous question be used to estimate the rotation period of the sun? Can these data be used to demonstrate that the sun rotates differentially? Examine images at a particular wavelength over a period of two weeks. From these observations, estimate the period of rotation of the sun.

3. Explore the Web to find photos and observations of aurora. How is auroral activity affected by solar activity?

4. What can you find on the Web about Earth-based efforts to generate energy through nuclear fusion? How do nuclear fusion power experiments attempt to trigger and control nuclear fusion? So-called "cold fusion" has been largely abandoned as a false trail. How did it resemble nuclear fusion?

Go to the Brooks/Cole Astronomy Resource Center (www.brookscole.com/astronomy) for critical thinking exercises, articles, and additional readings from InfoTrac College Edition, Brooks/Cole's online student library.

Perspective: Origins

Guidepost

This book is designed to concentrate on the origin and evolution of our solar system, and thus it does not include chapters on stars, galaxies, and the universe as a whole. Yet it is difficult to understand our solar system without understanding the nature of astronomical origins. Consequently, this short essay puts our solar system into cosmic perspective

We have three origins to discuss: the origin of stars, the origin of the chemical elements, and the origin of the universe. Those three topics tell the story of our own origin.

Figure P-1
The Horsehead Nebula stretches 1.8 ly from nose to ear and reminds us of the cycle of stellar birth and death. The dusty gas of the nebula is part of a larger cloud in which new stars are forming. Some of these newly born stars are hot and luminous, and they illuminate the nebula beyond the dark swirl that makes up the Horsehead. *(AURA/NOAO/NSF and Nigel Sharp)*

Everything has to come from somewhere. Every object has a history leading back into the past, not only to the creation of that object but to the formation of the atoms of which it is made. The atoms in our bodies are just a few billion years old, and the particles those atoms are made of had their birthday within minutes of the beginning of the universe over 10 billion years ago. Our universe is filled with history.

As we study our solar system, we must remember the history of the matter we are made of—the story of the birth and death of stars (Figure P-1). The carbon, oxygen, calcium, iron, and other atoms in our bodies

exist because stars have lived and died. We are no more isolated from the rest of the universe than a raindrop is isolated from the sea.

The Birth and Death of Stars

The stars around us seem permanent fixtures of the sky, but we know that stars are born and stars die. Their lives are long compared to a human life, but we can see

Figure P-2
The Great Nebula in Orion is visible to the unaided eye as a hazy patch in Orion's sword. It is, in fact, a glowing cloud of excited gas about 25 ly in diameter, inside which new stars are forming. *(Daniel Good)*

all of the stages of stellar birth, aging, and death around us, and thus we can tell the story of the stars, a story that begins in the darkness of interstellar space.

Although space seems empty, it is actually filled with thinly spread gas and dust, the interstellar medium. The gas atoms are mostly hydrogen, typically about a centimeter apart, and the dust is made of microscopic grains comprising only a few percent of the matter between the stars.

The interstellar medium is tenuous in the extreme, yet we have clear evidence that it exists. In some places, the interstellar medium is collected into great clouds of dusty gas that obscure the stars beyond. In other cases, a nearby hot star can ionize the gas and create a glowing nebula, such as the Orion Nebula, visible to the unaided eye in Orion (Figure P-2).

EGGs

EGGs

Figure P-3
Star clusters form deep inside dusty nebulae. (a) Hot stars out of the picture to the upper right are evaporating the Eagle Nebula and exposing dense cores of gas and dust called EGGs, objects that are the forerunners of stars. At least one seems to contain a newborn star (arrow). *(Jeff Hester and Paul Scowen, Arizona State University and NASA)* (b) Star cluster NGC2264, with an age of only a few million years, is one of the youngest star clusters known. Its stars still lie in the nebula from which they formed. *(Anglo-Australian Telescope Board)*

a

b

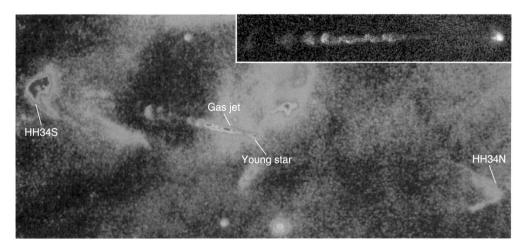

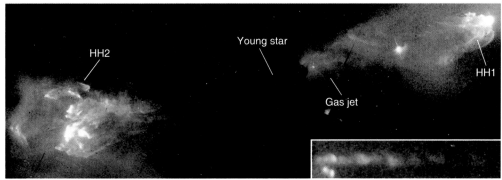

Figure P-4

In these two examples, a young star at the center emits jets of excited gas in opposite directions. Although the jets are visible themselves, they become dramatically evident where they have an impact on the surrounding interstellar gas and excite nebulae known as Herbig-Haro objects. Notice in the insets how detailed images of the jets show irregularities thought to be related to magnetic fields. *(HH34, Reinhard Mundt, Calar Alto 3.5-m telescope; HH1/HH2, J. Hester, Arizona State University, NASA)*

These nebulae not only adorn the sky, but they mark the birthplace of stars. Our sun was probably born in such a nebula about 5 billion years ago. If a cloud becomes dense enough, gravity can make the cloud collapse, and fragments of the cloud contract to form stars. It seems peculiar that something as cold as an interstellar gas cloud can give birth to hot stars, but gravity supplies the energy to heat the gas. As the atoms fall inward, they pick up speed, and, by the time they arrive near the center of a contracting clump of gas, they are traveling at high velocity. Temperature is a measure of the average velocity of the atoms in a material, so the contracting gas becomes hot, and eventually a star is born deep inside the gas cloud.

We don't see these protostars at first because they are hidden deep inside the dusty gas cloud, but they are easily visible in the infrared. As the more massive protostars contract to become hot, luminous stars, the light and gas flowing away from the newborn star blows the nebula away to reveal the new stars (Figure P-3). Thus, a single gas cloud can give birth to a cluster of stars.

At this stage in the formation of a star, something happens that is quite important to us planet dwellers—planets form. The evidence is very strong that a contracting gas cloud gives birth to a protostar at its center and also forms a spinning disk of gas and dust around the protostar. Such disks feed more gas into the growing protostar, and, by a process that isn't well understood, they eject powerful jets of gas along the axis of rotation. Astronomers are confident that these disks form because, for one thing, the jets of gas are easily visible at many wavelengths (Figure P-4). We will see, in Chapter 20 of this book, further evidence that many young stars form surrounded by these spinning disks. We will also see that planets form in these disks.

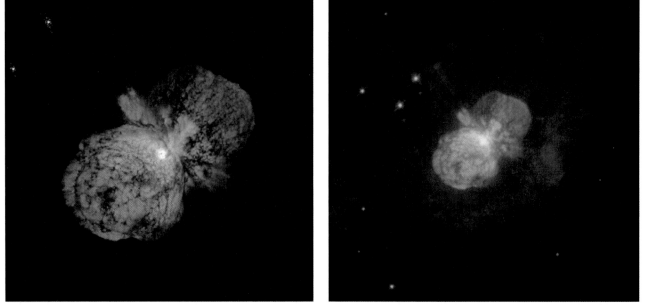

Figure P-5
The bright star Eta Carinae is a very-high-mass star system that is rapidly losing mass. It erupted in 1843, ejecting two lobes of gas. A more recent eruption ejected an expanding disk that sits like a plate pressed between two basketballs. *(Jon Morse/University of Colorado, NASA)* A deeper image (right) shows jets from recent eruptions and more distant clouds ejected in an earlier eruption. *(J. Hester/Arizona State University, NASA)*

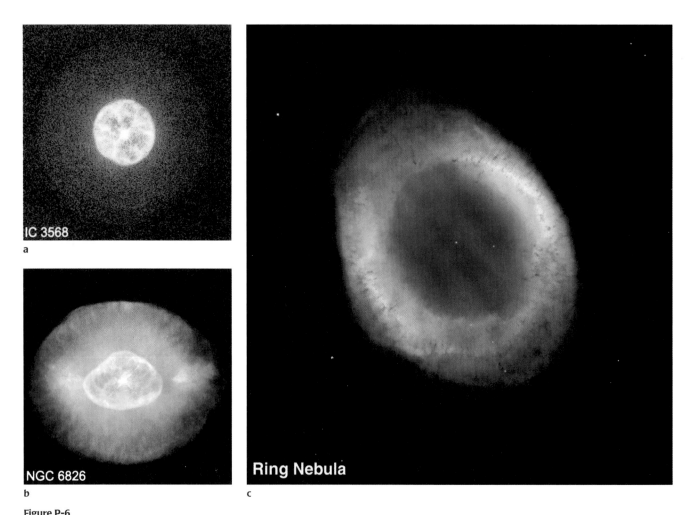

Figure P-6
Planetary nebulae are shells of gas expelled from dying stars. (a) IC3568 is highly symmetric and appears to be a spherical shell. (b) The red regions of NGC6826 are produced by some asymmetry in the ejected gas. (c) About a light-year in diameter, the Ring Nebula is bright enough to be visible in small telescopes. *(a and b, H. Bond, B. Ballick, and NASA; c, Hubble Heritage Team, AURA/STScI/NASA)*

The fundamental work of science is testing theories by comparing them with facts. As we think about science, we need to distinguish clearly between facts and theories. The facts are the evidence against which we test the theories.

Scientific facts are those observations or experimental results of which we are confident. An astronomer makes observations of stars and a botanist collects samples of related insects. A fact could be a precise measurement, such as the mass of a star expressed as a specific number, or it could be a simple observation, such as that a certain butterfly no longer visits a certain mountain valley. In each case, the scientist is gathering facts.

A theory, however, is a conjecture as to how nature works. If we are uncertain of the theory, we might call it a hypothesis. In any case, these conjectures are not facts; they are attempts to explain how nature works. In a sense, a theory or hypothesis is a story that scientists have made up to explain how nature works in some specific case. These stories can be wonderfully detailed and ingenious, but without evidence they are nothing more than hunches.

When one of these stories is tested against the facts and confirmed, we have more confidence that the story is more or less right. The more a story is tested successfully, the more confidence we have in it. The facts represent reality, and every theory or hypothesis must be repeatedly tested against reality.

We can't test one theory against another theory. Theories are not evidence; they are conjectures. If we were allowed to test one theory against another theory, we might fall into the trap of circular reasoning. "Elves make the flowers bloom. I know that elves exist because the flowers bloom." That is using two theories to confirm each other, and it leads us to nonsense. Only facts can be evidence.

Nor can we use a theory to deduce facts. We can use a theory to make predictions, which we can then test against facts, but the predictions themselves can never be certainties, so they can't be facts. The only way to arrive at facts is to consult nature and make direct measurements or observations.

As we study different problems in astronomy, we must carefully distinguish between facts and theories. Facts are the basic building blocks of all science, and they can come only from the careful study of nature. As Galileo would say, we must read the book of nature.

We might identify the birth of a star as the moment when it begins fusing hydrogen into helium. That nuclear fusion reaction releases energy, supports the star, and stops the contraction. Stars spend most of their lives fusing hydrogen, and, when the hydrogen is exhausted, they fuse helium into carbon. The more massive stars can fuse carbon into even heavier atoms, but iron is the limit beyond which no star can go.

The smallest stars are very common, but they are so faint they are hard to see. These small, cool stars, known as red dwarfs, probably outnumber any other kind of star. To ignite hydrogen fusion, a contracting star must have a mass of about 0.08 solar masses. Nature appears to make even lower mass objects, but these so-called brown dwarfs cannot ignite hydrogen fusion, so they slowly cool as they radiate away their heat. The most massive stars are over 100 times more massive than our sun, and they can easily ignite hydrogen fusion. However, they are so luminous that the outward flow of light makes them unstable, and they lose mass into space rapidly (Figure P-5). Such massive stars are very luminous, but they are also very rare. Nature makes very few of these high-mass stars.

This essay tells the story of the birth and death of stars in a few paragraphs, but modern astronomers know a great deal about the lives of the stars. Solving the mystery of the evolution of stars is one of the greatest accomplishments of modern astronomy. Astronomers have compared observational facts with theories and gradually unraveled the story of the stars (Window on Science P-1). Many details remain to be understood, but our quick summary here gives us background for our study of the origin of our solar system. Oddly enough, the birth of our solar system is linked in more than one way to the deaths of stars.

How a star dies depends on its mass. A lower-mass star like the sun can survive for billions of years, but it eventually exhausts its hydrogen and helium fuel and can't get hot enough to fuse carbon. It dies by puffing off its outer layers to produce an expanding nebula (Figure P-6). Such planetary nebulae were named for their planetlike appearance in a telescope, but they are in fact the remains of dying stars.

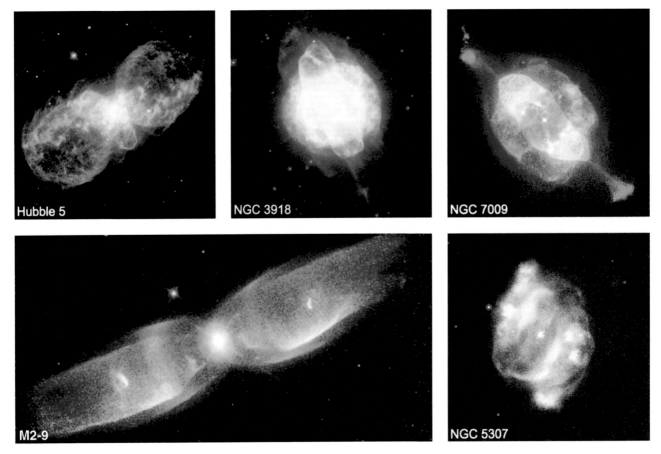

Figure P-7
When the outflowing gas creating a planetary nebula is focused in opposite directions it can produce the complex structures seen here. *(H. Bond, B. Ballick, and NASA)*

Modern telescopes have show that planetary nebulae are not all spherical shells around dying stars. The Hubble Space Telescope has imaged many planetary nebulae and found that they are often quite asymmetric (Figure P-7). In some cases, they appear to have been produced by gas ejected in opposite directions by the central star. Astronomers suspect this may be related to the rotation of the star, its magnetic field, the presence of a companion star, the presence of a disk of gas ejected when the star was a bit younger, or even by the presence of planets orbiting the star. Besides being as beautiful and as varied as the blossoms of wildflowers, planetary nebulae contain clues to the lives of the long-lived stars like the sun.

In contrast with stars like the sun, massive stars live very short lives, perhaps only millions of years, before they develop dead-end iron cores and explode as supernovae (Figure P-8). The core of such a dying massive star may form a neutron star or a black hole, but the outer parts of the star, newly enriched with the atoms made inside the star, are returned to the interstellar medium.

The violence of the supernova explosion can fuse atoms to build elements heavier than iron. Gold, silver, platinum, uranium, and other elements heavier than iron are rare and valuable because they are only made in the moments of a supernova explosion. The iodine in our thyroid glands was made in supernovae.

The atoms of which we are made had their birth inside stars. That process is common in the universe because stars are common. Our galaxy contains billions.

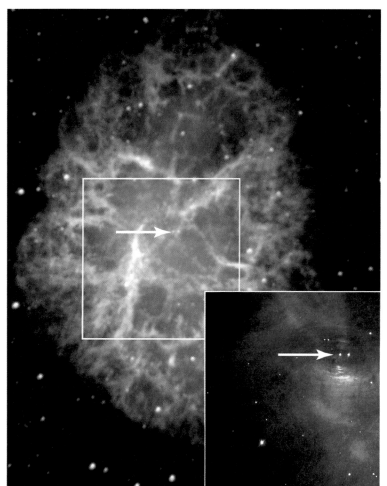

Figure P-8

(a) The Crab Nebula is a supernova remnant produced by the explosion of a star, an event seen by Chinese astronomers in AD 1054. Such expanding shells of gas mix elements manufactured inside stars back into the interstellar medium. The inset shows the boxed region of the nebula where a spinning neutron star, the remains of the star that exploded, energizes the nebula. *(Palomar Obs/Caltech. Inset: Jeff Hester and Paul Scowen, Arizona State University and NASA)* (b) Supernova 1987A (lower right), first seen in February 1987. It marks the explosive death of a massive star. The Tarantula Nebula is at upper left. *(AURA/NOAO/NSF)*

a

b

The Universe of Galaxies

The Milky Way that we see stretching across the sky is actually the disk of the galaxy we live in (Figure P-9). It contains about 100 billion stars, including our sun. We live about two-thirds of the way from the center of our galaxy to the edge.

Our galaxy is our home, but we are surrounded by a sea of galaxies. Some, like our Milky Way Galaxy, are disk-shaped spiral galaxies with graceful spiral arms marked by clouds of gas and bright newborn stars. But many galaxies are great swarms of stars with little remaining gas and dust (Figure P-10). Called elliptical galaxies, these range from dwarfs only a few hundred light-years in diameter to giants 100,000 light-years across.

Besides our Milky Way Galaxy, only one other galaxy is visible to our unaided eyes; the Andromeda Galaxy (see Figure P-9b) is visible as a patch of haze in the constellation Andromeda. (Use the star charts at the back of this book to locate the Andromeda Galaxy.) The Andromeda Galaxy is the most distant object you will ever see with your unaided eye. It is over 2 million light-years distant.

We might expect such titanic objects to be rare, but large telescopes reveal that the sky is filled with galaxies. Like grains of sand, galaxies fill the universe in every direction as far as we can see (Figure P-11). Grouped in clusters and superclusters, galaxies are the homes of the billions of stars that not only illuminate the universe, but that create the elements heavier than helium—elements that are critical to us because we are made of those atoms.

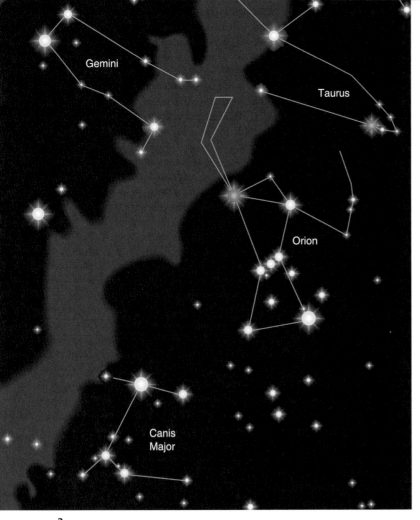

a

b

Figure P-9

(a) From the inside, we see our galaxy as a faintly luminous path that circles the sky. Invisible from the brightly lit skies of cities, the Milky Way is easily visible from a dark site. This artwork shows the location of a portion of the Milky Way near a few bright winter constellations. (b) If we could leave our galaxy and photograph it from a distance, it would look much like the Great Galaxy in Andromeda, a spiral galaxy a bit over 2 million ly from us. *(Bill Scheening and Vanessa Harvey, NSF REU, NOAO. © AURA, Inc.)* (See the star charts at the end of this book to locate the Milky Way throughout the year.)

a

b

Figure P-10

(a) Some galaxies like our own are spiral galaxies, beautiful disks of stars with the youngest stars located along spiral arms. The larger of such galaxies are typically 75,000 ly in diameter and contain 100 billion stars. *(Anglo-Australian Telescope Board)* (b) Many galaxies are great clouds of stars with little gas and dust and no spiral arms. Such elliptical galaxies are common. *(AURA/NOAO/NSF)*

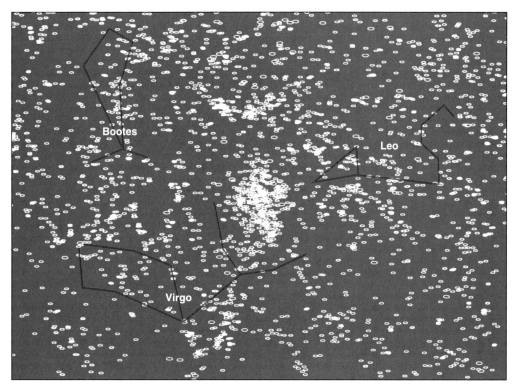

Figure P-11

The sky is filled with galaxies (marked by ovals in this diagram). The galaxies are grouped together in clusters and superclusters. The giant Virgo cluster lies at the center of this figure, and it is linked to other clusters to form the local supercluster in which our galaxy is located. Only the brighter galaxies are plotted in this diagram.

One of the most amazing discoveries of late 20th century astronomy was the recognition that galaxies can collide. When two galaxies collide, the stars swirl past each other without bumping, but the great clouds of gas and the magnetic fields in the galaxies do collide and compress each other. Furthermore, tidal forces twist and distort the colliding galaxies. Images of such colliding galaxies show their twisted shapes and far-flung streamers of stars and gas, but more detailed images reveal an even more impressive consequence of galaxy collisions—star formation (Figure P-12). The compression of the gas clouds can stimulate two colliding galaxies to form vast numbers of stars and massive star clusters. Such bursts of star formation salt the galaxies with newly formed heavy atoms. Our own Milky Way Galaxy has probably collided with more than one smaller galaxy, and the atoms of which we are made may have been formed because of those collisions.

A little less than a century ago, astronomers made an astonishing discovery. All of the galaxies in the sky are rushing away from each other. The spectrum of a galaxy is the combined spectrum of billions of stars, and the spectral lines visible in galaxy spectra are shifted toward the red end of the spectrum. This red shift is proportional to distance. That is, the farther away a galaxy is, the larger is the red shift that is visible in its spectrum (Figure P-13). This result, first discovered by the American astronomer Edwin Hubble in 1929, is clear evidence that the universe is expanding.

If the universe is expanding, we are forced to ask a question so challenging it seems at first like nonsense. Yet modern astronomy is beginning to find an answer. How did the universe begin?

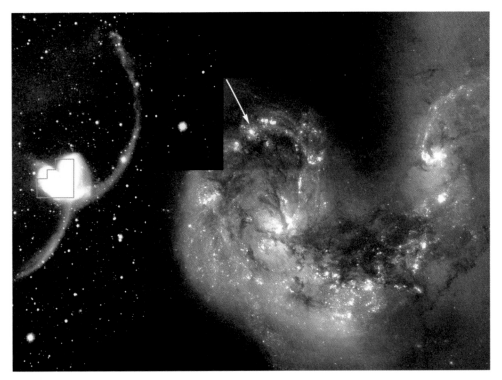

Figure P-12

The colliding galaxies NGC4038 and NGC4039 are known as the Antennae because their long curving tails resemble the antennae of an insect. Earth-based photos (left) show little detail, but a Hubble Space Telescope image (right) reveals the collision of two galaxies producing thick clouds of dust and raging star formation creating roughly a thousand massive star clusters such as the one at the top (arrow). Such collisions between galaxies are common. *(Brad Whitmore, STScI, and NASA)*

Figure P-13

Galaxies at different distances are shown at left, with their spectra shown at the right. The two vertical blue lines mark the unshifted location of the H and K lines of once-ionized calcium. The expansion of the universe causes a red shift that moves the spectral lines to longer wavelengths, as shown by the red arrows. More distant galaxies have larger red shifts, expressed as radial velocities in kilometers per second. *(Caltech)*

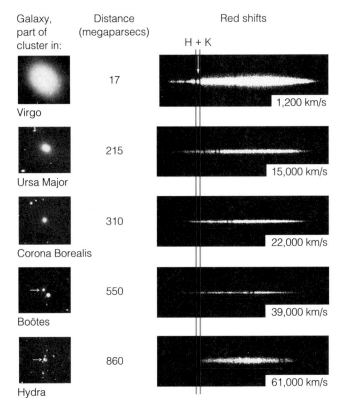

Galaxy, part of cluster in:	Distance (megaparsecs)	Red shifts
Virgo	17	1,200 km/s
Ursa Major	215	15,000 km/s
Corona Borealis	310	22,000 km/s
Boötes	550	39,000 km/s
Hydra	860	61,000 km/s

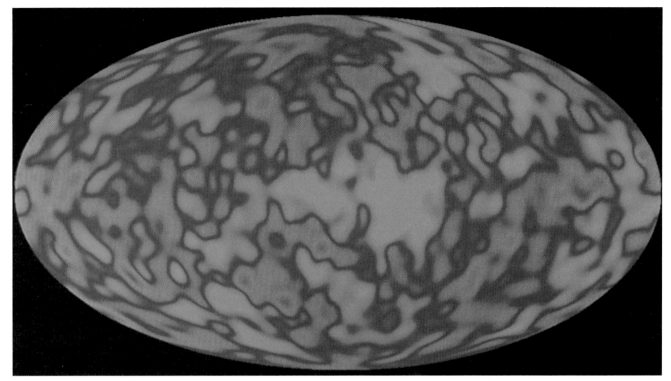

Figure P-14
This all-sky map projects the entire sky onto an oval. Colors show the temperature of the primordial background radiation, the glow from the big bang red-shifted into the far infrared. Red areas are very slightly hotter than blue areas, and these irregularities in the gas of the big bang are thought to have grown into the linked network of superclusters of galaxies that fill the universe. *(NASA)*

The Origin of the Universe

If the galaxies are all rushing away from each other, then we can imagine viewing a videotape running backward. We would see the galaxies rushing toward each other, and eventually we would see the galaxies pushing into each other, compressing and heating the gas until our video screen was filled with the glare of an intensely hot, fantastically dense gas filling the universe. This, the beginning of the universe, is a state called the big bang.

How can we know the big bang really happened? The evidence is dramatic—we can see it. To understand how we can see an event that occurred billions of years ago, we need to look once again at the galaxies around us. Nearby galaxies, such as the Andromeda Galaxy, are only a few million light-years away. The light we see has been traveling through space for only a few million years, and we see those galaxies as they were a few

million years ago. If we look at more distant galaxies, we see them as they were a billion years ago or more. When we look at the most distant galaxies, we look back in time and see them as they were nearly 10 billion years ago.

What do we see if we look at the empty places on the sky between the galaxies? We see a glow that was emitted by the hot, dense clouds of gas of the big bang. We can't see that glow with our eyes because the red shift is so great that the photons of light are shifted into the long-wavelength infrared part of the spectrum. Nevertheless, we can "see" the radiation with the proper detectors (Figure P-14). This primordial background radiation pours in on us from all directions, telling us that we are part of a universe that was very hot and very dense only a bit more than 10 billion years ago. We are part of the big bang.

No scientific theory or hypothesis can be proved correct. We can test a theory over and over by performing experiments or making observations, but we can never prove that the theory is absolutely true. It is always possible that we have misunderstood the theory or the evidence, and the next observation we make might disprove the theory. In that sense, we never learn anything new until we disprove a theory or hypothesis. Only at that point do we discover something we did not know before.

For example, we might propose the theory that the sun is mostly iron. We might test the theory by looking at the iron lines in the solar spectrum, and the strength of the lines would suggest our theory is right. Although our observation has confirmed our theory, we know nothing more than we did before. We might continue to confirm the theory over and over, but we still know nothing new, and there is always the danger that we have misunderstood the evidence. Finally, we might study the formation of atomic spectra and discover that iron atoms are very good absorbers of photons, so the strong iron lines in the solar spectrum do not mean that the sun is made mostly of iron. In fact, the hydrogen lines mean that the sun is mostly hydrogen atoms. Now that we have disproved our hypothesis, we have learned something new.

The nature of scientific hypothesis testing can lead to two kinds of mistakes. Sometimes nonscientists will say, "You scientists just want to tear everything down—you don't believe anything." It is the nature of science to test every hypothesis, not because the scientist wants to tear things down, but because the scientist wants to know what is the most dependable description of nature. Others will say, "You scientists are never sure of anything." Again, the scientist knows that no theory can ever be proved correct. That the sun will rise tomorrow is very likely, but in the end it is still a theory.

People will say of an idea they dislike, "That is only a theory," as if a theory were simply a random guess. In fact, a theory can be a well-tested truth in which all scientists have great confidence. Yet we can never prove that any theory is absolutely true.

Notice that the background radiation is visible in any direction we look. The big bang did not occur in a specific place, but it filled the entire volume of the universe. As the universe expanded, the total amount of volume increased, and the hot gases of the big bang cooled and formed galaxies. The expansion continues, and the galaxies continue to move away from each other. They are not fragments ejected from an explosion but parts of a whole that fills all of the volume of the universe and expands continuously as the total volume increases.

Thus we can't find a center to the expansion, or an edge. The universe may be infinitely large or it may contain only so much volume and no more, but it has no center. The expansion occurs as the total volume of the universe increases, but there is no edge pushing into unoccupied volume.

The big bang theory seems fantastic, but it has been tested and confirmed over and over, and modern astronomers have great confidence that there really was a big bang (Window on Science P-2). Of course, there are more details to understand. How did the first galaxies form? Why did the galaxies form in giant clusters? Will the universe expand forever? How old is the universe? We live in an exciting age when astronomers are able to ask these questions and expect to discover answers.

These ideas strain our imaginations and challenge the most sophisticated mathematics, but the lesson is a simple one. Our universe had a beginning not so long ago, and the matter we are made of had its birthday in that moment of cosmic beginnings. We are small, but we are part of something vast.

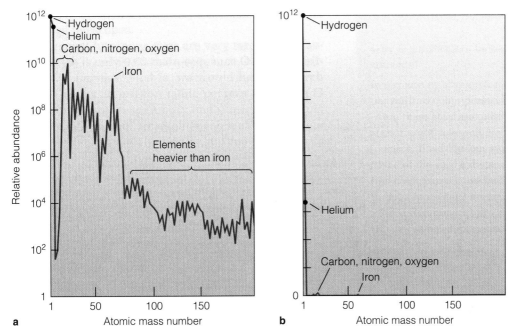

Figure P-15

The abundance of the chemical elements is usually plotted as in the graph at the left (a). Hydrogen is most common, followed by helium. Carbon, nitrogen, oxygen, and iron also seem abundant, but elements heavier than iron are quite rare because they are made only in supernova explosions. Plotting these abundances on a linear scale as in the graph at the right (b) gives us a more realistic impression. Most of the universe is hydrogen and helium. The atoms of which we are made are hardly more than a tiny impurity in the cosmos.

The Story of Matter

Astronomers can tell the story of the matter in the universe with only a few gaps. The matter was created in the big bang, collected into galaxies, and cooked in stars. Finally, some of that matter fell together to form our sun, our planet, and us.

Models of the big bang tell us that the first few minutes in the history of the universe were chaotic. Energy was so dense that it could create particles of matter, and that is when the first protons, neutrons, and electrons formed. Protons and electrons, even when they are not attached to each other, are hydrogen, so the first gas of the big bang was hydrogen. The temperature and density were so high that nuclear fusion reactions could occur, fusing hydrogen into heavier elements.

By the time the universe was 3 minutes old, the expansion had cooled its gases and nuclear fusion had stopped. In those first 3 minutes, about 25 percent of the mass had been converted into helium, but few heavier atoms could be made. Because there are no stable atomic nuclei with masses of 5 or 8 times that of hydrogen, the fusion process could not go past helium. Nearly all of the atoms of which we are made except for hydrogen and helium had to be made later inside stars (Figure P-15).

Figure P-16
An apparently empty spot on the sky only 1/30 the diameter of the full moon contains over 1500 galaxies in this extremely long time exposure known as the Hubble Deep Field. Presumably the entire sky is similarly filled with galaxies. *(R. Williams and the Hubble Deep Field Team, STScI, NASA)*

How the gases of the big bang made galaxies is one of the gaps in our story. New, giant telescopes such as the Hubble Space Telescope can see the faint traces of star clouds billions of light-years away. Thus, we see these objects as they were billions of years in the past. Perhaps these twisted clouds fell together to build the galaxies in the great clusters of galaxies that fill the universe today (Figure P-16).

We and our world are made of the ashes of the big bang cooked inside generations of stars and concentrated by the formation of Earth. Every object has a history that stretches back through the birth and death of stars to the first moment in time. We exist because stars have died.

CHAPTER TWENTY

The Origin of the Solar System

What place is this?

Where are we now?

Carl Sandburg
Grass

Guidepost

The preceding 19 chapters have described the origin, structure, and evolution of the physical universe, but they have neglected one important class of objects—planets. In this chapter, we can look back on what we have learned and find our place in the universe. We live on a planet. What does that mean? Where do we fit?

Each time we have studied a new object, we have asked how it formed and how it evolved to its present state. We have done that with stars and galaxies and the universe, so it is appropriate to begin our discussion of the solar system by considering its origin.

Another reason for discussing the origin of the solar system here is to give ourselves a framework into which we can fit the planets as we discuss them in the chapters that follow. Without a theoretical framework, science is nothing but a jumble of facts. With a good framework in hand, we will be ready to make sense of the solar system through the next six chapters.

Microscopic creatures live in the roots of your eyelashes. Don't worry. Everyone has them, and they are harmless.* They hatch, fight for survival, mate, lay eggs, and die in the tiny spaces around the roots of our eyelashes without doing us any harm. Some live in renowned places—the eyelashes of a glamorous movie star—but the tiny beasts are not self-aware; they never stop to say, "Where are we?" Humans are more intelligent; we have the ability and the responsibility to wonder where we are in the universe and how we came here.

In this chapter, we begin exploring our solar system: the sun and its family of planets. We must avoid, however, the great pitfall of science. We must avoid memorizing facts instead of understanding nature. Jupiter is 317.83 times more massive than Earth. That is a fact, and it is boring. The excitement of science comes from understanding nature. We want to understand how the sun and planets formed, and that will tell us *why* Jupiter is so massive.

Facts fit together to build understanding, so we begin our study of the solar system by developing a unifying hypothesis for the origin of the sun and planets. As we learn facts about the solar system we can use them to test and improve our hypothesis, and that will increase our understanding of nature. A unifying hypothesis is like a basket into which we can put facts. The facts are easier to remember because they make sense, and the combination of hypothesis and supporting facts represents real knowledge—real understanding—and not just memory.

We have no choice; we must understand our solar system out of self-preservation. We live on a planet with 6 billion other humans, and we are altering our world's environment in unplanned and unknown ways. We could make Earth uninhabitable. The deadly deserts of Mars and sulfuric acid fogs of Venus may help us understand our own more comfortable world.

We must also study planets as astronomers. It seems likely that roughly half of all stars have planetary systems, so there are more planets in the universe than stars. Also, the chemical reactions that give rise to life seem to need the moderate conditions of the surface of planets. To search for life in the universe, we must understand planets.

Above all, we study the solar system because it is our home in the universe. Because we are an intelligent species, we have the right and the responsibility to wonder what we are. Our kind have inhabited this solar system for at least a million years, but only within our lifetime have we begun to understand what a solar system is. Like sleeping passengers on a train, we waken, look out at the passing scenery, and mutter: "What place is this? Where are we now?"

Demodex folliculorum has been found in 97 percent of individuals with healthy skin.

20-1 The Great Chain of Origins

We are linked through a great chain of beginnings that leads backward through time to the first instant when the universe began 15 billion years ago. The gradual discovery of the links in that chain is one of the most exciting adventures of the human intellect. In earlier chapters, we discussed some of that story: the origin of the universe in the big bang, the formation of galaxies, the origin of stars, and the growth of the chemical elements. Here we will explore further to consider the formation of planets.

Early Hypotheses

The earliest theories for Earth's origin are myths and folktales that go back beyond the beginning of recorded history. In addition, almost all religions contain an account of the origin of the world. It was not until about the time of Galileo that philosophers began searching for rational explanations for natural phenomena. While people like Copernicus, Kepler, and Galileo tried to find logical explanations for the motions of Earth and the other planets, other philosophers began thinking about the origin of the planets.

The first rational theory for Earth's origin was proposed by the French philosopher and mathematician René Descartes (1596–1650). Because he lived and wrote before the time of Newton, Descartes did not have the concept of gravitation as the dominant force in the universe. Rather, he believed that force was communicated by contact between bodies and that the entire universe was filled with vortices of whirling particles. In 1644, he proposed that the sun and planets formed when a large vortex contracted and condensed. Thus, his hypothesis explained the general properties of the solar system known at the time.

A century later, in 1745, the French naturalist Georges-Louis de Buffon (1707–1788) proposed the alternative hypothesis that the planets were formed when a passing star collided with or passed close to the sun and pulled matter out of the sun and the star. The matter condensed to form the planets, and they fell into orbit around the sun (Figure 20-1a). This **passing star hypothesis** was popular off and on for two centuries, but it contains serious flaws. First, stars are very small compared to the distances between them, and thus they collide very infrequently. In the entire history of our galaxy, stars have probably collided only once. More important, the gas pulled from the sun and the star would be much too hot to condense to make planets. Furthermore, even if planets formed, they would not go into stable orbits.

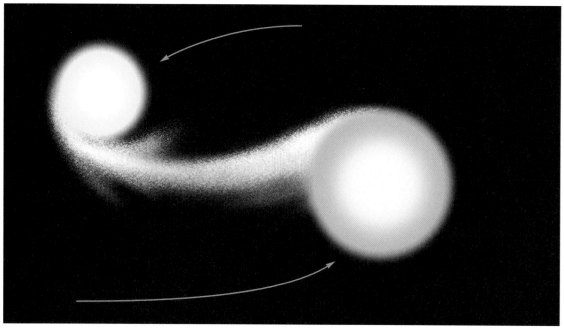

a

b

Figure 20-1

(a) The passing star hypothesis was catastrophic. It proposed that the sun was hit by or had a very close encounter with a passing star and that matter torn from the sun and the star formed planets orbiting the sun. The theory is no longer accepted. (b) Laplace's nebular hypothesis was evolutionary. It suggested that a contracting disk of matter conserved angular momentum, spun faster, and shed rings of matter that formed planets. The hypothesis could not explain the sun's low angular momentum and was never fully accepted, even though the modern theory shares some of its features.

The hypotheses of Descartes and Buffon fall into two broad categories. Descartes' hypothesis is **evolutionary** in that it calls upon common, gradual events to produce the sun and planets. If it is correct, stars with planets are very common. Buffon's hypothesis, on the other hand, is **catastrophic.** It calls upon unlikely, sudden events to produce the solar system, and thus it implies that solar systems are very rare. While our imagi-

nations may be tempted by colliding stars, modern hypotheses for the origins of the planets are evolutionary, with, as we will see, a few astonishing catastrophes thrown in (Window on Science 20-1).

Our modern theory of the origin of the solar system had its true beginning with Pierre-Simon de Laplace (1749–1827), the brilliant French astronomer and math-

Many theories in science can be classified as either evolutionary, in that they involve gradual processes, or catastrophic, in that they depend on specific, unlikely events. Scientists have generally preferred evolutionary theories. Nevertheless, catastrophic events do occur.

Something in us prefers catastrophic theories, perhaps because we like to see spectacular violence from a safe distance, which may explain the success of movies that include lots of car crashes and explosions. Also, cataclysmic theories resonate with Old Testament accounts of cata-

strophic events and special acts of creation. Thus, we have an understandable interest in catastrophic theories.

Nevertheless, most scientific theories are evolutionary. Such theories do not depend on unlikely events or special acts of biblical creation, and thus they are open to scientific investigation. For example, geologists much prefer theories of mountain building that are evolutionary, with the mountains being pushed up slowly as centuries pass. All the evidence of erosion and the folding of rock layers show that the process is gradual. Because most such

natural processes are evolutionary, scientists sometimes find it difficult to accept any theory that depends on catastrophic events.

We will see in this and later chapters that catastrophes do occur. Earth, for example, is bombarded by debris from space, and some of those impacts are very large. As you study astronomy or any other natural science, notice that most theories are evolutionary but that we must allow for the possibility of unpredictable catastrophic events.

ematician. In 1796, he combined Descartes' vortex with Newton's gravity to produce a model of a rotating cloud of matter contracting under its own gravitation and flattening into a disk—the **nebular hypothesis.** As the disk grew smaller, it had to conserve angular momentum and spin faster and faster. Laplace reasoned that, when it was spinning as fast as it could, the disk would shed its outer edge to leave behind a ring of matter. Then the disk could contract further, speed up again, and leave another ring. Thus, the contracting disk would leave behind a series of rings, each to become a planet circling the newborn sun at the center of the disk (Figure 20-1b).

According to the nebular hypothesis, the sun should be spinning very rapidly, or, to put it another way, the sun should have most of the angular momentum of the solar system. (Recall from Chapter 5 that angular momentum is the tendency of a rotating object to continue rotating.) As astronomers studied the planets and the sun, however, they found that the sun rotated slowly and that the planets moving in their orbits had most of the angular momentum in the solar system. In fact, the rotation of the sun contains only about 0.3 percent of the angular momentum of the solar system. Because the nebular hypothesis could not explain this **angular momentum problem,** it was never fully successful, and astronomers toyed with various versions of the passing star hypothesis for over a century.

In contrast to the astronomy of earlier centuries, 20th-century astronomy applied modern physics to the stars and galaxies and so illuminated our origins. We can now trace that great chain of origins that began with the big bang and led to the matter of which we are made. We will review the story of the origin of matter, and then we will add to it the story of how that matter formed our world.

A Review of the Origin of Matter

Look at your thumb. The matter in your thumb came into existence within minutes of the beginning of the universe. Astronomers have strong evidence that the universe began in an event called the big bang (Chapter 19), and by the time the universe was three minutes old, the protons, neutrons, and electrons in your thumb had come into existence. You are made of very old matter.

Although those particles formed quickly, they were not linked together as they are today. Most of the matter was hydrogen, and about 25 percent was helium. Very few heavier atoms were made in the big bang. Although helium is very rare in our bodies, your thumb contains many of those ancient hydrogen atoms unchanged since the universe began.

Within a billion years or so after the big bang, matter began to collect to form galaxies containing billions of stars. Astronomers understand that nuclear reactions inside stars combine low-mass atoms such as hydrogen to make heavier atoms (Chapter 12). Generation after generation of stars cooked the original particles, linking them together to build atoms such as carbon, nitrogen, and oxygen (Chapter 14). Look at your thumb. Even the calcium in the bone and the iron in the blood was assembled inside stars.

The atoms heavier than iron were created by rapid nuclear reactions that can only occur when a massive star explodes (Chapter 14). Gold and silver are rare in our bodies, but iodine is critical in your thyroid gland, and there are no doubt a few of those iodine atoms circulating through your thumb at this very moment thanks to the violent deaths of massive stars.

Our galaxy contains about 100 billion stars of which our sun is one. It formed from a cloud of gas and dust about 5 billion years ago, and the atoms in your thumb were part of that cloud. How the sun took shape, how the cloud gave birth to the planets, how those atoms in your thumb found their way onto Earth and into you is the story of this chapter. As we expore the origin of our solar system, keep in mind the great chain of origins that created the atoms. As the geologist Preston Cloud remarked, "Stars have died that we might live."

The Solar Nebula Hypothesis

The stars gave birth to the heavy atoms of which our world is made, and the modern theory of the origin of the planets is based on star formation. The **solar nebula hypothesis** proposes that the planets were formed from the disk of gas and dust that surrounded the sun as it formed (Figure 20-2).

As stars form in contracting clouds, they remain surrounded by cocoons of dust and gas, and the rotation of the cloud causes that dust and gas to form a spinning disk around the protostar. When the center of the star grows hot enough to ignite nuclear reactions, its surface quickly heats up, becomes more luminous, and blows away the gas and dust cocoon.

Modern theories suppose that planets form in the rotating disks of gas and dust around young stars. In-

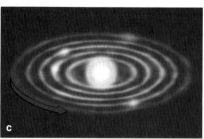

Figure 20-2
(a) Because the solar nebula was rotating, (b) it contracted into a disk, and (c) the planets formed with orbits lying in nearly the same plane.

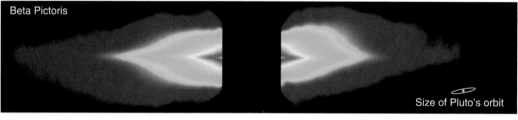
Beta Pictoris

Size of Pluto's orbit

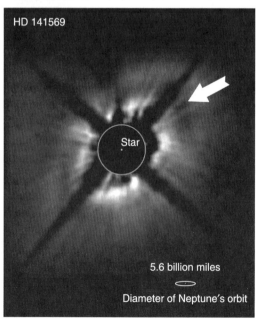
HD 141569

Star

5.6 billion miles

Diameter of Neptune's orbit

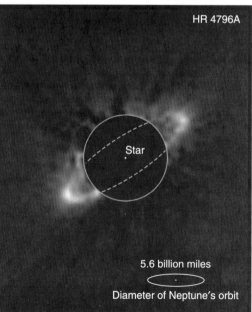
HR 4796A

Star

5.6 billion miles

Diameter of Neptune's orbit

Figure 20-3
The star Beta Pictoris (top) is hidden behind the black bar at center, and the dust disk around it is seen edge-on extending to right and left. *(Al Schultz, CSC/STScI, Sally Heap, GSFC, and NASA)* In the lower two images the stars are hidden behind the round masks with dust disks extending outward. The gap in the ring around HD141569 (arrow) may have been cleared by a planet. The narrowness of the ring around HR4796A suggests it is confined by planets. *(Alycia Weinberger, Eric Becklin, UCLA, Glenn Schneider, University of Arizona, and NASA)*

frared observations of T Tauri stars, for instance, show that some are surrounded by gas clouds rich in dust, and spectra show that these stars are blowing away their nebulae at speeds up to 200 km/s. Astronomers find many very young stars ejecting jets in opposite directions, bipolar flows (Chapter 12), evidence that the outflow was focused by a disk around the star. Our own planetary system probably formed in such a disk-shaped cloud around the sun. When the sun became luminous enough, the remaining gas and dust were blown away into space, leaving the planets orbiting the sun. Notice that the solar nebula hypothesis also suffers from the angular momentum problem. If we are to adopt the theory, we must understand why the sun now rotates so slowly, a puzzle we will consider later in this chapter.

If the solar nebula hypothesis is correct, then our Earth and the other planets of the solar system formed billions of years ago as the sun condensed from the interstellar medium. If that is true, then planets form as a by-product of star formation, and most stars should have planets. Before we go further, we should ask if there is any way to check this prediction against the evidence. Can we see planets orbiting other stars?

Planets Orbiting Other Stars

If the solar nebula theory is correct, then planets should be common, and we could test the theory by looking for planets orbiting other stars, objects astronomers call **extra-solar planets.**

Unfortunately, imaging an extra-solar planet is a bit like photographing a bug crawling on the bulb in a distant searchlight. Planets are small and shine by reflected light. Even a planet as big as Jupiter, 11 times Earth's diameter, reflects little light. If it orbited the nearest star to the Earth, it would be 4 ly away and would be only 22nd magnitude, a very faint image. Furthermore, it would be located only 4 seconds of arc from its star. Current telescopes can't image extra-solar planets directly.

Nevertheless, astronomers have found evidence of extra-solar planets by looking for the dust that accompanies planets and by looking for the motions that planets cause as they orbit their stars.

Infrared astronomers have found very cold, very low density disks of dust around other stars such as Beta Pictoris (Figure 20-3). The Beta Pictoris disk is about 100 times the diameter of our solar system and has an inner hole around the star which may be a place where planets have formed. These tenuous disks of cold dust are believed to be debris released from comets or produced by collisions among small bodies such as asteroids and comets. Our own solar system contains this kind of cold dust. Thus the cold, low-density dust disks are not evidence of active planet formation but are evidence that planetary systems have already formed.

Both visible and radio wavelength observations detect dense disks of gas orbiting young stars. For example, at least 50 percent of the stars in the Orion nebula are encircled by dense disks of gas and dust (Figure 20-4).

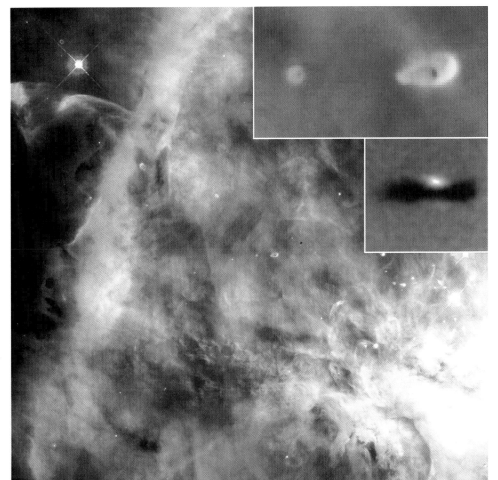

Figure 20-4
This Hubble Space Telescope image of the central part of the Orion Nebula reveals that many young stars there are surrounded by disks of gas and dust (insets). Most of the disks glow brightly because they are illuminated by the hottest stars in the nebula, but a few, such as the dark disk seen nearly edge-on (lower inset), are silhouetted against the bright nebula. Note that the central star in the silhouetted nebula is visible by light scattered by dust above and below the nebula. This particular disk is the largest detected, with a diameter about 17 times that of our solar system. *(C. R. O'Dell, Rice University, NASA; lower inset, Mark McCaughrean, Max-Planck Institute for Astronomy, C. R. O'Dell, and NASA)*

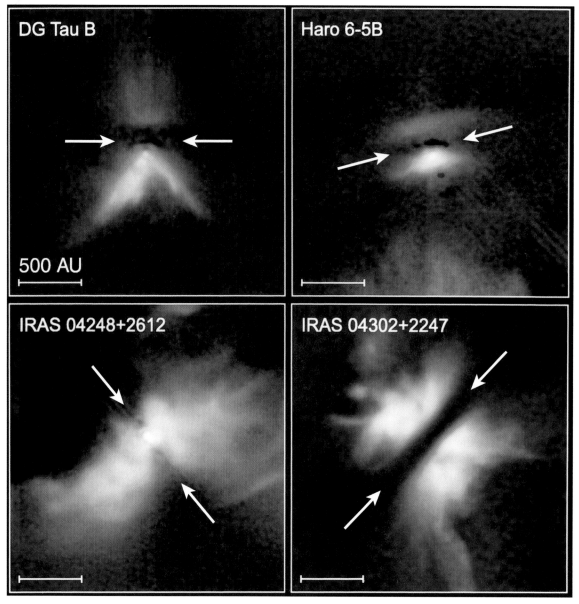

Figure 20-5

Dust disks around young stars are evident in these Hubble Space Telescope infrared images. The stars are so young that material is still falling inward and being illuminated by the light from the star. Dark bands across the nebulae (arrows) are caused by dense disks of gas and dust that orbit the stars. Note that the star at lower left is double. *(D. Padgett, IPAC/Caltech, W. Brandner, IPAC, K. Stapelfeldt, JPL, and NASA)*

A young star is visible at the center of each disk, and astronomers estimate that the disks contain at least a few times Earth's mass in a region a few times larger in diameter than our solar system. The disks are visible because they are brightly illuminated by the hot stars, in the Orion Nebula, and some of the disks are seen silhouetted against the bright nebula. The Orion star-forming region is only a few million years old, so it does not seem likely that planets could have formed in these disks yet. Nevertheless, the presence of so many disks is evidence that the conditions that create planets are common.

Hubble Space Telescope images of very young stars reveal that many are surrounded by glowing clouds of gas and dust bisected by a dark shadow caused by a dense disk of gas and dust circling the star (Figure 20-5). Thus astronomers have further evidence that newborn stars are commonly encircled by dense disks. These disks are related to the formation of bipolar flows in protostars (Figure 12-9). In contrast with the cold, low-density dust disks, these dense disks appear to be places where planets could be forming now.

In the middle 1990s, astronomers obtained further evidence of extra-solar planets. Although such a planet is too faint to be seen directly, it can be detected from

"Scientists are just a bunch of skeptics who don't believe in anything." That is a common complaint about scientists, but it misinterprets the fundamental characteristic of the scientist. Yes, scientists are skeptical about new ideas and discoveries, but scientists do hold strong beliefs about how nature works. Ultimately, scientists try to tell stories about how nature works, and skepticism is the scientists' guide.

Some people think that "telling a story" is telling a fib. The stories that scientists tell are exactly the opposite; perhaps we can call them antifibs, because they are as true as scientists can make them. The stories are explanations that have been tested over and over to make them accurate descriptions of nature. Skepticism is the tool scientists use to test every aspect of an explanation.

When planets were discovered orbiting 51 Pegasi, astronomers were skeptical—not because they thought the observations were wrong, but because that is how science works. Every observation is tested, and every discovery is confirmed. Only an idea that survives many tests begins to be accepted as a scientific truth.

To nonscientists, this makes scientists seem like irritable skeptics, but among scientists it is not bad manners to say, "Really, how do you know that?" or "Why do you think that?" or "Show me the evidence!"

Scientists try to tell stories about how nature works, stories that are sometimes called theories. A story by John Steinbeck can be a brilliant work of art invented to help us understand some aspect of our lives, and another story, such as a TV script, can be little more than chewing gum for

the mind. But scientists' stories are different in a critical way. They are true. To create such stories, scientists cannot create facts to fit their story. Rather, every link in a scientific story must be based on evidence and logic. Of course, evidence can be misunderstood, and logical errors happen. To test every aspect of their story, scientists must be continuously skeptical. That is the only way to discover those scientific truths that help us understand how nature works.

Skepticism is not a refusal to hold beliefs. Scientists often believe sincerely in their theories once those theories have been tested over and over. Rather, skepticism is the tool scientists use to find those natural principles worthy of belief.

---

the motion of the star it orbits. As the planet and the star orbit their common center of mass, the star wobbles a bit, just as a person walking an excited dog gets tugged about.

The first planet detected this way orbits the star 51 Pegasi. As the planet circles the star, the star wobbles slightly, and this very small motion of the star is detectable as Doppler shifts in the star's spectrum (Figure 20-6). From the motion of the star and estimates of the star's mass, astronomers can deduce that the planet has half the mass of Jupiter and orbits only 0.05 AU from the star. Half the mass of Jupiter amounts to 159 Earth masses, so this is a large planet. Note also that it orbits very close to its star.

Astronomers greeted the announcement of a planet orbiting 51 Pegasi with cheers and skepticism (Window on Science 20-2), but the discovery was confirmed, and more such planets were found. Roughly three dozen are known so far, including at least three orbiting the star Upsilon Andromidae—the first evidence astronomers have of a true planetary system.

Other extra-solar planets have been detected from high-precision studies of the brightness of certain stars. From Doppler shifts, astronomers knew that a planet orbited the star HD209458. Although they were not able to resolve the tiny disk of the planet, they were able to detect the small dip in brightness as the planet crossed in front of the star. That further confirmed the existence of the planet. Another planet was known to orbit the star Tau Boötis, but it could not be imaged directly.

Nevertheless, astronomers used the known period and Doppler shift in the spectrum of the star to detect oppositely directed Doppler shifts caused by light from the planet. Again, this confirms that the planet exists. Notice how the techniques used to detect these two planets resemble techniques used to observe eclipsing binaries and spectroscopic binaries (Chapter 10).

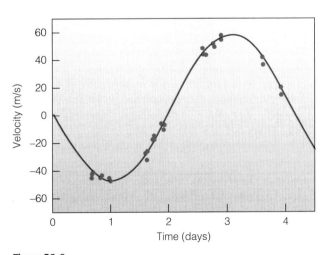

Figure 20-6
The velocity of the sunlike star 51 Pegasi can be deduced from its spectrum. The star moves with a period of 4.2 days and an amplitude of only 50 meters per second. This wobble in the star is apparently produced by a planet, the first planet found to orbit a normal star other than our sun. *(Adapted from data by Geoff Marcy and Paul Butler)*

The planets discovered so far tend to be massive and have short periods because lower-mass planets or longer-period planets are harder to detect. Low-mass planets don't tug on their stars very much, and present-day spectrographs can't detect the very small velocity changes that these gentle tugs produce. Planets with longer periods are harder to detect because Earth's astronomers have not been making high-precision observations for many years. Jupiter takes 11 years to circle the sun once, so it will take years for astronomers to see the longer-period wobbles produced by planets lying farther from their stars. Consequently we should not be surprised that most of the first planets discovered are massive and have short periods.

The new planets may seem odd for another reason. Our experience in our own solar system is that the large planets lie farther from the sun. How could big planets form so near their stars? Theorists find that planets that form in an especially dense disk of matter could spiral inward. Thus it is possible for a few planets to become the massive, short-period planets that we can detect most easily.

A few of the newly discovered extra-solar planets have elliptical orbits, and that seems odd compared to our solar system in which the planetary orbits are nearly circular. Theorists point out, however, that planets may interact in some young planetary systems and can be thrown into elliptical orbits. This is probably rare among planetary systems, but we find these extreme systems more easily because they tend to produce big wobbles.

The preceding paragraphs should reassure us that massive planets in small orbits or in elliptical orbits are not outrageous. At present we know only one planetary system well, our own. As we find more extra-solar planets, we will get a better understanding of what is normal and what is rare.

The discovery of extra-solar planets gives astronomers added confidence in the solar nebula theory. The theory predicts that planets are common, and astronomers are finding them orbiting many stars.

REVIEW Critical Inquiry

Why does the solar nebula theory imply planets are common?

Often, the implications of a theory are more important in its analysis than its own conjecture about nature. If the solar nebula theory is correct, the planets of our solar system formed from the disk of gas and dust that surrounded the sun as it condensed from the interstellar medium. If that is true, then it should be a common process. Most stars should form with disks of gas and dust around them, and so we would expect them to have planets.

Without knowing any more about the theory, the implication that planets are common suggests a way we could test the theory. If we look at a large number of stars and find lots of planets, the solar nebula theory is confirmed. But if we search and find few planets, it is contradicted. What observational evidence do we have so far on this issue?

So far only a few other stars are known to have planets. We can expect more such worlds to be discovered, and that raises the question of just how planets form. To answer that question, we begin by surveying our own solar system to discover the distinguishing characteristics of planetary systems. Those are the characteristics that our theory must eventually explain.

20-2 A Survey of the Solar System

To test our theories, we must search the present solar system for evidence of its past (Window on Science 20-3). In this section, we survey the solar system and compile a list of its most significant characteristics, potential clues to how it formed.

We begin with the most general view of the solar system. It is, in fact, almost entirely empty space (Figure 20-7). Imagine that we reduce the solar system until Earth is the size of a grain of table salt, about 0.3 mm (0.01 in.) in diameter. The moon is a speck of pepper about 1 cm (0.4 in.) away, and the sun is the size of a small plum 4 m (13 ft) from Earth. Mercury, Venus, and Mars are grains of salt. Jupiter is an apple seed 20 m (66 ft) from the sun, and Saturn is a smaller seed over 36 m (120 ft) away. Uranus and Neptune are slightly larger than average salt grains, and Pluto, the farthest planet (on average), is a speck of pepper over 150 m (500 ft) from the central plum. We would need a powerful microscope to detect the asteroids. These tiny specks of matter scattered around the sun are the remains of the solar nebula.

Revolution and Rotation

The planets revolve* around the sun in orbits that lie close to a common plane. The orbit of Mercury, the closest planet to the sun, is tipped 7° to Earth's orbit, and Pluto's orbit is tipped 17.2°. The rest of the planets' orbital planes are inclined by no more than 3.4°. Thus, the solar system is basically disk-shaped.

---

*Recall from Chapter 2 that English distinguishes between the words *revolve* and *rotate*. A planet revolves around the sun, but rotates on its axis. Cowboys in the old west didn't carry revolvers. They carried rotators.

Scientists often face problems in which they must reconstruct the past. Some of these reconstructions are obvious, such as an archaeologist excavating the ruins of a burial tomb, but others are less obvious. In each case, success requires the interplay of hypotheses and evidence to recreate a past that no longer exists.

The reconstruction of the past is obvious when we use the chemical abundance of stars to reconstruct the story of the formation of our galaxy, but a biologist studying a centipede is also reconstructing the past. How did this creature come to have a segmented body with so many legs? How did it develop the metabolism that allows it to move quickly and hunt prey? Although the problem might at first seem to be one of mere anatomy, the scientist must reconstruct an environment that no longer exists.

Geologists are fond of saying "the present is the key to the past," and that is true for any scientific problem that requires us to reconstruct the past. Because the past is over, we cannot observe it directly, so we must use the evidence of the present to test hypotheses about the nature of the past.

The grand interplay of evidence and theory is the distinguishing characteristic of science, and nowhere do we see it more clearly than when scientists reconstruct the past.

The rotation of the sun and planets on their axes also seems related to this disk shape. The sun rotates with its equator inclined only 7.25° to Earth's orbit, and most of the other planets' equators are tipped less than 30°. The rotations of Venus, Uranus, and Pluto are peculiar, however. Venus rotates backward compared with the other planets, and both Uranus and Pluto rotate on their sides (with their equators almost perpendicular to their orbits). We will discuss these planets in detail in Chapters 23 and 25, but later in this chapter we will try to understand how they could have acquired their peculiar rotations.

Apparently, the preferred direction of motion in the solar system—counterclockwise as seen from the

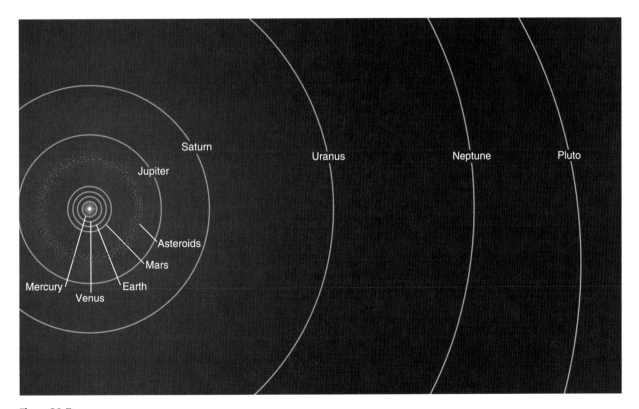

Figure 20-7
The solar system. The radii of the orbits are drawn to scale here. Note the eccentricity of Pluto's orbit. At this scale, only the sun would be visible. The planets are too small to be seen. Asteroids are discussed later in the chapter

north—is also related to its disk shape. All the planets revolve counterclockwise around the sun, and with the exception of Venus, Uranus, and Pluto they rotate counterclockwise on their axes. Thus, with only a few exceptions, most of which are understood, revolution and rotation in the solar system follow a disk theme.

Two Kinds of Planets

Perhaps the most important clue we have to the origin of the solar system comes from the division of the planets into two categories: terrestrial, or Earthlike, and Jovian, or Jupiter-like (Figure 20-8). The **terrestrial planets** are small, dense, rocky worlds with less atmosphere than the Jovian planets. The terrestrial planets—Mercury, Venus, Earth, and Mars—lie in the inner solar system. By contrast, the **Jovian planets** are large, gaseous, low-density worlds. The Jovian planets—Jupiter, Saturn, Uranus, and Neptune—lie in the outer solar system beyond the asteroids. Note that Pluto does not fit either category very well. It is small like the terrestrial planets but lies far from the sun and has a low density like the Jovian planets.

Although *terrestrial* means "Earthlike," Mercury is more like a moon. It is a small world, only 40 percent larger than Earth's moon, and thus it has held no atmosphere. Its gray, rocky surface is covered with thou-

Figure 20-8
The relative sizes of the planets compared with the disk of the sun illustrate the two kinds of planets. The four terrestrial planets are small, and the Jovian planets are large. Pluto is not included in the classification scheme.

Figure 20-9
Mercury is about 40 percent larger in diameter than Earth's moon, shown here to the same scale. Like the moon, Mercury lacks an atmosphere and is heavily cratered by the impact of large meteorites. *(Mercury, NASA; moon, Lick Observatory photograph)*

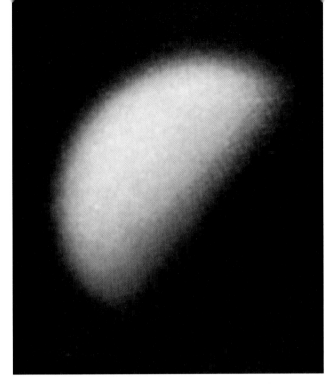

Figure 20-10
Venus is covered by a thick layer of clouds that never allows us to see its surface. In this photo, taken through an Earth-based telescope, the planet is visible at quarter phase. *(NASA)*

sands of overlapping craters, which, like the craters on Earth's moon, are the scars left behind by the impact of large meteorites (Figure 20-9). Such impact craters are very common in the solar system.

Venus is nearly as large as Earth, and its atmosphere is thick with heavy clouds that hide its surface (Figure 20-10). Radio signals can penetrate these clouds and bounce back from its rocky surface to give us a radar image of its terrain. These radar maps reveal that, although Venus is nearly the same size as Earth, it totally lacks Earth's oceans of water. Venus is a deadly, dry, desert world marked by impact craters, volcanoes, lava flows, and geological faults (Figure 20-11). In contrast, the atmosphere on Mars is thin, and we can see its surface clearly from space probes. These photos show that Mars, about half the diameter of Earth, is marked by impact craters and volcanoes.

Impact craters seem typical of all but the youngest surfaces in our solar system. The terrestrial planets, including Earth, are marked by such craters, and spacecraft have found craters on the moons of Mars, Jupiter, Saturn, Uranus, and Neptune. (There are no craters on the planets Jupiter, Saturn, Uranus, or Neptune because, as we will see, those planets do not have solid surfaces.)

Figure 20-11
Earth, Venus, and Mars to scale. Earth (left) is mostly a water-covered planet with large masses of exposed rocky surface and an atmosphere of changing cloud patterns. The surface of Venus is perpetually hidden below thick clouds, so we must depend on radar images to reveal its surface (right), a dry desert marked by impact craters, volcanism, lava flows, and faults. Mars (center) is a smaller world with impact craters and old volcanoes on its rocky surface. *(NASA)*

This suggests that craters are a characteristic of every object in the solar system with a surface capable of retaining such features.

The terrestrial planets have densities ranging from about 3 to 5 g/cm³. Typically, they contain cores of iron and nickel surrounded by a mantle of dense rock. In contrast, the Jovian planets are rich in hydrogen and helium, so their average density is low—less than 1.75 g/cm³. Saturn's very low density, 0.7 g/cm³, is less than the density of water, so the planet would float if we could find a bathtub big enough. Nevertheless, all the Jovian planets are massive. Jupiter is over 318 times as massive as Earth, and Saturn is about 95 Earth masses. Uranus and Neptune are smaller—15 and 17 Earth masses, respectively.

Although photographs of the Jovian planets reveal swirling cloud patterns (Figure 20-12), their atmospheres are not very deep compared to their radii. In fact, if these so-called gas giants were shrunk to a few centimeters in diameter, their gaseous hydrogen and helium atmospheres would be no deeper than the fuzz on a badly worn tennis ball.

Mathematical models predict that the interiors of Jupiter and Saturn are mostly hydrogen in two different states (Figure 20-13). Beneath the cloud belts of these planets are layers where pressure forces the hydrogen into a liquid state. Even deeper, the pressure is so high that the hydrogen atoms can no longer hold their electrons. With the electrons free to move about, the material becomes an excellent electrical conductor and is known as **liquid metallic hydrogen.** The centers of Jupiter and Saturn are believed to be hot cores of heavy elements slightly larger than Earth. These are often erroneously called "rocky cores"; the heat and pressure would not allow rock as it is known on Earth to exist.

Uranus and Neptune, being smaller, have less pressure inside and consequently cannot force hydrogen into the liquid metallic state. Rather, model calculations suggest they contain heavy-element cores surrounded by deep layers of icy slush mixed with rocky minerals and dissolved methane and ammonia. Above that is a very dense, deep atmosphere made mostly of hydrogen gas. Although Uranus and Neptune have low densities, hydrogen-rich compositions, and atmospheric features similar to those of Jupiter and Saturn, the two smaller Jovian worlds are clearly different kinds of planets.

Another characteristic of the Jovian planets is their large satellite systems. Jupiter has at least 16 known moons; 4 of them, the **Galilean satellites** (named after Galileo, who discovered them in 1610), are visible through small telescopes or even binoculars (Figure 20-14a). Saturn, Uranus, and Neptune also have systems of moons.

All of the Jovian planets have ring systems. Saturn has bright rings made of ice particles (Figure 20-14b), and Jupiter has a thin ring of dark, rocky dust. Uranus and Neptune also have thin rings composed of myriads of small particles.

Jupiter

Great Red Spot

Uranus

Neptune

Saturn

Figure 20-12
Of the Jovian planets (reproduced here to scale), Jupiter and Saturn are the largest. Jupiter is marked by dramatic cloud belts and the Great Red Spot, while Saturn has a bright ring system. Uranus looks featureless, but faint clouds are visible in computer-enhanced photos. The dark blue color of Neptune and the lighter blue of Uranus are caused by atmospheric methane, which absorbs red photons. *(Adapted from NASA photographs)*

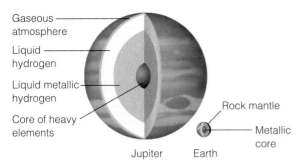

Gaseous atmosphere
Liquid hydrogen
Liquid metallic hydrogen
Core of heavy elements
Jupiter

Rock mantle
Metallic core
Earth

Figure 20-13
Cross sections of Earth and Jupiter illustrate the differences between the interiors of terrestrial and Jovian planets. Earth contains a metallic core with a rock mantle. Jupiter, 11.18 times larger in diameter, is composed mostly of liquid hydrogen. A core composed of heavy elements such as iron, nickel, and silicon lies at the center.

We will examine the satellites and rings of the Jovian planets in later chapters. Here it is sufficient to note that the Jovian planets have large numbers of satellites and that all four have ring systems. In contrast, the terrestrial planets have few satellites and no rings at all. These observations will help us understand how plan-

a

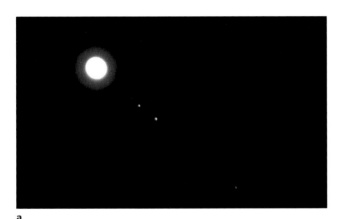

b

Figure 20-14
Jupiter and Saturn through a small telescope. (a) Jupiter has at least 16 satellites, but only the 4 Galilean moons are visible in small telescopes. In this photo, the planet was overexposed to show 3 of the moons. (b) The icy rings of Saturn are so reflective that they are visible through even a small telescope. *(Grundy Observatory)*

ets were formed in the solar nebula. For now, however, we turn our attention to the smaller members of our solar system.

Space Debris

The sun and planets are not the only remains of the solar nebula. The solar system is littered with three kinds of space debris: asteroids, comets, and meteoroids. Although these objects represent a tiny fraction of the mass of the system, they are a rich source of information about the origin of the planets.

The **asteroids,** sometimes called minor planets, are small rocky worlds, most of which orbit the sun in a belt between the orbits of Mars and Jupiter. Roughly 20,000 asteroids have been identified, of which about 500 to 1000 follow orbits that bring them into the inner solar system, where they can occasionally collide with a planet. Earth has been struck many times in its history. Some asteroids are located in Jupiter's orbit, some have been found beyond the orbit of Saturn, and a growing number are being discovered among and even beyond the outermost planets.

About 200 asteroids are more than 100 km (60 mi) in diameter, and over 2000 are more than 10 km (6 mi) in diameter. There are probably 500,000 that are larger than 1 km (0.6 mi) and billions that are smaller. Because even the largest are only a few hundred kilometers in diameter, Earth-based telescopes can detect no details on their surfaces.

We do however have clear evidence that asteroids are irregularly shaped cratered worlds. Spacecraft have flown past a few asteroids and transmitted photos back to Earth (Figure 20-15). Also, the Hubble Space Telescope

Figure 20-15
The asteroid Gaspra as photographed by the Galileo spacecraft. The asteroid is 20 by 12 by 11 km and rotates once every 7.04 hours. Its irregular shape and numerous craters testify that it has suffered many impacts. The color here is exaggerated to reveal subtle color variation. Gaspra would look gray to the human eye. *(NASA/Galileo Imaging Team)*

has been able to image some asteroids well enough to reveal shapes and major features. Even radio telescopes, used as radar, have imaged asteroids. All of these observations reveal that asteroids tend to be irregular in shape and heavily cratered. We will discuss these observations in detail in Chapter 26, but in our quick survey of the solar system we can note that all of the evidence leads to the conclusion that the asteroids have suffered many impacts from collisions with other asteroids.

Older theories proposed that the asteroids are the remains of a planet that broke up, but modern astronomers recognize the asteroids as the debris left over by a planet that failed to form at a distance of 2.8 AU from the sun. When we discuss the formation of planets later in this chapter, we must be prepared to explain why material in the solar nebula failed to form a planet at a distance of 2.8 AU.

In contrast to asteroids, the brightest **comets** are impressively beautiful objects (Figure 20-16a). Most comets are faint, however, and are difficult to locate even at their brightest. A comet may take months to sweep through the inner solar system, during which time it appears as a glowing head with an extended tail of gas and dust.

According to the **dirty snowball model** of comets, the tail, which can be greater than 1 AU in length for a bright comet, is produced by the nucleus, a ball of dirty ices (mainly water and carbon dioxide) no more than a few dozen kilometers in diameter. When the nucleus enters the inner solar system, the sun's radiation begins to vaporize the ices, releasing gas and dust. The pressure of sunlight and the solar wind pushes the gas and dust away, forming a long tail. The motion of the nucleus along its orbit, the pressure of sunlight, and the outward flow of the solar wind can create comet tails that are long and straight or gently curved, but in either case the tails of comets always point generally away from the sun (Figure 20-16b).

The nuclei of comets are icy bodies left over from the origin of the planets. Thus, we must conclude that at least some parts of the solar nebula were rich in ices. We will see later in this chapter how important ices were in the formation of the Jovian planets, and we will discuss comets in more detail in Chapter 26.

Unlike the stately comets, **meteors** flash across the sky in momentary streaks of light (Figure 20-17). They are commonly called "shooting stars." Of course, they are not stars but small bits of rock and metal falling into Earth's atmosphere and bursting into incandescent vapor about 80 km (50 mi) above the ground because of friction with the air. This vapor condenses to form dust, which settles slowly to Earth, adding about 40,000 tons per year to the planet's mass.

Technically, the word *meteor* refers to the streak of light in the sky. In space, before its fiery plunge, the object is called a **meteoroid,** and any part of it that sur-

a

b

Figure 20-16
(a) A comet may remain visible in the evening or morning sky for weeks as it moves through the inner solar system. Comet West was in the sky during March 1976. (b) A comet in a long, elliptical orbit becomes visible when the sun's heat vaporizes its ices and pushes the gas and dust away in a tail. *(Celestron International)*

Figure 20-17
A meteor is the streak of glowing gases produced by a bit of material falling into the earth's atmosphere. Friction with the air vaporizes the material about 80 km (50 mi) above Earth's surface. *(Daniel Good)*

vives its fiery passage to Earth's surface is called a **meteorite.** Most meteoroids are specks of dust, grains of sand, or tiny pebbles. Almost all the meteors we see in the sky are produced by meteoroids that weigh less than 1 g. Only rarely is one massive enough and strong enough to survive its plunge and reach Earth's surface.

Thousands of meteorites have been found, and we will discuss their particular forms in Chapter 26. We mention meteorites here for one specific clue they can give us concerning the solar nebula: Meteorites can tell us the age of the solar system.

The Age of the Solar System

If the solar nebula theory is correct, the planets should be about the same age as the sun. The most accurate way to find the age of a celestial body is to bring a sample into the laboratory and determine its age by analyzing the radioactive elements it contains.

When a rock solidifies, it incorporates known percentages of the chemical elements. A few of these elements are radioactive and can decay into another element, called the daughter element. For example, the

isotope of uranium ^{238}U can decay into an isotope of lead, ^{206}Pb. The **half-life** of a radioactive element is the time it takes for half of the atoms to decay. The half-life of ^{238}U is 4.5 billion years. Thus, the abundance of a radioactive element gradually decreases as it decays (Figure 20-18), and the abundance of the daughter element gradually increases. If we know the abundance of the elements in the original rock, we can measure the present abundance and find the age of the rock. For example, if we studied a rock and found that only 50 percent of the ^{238}U remained and the rest had become ^{206}Pb, we would conclude that one half-life must have passed and the rock was 4.5 billion years old.

Uranium isn't the only radioactive element used in radioactive dating. Potassium (^{40}K) decays with a half-life of 1.3 billion years to form calcium (^{40}Ca) and argon (^{40}Ar). Rubidium (^{87}R) decays to strontium (^{87}Sr) with a half-life of 47 billion years. Any of these elements can be used as a radioactive clock to find the age of mineral samples.

Of course, to find a radioactive age, we need a sample in the laboratory, and the only celestial bodies from which we have samples are Earth, the moon, Mars, and meteorites.

The oldest Earth rocks so far discovered and dated are northern Canadian minerals, about 4.0 billion years old. That does not mean that Earth formed 4.0 billion years ago. The surface of Earth is active, and the crust is continually destroyed and reformed from material welling up from beneath the crust (see Chapter 21). Thus, the age of these oldest rocks tells us only that Earth is *at least* 4.0 billion years old.

One of the most exciting goals of the Apollo lunar landings was bringing lunar rocks back to Earth's laboratories, where they could be dated. Because the moon's surface is not being recycled like Earth's, some parts of it might have survived unaltered since early in the history of the solar system. Dating the rocks showed the oldest to be 4.48 billion years old. Thus, the solar system must be *at least* 4.48 billion years old.

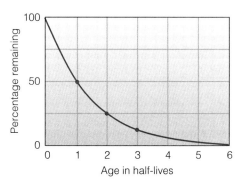

Figure 20-18
The number of nuclei remaining in a radioactive sample decreases such that 50 percent are left after one half-life, 25 percent after two half-lives, 12.5 percent after three half-lives, and so on.

Table 20-1 Characteristic Properties
of the Solar System

1. Disk shape of the solar system
   Orbits in nearly the same plane
   Common direction of rotation and revolution
2. Two planetary types
   Terrestrial—inner planets; high density
   Jovian—outer planets; low density
3. Planetary ring systems for Jupiter, Saturn, Uranus, and Neptune
4. Space debris—asteroids, comets, and meteors
   Composition
   Orbits
5. Common ages of about 4.6 billion years for Earth, the moon, Mars, meteorites, and the sun

Although no one has yet been to Mars, a few meteorites found on Earth have been identified by their chemical composition as having come from Mars. Most of these have ages of only 1.3 billion years, but one has an age of about 4.5 billion years. Thus, Mars must be at least that old.

Another important source for determining the age of the solar system is meteorites. Radioactive dating of meteorites yields a range of ages, with the oldest about 4.6 billion years old. This figure is widely accepted as the age of the solar system.

One last celestial body deserves mention: the sun. Astronomers estimate the age of the sun to be about 5 billion years, but this is not a radioactive date because they cannot obtain a sample of solar material. Instead, they estimate the sun's age from the radioactive ages of Earth, the moon, and meteorites. Computer models of the sun give only approximate ages, but they generally agree with the age of Earth.

Apparently, all the bodies of the solar system formed at about the same time some 4.6 billion years ago. This completes our survey of the solar system. We have found five significant characteristics (Table 20-1). We must now try to understand how the formation of the planets in the solar nebula explains these significant characteristics.

REVIEW Critical Inquiry

In what ways is the solar system a disk?

First, the general shape of the solar system is that of a disk. The planets follow orbits that lie in nearly the same plane. The orbit of Mercury is inclined 7° to the plane of Earth's orbit, and the orbit of Pluto is inclined a bit more than 17°. Thus, the planets follow orbits confined to a thin disk with the sun at its center.

Second, the motions of the sun and planets also follow this disk theme. The sun and most of the planets rotate in the same direction, counterclockwise as seen from the north, with their equators near the plane of the solar system. Also, all of the planets revolve around the sun in that same direction. Thus, our solar system seems to prefer motion in the same direction, which further reflects a disk theme.

One of the basic characteristics of our solar system is its disk shape, but another dramatic characteristic is the division of the planets into two groups. What are the distinguishing differences between the terrestrial and Jovian planets?

We have completed our survey of the solar system and gathered the preliminary evidence we can use to evaluate the solar nebula hypothesis. Now it is time to ask how the planets formed from the solar nebula and took on the characteristics we see today.

20-3 The Story of Planet Building

The challenge for modern planetary astronomers is to compare the characteristics of the solar system with the solar nebula theory and tell the story of how the planets formed.

The Chemical Composition of the Solar Nebula

Everything we know about the solar system and star formation suggests that the solar nebula was a fragment of an interstellar gas cloud. Such a cloud would have been mostly hydrogen with some helium and tiny traces of the heavier elements.

This is precisely what we see in the composition of the sun (see Table 7-3). Analysis of the solar spectrum shows that the sun is mostly hydrogen, with 25 percent of its mass being helium and only about 2 percent being heavier elements. Of course, nuclear reactions have fused some hydrogen into helium, but this happens in the sun's core and has not affected its surface composition. Thus, the composition revealed in its spectrum is essentially the composition of the gases from which it formed.

This must have been the composition of the solar nebula, and we can see that composition reflected in the chemical compositions of the planets. One factor, however, can dramatically alter the composition of a planet. We must recognize that planets grow by two processes. Planets begin growing by the sticking together of solid bits of matter. Only when a planet has grown to a mass of about 15 Earth masses does it have enough gravitation to begin capturing gas directly from the solar nebula in a process called **gravitational collapse.** The Jovian planets began forming by the aggregation of bits of rock and ice. Once they accumulated enough mass, they began growing by gravitational collapse. That is, they could capture large amounts of gas, mostly hydrogen and helium, directly from the solar nebula. Jupiter and Saturn grew rapidly and captured the most hydrogen and helium, while Uranus and Neptune grew more slowly and didn't capture gas as rapidly as Jupiter and Saturn. Thus, the Jovian worlds grew to be low-density, hydrogen-rich planets, with Jupiter and Saturn the largest and least dense.

The terrestrial planets, in contrast, contain very little hydrogen and helium. These small planets have such low masses they were unable to keep the hydrogen and helium from leaking into space. Indeed, because of their small masses, they were probably unable to capture very much of these lightweight gases from the solar nebula. The terrestrial planets are dense worlds because they are composed of the heavier elements from the solar nebula.

Although we can see how the chemical composition of the solar nebula is reflected in the present composition of the sun and planets, a very important question remains: How did gas and dust in the solar nebula come together to form the solid matter of the planets? We must answer that question in two stages. First, we must understand how gas and dust formed billions of small solid particles, and then we must explain how these particles built the planets.

The Condensation of Solids

The key to understanding the process that converted the nebular gas into solid matter is the variation in density among solar system objects. We have already noted that the four inner planets are high-density, terrestrial bodies, whereas the outermost planets are low-density, giant planets (except for Pluto, which is low in density but not a giant). This division is due to the different ways gases condensed into solids in the inner and outer regions of the solar nebula.

Even among the four terrestrial planets, we find a pattern of subtle differences in density. Merely listing the observed densities of the terrestrial planets does not reveal the pattern, because Earth and Venus, being more massive, have stronger gravity and have squeezed their interiors to higher densities. We must look at the **uncompressed densities**—the densities the planets would have if their gravity did not compress them. These densities (Table 20-2) show that the closer a planet is to the sun, the higher its uncompressed density.

This density variation probably originated when the solar system first formed solid grains. The kind of matter that condensed in a particular region would depend on the temperature of the gas there. In the inner

Table 20-2 Observed and Uncompressed Densities

Planet	Observed Density (g/cm³)	Uncompressed Density (g/cm³)
Mercury	5.44	5.4
Venus	5.24	4.2
Earth	5.50	4.2
Mars	3.94	3.3
(Moon)	3.36	3.35

regions, the temperature may have been 1500 K or so. The only materials that could form grains at this temperature are compounds with high melting points, such as metal oxides and pure metals, which are very dense. Farther out in the nebula it was cooler, and silicates (rocky material) could condense. These are less dense than metal oxides and metals. In the cold outer regions, ices of water, methane, and ammonia could condense. These are low-density materials.

The sequence in which the different materials condense from the gas as we move away from the sun is called the **condensation sequence** (Table 20-3). It suggests that the planets, forming at different distances from the sun, accumulated from different kinds of materials. Thus, the inner planets formed from high-density metal oxides and metals, and the outer planets formed from low-density ices.

We must also remember that the solar nebula did not remain the same temperature throughout the formation of the planets but may have grown progressively cooler. Thus, a particular region of the nebula may have begun by producing solid particles of metals and metal

Table 20-3 The Condensation Sequence

Temperature (K)	Condensate	Planet (Estimated Temperature of Formation; K)
1500	Metal oxides	Mercury (1400)
1300	Metallic iron and nickel	
1200	Silicates	
1000	Feldspars	Venus (900)
680	Troilite (FeS)	Earth (600)
		Mars (450)
175	H_2O ice	Jovian (175)
150	Ammonia–water ice	
120	Methane–water ice	
65	Argon–neon ice	Pluto (65)

oxides but, after cooling, began producing particles of silicates. Allowing for the cooling of the nebula makes our theory much more complex, but it also makes the processes by which the planets formed much more understandable.

The Formation of Planetesimals

In the development of a planet, three groups of processes operate. First, grains of solid matter grow larger, eventually reaching diameters ranging from a few centimeters to kilometers. The larger of these objects, called **planetesimals,** are believed to be the bodies that the second group of processes collects into planets. Finally, a third set of processes clears away the solar nebula. The study of planet building is the study of these three groups of processes.

According to the solar nebula theory, planetary development in the solar nebula began with the growth of dust grains. These specks of matter, whatever their composition, grew from microscopic size by two processes: condensation and accretion.

A particle grows by **condensation** when it adds matter one atom at a time from a surrounding gas. Thus, snowflakes grow by condensation in Earth's atmosphere. In the solar nebula, dust grains were continuously bombarded by atoms of gas, and some of these stuck to the grains. A microscopic grain capturing a gas atom increases its mass by a much larger fraction than a gigantic boulder capturing a single atom. Thus, condensation can increase the mass of a small grain rapidly, but as the grain grows larger, condensation becomes less effective.

The second process is **accretion,** the sticking together of solid particles. In building a snowman, we roll a ball of snow across the snowy ground so that it grows by accretion. In the solar nebula, the dust grains were, on the average, no more than a few centimeters apart, so they collided frequently. Their mutual gravitation was too small to hold them to each other, but other effects may have helped. Static electricity generated by their passage through the gas could have held them together, as could compounds of carbon that might have formed a sticky surface on the grains. Ice grains might have stuck together better than some other types. Of course, some collisions might have broken up clumps of grains; on the whole, however, accretion must have increased grain size. If it had not, the planets would not have formed.

There is no clear distinction between a very large grain and a very small planetesimal, but we can consider an object a planetesimal when its diameter becomes a kilometer or so. Objects this size and larger were subject to new processes that tended to concentrate them. One important effect may have been that the growing planetesimals collapsed into the plane of the solar nebula. Dust grains could not fall into the plane

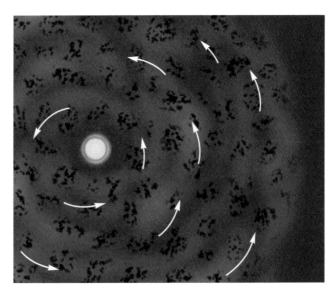

Figure 20-19
Gravitational instabilities in the rotating disk of planetesimals may have forced them to collect in clumps, accelerating their growth.

because the turbulent motions of the gas kept them stirred up, but the larger objects had more mass, and the gas motions could not have prevented them from settling into the plane of the spinning nebula. This would have concentrated the solid particles into a thin plane about 0.01 AU thick and would have made further planetary growth more rapid.

This collapse of the planetesimals into the plane is analogous to the flattening of a forming galaxy. However, an entirely new process may have become important once the plane of planetesimals formed. Computer models show that the rotating disk of particles should have been gravitationally unstable and would have broken up into small clouds (Figure 20-19). This would further concentrate the planetesimals and help them coalesce into objects up to 100 km (60 mi) in diameter. Thus, the theory predicts that the nebula became filled with trillions of solid particles ranging in size from pebbles to tiny planets. As the largest began to exceed 100 km in diameter, new processes began to alter them, and a new stage in planet building began, the growth of protoplanets.

The Growth of Protoplanets

The coalescing of planetesimals eventually formed **protoplanets,** massive objects destined to become planets. As these larger bodies grew, new processes began making them grow faster and altered their physical structure.

If planetesimals collided at orbital velocities, it is unlikely that they would have stuck together. The average orbital velocity in the solar system is about 30 km/s (67,000 mph). Head-on collisions at this velocity would have pulverized the material. However, the planetesimals were moving in the same direction in the nebular plane and thus avoided head-on collisions. Instead, they merely rubbed shoulders at low relative velocities. Such collisions would have been more likely to fuse them than to shatter them.

In addition, some adhesive effects probably helped. Sticky coatings and electrostatic charges on the surfaces of the smaller planetesimals probably aided formation of larger bodies. Collisions would have fragmented some of the surface rock, but if the planetesimals were large enough, their gravity would have held on to some fragments, forming a layer of soil composed entirely of crushed rock. Such a layer on the larger planetesimals may have been effective in trapping smaller bodies.

The largest planetesimals would grow the fastest because they had the strongest gravitational field. Not only could they hold on to a cushioning layer to trap fragments, but their stronger gravity could also attract additional material. These planetesimals probably grew quickly to protoplanetary dimensions, sweeping up more and more material. When massive enough, they trapped some of the original nebular gas to form primitive atmospheres. At some point, they crossed the boundary between planetesimals and protoplanets.

To trace the growth of a protoplanet, we can think of the formation of Earth. In its simplest form, our theory of protoplanet growth supposes that all the planetesimals had about the same chemical composition. The planetesimals accumulated gradually to form a planet-size ball of material that was of homogeneous composition throughout. Once the planet formed, heat began to accumulate in its interior from the decay of short-lived radioactive elements, and this heat eventually melted the planet and allowed it to differentiate. **Differentiation** is the separation of material according to density. When the planet melted, the heavy metals such as iron and nickel settled to the core, while the lighter silicates floated to the surface to form a low-density crust. The story of planet formation from planetesimals of similar composition is shown in the left half of Figure 20-20.

In this simple form, the theory predicts that Earth's present atmosphere was not its first. The first atmosphere consisted of gases trapped from the solar nebula—mostly hydrogen and helium. They were later driven off by the heat, aided perhaps by outbursts from the infant sun, and new gases that baked from the rocks formed a secondary atmosphere. This creation of a planetary atmosphere from the planet's interior is called **outgassing.**

We can improve this simple theory of planet formation in two ways. First, it seems likely that the solar nebula cooled during the formation of the planets, so they did not accumulate from planetesimals of common composition. As planet building began, the first particles to condense in the inner solar system were metals and metal oxides, so the protoplanets may have begun by accreting metallic cores. Later, as the nebula cooled,

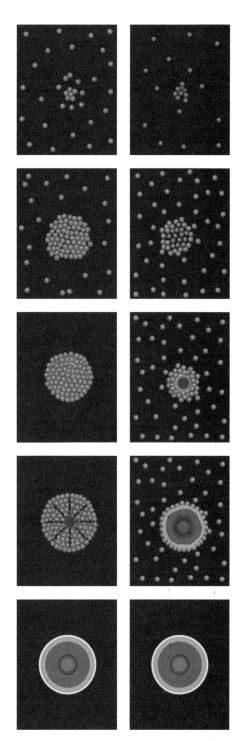

Figure 20-20
At left, a planet grows slowly by the accretion of planetesimals of similar composition. After the protoplanet has formed, the accumulation of heat from radioactive decay in its interior melts it and allows it to differentiate. An improved theory, right, proposes that the solar nebula cooled as the planets formed. The first particles to accumulate were metals, but later particles were silicates. Also, it seems likely that the planets formed fast enough to prevent the loss of the heat of formation, which melted the protoplanets and allowed them to differentiate as they formed.

more silicates could form, and the protoplanets added silicate mantles. A second improvement in the theory proposes that the planets grew so rapidly that the heat released by the in-falling particles, the **heat of formation,** did not have time to escape. This heat rapidly accumulated and melted the protoplanets as they formed. If this is true, then the planets would have differentiated as they formed. This improved story of planet formation is shown in the right half of Figure 20-20.

Our improved theory suggests that our present atmosphere was not produced entirely by outgassing. If Earth formed in a molten state, then there was never a time when it had a primitive atmosphere of hydrogen and helium accumulated from the solar nebula. Rather, the gases of the atmosphere were released by the molten rock as the protoplanet grew. Those gases would not have included much water, however, so some astronomers now think that Earth's water and much of its present atmosphere accumulated late in the formation of the planets as Earth swept up volatile-rich planetesimals forming in the cooling solar nebula. Such icy planetesimals may have formed in the outer parts of the solar nebula and been scattered by encounters with the Jovian planets in a bombardment of comets.

The Jovian planets grew to their present size in about 10 million years. How do astronomers know this? The T Tauri stars that have gas disks are all younger than 10 million years; older T Tauri stars have blown their gas disks away. The Jovian planets grew mainly from the gas in the solar nebula, so they must have finished forming by the time the sun blew its gas away. Jupiter and Saturn appear to have formed quickly, so they captured lots of gas. Uranus and Neptune, however, seem to have grown more slowly, perhaps because the solar nebula was less dense in its outer parts and also because orbital velocities were slower and material collided less often in the outer nebula. In any case, by the time Uranus and Neptune grew massive enough to capture gas from the solar nebula, the sun had blown much of the gas away. Thus Uranus and Neptune contain less hydrogen than Jupiter and Saturn.

The terrestrial planets probably took longer to form because they did not grow from the gas in the nebula but had to accrete from planetesimals. Nevertheless, the terrestrial planets must have been nearly complete by the time the sun reached the main sequence at an age of about 30 million years. Models suggest that the planets were fully formed by about 100 million years.

Traditionally astronomers have assumed that the planets formed at about their present distances from the sun, but new mathematical models suggest that Uranus and Neptune may have formed at about the distance of Jupiter. Gravitational interactions with massive Jupiter could have moved them outward. This is an exciting new insight, but little evidence exists to test the hypothesis at this point. Some extra-solar planets appear to have had their orbits altered by interactions with other planets, so we may learn more about the formation of our own solar system as we discover more extra-solar planets and begin to understand what is normal and what is peculiar in planetary systems.

The details of planet formation are still being unraveled, although it seems clear that the solar nebula hypothesis can explain the formation of the planets around the infant sun. But can it explain the distinguishing characteristics of the solar system that we compiled earlier in this chapter?

Explaining the Characteristics of the Solar System

Table 20-1 contains the list of distinguishing characteristics of the solar system. Any theory of the origin of the solar system should explain these characteristics.

The disk shape of the solar system is inherited from the solar nebula. The sun and planets revolve and rotate in the same direction because they formed from the same rotating gas cloud. The orbits of the planets lie in the same plane because the rotating solar nebula collapsed into a disk, and the planets formed in that disk.

The solar nebula hypothesis is evolutionary in that it calls on continuing processes to gradually build the planets. To explain the rotation of Venus, Uranus, and Pluto, however, we must introduce catastrophic events. Apparently, these planets acquired their peculiar rotations when they were struck by very large planetesimals late in their formation. Such impacts may be unusual, but some planets probably experienced them, and an off-center impact by a planetesimal as large as Mars might have made Venus rotate backward. A similar event might have forced Uranus to rotate on its side. In Chapter 25, we will discuss Pluto, and we will see that such a catastrophic event may have been involved in its formation as well.

The second item in Table 20-1, the division of the planets into terrestrial and Jovian worlds, can be understood through the condensation sequence. The terrestrial planets formed in the inner part of the solar nebula, where the temperature was high and most compounds remained gaseous. Only compounds such as the metals and silicates could condense to form solid particles. Planets must begin forming by the sticking together of solid particles because a small blob of matter does not have enough gravitation to capture gas. Thus, the planets that began growing in the inner solar system had to form mainly from metals and silicates. They became the small, dense terrestrial planets.

In contrast, the Jovian planets formed in the outer solar nebula, where the lower temperature allowed the gas to form large amounts of ices, perhaps three times more ices than silicates. Thus, the Jovian planets grew rapidly and became massive.

The heat of formation (the energy released by infalling matter) was tremendous for these massive planets. Jupiter must have grown hot enough to glow with a luminosity of about 1 percent that of the present sun. However, because it never got hot enough to generate nuclear energy as a star would, it never generated its own energy. Jupiter is still hot inside. In fact, both Jupiter and Saturn radiate more heat than they absorb from the sun, so they are evidently still cooling.

Although planets must begin growing by the accretion of solid particles, once they become about 15 times more massive than Earth they have enough gravity to capture gas directly from the solar nebula. Thus, the Jovian planets grew by attracting vast amounts of nebular gas and grew rich in light gases such as hydrogen and helium. The terrestrial planets could not do this because they never became massive enough and because the gas in the inner nebula was hotter and more difficult to capture.

When planets were first discovered orbiting other stars, some astronomers wondered if the solar nebula theory was quite right. The newly discovered planets are massive, so they are presumably Jovian, but they are close to their stars where a terrestrial planet should be. Also, some of the new planets follow elliptical orbits. Models of planet formation resolved the problem. If a Jovian planet forms in a disk rich in planetesimals, it may sweep up mass so fast it migrates inward toward its star. Also, interactions among massive planets could throw a planet into an elliptical orbit. Thus the newly discovered planets do not pose a problem for the solar nebula explanation of terrestrial and Jovian planets.

A glance at the solar system suggests that we should expect to find a planet between Mars and Jupiter at the present location of the asteroid belt. Evidently, we have asteroids there, and not a planet, because Jupiter grew into such a massive planet it was able to gravitationally disturb the motion of nearby planetesimals. Thus, the bodies that should have formed a planet just inward in the solar system from Jupiter were broken up, thrown into the sun, or ejected from the solar system. The asteroids we see today are the last remains of those objects.

The large satellite systems of the Jovian worlds may contain two kinds of moons. Some moons may have formed in orbit around the forming planet in a miniature of the solar nebula. But some of the smaller moons may be captured planetesimals and asteroids. The large masses of the Jovian planets would have made it easier for them to capture satellites.

In Table 20-1, we noted that all four Jovian worlds have ring systems, and we can understand this by considering the large mass of these worlds and their remote location in the solar system. A large mass makes it easier for a planet to hold onto orbiting ring particles; and being farther from the sun, the ring particles are not as easily swept away by the pressure of sunlight and the solar wind. It is hardly surprising, then, that the terrestrial planets, low-mass worlds located near the sun, have no planetary rings.

The last entry in Table 20-1 is the common ages of solar system bodies, and the solar nebula hypothesis has no difficulty explaining that characteristic. If the hypothesis is correct, then the planets formed at the same time as the sun and thus should have roughly the same age. Although we have mineral samples for radioactive dating only from Earth, the moon, Mars, and meteorites, so far the ages seem to agree.

The solar nebula hypothesis can account for the distinguishing characteristics of the solar system, but there is yet another test we should apply to the hypothesis. What about the problem that troubled Laplace and his nebular hypothesis? What about the angular momentum problem? If the sun and planets formed from a contracting nebula, the sun should have been left spinning very rapidly. That is, it should have most of the angular momentum in the solar system, and instead it has very little. The solar nebula hypothesis allows us to explain the angular momentum problem by thinking about the relationship between the origin of planetary systems and star formation. What we know about star formation tells us that the sun must have had a strong stellar wind when it was young. Also, a rapidly spinning star is likely to generate a strong magnetic field. Like a great paddle wheel, the sun's magnetic field must have stuck out into space, and gas flowing away from the sun would have been dragged around with the rotating magnetic field. This would have slowed the sun and left it rotating slowly. Astronomers now believe that the angular momentum problem is no longer an objection to the solar nebula hypothesis.

Our general understanding of the origin of the solar system also explains the origin of asteroids, meteors, and comets. They appear to be the last of the debris left behind by the solar nebula. These objects are such important sources of information about the history of our solar system that we will discuss them in detail in Chapter 26. But for now we ask: What happened to the solar nebula?

Clearing the Nebula

The planets appear to have grown in the solar nebula at the same time that the sun was forming, about 4.6 billion years ago. Indeed, much of the planetary growth may have taken place in darkness, because the sun may not have become luminous yet. But once the sun became hot and luminous, it began to clear the nebula of gas and dust and brought planet building to a halt.

Four effects helped clear the nebula. The most important was **radiation pressure.** When the sun became a luminous object, light streaming from its surface pushed against the particles of the solar nebula. Large bits of matter like planetesimals and planets were not affected, but low-mass specks of dust and individual gas atoms were pushed outward and eventually driven from the system. This is not a sudden process, and it

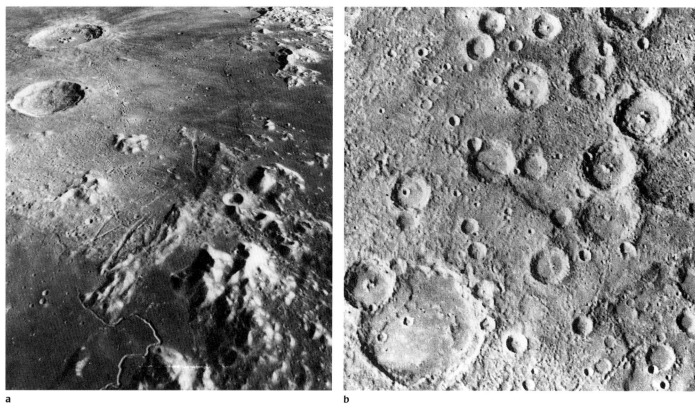

a   b

Figure 20-21
Every old, solid surface in the solar system is scarred by craters. (a) Earth's moon is scarred by craters ranging
from basins hundreds of kilometers in diameter down to microscopic pits. (b) The surface of Mercury as photographed by a passing spacecraft shows vast numbers of overlapping craters. *(NASA)*

may not have occurred at the same time everywhere in
the nebula. Before sunlight could begin clearing the
outer nebula, it first had to push its way through the
inner nebula.

The second effect that helped clear the nebula was
the solar wind, the flow of ionized hydrogen and other
atoms away from the sun's upper atmosphere. This flow
is a steady breeze that rushes past Earth at about 400
km/s (250 mi/s). When the sun was young, it may have
had an even stronger solar wind, and irregular fluctuations in its luminosity, like those observed in T Tauri
stars, may have produced surges in the wind that helped
push dust and gas out of the nebula.

The third effect for clearing the nebula was the
sweeping up of space debris by the planets. All of the
old, solid surfaces in the solar system are heavily cratered by meteorite impacts (Figure 20-21). Earth's moon,
Mercury, Venus, Mars, and most of the moons in the
solar system are covered with craters. A few of these
craters have been formed recently by the steady rain of
meteorites that falls on all the planets in the solar system, but most of the craters we see appear to have been
formed roughly 4 billion years ago in what is called the
heavy bombardment, as the last of the debris in the
solar nebula was swept up by the planets. We will find

more evidence of this bombardment in later chapters.
Thus, many of the last remaining solid objects in the
solar nebula were swept up by the planets soon after
they formed.

The fourth effect was the ejection of material from
the solar system by close encounters with planets. If
a small object such as a planetesimal passes close to a
planet, it can gain energy from the planet's gravitational
field and be thrown out of the solar system. Ejection is
most probable in encounters with massive planets, so
the Jovian planets were probably very efficient at ejecting the icy planetesimals that formed in their region of
the nebula.

Together these four effects cleared away the solar
nebula. This may have taken considerable time, but
eventually the solar system was relatively clear, the
planets could no longer gain mass, and planet building
ended.

R E V I E W Critical Inquiry

Why are there two kinds of planets in our solar system?

Planets begin forming from solid bits of matter, not from
gas. Consequently, the kind of planet that forms at a

given distance from the sun depends on the kind of compounds that can condense out of the gas to form solid particles. In the inner parts of the solar nebula, the temperature was so high that most of the gas could not condense to form solids. Only metals and silicates could form solid grains, and thus the innermost planets grew from this dense material. Much of the mass of the solar nebula consisted of hydrogen, helium, water vapor, and other gases, and they were present in the inner solar nebula but couldn't form solid grains. The small terrestrial planets couldn't grow from these gases, so the terrestrial planets are small and dense.

In the outer solar nebula, the composition of the gas was the same, but it was cold enough for water vapor to condense to form ice grains, and because hydrogen and oxygen are so abundant there was lots of ice available. Thus, the outer planets grew from solid bits of metal and silicate combined with large amounts of ice. The outer planets grew so rapidly that they became big enough to capture gas directly, and thus they became the hydrogen- and helium-rich Jovian worlds.

The condensation sequence combined with the solar nebula hypothesis gives us a way to understand the formation of the planets. Can we extend our insight to other stars? What would a planetary system be like if it formed around a star that was slightly hotter than the sun?

The great chain of origins leads from the first instant of the big bang through the birth of galaxies, the formation of stars, the origin of our solar system, and finally to us. One of the most elegant and complex links in that chain is planet building. As we explore the solar system in detail in the following chapters, we must stay alert for further clues to the birth of the planets.

Summary

René Descartes proposed that the solar system formed from a contracting vortex of matter, and Buffon later suggested that a passing star pulled matter out of the sun to form the planets. While Buffon's hypothesis was catastrophic, Laplace's nebular hypothesis was evolutionary. The nebular hypothesis required a contracting nebula to leave behind rings that formed each planet, but it could not explain the low angular momentum of the sun. The modern solar nebula hypothesis suggests that the sun was slowed by its magnetic field and mass loss. Disks of gas and dust have been found around young stars, so astronomers suspect that planetary systems are common. Roughly three dozen planets have been found orbiting other stars.

The solar nebula hypothesis proposes that the solar system began as a contracting cloud of gas and dust that flattened into a rotating disk. The center of this cloud eventually became the sun, and the planets formed in the disk of the nebula. The properties of this nebula and the processes that built the planets explain many of the characteristic properties of the solar system.

One of the most striking properties of the solar system is its disk shape. The orbits of the planets lie in nearly the same plane, and they all revolve around the sun in the same direction, counterclockwise as seen from the north. This is also true of many of the satellites of the planets. With only three exceptions, the planets rotate counterclockwise around axes roughly perpendicular to the plane of the solar system. This disk shape and the motion of the planets appear to have originated in the solar nebula.

Another striking feature of the solar system is the division of the planets into two families. The terrestrial planets, which are small and dense, lie in the inner part of the system. The Jovian planets are large, low-density worlds in the outer part of the system. In general, the closer a planet lies to the sun, the higher its uncompressed density.

The solar system is now filled with smaller bodies such as asteroids, comets, and meteors. The asteroids are small, rocky worlds, most of which orbit the sun between Jupiter and Mars. They appear to be material left over from the formation of the solar system.

Another important characteristic of the solar system bodies is their similar ages. Radioactive dating tells us that Earth, the moon, Mars, and meteorites are no older than about 4.6 billion years. Thus, it seems our solar system took shape about 4.6 billion years ago.

According to the condensation sequence, the inner part of the nebula was so hot that only high-density minerals could form solid grains. The outer regions, being cooler, condensed to form icy material of lower density. The planets grew from these solid materials, with the denser planets forming in the inner part of the nebula and the lower-density Jovian planets forming farther from the sun.

Planet building began as dust grains growing by condensation and accretion into planetesimals ranging from a kilometer to hundreds of kilometers in diameter. These planetesimals settled into a thin plane around the sun and accumulated into larger bodies, the largest of which grew the fastest and eventually became protoplanets.

Once a planet had formed from a large number of planetesimals, heat from radioactive decay could have melted it and allowed it to differentiate into a dense metallic core and a lower-density silicate crust. In fact, it is possible that the solar nebula cooled as the protoplanets grew so that the first planetesimals were metallic and later additions were silicate. It is also likely that the planets grew rapidly enough that the heat of formation released by the in-falling material melted the planets and allowed them to differentiate as they formed.

The Jovian planets probably grew rapidly from icy materials and became massive enough to attract and hold vast amounts of nebular gas. Heat of formation raised their temperatures very high when they were young, and Jupiter and Saturn still radiate more heat than they absorb from the sun.

Once the sun became a luminous object, it cleared the nebula as its light and solar wind pushed material out of the system. The planets helped by absorbing some planetesimals and ejecting others from the system. Once the solar system was clear of debris, planet building ended.

New Terms

passing star hypothesis	meteoroid
evolutionary hypothesis	meteorite
catastrophic hypothesis	half-life
nebular hypothesis	gravitational collapse
angular momentum problem	uncompressed density
solar nebula hypothesis	condensation sequence
extra-solar planets	planetesimal
terrestrial planet	condensation
Jovian planet	accretion
liquid metallic hydrogen	protoplanet
Galilean satellites	differentiation
asteroid	outgassing
comet	heat of formation
dirty snowball model	radiation pressure
meteor	heavy bombardment

Review Questions

1. What were the objections to the passing star hypothesis? to the nebular hypothesis?
2. What produced the helium now present in the sun's atmosphere? in Jupiter's atmosphere? in the sun's core?
3. What produced the iron in Earth's core and the heavier elements like gold and silver in Earth's crust?
4. What evidence do we have that disks of gas and dust are common around young stars?
5. According to the solar nebula theory, why is the sun's equator nearly in the plane of Earth's orbit?
6. Why does the solar nebula theory predict that planetary systems are common?
7. Why do we think the solar system formed about 4.6 billion years ago?
8. If you visited another planetary system, would you be surprised to find planets older than Earth? Why or why not?
9. Why is almost every solid surface in our solar system scarred by craters?
10. What is the difference between condensation and accretion?
11. Why don't terrestrial planets have rings and large satellite systems like the Jovian planets?
12. How does the solar nebula theory help us understand the location of asteroids?
13. How does the solar nebula theory explain the dramatic density difference between the terrestrial and Jovian planets?

14. If you visited some other planetary system in the act of building planets, would you expect to see the condensation sequence at work, or was it unique to our solar system?
15. Why do we expect to find that planets are differentiated?
16. What processes cleared the nebula away and ended planet building?

Discussion Questions

1. In your opinion, should all solar systems have asteroid belts? Should all solar systems show evidence of an age of heavy bombardment?
2. If the solar nebula hypothesis is correct, then there are probably more planets in the universe than stars. Do you agree? Why or why not?

Problems

1. If you observed the solar system from the nearest star (1.3 pc), what would the maximum angular separation be between Jupiter and the sun? (*Hint:* Use the small-angle formula.)
2. The brightest planet in our sky is Venus, which is sometimes as bright as apparent magnitude −4 when it is at a distance of about 1 AU. How many times fainter would it look from a distance of 1 parsec (206,265 AU)? What would its apparent magnitude be? (*Hints:* Remember the inverse square law, Chapter 5; see Chapter 2.)
3. What is the smallest-diameter crater you can identify in Figure 20-9? (*Hint:* See the table in the Appendix, Properties of the Planets, to find the diameter of Mercury in kilometers.)
4. A sample of a meteorite has been analyzed, and the result shows that out of every 1000 nuclei of ^{40}K originally in the meteorite, only 100 have not decayed. How old is the meteorite? (*Hint:* See Figure 20-18.)
5. In Table 20-2, which object's observed density differs least from its uncompressed density? Why?
6. What composition might we expect for a planet that formed in a region of the solar nebula where the temperature was about 100 K?
7. Suppose that Earth grew to its present size in 1 million years through the accretion of particles averaging 100 g each. On the average, how many particles did Earth capture per second? (*Hint:* See Appendix A to find Earth's mass.)
8. If you stood on Earth during its formation as described in Problem 7 and watched a region covering 100 m², how many impacts would you expect to see in an hour? (*Hints:*

Assume that Earth had its present radius. The surface area of a sphere is $4\pi r^2$.)

9. The velocity of the solar wind is roughly 400 km/s. How long does it take to travel from the sun to Pluto?

Critical Inquiries for the Web

1. How does our solar system compare with the others that have been found? Search the Internet for sites that give information about planetary systems around other stars. What kinds of planets have been detected by these searches so far? Discuss the selection effects (see Window on Science 13-2) that must be considered when interpreting these data.

2. The process of protoplanetary accretion is still not well understood. Search the Web for current research in this field. From the results of your search, outline the basic steps in the formation of a protoplanet through accretion. What specific factors are important in these models of planet building? Do these models produce planetary systems similar to the ones we know to exist?

3. How is radioactive dating carried out on meteorites and rocks from surfaces of various bodies in the solar system? Look for Web sites on the details of radioactive dating, and summarize the methods used to uncover the abundances of radioactive elements in a particular sample. (*Hint:* Try looking for information on how a particular meteorite—for example, the Martian meteorite ALH84001—was studied, what age range was determined, and what radioactive elements were used to arrive at the age.)

Exploring *The Sky*

1. Use *The Sky* to observe and describe the shape of the solar system as defined by

   a. The planets
   How to proceed: Enter the **3D Solar System Mode.** Simulate the motion of the terrestrial planets (planets out to Mars) with a time increment of one day. Use the up and down arrow keys to view the orbits from different angles. Note the general "flatness" of the inner solar system. Which of the orbits of the terrestrial planets appears to have the greatest inclination to Earth's orbit? Now zoom out to show all the planets, and again observe the motion of the planets from different angles. Again note the relative "flatness" of the outer solar system. Which of the orbits has the largest inclination to the plane of the solar system?

   b. Comets
   How to proceed: Zoom in so the orbit of Jupiter fills the screen, and run the simulation with a time increment of 10 days. Note that aside from showing you the planets, *The Sky* shows a number of preinstalled comets and asteroids. What can you say about the inclination of the orbits of the comets?

 Go to the Brooks/Cole Astronomy Resource Center (www. brookscole.com/astronomy) for critical thinking exercises, articles, and additional readings from InfoTrac College Edition, Brooks/Cole's online student library.

CHAPTER TWENTY-ONE

Planet Earth

Guidepost

Astronomy has been described as the science of everything above the clouds. Planetary astronomers, however, must also think about what lies below the clouds, because Earth is the basis for comparison with all other Earthlike planets. We know Earth well, and we can apply what we know about Earth to other worlds.

There is another reason for studying Earth in an astronomy course. Astronomy is really about us. Astronomy is exciting and fascinating because it helps us understand what we are and where we are in the universe. Thus, we cannot omit Earth from our discussion—it is where we are.

The next two chapters will discuss the Earthlike planets, but that will not end our thoughts on the Earth. The moons of the giant outer worlds will seem Earthlike in strange ways, and our discussion of the smaller bodies of our solar system will alert us to the dangers Earth faces. Throughout the rest of this book, we will remain painfully aware of the fragile beauty of our planet.

Stars die, but planets don't. That should give us a feeling of security, but we know that the survival of our planet is limited by the life expectancy of our star. Furthermore, although planets don't die, they do evolve, and our planet has changed dramatically since it formed. Mountains rise and weather away, the climate warms and then cools, and the continents themselves drift and grind against each other. To fully understand how our own planet works, we need to know how planets in general evolve. Why is Venus hot and Mars cold? Why does Mercury have many craters and Earth only a few?

To answer such questions, we must use **comparative planetology,** the study of planets through comparison and contrast, and Earth is our basis of comparison. One reason why we begin with Earth is that it is the planet we know best. We can study it carefully and then draw comparisons to other worlds. But another reason is that our home is a planet of extremes (Data File Two). Its interior is molten and generates a magnetic field. Its crust is active, with moving sections that push against each other and trigger earthquakes, volcanoes, and mountain building. Even Earth's atmosphere is extreme. Processes have altered Earth's air from its original composition to a highly unusual, oxygen-rich sea of gas. Once we understand Earth's complex properties, the remaining planets in our system should be easier to comprehend.

In our study of Earth, we will find a four-stage history of planetary development. The moon and all the terrestrial planets have passed through these stages, although differences in the way the planets were altered by these stages have produced dramatically different worlds. The moon, for example, is much like Earth, but its evolution has been dramatically altered by its smaller size. Thus, as we explore the solar system, we will discover not entirely new processes but rather familiar effects working in slightly different ways.

21-1 The Early History of Earth

Like all the terrestrial planets, Earth formed from the inner solar nebula about 4.6 billion years ago. As it took form, it began to change, passing through four developmental stages (Figure 21-1). Processes that occurred during these four stages determined the present condition of Earth's interior, crust, and atmosphere.

Four Stages of Planetary Development

The first stage of planetary evolution is *differentiation,* the separation of material according to density. Earth now has a dense core and a lower-density crust, and that structure must have originated very early.

Differentiation would have occurred easily if Earth was molten when it was young. Two sources of heat

Earth   Data File Two

A shaded relief map of Earth with the oceans removed reveals the geological features on the continents and under the oceans. *(National Geophysical Data Center)*

Average distance from the sun	1.00 AU (1.495979×10^8 km)
Eccentricity of orbit	0.0167
Maximum distance from the sun	1.0167 AU (1.5210×10^8 km)
Minimum distance from the sun	0.9833 AU (1.4710×10^8 km)
Inclination of orbit to ecliptic	0°
Average orbital velocity	29.79 km/s
Orbital period	1.00 y (365.26 days)
Period of rotation (with respect to the sun)	$24^h00^m00^s$
Inclination of equator to orbit	23°27′
Equatorial diameter	12,756 km
Mass	5.976×10^{24} kg
Average density	5.497 g/cm³ (4.2 g/cm³ uncompressed)
Surface gravity	1.0 Earth gravities
Escape velocity	11.2 km/s
Surface temperature	−50° to 50°C (−60° to 120°F)
Average albedo*	0.39
Oblateness	0.0034

*Albedo is the fraction of light striking an object that the object reflects. The albedo of a perfect reflector is 1.

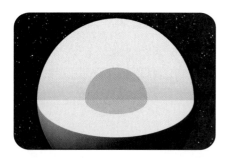

Figure 21-1
The four stages of planetary development. First: Differentiation into core and crust as dense material sinks to the center and low-density material floats to the surface. Second: Cratering of the newly formed crust by the heavy bombardment of meteorites in the young, debris-filled solar nebula. Third: Flooding of lowlands by lava from the interior and later by water from the cooling atmosphere. Fourth: Slow surface evolution due to geological activity and erosion.

could have heated Earth. First, heat of formation would be created by in-falling material. A meteorite hitting Earth at high velocity converts most of its energy of motion into heat, and the in-falling of a large number of meteorites could release tremendous heat. If Earth formed rapidly, this heat would have accumulated much more rapidly than it could leak away, and Earth may have been molten when it formed. A second source of heat requires more time to develop. The decay of radioactive elements trapped in the Earth releases heat

gradually, but as soon as Earth formed, that heat would have begun to accumulate and could have helped melt Earth to facilitate differentiation. Most of Earth's radioactive elements are now concentrated in the crust, where they continue to warm and soften the rock layers.

Earth formed by material falling together, but meteorites could have left no trace until a crust solidified. Once Earth had a hard surface, the meteorites could form craters. This second stage in planetary evolution, *cratering,* was violent. The heavy bombardment was intense because the solar nebula was filled with rocky and icy debris, and the young Earth was battered by meteorites that pulverized the newly forming crust. The largest meteorites blasted out crater basins hundreds of kilometers in diameter. As the solar nebula cleared, the amount of debris decreased, and the level of cratering fell to its present low level. Although meteorites still occasionally strike Earth and dig craters, cratering is no longer the dominant influence on Earth's geology. As we compare other worlds with Earth, we will discover traces of this intense period of cratering, the heavy bombardment, on every old surface in the solar system.

The third stage, *flooding,* no doubt began while cratering was still intense. The fracturing of the crust and the heating caused by radioactive decay allowed molten rock to well up through fissures and flood the deeper basins. We will discuss such flooded basins on other worlds, such as the moon, but all traces of this early lava flooding have been destroyed by later geological activity in Earth's crust. On Earth, flooding continued as the atmosphere cooled and water fell as rain, filling the deepest basins to produce the first oceans. Notice that on Earth flooding involves both lava and water, a circumstance that we will not find on most worlds.

The fourth stage, *slow surface evolution,* has continued for the last 3.5 billion years or more. Earth's surface is constantly changing as sections of crust slide over each other, push up mountains, and shift continents. Also, moving air and water erode the surface and wear away geological features. Almost all traces of the first billion years of Earth's geology have been destroyed by the active crust and erosion.

Earth as a Planet

We are studying Earth in this chapter in order to use it as a basis for comparison with other worlds. Of course, four of the worlds in our solar system, the Jovian planets, are dramatically different from Earth, so we will not try to compare them with Earth. But the four terrestrial planets are rocky worlds much like Earth, and many of the moons in the solar system have geology that we can compare with Earth and its four stages of development.

All terrestrial planets pass through these four stages, so in that respect Earth is a good basic reference planet for comparative planetology. Some planets have em-

a

b

Figure 21-2
(a) A typical place on Earth. Most of Earth is covered by liquid water. *(William K. Hartmann)* (b) Moving water in the form of liquid or solid can erode entire mountain ranges to nothing in a tiny fraction of Earth's total age. *(Janet Seeds)*

phasized one stage over another, and some planets have failed to progress fully through the four stages. Nevertheless, Earth is a good standard of comparison. Every major process on any rocky world in our solar system is represented in some form on Earth. Clearly, Earth is a highly complex world; for that reason, it makes a good standard of comparison.

On the other hand, Earth is peculiar in two ways. First, it has large amounts of liquid water on its surface (Figure 21-2a). Fully 75 percent of its surface is covered by this liquid, and no other planet in our solar system is known to have such extensive liquid water. Not only does water fill the oceans, but it evaporates into the atmosphere, forms clouds, and then falls as rain. Water falling on the continents flows downhill to form rivers that flow back to the sea, and, in so doing, the water produces extensive erosion (Figure 21-2b). Entire mountain ranges can literally dissolve and wash away in only a few tens of millions of years, less than 1 percent of Earth's total age. We will not see such intense erosion on most worlds. Liquid water is, in fact, a rare material on most planets.

Our planet is special in a second way. Some of the matter on the surface of this world is alive, and a small part of that living matter is aware. We do not know how the presence of living matter has affected the evolution of Earth, but this process seems to be totally missing from other worlds in our solar system. Furthermore, the thinking part of the life on Earth, humankind, is actively altering our planet. It does not seem likely that we have made major changes in Earth yet, but our study of Earth as a standard of comparison with other worlds will also give us new insight into our own planet and how we may be altering it.

R E V I E W Critical Inquiry

Why do we think Earth went through a stage of cratering?

Recall from the previous chapter that the planets formed by the accretion of planetesimals from the solar nebula. The protoearth may have been molten as it formed, but as soon as it grew cool enough to form a solid crust, the remaining planetesimal impacts would have formed craters. Thus, we can reason from the solar nebula hypothesis that Earth should have been cratered. But we can't use a theory as evidence to support some other theory. To find real observational evidence, we need only look at the moon. The moon has craters, and so does every other old surface in our solar system. There must have been a time, when the solar system was young, when there were large numbers of objects striking all the planets and moons and blasting out craters. If it happened to other worlds in our solar system, it must have happened to Earth, too.

Of course, we don't see those craters on Earth today. Why not?

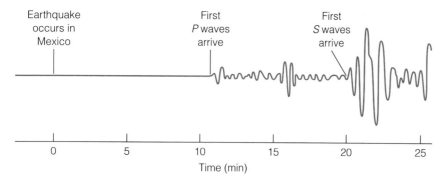

First *P* waves arrive

First *S* waves arrive

Time (min)

Figure 21-3
A seismograph in northern Canada made this record of seismic waves from an earthquake in Mexico. The first vibrations, *P* waves, arrived 11 minutes after the quake, but the slower *S* waves took 20 minutes to make the journey.

As we explore our solar system in the chapters that follow, Earth will be our standard of comparison. Our goal in this chapter is to study Earth's interior, crust, and atmosphere in order to establish a basis for comparison with other planets. Only by understanding Earth in detail can we understand planets in general.

21-2 The Solid Earth

Although we might think of Earth as made of solid rock, it is in fact neither entirely solid nor entirely rock. The thin crust seems solid, but it floats and shifts on a semiliquid layer of molten rock just below the crust. Below that lies a deep, rocky mantle surrounding a core of liquid metal. Much of what we see on the surface of Earth is determined by its interior.

Earth's Interior

The theory of the origin of planets from the solar nebula predicts that Earth should have melted and differentiated into a dense metallic core with a low-density silicate crust. But did it differentiate? Clearly, Earth's surface is made of silicates, but what of the interior?

High temperature and tremendous pressure in Earth's interior make any direct exploration impossible. Even the deepest oil wells extend only a few kilometers down and don't reach through the crust. It is quite impossible to drill far enough to sample Earth's core. Yet Earth scientists have studied the interior and found clear proof that Earth did differentiate.

This analysis of Earth's interior is possible because earthquakes produce vibrations called **seismic waves,** which travel through the interior and eventually register on sensitive detectors, known as **seismographs,** all over the world (Figure 21-3). Two kinds of seismic waves are important to us here. The *P,* or **pressure, waves** are much like sound waves in that they travel as a region of compression. As a *P* wave passes, particles of matter vibrate back and forth parallel to the direction of wave travel (Figure 21-4a). In contrast, the *S,* or **shear, waves** move as displacements of particles perpendicular to

the waves' direction of travel (Figure 21-4b). Thus, *S* waves distort the material but do not compress it. Normal sound waves are pressure waves, whereas the vibrations in a bowl of jelly are shear waves. Because *P* waves are compression waves, they can move through a liquid, but *S* waves cannot. A glass of water can't

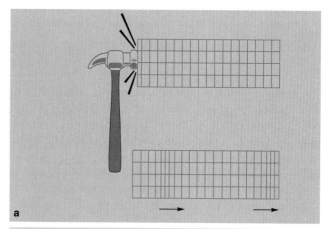

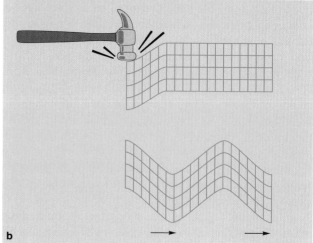

Figure 21-4
(a) *P,* or pressure, waves, like sound waves in air, travel as a region of compression. (b) *S,* or shear, waves, like vibrations in a bowl of jelly, travel as displacements perpendicular to the direction of travel. *S* waves tend to travel more slowly than *P* waves and cannot travel through liquids.

shimmy like jelly because a liquid does not have the rigidity required to transmit *S* waves.

The *P* and *S* waves caused by an earthquake do not travel in straight lines or at constant speed within Earth. The waves may reflect off boundaries between layers of different density, or they may be refracted as they pass through such a boundary. In addition, the gradual increase in density toward Earth's center means the speed of sound is greater toward the center. Sound waves approaching such a region tend to curve away (refract), so seismic waves do not travel in straight lines through Earth's interior but are refracted to follow curved paths. Earth scientists can use the arrival times of reflected and refracted seismic waves from distant earthquakes to construct a picture of Earth's interior.

Such studies show that the interior consists of three parts: a central core, a thick mantle, and a thin crust. *S* waves give us an important clue to the nature of the core, because no such waves are observed to travel through the core. When an earthquake occurs, no direct *S* waves register on seismographs on the opposite side of Earth, as if the core were casting a shadow (Figure 21-5). This shows that the core is mostly liquid, and the size of the shadow fixes the size of the core at about 55 percent of Earth's radius. Theoretical calculations predict that the core is hot (about 6000 K), dense (about 14 g/cm^3), and composed of iron and nickel.

Although the core is hot, it is not entirely molten. Nearer the center the material is under higher pressure, which in turn raises the melting point so high that the material cannot melt at the existing temperature (Figure 21-6). Thus, there is an inner core of solid iron and nickel. Estimates suggest the inner core's radius is about 22 percent that of Earth.

The **mantle** is the layer of dense rock and metal oxides that lies between the molten core and the crust. The paths of seismic waves in the mantle show that it is not molten, but it is not precisely solid either. Mantle material behaves like a **plastic,** a material with the properties of a solid but capable of flowing under pressure. The asphalt used in paving roads is a common ex-

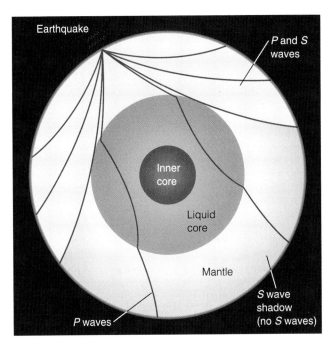

Figure 21-5
P and *S* waves give us clues to Earth's interior. That no direct *S* waves reach the far side shows that Earth's core is liquid. The size of the *S* wave shadow tells us the size of the outer core.

ample of a plastic. It shatters if struck with a sledge-hammer, but it bends under the weight of a heavy truck. Just below Earth's crust, where the pressure is less than at greater depths, the mantle is most plastic.

Earth's rocky crust is made up of low-density rocks and floats on the denser mantle. The crust is thickest under the continents, up to 60 km thick, and thinnest under the oceans, where it is only about 10 km thick. Unlike the mantle, the crust is brittle and breaks much more easily than the plastic mantle.

Perhaps no region is more immediate yet more inaccessible than Earth's core. Only a few thousand kilometers from where we sit, Earth's interior is beyond our immediate reach. Yet we can send our imagination

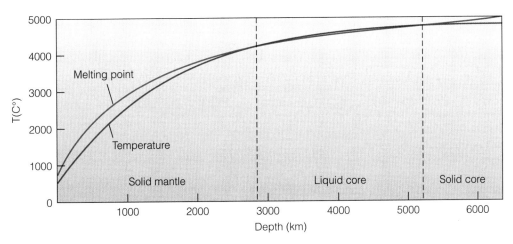

Figure 21-6
Theoretical models combined with observations of the velocity of seismic waves reveal the temperature inside Earth (blue line). The melting point of the material (red line) is determined by its composition and by the pressure. In the mantle and in the inner core, the melting point is higher than the existing temperature, and the material is not molten.

One of the most fascinating aspects of science is its power to show us the unseen. That is, it gives us an opportunity to learn about regions we can never visit. We saw this in earlier chapters when we discussed the inside of the sun and stars, the surface of neutron stars, the event horizon around black holes, the cores of active galaxies, and more. In this chapter, we have seen Earth's core.

Not only does science take us to places we can never visit, but it takes us to scales we can never explore. For example, physicists can explore the inside of an atom, and biologists can study the structure of a virus. Not only can we stretch the scale of space, but we can also stretch the scale of time. Geologists can watch mountain ranges rising slowly from the plains and eroding away to grit. Astronomers can watch the 200-million-year-long whirl of two galaxies colliding.

In this respect, astronomy, geology, and psychology are particularly similar. All three sciences study realms that cannot be explored in person. The insides of stars, Earth's ancient history, and the workings of the human mind lie beyond our immediate experience. Only by careful observation and careful thought can scientists learn to understand these aspects of the natural world.

Science can reveal the unseen because it can combine human imagination with the predictive power of the laws of nature and the precision of mathematics. This calls on scientists to be as creative as artists and as precise as surgeons to produce reliable theories and models, but the reward is one of the great thrills of science—exploring beyond the limits of normal human experience.

where our bodies can never go (Window on Science 21-1). Earth's seismic activity gives us evidence of Earth's innermost secrets. But there is another source of evidence about Earth's interior—its magnetic field.

The Magnetic Field

Apparently, Earth's magnetic field is a direct result of its rapid rotation and its molten metallic core. The internal heat forces the liquid core to churn with convection while Earth's rotation turns the core about an axis. This core is a highly conductive iron–nickel alloy, a better electrical conductor than copper, the material commonly used for electrical wiring. The rotation of this convecting, conducting liquid generates Earth's magnetic field in a process called the dynamo effect (Figure 21-7). This is believed to be the same process that generates the solar magnetic field in the convective layers of the sun, and we will meet it again when we explore other planets.

Earth's magnetic field protects it from the solar wind. Blowing outward from the sun at about 400 km/s, the solar wind consists of ionized gases carrying a small part of the sun's magnetic field. When the solar wind encounters Earth's magnetic field, it is deflected like water flowing around a boulder in a stream. The surface where the solar wind is first deflected is called the **bow shock,** and the cavity dominated by Earth's magnetic field is called the **magnetosphere** (Figure 21-8a). High-energy particles from the solar wind leak into the magnetosphere and become trapped within Earth's magnetic field to produce the **Van Allen belts** of radiation. We will see in later chapters that all planets that have magnetic fields have bow shocks, magnetospheres, and radiation belts.

One dramatic result of Earth's magnetic field is the aurora, glowing rays and curtains of light in the upper atmosphere (Figure 21-8b). The solar wind carries charged particles past Earth's extended magnetic field, and this generates tremendous electrical currents that flow into Earth's atmosphere near the north and south

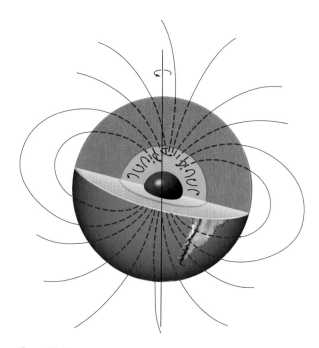

Figure 21-7
The dynamo effect couples convection in the liquid core with Earth's rotation to produce electric currents that are believed to be responsible for Earth's magnetic field.

magnetic poles. The currents ionize gas atoms in Earth's atmosphere, and when the ionized atoms capture electrons and recombine, they emit light as if they were part of a vast "neon" sign. Thus, the spectrum of an aurora is an emission spectrum.

Although we are confident that Earth's magnetic field is generated within its molten core, many mysteries remain. For example, rocks retain traces of the magnetic field in which they solidify, and some contain fields that point backward. That is, they imply that Earth's magnetic field was reversed at the time they solidified. Careful analysis of such rocks indicates that Earth's field has reversed itself every million years or so, with the north magnetic pole becoming the south magnetic pole and vice versa. These reversals are poorly understood, but they may be related to changes in the core convection.

Convection in Earth's core is important because it generates the magnetic field. As we will see in the next section, convection in the mantle constantly remakes Earth's surface.

Earth's Crust

Earth's surface is active. It is continually destroyed and renewed as large sections of crust move about. Geologists refer to this phenomenon as **plate tectonics.** *Plate* refers to the sections of moving crust, and *tectonics* comes from the Greek word for "builder." Plate tectonics is the builder of Earth's surface, because interactions between plates destroy old terrain, push up mountains, and create new crust.

The plates consist of an upper layer of very low density rock plus the uppermost layers of the mantle. These layers make up the **lithosphere,** literally the rock-sphere at Earth's surface. At the base of the lithosphere, at a depth of about 100 km, the melting point is low enough that the rock becomes highly plastic. That is, it deforms and flows easily. Thus, the sections of lithosphere—the tectonic plates—move easily on this plastic layer.

Two features of the ocean floor give us dramatic evidence of plate tectonics. In many places, the ocean floor is marked by a long underwater mountain range called a **midocean rise,** which is split by a **midocean rift** where the ocean crust is pulling apart at the rate of 2 to 4 cm per year and lava flows are welling up to form new ocean crust. (You may have heard this called "seafloor spreading.") In other places, sections of ocean floor are pushed below continents into **subduction zones** in a process that builds mountains and triggers volcanism (Figure 21-9).

The evidence is clear that midocean rises are young. The rocks near a midocean rise are geologically young **basalt,** rock formed by solidified lava, and the further the basalt is from the rift the older it is. Also, the ocean floor farther from a rift has accumulated more sediment than the ocean floor near the rift. Further evidence lies in the magnetic properties of the ocean floor. As molten rock wells up along a midocean rift and hardens, it traps a record of Earth's magnetic field. As the ocean crust spreads and Earth's magnetic field periodically reverses, the newly created basalt records the changing direction of Earth's magnetic field in a series of bands that run parallel to the rift and mirror each other on opposite sides of the rift. This

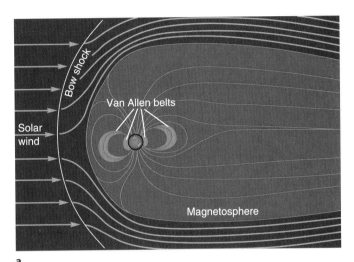

a

b

Figure 21-8
(a) Earth's magnetic field dominates space around Earth by deflecting the solar wind around a region called the magnetosphere and by trapping high-energy particles to form radiation belts. (b) Electrical currents in Earth's magnetic field flow downward into Earth's atmosphere and stimulate atmospheric atoms to emit photons, which we see as auroral displays. The general red glow in this photograph and the "curtain" effect are both examples of auroras. *(b, Courtesy David Miller)*

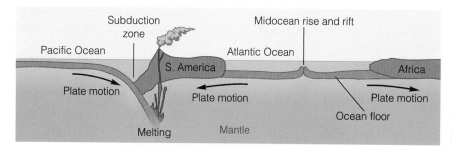

Figure 21-9

A cross section of Earth's crust shows the thick crust of the continents and the thinner crust of the sea floors. The motion of crustal plates allows rising magma to build new crust at the midocean rift, and old sea floor descends under the continents in subduction zones where it triggers mountain building and volcanism.

was first detected near Iceland on the midocean rift that runs from the South Atlantic into the North Atlantic (Figure 21-10).

Like a conveyor belt, ocean crust is created in midocean rifts, spreads outward, and disappears down subduction zones to be remelted by the heat of Earth's mantle. Where ocean crust slides downward under a continent, it crumples the crust at the edge of the continent to build mountains, and the melting of the ocean crust releases low-density magma that rises to the surface to create volcanoes. The Andes Mountains and associated volcanism along the west coast of South

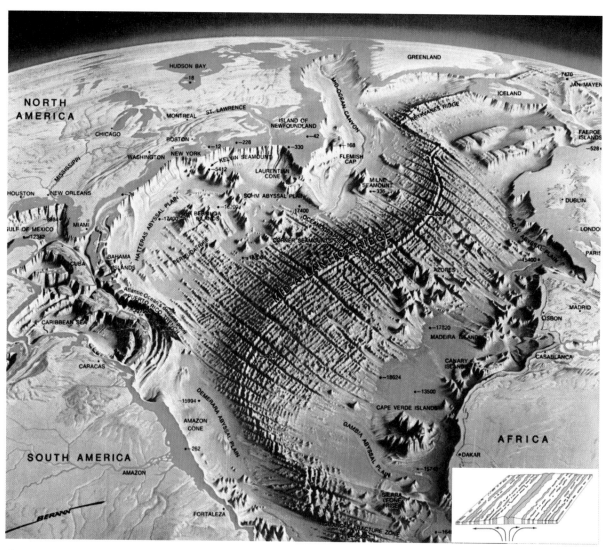

Figure 21-10

If we could drain the Atlantic Ocean, we would see the midocean rift snaking along the crest of the midocean rise from the South Atlantic all the way to Iceland and beyond. The ocean floor spreads open along the rift as the continents drift apart. Magma rising into the rift solidifies, recording the alternation of Earth's magnetic field (inset) in a symmetric pattern of normal (dark) and reversed (white) fields. *(Alcoa)*

America are caused by the ocean floor sliding under the continent. The coastal volcanoes (including Mount St. Helens) extending from northern California into Alaska are caused by a tectonic plate plunging below North America. Detailed images of Earth's crust created from satellite data show the worldwide network of midocean ridges and subduction zones (Figure 21-11).

One of the telltale marks of plate tectonics is **folded mountain chains,** long, linear ranges of mountains that look much like folds in a rug that has been pushed up against a wall. These mountain ranges are caused by the collision of plates and the compression of Earth's crust. The Allegheny Mountains are typical folded mountain ranges. We will search for such folded mountain ranges on other planets.

When a seafloor plate collides with a continental plate, it does not always slip below the continent. The floor of the Atlantic Ocean is locked to North and South America, and, as the Atlantic ocean spreads, North and South America are drifting westward. This continental drift can be measured by radio astronomers who use pulsars to calculate the distance between radio antennas in Europe and North America. The radio telescopes are moving apart at roughly 3 cm per year.

Crustal plates can also split apart to open new seas. The first sign of such a splitting is a long, straight, deep depression called a **rift valley.** Africa has recently split from Arabia, opening a rift valley now filled by the Red Sea. Other rift valleys extend southward from the Red Sea across eastern Africa. We will stay alert for rift valleys on other worlds.

The crustal plates on Earth have drifted about for much of the history of our planet, sometimes coming together to form very large continents and sometimes separating. Roughly 250 million years ago, the continents were all part of a single large continent called

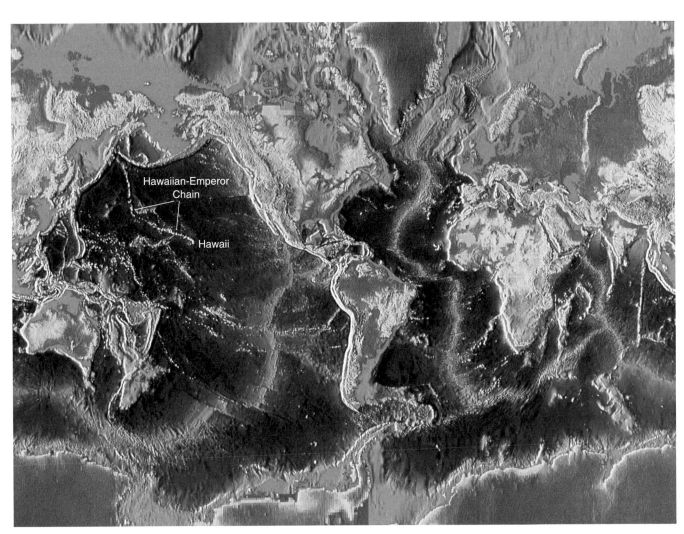

Figure 21-11
This color-coded map of Earth reveals a worldwide network of midocean rises and subduction zones such as those near Japan and the west coast of South America. The Hawaiian-Emperor chain of extinct undersea volcanoes extends northwest from Hawaii and then turns north, evidently because the direction of plate motion changed. (Compare with Figure 21-10.) *(National Geophysical Data Center)*

Pangaea. The east coast of North America was pushed against Africa, which pushed up great mountain ranges as high as the Himalayas. We see the remains of these as the Allegheny Mountains. Beginning roughly 200 million years ago, the plates spread apart to open new oceans (Figure 21-12). The continents continue to move, as demonstrated by India colliding with Asia and pushing up the Himalaya Mountains.

Earthquakes outline the edges of the plates (Figure 21-13). Small shallow earthquakes mark the midocean rifts as they pull apart, and major earthquakes occur where plates collide as in northern India. In addition, sections of crust in direct contact may slip past each other, as in the case of the Pacific plate, which is carrying part of southern California northward along the San Andreas Fault and generating frequent earthquakes. Volcanism too outlines the plates along the midocean rifts and subduction zones. The volcanism surrounding the Pacific Ocean has been called the Ring of Fire.

In a few places, such as Hawaii, volcanism occurs in the middle of a plate because a rising current of hot magma in the mantle punches through the plate to form volcanoes. The movement of the plate can result in a chain of volcanoes, as shown in Figure 21-11.

All of the tectonic activity we see on our world is driven by heat rising from the interior. Just as heat flowing out of the sun causes active regions and eruptions on the sun's surface, so too does the heat flowing out of Earth's interior produce plate motions, mountain building, volcanism, and earthquakes.

The Astronomer's Geology

As astronomers, we want to use Earth as a basis for comparison with other worlds, and thus we must take an astronomer's view of geology. As we study geological features on other worlds, we will discover that Earth is astonishing for its geological activity, both through plate tectonics and erosion.

Earth's surface seems to change very slowly, but that is only because we live such short lives. Fifteen minutes at a bus stop seem an eternity, and a night sleeping in an airport can be an eon. But, in truth, a human lifetime is less than an eye blink in geological terms.

The Grand Canyon (Figure 21-14a) seems a vast and imposing geological feature, and we are not surprised to hear that the Colorado River took 10 million years to dig the canyon. The bottom of the canyon is over a mile (1.6 km) below the rim, and it takes several hours to hike to the bottom. The Grand Canyon is a geological paradise, but the astronomer sees it as a flicker in Earth's history. When we say that the Grand Canyon is 10 million years old, it sounds imposing, but remember that Earth is 4.6 billion years old. If we express 10 million in units of billions, we discover that the river began carving the canyon only 0.01 billion years ago, only a tiny fraction of Earth's history (Figure 21-14b). Although the human race has dammed the river and temporarily reduced its erosive power, the dam cannot survive more than a few centuries, and once humanity is done playing with the river, it will resume its business in the canyon. The entire canyon will probably be gone in a few times 0.01 billion years.

The astronomer takes a long view of geology. Entire mountain ranges can be pushed up and eroded away to sand in a few times 0.01 billion years. The continents drift and jostle against each other, and we study the breakup of Pangaea 200 million years ago. That is only 0.2 billion years ago. The oldest horizontal sedimentary layers at the bottom of the Grand Canyon were laid down 570 million years ago, and they lie on top of

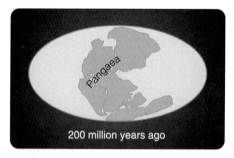

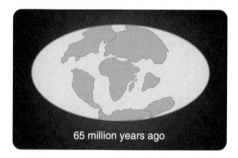

Figure 21-12
Continental drift has broken Pangaea first into Laurasia and Gondwanaland and eventually into the continents we know. *(Adapted from R. S. Dietz and J. C. Holden.* Scientific American *223, April 1970)*

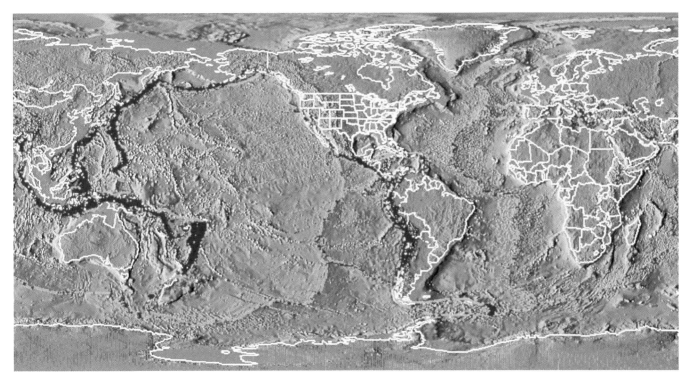

Figure 21-13
Earthquakes, plotted as red dots, outline the edges of Earth's crustal plates (yellow). Volcanism follows the same pattern, and much of the Pacific is ringed by volcanic peaks that extend from Chile through Central America, Alaska, and south through Japan—the "ring of fire." Earthquakes and volcanism in Hawaii are produced where a hot spot in the mantle penetrates upward through the middle of a plate. *(National Geophysical Data Center)*

the remains of tilted and eroded layers that were deposited over a hundred million years before and then almost entirely eroded away. All of that happened within the last 1 billion years.

As we visit the other terrestrial planets, we will discover dry riverbeds 2 billion years old, lava flows 3 billion years old, and craters that are 4 billion years old. Almost nowhere on Earth can we find geological

a

Figure 21-14
(a) The Grand Canyon illustrates the impermanence of geological features on Earth. The Colorado River began cutting the canyon only 0.01 billion years ago, and the oldest horizontal sedimentary layers at the bottom of the canyon were laid down in the bottom of an ocean 0.57 billion years ago. *(M. A. Seeds)* (b) On a time line of Earth's geological history, the cutting of the Grand Canyon can be represented as a narrow line. Many of the geological features we know on Earth and the rise of complex living things have been recent events.

Formation of Grand Canyon
Age of dinosaurs
Breakup of Pangaea
First animals emerge on land

Formation of Earth
Heavy bombardment
?
Oldest fossil life

4.6   4   3   2   1   Now
Billions of years ago

b

features with such great ages. The oldest surviving portions of Earth's crust, including parts of Canada and parts of South Africa and Australia, are about 4 billion years old. Most of Earth's crust is much younger. The constant churning of Earth's surface has wiped away all record of older crust.

Plate tectonics also explains why Earth contains so few impact craters. The moon is richly cratered, but Earth contains only about 150 impact craters. Plate tectonics has subducted and melted all but the more recent craters on Earth.

As astronomers, we see Earth's geology dominated by two dramatic forces. Just below the thin crust of solid rock lies a churning molten interior that rips the crust to fragments and pushes the pieces about like bits of algae on a pond. The second force modifying the crust is water. It falls as rain and snow and tears down the mountains, erodes the river valleys, and washes any raised ground into the sea. Like angry children, tectonics and erosion build mountains and continents and then rip them to nothing.

R E V I E W Critical Inquiry

What evidence do we have that Earth has a molten metallic core?

In science, the critical analysis of ideas must eventually return to evidence. In this case, the evidence is indirect because we can never visit Earth's core. Seismic waves from distant earthquakes pass through Earth, but a certain kind of wave, the *S* waves, cannot pass through the core. Because the *S* waves cannot propagate through a liquid, we suspect that Earth's core is a liquid. Earth's magnetic field gives us further evidence of a metallic core. The theory for the generation of magnetic fields, the dynamo effect, requires a liquid, convective, conducting core. If the core were not a liquid, it would not be able to generate a magnetic field. Thus, we have two different kinds of evidence to tell us that our planet has a liquid core.

Every theory must be supported by evidence. What evidence do we have to support the theory of plate tectonics?

For the astronomer, accustomed to the longest view of cosmic time, Earth's geology seems not so much the history of slow-moving rock as it does the churning surface of a pot of boiling soup. And just as steam rises from boiling soup, vapors rise from Earth's crust. Our atmosphere not only acts to erode the rocky surface, but some of it came from the crust as a by-product of geological activity.

21-3 The Atmosphere

We cannot complete our four-stage history of Earth without mentioning the atmosphere. Not only is it necessary for life, but it is also intimately related to the crust. It affects the surface by eroding geological features through wind and water, and in turn the chemistry of Earth's surface affects the composition of the atmosphere.

Origin of the Atmosphere

Until recent years, Earth scientists believed that Earth's atmosphere had developed slowly by outgassing, the release of gases from a planet's interior. According to this picture, Earth formed by the slow accumulation of planetesimals into a large, cold ball of matter whose composition was uniform from surface to center (Figure 20-20). Early Earth might have attracted gases such as hydrogen, helium, and even methane and ammonia from the solar nebula to form a **primeval atmosphere.** The slow decay of uranium, thorium, and potassium heated Earth's interior, melted it, caused it to differentiate into core, mantle, and crust, and drove out gases that diluted the primeval atmosphere. These products of outgassing became a **secondary atmosphere** rich in CO_2 and H_2O.

A more recent understanding of how the solar system formed suggests that Earth formed rapidly and that the heat of formation released by the in-falling planetesimals melted Earth as it formed. Thus, any volatiles in the rock, such as CO_2 and H_2O, would have been driven out immediately, and the atmosphere would have grown rich in these gases from the beginning.

For some years, astronomers have suspected that the water on Earth arrived late in the formation process as a bombardment of volatile-rich planetesimals. These icy bodies, the theory goes, were scattered by the growing mass of the outer planets, and the inner solar system was bombarded by a storm of comets. This theory now faces a serious objection. Spectrographic studies of comets reveal that the ratio of deuterium to hydrogen in comets does not match the ratio in the water on Earth. This seems to tell us that the water in our oceans was never part of the icy bodies of the outer solar system. The story of the origin of Earth's atmosphere and water is yet to be fully understood.

However the atmosphere originated, the mix of gases must have changed. The young atmosphere must have been rich in water vapor, carbon dioxide, and other volatiles. As the atmosphere cooled, the water condensed to form the first oceans, and as the oceans grew bigger, carbon dioxide began to dissolve in the water. Carbon dioxide is easily soluble in water—which is why carbonated beverages are so easy to manufacture—

but the first oceans could not have absorbed all of the atmospheric carbon dioxide if the gas had not reacted with dissolved substances in the seawater to form silicon dioxide, limestone, and other mineral sediments. Thus, chemical reactions in the oceans transferred the carbon dioxide from the atmosphere to the seafloor.

Any methane or ammonia in the atmosphere would have been broken up by ultraviolet radiation from the sun. Earth's lower atmosphere is now protected from ultraviolet radiation by an **ozone layer** about 25 km above the surface. However, the atmosphere of the young Earth did not contain free oxygen, so an ozone layer could not form, and the sun's ultraviolet radiation penetrated deep into the atmosphere. The energetic ultraviolet photons broke up water, methane (CH_4), and ammonia (NH_3), and the hydrogen escaped to space. Carbon and oxygen formed carbon dioxide and dissolved in the oceans, leaving nitrogen to accumulate in the atmosphere (Table 21-1).

The origin of the atmospheric oxygen is linked to the origin of life, the subject of Chapter 27, but it is sufficient here to note that life must have originated within a billion years of Earth's formation. This life did not significantly alter the atmosphere, however, until nature invented photosynthesis about 3.3 billion years ago. Photosynthesis absorbs carbon dioxide from the air or water and utilizes it for plant growth, releasing oxygen. Because oxygen is a very reactive element, it combines easily with other compounds, and thus the oxygen abundance grew slowly at first. Apparently, the development of large, shallow seas along the continental margins half a billion years ago allowed ocean plants to manufacture oxygen faster than chemical reactions could consume it. Atmospheric oxygen then increased rapidly, and it is still increasing at the rate of about 1 percent every 36 million years.

Earth's climate is critically sensitive to many different factors. For example, the **albedo** of a world is the fraction of the sunlight that hits it that gets reflected away. An albedo of 1 would be a perfectly white world, and an albedo of 0 would be perfectly black. Earth's overall albedo is 0.39, meaning it reflects back into space 39 percent of the sunlight that hits it. Much of this reflection is caused by clouds, and the formation of clouds depends critically on the presence of water vapor in the upper atmosphere, the temperature of the upper atmosphere, and the patterns of atmospheric circulation. Even a small change in any of these factors could change Earth's albedo and thus its climate. We are just beginning to understand all of the factors that have combined to assure our comfortable lives on the surface of our world.

Human Effects on Earth's Atmosphere

Earth is comfortable for human life because of our atmosphere. The temperature is moderate at sea level, but if we climb to the top of a high mountain, we find the temperature to be very low (Figure 21-15). Most clouds form at such altitudes. Higher still we find the air much colder and so thin it cannot protect us from the intense ultraviolet radiation in sunlight. We can walk Earth's surface in safety because our atmosphere protects us, but we are altering our atmosphere in at least two serious ways: We are adding carbon dioxide (CO_2), and we are destroying ozone.

The concentration of CO_2 in our atmosphere is important to us because CO_2 can trap heat in a process called the **greenhouse effect** (Figure 21-16a). When sunlight shines through the glass roof of a greenhouse, it heats the benches and plants inside. The warmed interior radiates infrared radiation, but the infrared cannot

Table 21-1 Earth's Atmosphere

Gas	Percent by Weight
N_2	75.5
O_2	23.1
Ar	1.29
CO_2	0.05
Ne	0.0013
He	0.00007
CH_4	0.0001
Kr	0.0003
H_2O (vapor)	1.7–0.06

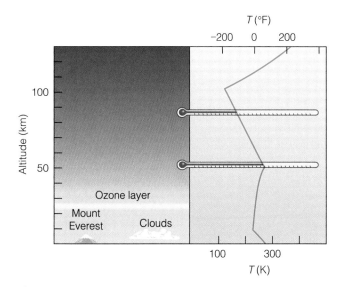

Figure 21-15
Thermometers placed in Earth's atmosphere at different levels would register the temperatures shown in the graph at the right. The lower few kilometers where we live is comfortable, but higher in the atmosphere the temperature is quite low. The ozone layer lies about 25 km above Earth's surface.

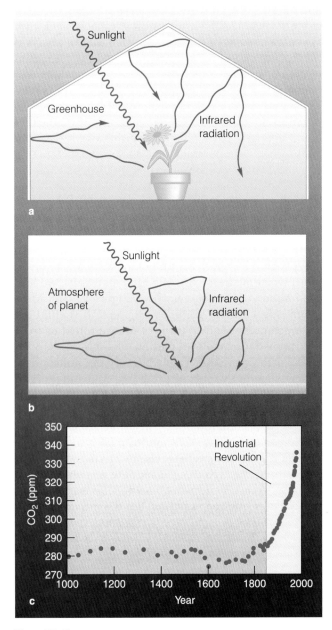

Figure 21-16
The greenhouse effect. (a) Short-wavelength light can enter a greenhouse and heat its contents, but the longer-wavelength infrared radiated by the warm interior cannot get out. (b) The same process can heat a planet's surface if its atmosphere contains CO_2. (c) The concentration of CO_2 in Earth's atmosphere as measured in Antarctic ice cores remained roughly constant until the Industrial Revolution. *(Adapted from a figure by Etheridge, Steele, Langenfelds, Francey, Barnola, and Morgan.)*

get out through the glass. Heat is trapped within the greenhouse, and the temperature climbs until the glass itself grows warm enough to radiate heat away as fast as the sunlight enters. In Earth's atmosphere, the CO_2 admits sunlight, and the surface grows warm. However, the CO_2 is not transparent to infrared radiation, so some of the heat is trapped, and that helps keep our planet warm (Figure 21-16b).

CO_2 is not the only greenhouse gas. Water vapor, methane, and other gases in our atmosphere contribute to warming Earth. But CO_2 is the most important greenhouse gas, and it is the one that the human race is changing. We are adding CO_2 to our atmosphere and making Earth warmer. For over 4 billion years, natural processes on Earth have removed CO_2 from the atmosphere and buried the carbon in the form of coal, oil, and natural gas. Since the beginning of the industrial revolution, our civilization has been digging up this carbon, burning it to get energy, and releasing CO_2 back into the atmosphere. At the same time, we are cutting down the great forests that could absorb large amounts of CO_2. The result has been a steady growth in the amount of CO_2 in our atmosphere (Figure 21-16c).

The increasing levels of CO_2 in our atmosphere have been blamed for a warming of Earth through the greenhouse effect. Studies of the growth rings in very old trees show that Earth's climate has been cooling for the last 1000 years, but the 20th century reversed that trend. The global climate has warmed by 0.3° to 9.6° C in the last century. It is difficult to be sure how much of the day-to-day weather changes we see are linked to greenhouse warming, but it is worrisome that the warmest decade in history was the 1990s, with 1998 being the hottest year of the 20th century. The second hottest year was 1990.

To avoid a planetwide disaster, governments will need to limit the amount of CO_2 released to the atmosphere, and that will mean using energy sources that are more efficient and cleaner than the burning of fossil fuels. We must also conserve our forests, because they cleanse the air of excess CO_2.

In addition to the CO_2 problem, we are destroying ozone in Earth's atmosphere. Ozone is an unstable molecule formed of three oxygen atoms (O_3). Ozone is produced in auto emissions, and it is a pollutant in city air. But ozone is produced naturally in our atmosphere at an altitude of about 25 km, as we have seen, and there it is beneficial. This ozone layer absorbs ultraviolet photons and prevents them from reaching the ground.

Certain chemicals called chlorofluorocarbons (CFCs), used in industrial processes and in refrigeration and air conditioning, can destroy ozone. As these CFCs escape into the atmosphere, they become mixed into the ozone layer and convert the ozone into normal oxygen molecules. Oxygen does not block ultraviolet radiation, so depleting the ozone layer will cause an increase in ultraviolet radiation at Earth's surface. In small doses, ultraviolet radiation can produce a suntan, but in larger doses it can cause skin cancers.

The ozone layer over the Antarctic may be especially sensitive to CFCs. Since the late 1970s, the ozone concentration has been falling over the Antarctic, and a hole in the ozone layer now develops over the continent each October—spring for Antarctica (Figure 21-17). Satellite and ground-based measurements show that the same thing is happening at higher northern latitudes

Ozone hole

Dobson units

Dark grey < 100, red > 500 DU

Figure 21-17
Data from the Earth Probe satellite maps the ozone layer over Earth each day in Dobson units, a measure of ozone density. The lowest density regions of the ozone layer are shown in gray and pink and the highest in reds. White areas are caused by missing data. *(Ozone Processing Team, Goddard Space Flight Center, NASA)*

and that the amount of ultraviolet radiation reaching the ground is increasing. This is an early warning that human civilization is modifying Earth's atmosphere in a potentially dangerous way.

The CO_2 problem and the ozone problem in Earth's atmosphere are paralleled on Venus and Mars. When we study Venus in Chapter 23, we will discover a runaway greenhouse effect that has made the planet into a thermal Hades. On Mars we will discover an atmosphere without an ozone layer. A few minutes of sunbathing on Mars would kill you. Once again, we will learn more about our own planet by studying other planets.

REVIEW Critical Inquiry

Why is our atmosphere poor in carbon dioxide and rich in oxygen?

We might expect our atmosphere to be very rich in carbon dioxide, CO_2. Outgassing from Earth's crust releases mostly CO_2 and water vapor. Luckily for us, CO_2 is highly soluble in water, and Earth's surface temperature is just right for it to be covered with liquid water. The CO_2 dissolves in the oceans and combines with minerals in seawater to form deposits of silicon dioxide, limestone, and other mineral deposits. Thus, the CO_2 is removed from our atmosphere and buried in Earth's crust. Oxygen, in contrast, is highly reactive and forms oxides so easily it could not remain in our atmosphere very long unless it was continually replenished.

We depend on life in the form of green plants to release oxygen into our atmosphere faster than chemical reactions can remove it. Were it not for liquid-water oceans, we would be choking in a thick CO_2 atmosphere with no free oxygen.

Why would an excess of CO_2 and a deficiency of free oxygen be harmful to all life on Earth in ways that go beyond mere respiration?

Earth is a rich and beautiful world, and when we visit Venus and Mars, both of which have CO_2-rich atmospheres, we will appreciate Earth even more. But before we rush to compare Earth with planets that have atmospheres, we should examine two worlds that have no atmosphere at all—the moon and Mercury. As usual, the comparison of worlds will be revealing.

Summary

Comparative planetology is the study of planets by comparison and contrast. We use Earth as the basis for our study of planets because we know it best and because it is a planet of extremes. It has an active crust and a water-rich atmosphere.

Like all terrestrial planets, Earth has passed through a four-stage history: (1) differentiation; (2) cratering; (3) flooding by lava, water, or both; and (4) slow surface evolution.

We know that Earth differentiated because seismic waves traveling through its interior reveal a core of nickel and iron, a thick mantle of dense rock, and a thin crust of low-density rock. Except for the very center, the core is molten, and circulation within the conducting liquid combined with Earth's rotation generates the planet's magnetic field through the dynamo effect.

Circulation in the plastic rock of the upper mantle moves plates of crustal rock and makes Earth's surface active. The crust splits open in midocean rifts where upwelling lava makes new rock, and old crust plunges into deep trenches along the coasts of some continents, where it is remelted. This plate tectonics causes volcanism and earthquakes and constantly recycles Earth's crust. Thus, we see few old rocks and few impact craters.

Earth's atmosphere has developed from gases and ices that were trapped in the in-falling planetesimals that built Earth. Other gases have been outgassed by volcanism, and the oxygen has been added by plant life.

Lowland regions must have been flooded by molten lava when Earth was young, but on Earth, water can form a liquid, so the lowlands are now filled with oceans.

The human race is altering Earth's atmosphere in two dramatic ways: We are increasing the greenhouse effect, and we are reducing the ozone layer. The greenhouse effect is produced by carbon dioxide in the atmosphere, which traps heat and makes Earth warmer. As we burn fossil fuels and cut down forests, we add carbon dioxide to the atmosphere and reduce the plant life that cleans the air. Thus, we appear to be making Earth a warmer planet, and that could eventually alter our climate, melt the polar ice caps, flood low regions, and reduce our ability to grow food.

Certain waste gases called chlorofluorocarbons are thinning the ozone layer 25 km above Earth's surface. Ozone absorbs ultraviolet radiation from the sun, and as we thin the ozone layer, we admit more ultraviolet radiation. Such high-energy photons can cause skin cancers.

New Terms

comparative planetology	midocean rise
seismic waves	midocean rift
seismograph	subduction zone
pressure (P) waves	basalt
shear (S) waves	folded mountain chain
mantle	rift valley
plastic	primeval atmosphere
bow shock	secondary atmosphere
magnetosphere	ozone layer
Van Allen belts	albedo
plate tectonics	greenhouse effect
lithosphere	

Review Questions

1. How do we know that Earth differentiated?

2. What evidence do we have that Earth's metallic core is molten?

3. How are earthquakes near midocean rifts different from those in southern California?

4. If Earth's core were not an electrical conductor, how would the magnetic field differ?

5. What evidence do we have that the Atlantic Ocean is growing wider?

6. How is our concept of Earth's first atmosphere related to the speed with which Earth formed from the solar nebula?

7. What has produced the oxygen in our atmosphere?

8. Why would we expect an increase in CO_2 in Earth's atmosphere to cause a rise in Earth's temperature?

9. Why would a decrease in the density of the ozone layer cause public health problems?

Discussion Questions

1. If we visited a planet in another solar system and discovered oxygen in its atmosphere, what might we guess about its surface?

2. If liquid water is rare on the surface of planets, then most terrestrial planets must have CO_2-rich atmospheres. Why?

Problems

1. In Figure 21-3, the earthquake occurred 7440 km from the seismograph. How fast did the P waves travel in km/s? How fast did the S waves travel?

2. What percentage of Earth's volume is taken up by its metallic core?

3. If the Atlantic seafloor is spreading at 3 cm/year and is now 6400 km wide, how long ago were the continents in contact?

4. The Hawaiian-Emperor chain of undersea volcanoes is about 7500 km long. If the Pacific plate is moving 9.2 cm a year, how old is the oldest detectable volcano in the chain? What has happened to older volcanoes in the chain?

5. From Hawaii to the bend in the Hawaiian-Emperor chain is about 4000 km. Use the speed of plate motion given in Problem 4 to estimate how long ago the direction of plate motion changed. (The San Andreas Fault in southern California became active at about the same time!)

6. Calculate the age of the Grand Canyon as a percent of Earth's age.

Critical Inquiries for the Web

1. Search the Web to find the location of the most recent earthquakes, and plot them on Figure 21-13. The Earth quakes somewhere each day. Search for seismographs that will give you information about the most recent earthquakes.

2. Search for information about the arrangement of land masses on Earth before Pangaea. Were there previous supercontinents? What was Rodinia? What is the evidence?

3. Earth's magnetic poles are not coincident with its axis of rotation, nor are their positions fixed. Search for information about the location of the magnetic poles and their motions. Why do the positions change?

4. What are the oldest known rocks on Earth? Where do they come from and how were their ages determined?

5. What evidence can you find about our changing climate? Is the rise in Earth's temperature related to the greenhouse effect? Scientific, industrial, and political leaders do not agree on this issue, so be prepared to analyze arguments with great care.

6. Can you locate an image of the Antarctic ozone hole? Find data for the ozone concentration in the upper atmosphere over the place where you live. What is the latest news concerning ozone depletion?

Go to the Brooks/Cole Astronomy Resource Center (www. brookscole.com/astronomy) for critical thinking exercises, articles, and additional readings from InfoTrac College Edition, Brooks/Cole's online student library.

CHAPTER TWENTY-TWO

The Moon and Mercury: Airless Worlds

That's one small step for man . . .

one giant leap for mankind.

Neil Armstrong
On the Moon

Beautiful, beautiful. Magnificent

desolation.

Edwin E. Aldrin, Jr.
On the Moon

Guidepost

The two preceding chapters have been preparation for the exploration of the planets. In this chapter, we begin that detailed study with two goals in mind. First, we search for evidence to test the solar nebula hypothesis for the formation of the solar system. Second, we search for an understanding of how planets evolve once they have formed.

The moon is a good place to begin because people have been there. This is an oddity in astronomy in that astronomers are accustomed to studying objects at a distance. In fact, many of the experts on the moon are not astronomers but geologists, and much of what we will study about the moon is an application of earthly geology.

While no one has visited Mercury, we will recognize it as familiar territory. It is much like the moon, so our experience with lunar science will help us understand Mercury as well as the other worlds we will visit in the chapters that follow.

When Neil Armstrong stepped out onto the moon on July 20, 1969, he spoke words for the history books. Edwin Aldrin, the second man to walk on the moon, spoke more directly about what he saw. The lunar landscape is indeed "magnificent desolation." Had Armstrong and Aldrin landed on Mercury, they would have had much the same view: an airless, rocky surface blasted to dust by over 4 billion years of impacts.

In this chapter, we begin our study of other worlds by applying comparative planetology to the study of the moon and Mercury. As we compare them with each other and with Earth, we will discover three important themes of planetary astronomy. First, we will see the importance of impact cratering. The surfaces of the moon and Mercury have been scarred by the impact of uncountable meteorites in a process that must have affected all the planets. Second, we will observe the importance of internal heat in worlds smaller than Earth. Third, we will discover the possibility of giant impacts so large they could shatter planets. By concentrating on these three principles, we will be able to organize the flood of details astronomers have learned about the moon and Mercury.

Our ultimate goal in the study of any world is to tell its story. How did the planet form, how has it evolved, and why is it the way it is? Each time we study a new planet, we will conclude our discussion by trying to combine what we have learned to tell the planet's story. For the moon and Mercury, we will use our three themes—cratering, internal heat, and giant impacts—to try to understand how small worlds can become such magnificently desolate places.

A purist might argue that the moon and Mercury can't be similar because one is a satellite and one is a planet. But both are worlds. They have had complex geological histories that we can compare with each other and with Earth's history. The moon and Mercury illustrate the fate of small, airless worlds.

22-1 The Moon

Only 12 people have stood on the moon, but planetary astronomers know it well. The photographs, measurements, and samples brought back to Earth paint a picture of an airless, ancient, battered crust.

The View from Earth

A few billion years ago, the moon must have rotated faster, but Earth is four times larger in diameter than the moon (Data File Three), and its tidal forces on the moon are strong. Earth's gravity raised tidal bulges on the moon, and friction in the bulges slowed the moon until it now rotates once each orbit to keep the same side facing Earth in what is called **tidal coupling**. This means that we who live on Earth always see the same

The Moon Data File Three

The moon's diameter is roughly 27 percent that of Earth.

Average distance from Earth	384,400 km (center to center)
Eccentricity of orbit	0.055
Maximum distance from Earth	405,500 km
Minimum distance from Earth	363,300 km
Inclination of orbit to ecliptic	5°9′
Average orbital velocity	1.022 km/s
Orbital period (sidereal)	27.321661 days
Orbital period (synodic)	29.5305882 days
Inclination of equator to orbit	6°41′
Equatorial diameter	3476 km
Mass	7.35×10^{22} kg (0.0123 $M_{\oplus}$)
Average density	3.36 g/cm³ (3.35 g/cm³ uncompressed)
Surface gravity	0.167 Earth gravities
Escape velocity	2.38 km/s (0.21 $V_{\oplus}$)
Surface temperature	−170° to 130°C (−274° to 266°F)
Average albedo	0.07

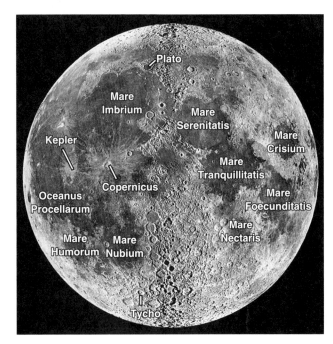

Figure 22-1

The side of the moon that faces Earth is a familiar sight. Craters have been named for famous scientists and philosophers, and the dark, smooth areas have been called seas (*mare* in Latin). Mare Imbrium is the Sea of Rains and Mare Tranquillitatis is the Sea of Tranquillity. There is, in fact, no water on the moon. *(UCO/Lick Observatory photograph)*

a

b

Figure 22-2

(a) The edge of Mare Imbrium is marked by high mountains and the sinuous rille called Hadley Rille (arrow). Such rilles were caused by lava flowing through a channel. Other rilles, straight faults, and many small craters are visible in the surface of the mare. (b) Apollo 15 landed near the 1200-foot-deep rille (circle in part a) and used a battery-powered rover to visit the rille and nearby mountains and craters. *(NASA)*

side of the moon; we never see the back. The moon's familiar face has shone down on Earth since long before there were humans (Figure 22-1).

Even from Earth, we can tell that the moon is an airless world. How do we know? First, our understanding of gravity tells us that a world as small as the moon must have a low escape velocity (Data File Three), and gas atoms near its surface escape easily into space. Furthermore, we can see dramatic shadows near the **terminator,** the dividing line between daylight and darkness. There is no air on the moon to scatter light and soften shadows. We also see stars disappear behind the **limb** of the moon—the edge of its disk—without dimming. Clearly, the moon is an airless (and soundless) world.

The most dramatic features on the moon, regions that are visible even to our unaided eyes, are **mare,** Latin for "sea." The dark maria (plural of *mare*) that make up the face of the man in the moon were so named because early observers thought the great smooth expanses were seas. Even a small telescope will show that the maria are marked by ridges, faults, smudges, and craters and therefore can't be water. The maria are actually great lowland plains covering 17 percent of the moon's surface.

Not many decades ago, astronomers argued about the nature of the maria. Some said that they were great solidified lava flows, but others argued that they were vast oceans of dust blasted from the lunar mountains by billions of years of meteorites and drawn to the low-

lands by the lunar gravity. Science fiction writers wrote fascinating stories about the dust oceans. For example, *A Fall of Moon Dust* by Arthur C. Clarke tells of tourists in a lunar sightseeing boat trapped when their boat sinks in the dust to the bottom of a mare. Although it is a wonderful rescue story, it can never happen. We know now that the maria are great lava flows with only a few centimeters of loose dust on their surfaces.

These lava flows suggest volcanism, but only small traces of past volcanic activity are visible from Earth. No major volcanic peaks are visible, and no active volcanism has been detected. The lava flows that created the maria happened long ago and were much too fluid to build peaks. We can see a few small domes pushed up by lava below the surface, and we can see long, winding channels called **sinuous rilles** (Figure 22-2), often found near the edges of the maria. These channels were

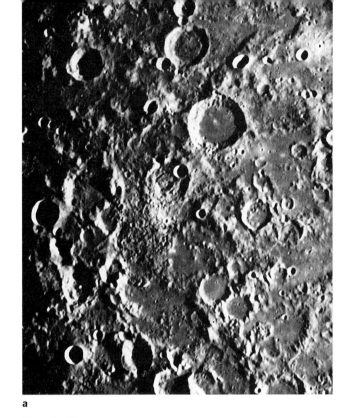

a

b

Figure 22-3
(a) The lunar highlands are the oldest parts of the moon and are heavily battered by craters. (b) The lunar low-lands are filled by solidified lava flows containing few craters and long wrinkle ridges. The lava flows occurred after the end of the heavy bombardment and thus escaped most cratering impacts. *(Lick Observatory)*

evidently cut by flowing lava. In some cases, such a channel may have had a roof of solid rock, forming a lava tube. When the lava drained away, meteorite impacts collapsed the roof to form a sinuous rille. Thus, the view from Earth gives us only hints of ancient volcanic activity associated with the maria.

Besides the maria, the most striking feature of the moon is the millions of impact craters. The lighter parts of the moon's face are heavily cratered highlands. Although they look mountainous at first glance, there are no folded mountain ranges on the moon. Its crust has never wrinkled into folded mountains as Earth's crust has. The mountainous regions of the highlands are actually millions of meteorite craters, one on top of the other (Figure 22-3a). We will discuss the process that forms craters in the next section.

The crater density in different regions of the moon gives us a way of estimating the ages of different features, important information if we are to tell the story of the moon. The highlands must be old to have accumulated so many craters, but the maria contain only a few craters, so they can't be that old (Figure 22-3b). Most of the craters we see on the moon were produced during the first half billion years of the moon's existence by the heavy bombardment at the end of planet formation (Chapter 20). As the debris left over from the solar nebula was gradually swept up, the rate of cratering fell rapidly, as shown by the curve in Figure 22-4. Notice that the vertical axis in the graph is exponential (Win-

dow on Science 9-1). The cratering rate fell very rapidly and only half a billion years after the formation of the solar system, the cratering rate had fallen by a factor of 10,000.

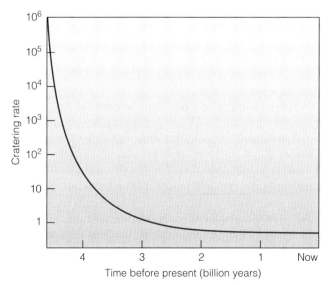

Figure 22-4
The cratering rate was high when the solar system first formed, but it fell rapidly as the debris in the solar nebula was swept up. Although the cratering curve is drawn as a smooth line here, it may have been an irregular decline with episodes of cratering.

Like any technical subject, science includes a mass of details, facts, figures, measurements, and observations. It is easy to be overwhelmed by the flood of details, but one of the most important characteristics of science comes to our rescue. The goal of science is not to discover more details but to explain the details with a unifying hypothesis or theory. A good theory is like a basket that makes it easier for us to carry a large assortment of details.

This is true of all the sciences. When a psychologist begins studying the way the human eye and brain respond to moving points of light, the data are a sea of detailed measurements and observations. Once the psychologist forms a hypothesis about the way the eye and brain interact, the details fall into place as parts of a log-

ical story. If we understand the hypothesis, the details all fit together and make sense, and thus we can remember the details without blindly memorizing tables of facts and figures. The goal of science is understanding, not memorization.

Scientists are in the story-telling business. The stories are often called hypotheses or theories, but they are really just stories to explain how nature works. The difference between scientific stories and works of fiction lies in the use of facts. Scientific stories are constructed to fit all known facts and are then tested over and over against new facts obtained by observation and experiment.

When we try to tell the story of each planet in our solar system, we pull together all the hypotheses and theories

and try to make them into a logical history of how the planet got to be the way it is. Of course, our stories will be incomplete because astronomers don't understand all the factors in planetary astronomy. Nevertheless, our story of each planet will draw together the known facts and details and attempt to make them a logical whole.

Memorizing a list of facts can give us a false feeling of security, just as when we memorize the names of things without understanding them (Window on Science 2-2). Rather than memorizing facts, we should search for the unifying hypothesis that pulls the details together into a single story. Our goal in studying science is understanding nature, not just remembering facts.

Planetary astronomers argue whether the cratering rate fell smoothly as the debris in the solar system was swept up or whether there were bursts of cratering. A major impact between large planetesimals might shower the solar system with fragments and produce a storm of crater formation. The final formation of Uranus and Neptune in the outer solar system might have flung icy planetesimals through the solar system like giant shrapnel. Nevertheless, we can be sure that the craters produced in the early heavy bombardment were probably obliterated by impacts during the late heavy bombardment, and that the smooth lava flows formed the maria after the last of the heavy bombardment.

A few minutes at a telescope will reveal clear cases of craters that partially cover other craters. Evidently, older craters were partially destroyed by newer craters. We can also find craters that have been partially buried by maria (see Figure 22-3b). Evidently, these craters are older than the maria. These are called **relative ages** because we can tell which feature formed first but we can't give either feature an age in years. By comparing the number of craters on a given surface with the rate of cratering as given in Figure 22-4, planetary astronomers can estimate **absolute ages** in years. Of course, an even more accurate way to find the ages of lunar features is to go to the moon and bring back rocks that can be dated radioactively. We will discuss these radioactive ages later.

The view from Earth tells us that the moon has an ancient crust that was heavily cratered when the solar system was young. The maria are slightly younger than the highlands, but all parts of the moon are old and scarred by craters. Our goal is to tell the story of the moon (Window on Science 22-1), and to do that we must understand how craters are formed.

Impact Craters

When a meteorite strikes a planet, it delivers tremendous energy and can excavate an impact crater many times its own diameter. Meteorites strike the surface of a planet with velocities that range from 15 to 60 km/s, and even a small meteorite can carry tremendous energy. A meteorite with the mass of a house would hit with energy equivalent to a large nuclear bomb.

When a meteorite strikes, it penetrates the surface, deforms the rock layers, and converts some of its energy into enough heat to vaporize itself and nearby rock (Figure 22-5). The deformed rock rebounds, and shock waves move upward and outward, driven in part by the heat of the impact. These rebounding shock waves blast away the surface to create a crater. The resulting scar is a circular crater over ten times the diameter of the impacting meteorite and about 10 percent as deep as it is wide. Rock blasted out of the crater falls to form a rim that rises slightly above the surrounding surface.

When we look at the moon through even a small telescope, we see many large craters that contain a cen-

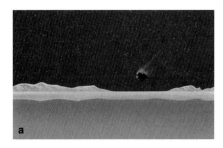

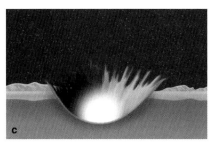

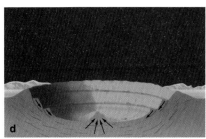

Figure 22-5
The formation of a crater. An approaching meteorite (a) strikes the lunar surface (b) and penetrates deeply. The meteorite vaporizes, and shock waves fracture the rock and blast the surface away (c) to form a circular crater. (d) Ejected material forms a raised crater rim. The rebound of the rock can form a raised central peak, and slumping along the crater wall can form terraces. The rock below the crater is severely fractured.

tral mountain and terraces along the inner walls of the crater. The central mountains are produced by the rebound of the fractured rock under the center of the crater. The terraces are produced by slumping as material slides back down into the crater from the inner walls. In even larger craters, we see an inner ring of mountains, and astronomers refer to these features as **multi-ringed basins.** We will find multiringed basins on the moon and other worlds.

Debris blasted out of the crater is called **ejecta;** it falls back to the surface to create an ejecta blanket sur-

rounding the crater. Some ejecta falls back into the crater, but some can be blasted far across the surface. On the moon, we can see the ejecta blankets around young craters because the rough surface scatters sunlight and looks bright. Ejecta shot out along specific directions can fall back to the surface and form bright **rays.** The crater Copernicus is the center of a vast ray system (Figure 22-1), and rays from the crater Tycho extend halfway around the moon. Older craters must have had rays, but exposure to sunlight and a continuous blasting from micrometeorites smaller than grains of sand gradually darken and smooth the lunar surface and obliterate the rays. Thus, the presence of bright, rayed craters shows that at least a few craters have been formed recently. Tycho, for example, may be only 100 million years old.

Impacts on the moon are tremendously powerful. Chunks of ejecta can travel large distances and fall back to the surface to form more craters called **secondary craters.** Some secondary craters are even big enough to form their own secondary craters. Some lunar impacts have been violent enough to blast rocks completely off the moon, and a few such rocks have been found on Earth. They must have been ejected from the moon and fallen to Earth as meteorites (Figure 22-6).

In the introduction to this chapter, we identified three themes of planetary astronomy that we would meet in studying the moon and Mercury. The first was impact cratering, and the blasted face of the moon tells us of a time when meteorites rained down on the lunar surface. We will see evidence of this early bombardment on other worlds.

The view from Earth is stark and fascinating, but we can learn only so much by looking. The only way to really understand the moon is to go there.

Figure 22-6
The meteorite ALHA81005 was found in Antarctica in January 1982 and later identified as a fragment of the lunar surface. *(NASA, Ursula B. Marvin)*

Figure 22-7
The Apollo astronauts took two spaceships to the moon. While one astronaut stayed in orbit in the command module, the other two rode the lunar lander down to the surface. *(NASA)*

The Apollo Missions

On May 25, 1961, President John Kennedy addressed Congress and committed the United States to landing a human being on the moon by 1970. The reasons for that decision are complicated and related more to economics and the stimulation of technology than to science, but the Apollo program became a fantastic adventure in science, a flight to the moon that changed how we think about Earth.

Flying to the moon is not particularly difficult; with powerful enough rockets and enough food, water, and air, it is a simple trip. Landing on the moon is more difficult but not impossible. The moon's gravity is only one-sixth that of Earth, and there is no atmosphere to disturb the trajectory of the spaceship. Getting to the moon is simple and landing is possible; the difficulty is doing both on one trip. The spaceship must carry food, water, and air for a number of days in space plus fuel and rockets for midcourse corrections and for a return to Earth. All of this adds up to a ship that is too massive to make a safe landing on the lunar surface. The solution was to take two spaceships to the moon, one to ride in and one to land in (Figure 22-7).

The command module was the long-term home and command center for the trip. Three astronauts had to live in it for a week, and it had to carry all the life-support equipment, navigation instruments, computers, power packs, and so on for a week's jaunt in space. The lunar landing module (LM for short) was tacked to the front of the command module like a bicycle strapped to the front of the family camper. It carried only enough fuel and supplies for the short trip to the lunar surface, and it was built to minimize weight and maximize maneuverability.

The weaker gravity of the moon made the design of the LM simpler. Landing on Earth requires reclining couches for the astronauts, but the trip to the lunar surface involved smaller accelerations. In an early version of the LM, the astronauts sat on what looked like bicycle seats, but these were later scrapped to save weight. The astronauts had no seats at all in the LM, and once they began their descent and acquired weight, they stood at the controls held by straps, riding the LM like two daredevils riding a rocket surfboard.

Lifting off from the lunar surface, the LM saved weight by leaving the larger descent rocket and support stage behind. Only the astronauts, their instruments, and the cargo of rocks returned to the command module orbiting above. Again, the astronauts in the LM blasted up from the lunar surface standing at the controls. The rocket engine that lifted them back into lunar orbit was not much bigger than a dishwasher.

The most complicated parts of the trip were the rendezvous and docking between the tiny remains of the LM and the command module, aided by radar systems and computers. The astronauts rejoined their colleague in the command module, transferred their moon rocks, and jettisoned the LM. Only the command module returned to Earth.

The first manned lunar landing was made on July 20, 1969. While Michael Collins waited in orbit, Neil Armstrong and Edwin Aldrin, Jr., took the LM down to the surface. Although much of the descent was controlled by computers, the astronauts had to override a number of computer alarms and take control of the LM

Table 22-1 Apollo Lunar Landings

Apollo Mission*	Astronauts: Commander LM Pilot CM Pilot	Date	Mission Goals	Sample Weight (kg)	Typical Samples	Ages (10⁹ y)
11	Armstrong Aldrin Collins	July 1969	First manned landing; Mare Tranquillitatis	21.7	Mare basalts	3.48–3.72
12	Conrad Bean Gordon	November 1969	Visit Surveyor 3; sample Oceanus Procellarum (mare)	34.4	Mare basalts	3.15–3.37
14	Shepard Mitchell Roosa	February 1971	Fra Mauro, Imbrium ejecta sheet	42.9	Breccia	3.85–3.96
15	Scott Irwin Worden	July 1971	Edge of Mare Imbrium and Apennine Mountains, Hadley Rille	76.8	Mare basalts; highland anorthosite	3.28–3.44 4.09
16	Young Duke Mattingley	April 1972	Sample highland crust; Cayley formation (ejecta); Descartes	94.7	Highland basalt; breccia	3.84 3.92
17	Cernan Schmitt Evans	December 1972	Sample highland crust; dark halo craters; Taurus–Littrow	110.5	Mare basalt; highland breccia; fractured dunite	3.77 3.86 4.48

*The Apollo 13 mission suffered an explosion on the way to the moon and did not land.

to avoid a boulder-strewn crater bigger than a football field. Climbing down the ladder, Armstrong and then Aldrin stepped onto the lunar surface.

Between July 1969 and December 1972, 12 people reached the lunar surface and collected 380 kg (840 lb) of rocks and soil (Table 22-1). The flights were carefully planned to visit different regions and thus develop a comprehensive history of the lunar surface.

The first flights went into relatively safe landing sites (Figure 22-8a)—Mare Tranquillitatis for Apollo 11 and Oceanus Procellarum for Apollo 12. Apollo 13 was aimed at a more complicated site, but an explosion in an oxygen tank on the way to the moon ended all chances of a landing and nearly cost the astronauts their lives. They succeeded in using the life support in the LM to survive.

The last four Apollo missions, 14 through 17, sampled geologically important places on the moon. Apollo 14 visited the Fra Mauro region, which is covered by ejecta from the impact that dug the basin now filled by Mare Imbrium. Apollo 15 visited the edge of Mare Imbrium at the foot of the Apennine Mountains and examined Hadley Rille (Figure 22-2b). Apollos 16 and 17 visited highland regions to sample the older parts of the lunar crust (Figures 22-8b and 22-9). Almost all of the lunar samples from these six landings are now held at the Planetary Materials Laboratory at the Johnson Space Center in Houston. They are a national treasure containing clues to the beginnings of our solar system.

Moon Rocks

Of all of the rock samples that the Apollo astronauts carried back to Earth, every one is igneous. That is, they formed by the solidification of molten rock. No sedimentary rocks were found, which is consistent with the moon never having had liquid water on its surface. In addition, the rocks were extremely dry. Almost all Earth rocks contain 1 to 2 percent water, either as free water trapped in the rock or as water molecules chemically bonded with certain minerals. But moon rocks contain no water at all.

Rocks from the lunar maria are dark-colored, dense basalts much like the solidified lava produced by the Hawaiian volcanoes. These rocks are rich in heavy elements such as iron, manganese, and titanium, which give them their dark color. Some of the basalts are **vesicular,** meaning that they contain holes caused by bubbles of gas in the molten rock (Figure 22-10a). Like bubbles in a carbonated beverage, these bubbles do not form while the magma is under pressure. Only when the rock flows out onto the surface, where the pressure is low, do bubbles appear. Thus, the presence of vesicular basalts assures us that these rocks formed in lava flows that reached the surface and did not solidify underground.

a

b

Figure 22-8
(a) Edwin E. Aldrin, Jr., one of the three Apollo 11 astronauts, stands on the dusty surface of Mare Tranquillitatis in the lunar lowlands, site of the first lunar landing. The flat lava plain is almost featureless, and the horizon is straight. (b) In the lunar highlands at Taurus–Littrow, the surface is mountainous and irregular. Note the horizon. Harrison Schmitt, a scientist–astronaut on Apollo 17, is shown passing huge boulders. *(NASA)*

The ages of the mare basalts can be found from the radioactive atoms they contain (Chapter 20), and these ages range from about 3.1 to 3.8 billion years. Thus, the lava flows that formed these rocks must have happened some time after the end of the heavy bombardment.

Rocks from the highlands are light-colored, lower-density **anorthosite.** This rock is rich in calcium, aluminum, and oxygen and is the kind of rock we would expect to form first and float to the top of cooling molten magma. It seems that the anorthosite is the true crustal rock of the moon, while the basalts are lava flows that filled the lowlands. The ages of the anorthosite range from 4.0 to 4.5 billion years, significantly older than the mare basalts.

A large fraction of the lunar rocks are **breccias,** rocks that are made up of fragments of earlier rocks cemented together by heat and pressure (Figure 22-10b). Evidently, meteorite impacts have broken up the rocks and fused them together time after time.

Both the highlands and the lowlands of the moon are covered by a layer of powdered rock and crushed fragments called the **regolith.** It is about 10 m deep on the maria but over 100 m deep in certain places in the highlands. About 1 percent of the regolith is meteoric fragments; the rest is the smashed remains of moon rocks that have been ground down by the constant blasting of meteorites. The smallest meteorites, the **micrometeorites,** do the most damage by their constant sandblasting of the lunar surface (Figure 22-10c). Thus, we see that impact cratering dominates the lunar surface and is even responsible for the lunar regolith.

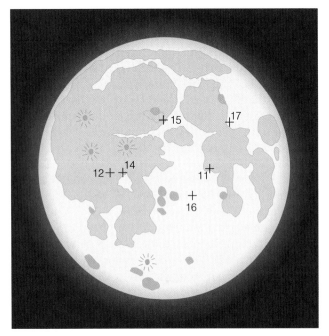

Figure 22-9
The landing sites of the six Apollo missions that reached the moon. For the names of lunar features, see Figure 22-1.

One of the best ways to think about a scientific problem is to follow the energy. According to the principle of cause and effect, every effect must have a cause, and every cause must involve energy (Window on Science 5-3). Energy moves from regions of high concentration to regions of low concentration and in doing so produces changes. For example, we may burn coal to make steam in a power plant, and the steam passes through a turbine and then escapes into the air. In flowing from the burning coal to the atmosphere, the heat spins the turbine and makes electricity.

It is surprising how commonly scientists use energy as a key to understanding nature. A biologist might ask where certain birds get the energy to fly thousands of miles, and an economist might ask where

the economic energy to support the creation of new investments comes from. Energy is everywhere, and when it moves, whether it is birds, money, or molten magma, it causes change. Energy is the cause in cause and effect.

In earlier chapters, we used the flow of energy from the inside of a star to its surface to understand how stars, including the sun, work. We saw that the outward flow of energy supports the star against its own weight, drives convection currents that produce magnetic fields, and causes surface activity such as spots, prominences, and flares. We were able to understand stars because we could follow the flow of energy outward from their interiors.

We can think of a planet by following the energy. The heat in the interior of a

planet may be left over from the formation of the planet, or it may be heat generated by radioactive decay, but it must flow outward toward the cooler surface, where it is radiated into space. In flowing outward, the heat can cause convection currents in the mantle, magnetic fields, plate motions, quakes, faults, volcanism, mountain building, and more.

When you think about any world, be it a small asteroid or a giant planet, think of it as a source of heat that flows outward through the planet's surface into space. If we can follow that energy flow, we can understand a great deal about the world. A planetary astronomer once said, "The most interesting thing about any planet is how its heat gets out."

The moon rocks are old, dry, igneous, and badly shattered by impacts. We can use these facts, combined with what we know of lunar features, to reconstruct the moon's past.

The History of the Moon

The ultimate goal in the study of any world is to describe its history, to explain how it evolved from its formation to its present state (Window on Science 22-1). We have seen that the surface of the moon has been se-

verely battered by impact craters, but the maria tell of a different theme in lunar history—the moon's internal heat. Like Earth, the moon formed with heat trapped in its interior, and radioactive decay released more heat. Unlike Earth, the moon is small, and that has sealed its fate. Because it is small, it could not retain its internal heat, it aged quickly, and it is now a dead world (Window on Science 22-2).

The interior of the moon appears to contain little heat. Studies of the shape of the moon's gravitational field show that the anorthosite crust is thicker on the

a   b   c

Figure 22-10
Moon rocks. (a) A vesicular basalt contains numerous bubbles frozen into the rock when it solidified from a lava flow on the surface. (b) A breccia is formed from two kinds of rocks cemented together. (c) Rocks exposed on the surface of the moon for long periods become pitted with small micrometeorite craters. *(NASA)*

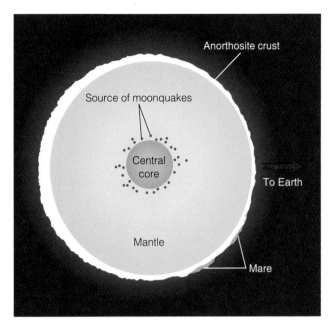

Figure 22-11
A cross section of the moon shows an anorthosite crust thinner on the side facing Earth, depressions filled with lava flows to form the maria, and a deep mantle of brittle rock. The core of the moon is small and may contain some iron.

far side than on the near side. The moon does not have a magnetic field of its own, so it cannot contain circulating molten iron or rock. The mantle appears to be deep and cold, and studies of the interaction of the moon with Earth's magnetic field suggest it has a very small core that contains at least some iron (Figure 22-11). Seismographs left on the lunar surface by the Apollo astronauts detected moonquakes about as large as big firecrackers deep in the moon's interior. These could be due to tidal stresses.

The moon is now a cold, dead, geologically inactive world, but it began as a sea of molten rock. The lack of rocks older than about 4.4 billion years and the chemical composition of the anorthosite have convinced planetary astronomers that the surface of the moon was covered by a sea of magma when it was very young. The energy to melt the rock probably came from the heat of formation (Chapter 20) released by in-falling meteorites.

In the first of the four stages of planetary evolution, the moon cooled and differentiated as the denser minerals sank into the interior and the lower-density minerals floated to the top, like slag in an open-hearth furnace. Given the present chemical composition of the moon, we would expect this slag to cool and form an anorthosite crust within roughly a hundred million years of the moon's formation.

The second stage of planetary evolution is cratering, and the moon was heavily battered by meteorites. The early heavy bombardment must have blasted the newly formed crust over and over. Most rocks older than 4 billion years are smashed to fragments, and the entire crust of the moon is fractured to a depth of a few kilometers. The largest meteorites excavated vast basins. The mare Oceanus Procellarum may overlie one of the oldest and largest basins on the moon, but later impacts and lava flows have partially obliterated its circular outline. Mare Imbrium was excavated by the impact of a meteorite the size of Rhode Island (roughly 65 km across). The resulting basin was 1300 km across, and the impact smashed the crust into mountain ranges up to 5 km high. Ejecta from this impact covers 16 percent of the lunar surface (Figure 22-12). Apollo 14 landed at Fra Mauro to sample it.

Figure 22-12
Apollo 14 landed on rugged terrain suspected of being ejecta from the Imbrium impact. The large boulder here is ejecta that, at some time in the past, fell here from an impact beyond the horizon. *(NASA)*

Our story of the moon might suggest that it was a violent place during this cratering phase, but large impacts were in fact rare and, except for a continuous rain of smaller meteorites, the moon was, for the most part, a peaceful place even during the heavy bombardment. Centuries might pass between major impacts. Of course, a large impact might have threatened to bury us under ejecta or jolted us by seismic shocks. We could have felt the Imbrium impact anywhere on the moon, but had we been standing on the side of the moon opposite this impact, we would have been at the focus of seismic waves traveling around the moon from different directions. When the waves met under our feet, the surface would have jerked up and down by as much as 10 m. The place on the moon opposite the Imbrium Basin is a strangely disturbed landscape called **jumbled terrain.** We will find similar effects of large impacts on other worlds.

The formation of large, circular impact basins broke the crust to such depth that pockets of rock melted by the decay of radioactive elements flowed out onto the surface (Figure 22-13). Flow after flow filled the basins to produce the maria. This process began before the end of heavy bombardment, but the older lava flows were destroyed by impacts. Those great basins that were flooded from about 3.8 to 3 billion years ago escaped the cratering, and we see them as maria.

The flooding of the maria, the third stage of planet formation, produced distinctive features. Some maria such as Mare Imbrium, Mare Serenitatis, and Mare Crisium retain their round shapes, but others are irregular because the lava overflowed the edges of the basin or because the shape of the basin was modified by further cratering. The moving lava formed channels that we see as sinuous rills (Figure 22-2). The weight of the maria pressed the crater basins downward, and the solidified lava was compressed and formed wrinkle ridges visible even in small telescopes (Figure 22-3b). The tension at the edges of the maria broke the hard lava to produce straight fractures and faults (Figure 22-2a). Later cratering and overlapping lava floods further modified the maria. Thus the maria are accumulations of features reflecting a complex history (Figure 22-14).

Almost all of this flooding occurred on the side of the moon facing Earth. The back side of the moon is heavily cratered, but only a few of the largest craters contain any lava flows (Figure 22-15). Even the multiringed basin Mare Orientale, near the west limb, is only slightly flooded. The difference

Mare Imbrium

Mare Serenitatis

Mare Crisium

a

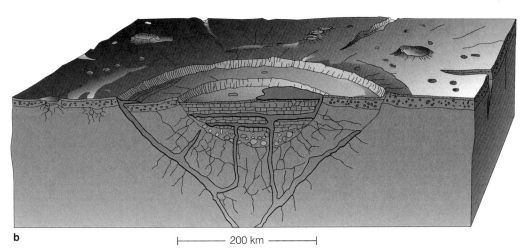

b

⊢ 200 km ⊣

Figure 22-13
(a) A topographic map of the moon's near side shows low regions as blue and high regions as yellow and red. The maria are clearly lowlands, and many have the circular outline of large impact basins. *(NASA/JPL)* (b) Major impacts produce multiringed basins and may later be flooded by lava following fractures upward. *(Adapted from William Hartmann,* Astronomy: The Cosmic Journey, *Belmont, CA: Wadsworth Publishing Co., 1978)*

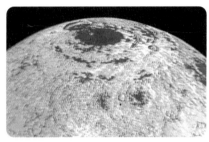

Figure 22-14
The evolution of the lunar surface. Mare Imbrium began when a meteorite as large as Rhode Island excavated a great crater basin. Lava welled up to flood the basin, starting about 3.8 billion years ago and lasting until no less than 2 billion years ago. Since then only scattered meteorites have modified Mare Imbrium. *(Don Davis)*

Figure 22-15
Lava flows occurred repeatedly on the lunar near side but flooded only a few of the small basins on the lunar far side. Nevertheless, the far side of the moon is marked by some very deep basins including the South Pole–Aitken basin (purple). Although it is very deep, it has an old, cratered floor and appears to have suffered little flooding. *(NASA/JPL)*

impact that formed the Oceanus Procellarum Basin and weakened the near-side crust.

The moon is small, and small worlds cool rapidly because they have a large ratio of surface area to volume, just as a small cupcake cools more rapidly than a large cake. The moon has lost much of its internal heat, and it is the outward flow of heat that produces geological activity. The crust of the moon rapidly grew thick and never divided into moving plates. There are no rift valleys or folded mountain chains on the moon. The lava flows on the moon ended one or two billion years ago when the moon's temperature fell too low to maintain subsurface lava.

The moon is now in the fourth stage of planetary evolution, slow surface evolution. It has no atmosphere, so there is no liquid water to cause erosion. Some observations from spacecraft suggest there may be water trapped at the moon's north and south poles in areas of the regolith that never receive sunlight. That water may have been brought to the moon by rare impacts of comets, which would leave the moon temporarily surrounded by a cloud of water vapor. Some of that water could be trapped at the poles. Nevertheless, this water would not affect the evolution of the surface.

is due to the thickness of the crust. On the side toward Earth, the lunar crust is only about 60 km thick, but on the far side it is more than 100 km thick (Figure 22-11). This asymmetry may have resulted from the immense

The surface of the moon is constantly blasted by micrometeorites that grind the surface rocks down to dust. Larger impacts are rare; no new crater is known to have appeared on the moon since astronomers began observing it with telescopes in the 17th century. Thus there is little change on the moon. The footprints of the Apollo astronauts will last for millions of years, but they will eventually be erased by the constant flux of micrometeorites.

The overall terrain on the moon is almost fixed. A billion years from now, plate tectonics will have totally changed the map of Earth, erosion will have worn away the Rocky Mountains, the descent stage of the Apollo 17 lunar lander will be nothing but a peculiar chemical contamination in the regolith at Taurus–Littrow, but the lunar mountains will look much the same.

Our story of the moon's history is as complete as comparative planetology can make it, but it omits one initial detail. Where did Earth get such a large satellite?

The Origin of the Moon

The Apollo missions to the moon told us a great deal, but they failed to answer the most basic question: How did the moon form? The three traditional theories do not work, but a newer hypothesis may solve the problem.

The first of the three traditional theories, the **fission hypothesis,** supposes that the moon formed by the fission of Earth. If the young Earth spun fast enough, tides raised by the sun might make it break into two parts (Figure 22-16a). If this separation occurred after Earth differentiated, the moon would have formed from crust material. Thus, the theory explains the moon's low density.

But the fission theory has problems. No one knows why the young Earth should have spun so fast, nearly ten times faster than today, nor where all that angular momentum went after the fission. In addition, the moon's orbit is not in the plane of Earth's equator, as it would be if it had formed by fission.

The second traditional theory is the **co-accretion** (or double-planet) **hypothesis.** It supposes that Earth and the moon condensed as a double planet from the same cloud of material (Figure 22-16b). However, if they formed from the same material, they should have the same chemical composition and density, which they don't. The moon is very poor in iron and related elements such as titanium, but it also contains almost exactly the same ratios of oxygen isotopes as does Earth's mantle. Co-accretion cannot explain these compositional differences.

The third theory is the **capture hypothesis.** It supposes that the moon formed somewhere else and was later captured by Earth (Figure 22-16c). We might imagine the moon forming inside the orbit of Mercury, where the heat would prevent the condensation of solid metallic grains and only high-melting-point metal oxides could have solidified. A later encounter with Mercury could "kick" the moon out to Earth.

The capture theory is not popular, because it requires highly unlikely events involving interactions with Mercury and Earth to move the moon from place to place. Scientists are always suspicious of explanations that require a chain of unlikely coincidences. Also, upon encountering Earth, the moon would have been moving so rapidly that Earth's gravity would have been unable to capture it without ripping the moon to fragments through tidal forces.

Until recently, astronomers were thus left with no acceptable theory to explain the origin of the moon, and they occasionally joked that the moon could not exist. But during the 1980s, planetary astronomers developed a new theory that combines the best aspects of the fission hypothesis and the capture hypothesis.

The **large-impact hypothesis** supposes that the moon formed from debris ejected into a disk around Earth by the impact of a large body (Figure 22-17). The impacting body may have been twice as large as Mars. Instead of saying that Earth was hit by a large body, we may be more nearly correct in saying that Earth and the moon resulted from the collision and merger of two very large planetesimals. The resulting large body became Earth, and the ejected debris formed the moon.

This hypothesis would explain a number of things. The collision had to occur at a steep angle to eject enough matter to make the moon. That is, the objects did not collide head-on. Such a glancing collision would have spun the resulting material rapidly, which would explain the

a

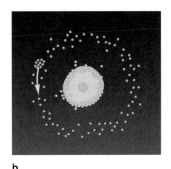
b

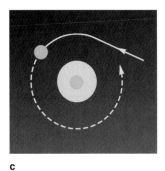
c

Figure 22-16
Three traditional theories for the moon's origin. (a) Fission theories suppose that Earth and the moon were once one body and broke apart. (b) Co-accretion theories suppose that the moon formed from material near Earth. (c) Capture theories suggest that the moon formed elsewhere and was captured by Earth. None of these theories explains all the facts.

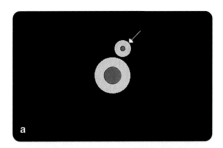

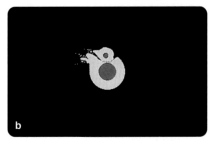

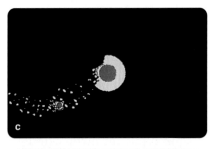

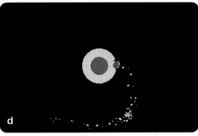

Figure 22-17
A Mars-sized planetesimal strikes the young, molten Earth, blasting iron-poor mantle material into orbit. In this simulation, the iron core of the planetesimal falls back into Earth within 4 hours, and the moon begins to form within about 10 hours.

angular momentum in the Earth–moon system. If the colliding planetesimals had already differentiated, the ejected material would be mostly iron-poor mantle and crust. Calculations show that the iron core of the impacting body would have fallen into the Earth. This

would explain why the moon is so poor in iron and why the abundances of other elements are so similar to rocks from Earth's mantle. Also, the material that eventually became the moon would have remained in a disk long enough for volatile elements, which the moon lacks, to be lost to space.

Thus, the moon may be the remains of a giant impact. Until recently, astronomers have been reluctant to consider such catastrophic events, but a number of lines of evidence suggest that planets have been affected by giant impacts. Consequently, the third theme we identified in the introduction to this chapter, giant impacts, has the potential to help us understand other worlds.

REVIEW Critical Inquiry

If the moon was intensely cratered by the heavy bombardment and then formed great lava plains, why didn't the same thing happen on Earth?

In fact, the same thing did happen on Earth. Although the moon has more craters than Earth, the moon and Earth are the same age, and both were battered by meteorites during the heavy bombardment. Some of those impacts on Earth must have been large and dug giant multiringed basins. Lava flows must have welled up through Earth's crust and flooded the lowlands to form great lava plains much like the lunar maria.

Earth, however, is different from the moon in one important respect. It is larger. Because it is a larger world, Earth has more internal heat, and that heat escapes more slowly than the moon's heat did. The moon is now geologically dead, but Earth is very active, with heat flowing outward from the interior to drive plate tectonics. The moving plates long ago erased all evidence of the cratering and lava flows dating from Earth's youth. Earth's surface was almost certainly cratered and flooded by lava flows, as was the moon's, but our planet has continued to evolve, and the moon has not.

Comparative planetology is a powerful tool in that it allows us to see similar processes occurring under different circumstances. For example, suppose you visited a world in our solar system that was even smaller than the moon and so had lost its heat even faster. What kind of surface features would you find, and what kinds of rocks would you pick up? More important, what features and rocks would be missing? We will find such small worlds among the asteroids.

The moon's evolution has been restricted by its size. It is too small to have an atmosphere, and it cooled so rapidly that it never developed plate motion. Another world in our solar system has suffered from its diminutive size. Mercury, like the moon, is a small, rocky world, cratered by impacts and flooded by ancient lava flows driven by its internal heat.

22-2 Mercury

Earth's moon and Mercury are similar in a number of ways. Their rotation has been altered by tides, their surfaces are heavily cratered, their lowlands are flooded in places by ancient lava flows, and both are small and airless and have ancient, inactive surfaces. Yet the differences between them are impressive and will help us understand the nature of these airless worlds.

Mercury is the innermost planet in the solar system (Data File Four), and thus its orbit keeps it near the sun in the sky. It is sometimes visible near the horizon in the evening sky after sunset or in the dawn sky just before sunrise. An Earth-based telescope shows Mercury passing through phases like the moon, but little detail is visible (Figure 22-18). Most of what we know about the surface of Mercury is based on photographs and measurements taken by the Mariner 10 spacecraft when it looped through the inner solar system in 1974 and 1975. Until then, Mercury was a mystery world.

Rotation and Revolution

During the 1880s, the Italian astronomer Giovanni Schiaparelli sketched the faint features he thought he saw on the disk of Mercury and concluded that the planet was tidally coupled to the sun just as the moon is tidally coupled to Earth. That is, he concluded that Mercury rotated on its axis to keep the same side facing the sun throughout its orbit. This was actually a

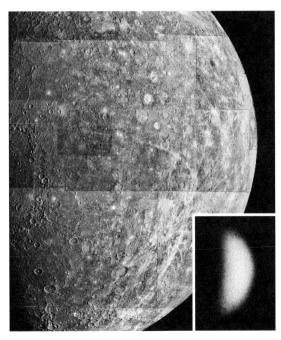

Figure 22-18
Photomosaic of Mercury made by the Mariner 10 spacecraft. Caloris Basin is in shadow at left. Almost no surface detail is visible from Earth (inset). *(NASA, inset courtesy Lowell Observatory)*

Mercury Data File Four

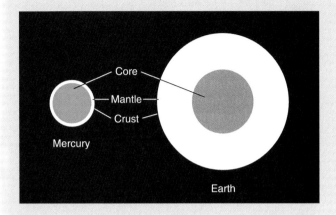

In proportion to its size, Mercury has a very large metallic core.

Average distance from the sun	0.387 AU (5.79 × 10⁷ km)
Eccentricity of orbit	0.2056
Maximum distance from the sun	0.467 AU (6.97 × 10⁷ km)
Minimum distance from the sun	0.306 AU (4.59 × 10⁷ km)
Inclination of orbit to ecliptic	7°00′16″
Average orbital velocity	47.9 km/s
Orbital period	87.969 days (0.24085 y)
Period of rotation	58.646 days (direct)
Inclination of equator to orbit	0°
Equatorial diameter	4878 km (0.382 $D_\oplus$)
Mass	3.31 × 10²³ kg (0.0558 $M_\oplus$)
Average density	5.44 g/cm³ (5.4 g/cm³ uncompressed)
Surface gravity	0.38 Earth gravities
Escape velocity	4.3 km/s (0.38 $V_\oplus$)
Surface temperature	−173° to 330°C (−279° to 620°F)
Average albedo	0.1
Oblateness	0

very good guess because, as we will see, tidal coupling between rotation and revolution is common in the solar system. But the rotation of Mercury is more complex than Schiaparelli thought.

In 1962, radio astronomers detected radio emissions from the planet and concluded that the dark side was not as cold as it should have been if the planet kept one side perpetually in darkness. In 1965, radio astronomers made radar contact with Mercury. They used the 305-m Arecibo dish (Figure 6-22) to transmit a pulse of radio energy at Mercury and then waited for the reflected signal to return. Doppler shifts in the reflected radio pulse showed that the planet was rotating with a period of only about 59 days, much shorter than the orbital period of 87.969 days.

Apparently, Mercury is tidally coupled to the sun such that it rotates not once per orbit but 1.5 times per orbit (Figure 22-19). Thus, its period of rotation is two-thirds its orbital period, or 58.65 days. This means that a mountain on Mercury directly below the sun at one place in its orbit will point away from the sun one orbit later and toward the sun after the next orbit.

If you flew to Mercury and landed your spaceship in the middle of the day side, the sun would be high overhead, and it would be noon. Your watch would show that almost 44 Earth days passed before the sun set in the west, and a total of 88 Earth days would have passed before the sun reached the midnight position. In those 88 Earth days, Mercury would have completed one orbit around the sun (Figure 22-19e). It would require another entire orbit of Mercury for the sun to return to the noon position overhead. So a full day on Mercury is two Mercury years long!

The complex tidal coupling between the rotation and revolution of Mercury is an important illustration of the power of tides. Just as the tides in the Earth–moon system have slowed the moon's rotation and locked it to Earth, so have the sun–Mercury tides slowed the rotation of Mercury and coupled its rotation to its revolution. We can refer to such a relationship as a **resonance.** We will see many such resonances as we explore the solar system.

Like its rotation, Mercury's orbital motion is also peculiar. In Chapter 5, we discussed the orbital motion of Mercury and saw that its elliptical orbit precesses faster than can be explained by Newton's laws but at just the rate predicted by Einstein's theory of general relativity. Thus, the orbital motion of Mercury is taken as a confirmation of the curvature of space-time as predicted by general relativity.

The Surface of Mercury

Because the days are so long on Mercury and because it is close to the sun and too small to keep an atmosphere, the temperatures on Mercury are extreme. The highest daytime temperatures can exceed 600 K (330°C),

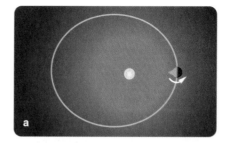

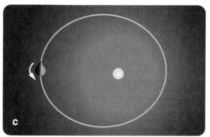

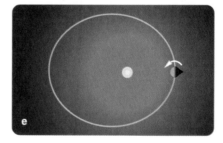

Figure 22-19
Mercury's rotation is tidally coupled to the sun, but unlike the moon, which rotates to keep the same face toward Earth, Mercury rotates 1.5 times during each orbit. A mountain facing the sun (a) will face away from the sun after one orbit (e) and will face the sun again after the next orbit.

although 500 K is a more usual high temperature. In the dark nights, with no atmosphere to retain heat, the surface can cool to 100 K (−173°C). Thus, the surface of Mercury cycles from very high temperatures to very low temperatures each Mercury day.

In appearance, Mercury looks much like Earth's moon. It is heavily battered, with craters of all sizes, including some large basins. Some craters are obviously old and degraded; others seem quite young and have bright rays of ejecta. However, a quick glance at photos of Mercury shows no large, dark maria like the moon's flooded basins.

The largest basin on Mercury is called Caloris Basin after the Latin word for "heat," recognition of its location at one of the two "hot poles," which face the sun at alternate perihelions. At the times of the Mariner encounters, the Caloris Basin was half in shadow (Figure 22-20a). Although half cannot be seen, the low angle of illumination is ideal for the study of the lighted half because it produces dramatic shadows.

The Caloris Basin is a gigantic impact basin 1300 km in diameter with concentric mountain rings up to 3 km high. The impact threw ejecta 600 to 800 km across the planet, and the focusing of seismic waves on the far side produced peculiar terrain (called lineated terrain) that looks much like the jumbled surface of the moon (Figure 22-20b and c). The Caloris Basin is partially filled with lava flows. Some of this lava may be material melted by the energy of the impact, but some may be lava from below the crust that leaked up through cracks. The weight of this lava and the response of the crust have produced deep cracks in the central lava plains. The geophysics of such large, multiringed crater basins is not well understood at present, but Caloris Basin seems to be the same kind of structure as the Imbrium Basin on the moon, but it has not been as deeply flooded.

When planetary scientists began looking at the Mariner photographs in detail, they discovered something not seen on the moon. Mercury is marked by great curved cliffs called **lobate scarps** (Figure 22-21). These seem to have formed when the planet cooled and shrank by a few kilometers, wrinkling its crust as a drying apple wrinkles its skin. Some of these scarps are as high as 3 km and reach hundreds of kilometers across the surface. Other faults in Mercury's crust are straight and may have been produced by tidal stresses generated when the sun slowed Mercury's rotation.

The Plains of Mercury

The most striking difference between Mercury and the moon is that Mercury lacks the great dark lava plains so obvious on the moon. Under careful examination, the Mariner 10 photographs show that Mercury has plains, two different kinds, in fact, but they are different from the moon's. Understanding the differences is the key to understanding the history of Mercury.

Much of Mercury's surface is old, cratered terrain (Figure 22-22), but other areas called **intercrater plains** are less heavily cratered. These plains are marked by meteorite craters less than 15 km in diameter and secondary craters produced by chunks of ejecta from larger

a

b

c

Figure 22-20

(a) Only part of the giant Caloris Basin on Mercury is visible in this Mariner 10 photomosaic. The center of the basin is in darkness just beyond the lower left corner of the photo, and the outer ring is marked by a dashed line. This giant crater basin was produced by a large impact and subsequent lava flows. (b) Seismic waves from the impact were focused on the far side of Mercury and produced the jumbled terrain shown in part c. *(NASA)*

Figure 22-21
A lobate scarp (arrow) crosses craters, indicating that Mercury cooled and shrank, wrinkling its crust, after many of its craters had formed. *(NASA)*

impacts. Unlike the heavily cratered regions, the intercrater plains are not totally saturated with craters. This suggests that the intercrater plains were produced by lava flows that buried older terrain.

In contrast, smaller regions called **smooth plains** appear to be even younger than the intercrater plains. They have fewer craters and appear to be ancient lava flows. Much of the region around the Caloris Basin is composed of these smooth plains (Figure 22-22), and they appear to have formed soon after the Caloris impact.

Given the available evidence, most planetary astronomers believe that the plains of Mercury are solidified lava flows, in which case they are much like the maria on the moon. Unlike the maria, Mercury's lava plains are not significantly darker than the rest of the crust. Except for a few bright crater rays, Mercury's surface is a uniform gray with an albedo of only about 0.1. Thus, Mercury's lava plains are not as dramatically obvious on photographs as the much darker maria on our own moon.

The Interior of Mercury

One of the most striking differences between Mercury and the moon is the difference between their interiors. We have seen that the moon is a low-density world that contains no more than a small core of metals. Yet evidence suggests that Mercury has a large metallic core.

Mercury is over 60 percent denser than the moon. Yet Mercury's surface appears to be normal rock, so we must conclude that its interior contains a large core of dense metals, probably iron. In proportion to its size, Mercury must have a larger metallic core than Earth (see diagram in Data File Four).

If Mercury has a large metallic core that remains molten, then the dynamo effect should generate a magnetic field (Chapter 21). The Mariner 10 spacecraft, however, found a magnetic field only about 0.5 percent as strong as Earth's, and this weak field makes it difficult to understand the nature of the interior. Because Mercury is a relatively small world, about halfway between the size of Earth and the size of the moon, it should have lost most of its internal heat long ago. But if its metallic core is not molten, we would not expect to detect a magnetic field as strong as 0.5 percent of Earth's field. The verdict is still out on the nature of Mercury's large metallic core and weak magnetic field.

Similarly, it is difficult to explain the proportionally large size of the metallic core inside Mercury. In Chapter 20, we saw that planets that formed near the sun should incorporate more metals and thus be denser, but some models suggest that this cannot account for all of Mercury's large metallic core. One hypothesis that has been proposed involves a giant impact when Mercury was young, an impact much like the planet-shattering impact proposed to explain the origin of the moon. If the forming planet had differentiated and was then struck by a large planetesimal, the impact could have shattered the crust and mantle and blasted much of the lower-density material into space. The denser core could have survived, re-formed, and attracted some of the lower-density debris to form a thin mantle and so leave Mercury with a deficiency of low-density crust and mantle rock. Thus, both Mercury and the moon may be products of giant impacts.

Figure 22-22
A geological map of part of Mercury shows the Caloris Basin at left and the hilly, broken terrain associated with that impact. The intercrater plains are interrupted by smooth plains and more heavily cratered terrain.

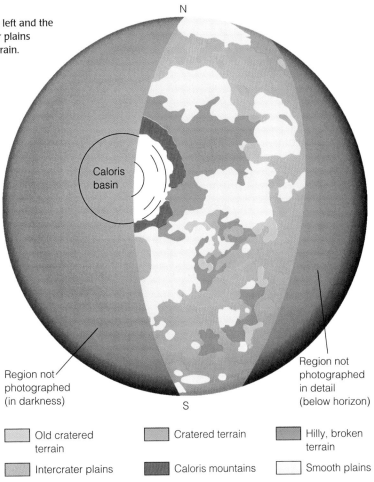

Region not photographed (in darkness)

Region not photographed in detail (below horizon)

Caloris basin

N

S

| | Old cratered terrain | | Cratered terrain | | Hilly, broken terrain |
| | Intercrater plains | | Caloris mountains | | Smooth plains |

A History of Mercury

Having reviewed the evidence, we must now try to tell the story of Mercury. Presumably, it formed in the innermost part of the solar nebula, and, as we have seen, a giant impact may have robbed it of some of its lower-density rock and left it a small, dense world with a large metallic core.

Like the moon, Mercury suffered heavy cratering by debris in the young solar system. We don't know the absolute ages of features on Mercury, but we can assume that cratering, the second stage of planetary evolution, occurred over the same period as did the cratering on the moon—during the heavy bombardment that lasted from formation until about 4 billion years ago. This intense cratering declined rapidly as the planets swept up the last of the debris left over from planet building.

The cratered surface of Mercury is not exactly like that of the moon. Because of Mercury's stronger gravity, the ejecta from a crater on Mercury would be thrown only about 65 percent as far as on the moon, and thus the ejecta from an impact does not blanket as much of the surface. Also, the intercrater plains appear to have formed when lava flows occurred during the early bombardment, burying the older surface, and then accumulated more craters. Sometime near the end of cratering, a planetesimal over 100 km in diameter smashed into the planet and blasted out the great multiringed Caloris Basin. Only parts of that basin have been flooded by lava flows. Finally, as cratering rapidly declined, the cooling core contracted, and the crust broke to form the lobate scarps.

The third stage, flooding, began with lava flows filling some lowlands. The Caloris impact may have been so big it fractured the crust and triggered still more outpourings of molten lava. These later lava flows formed the smooth plains. Nevertheless, the age of lava flooding ended quickly, perhaps because the shrinkage of the planet squeezed off the lava channels to the surface. Mercury lacks a true atmosphere, so we would not expect flooding by water, but radar images show a bright spot at the planet's north pole (Figure 22-23) that may be caused by ice trapped in perpetually shaded crater floors where the temperature never exceeds 60 K (−350°F).

The fourth stage in the story of Mercury, slow surface evolution, is now limited to micrometeorites, which grind the surface to dust; rare larger meteorites, which

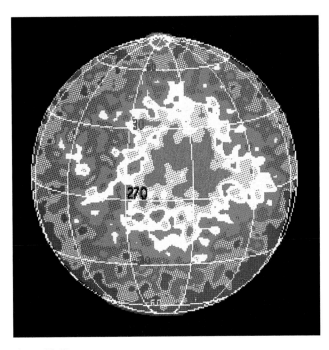

Figure 22-23
A radar map of Mercury reveals a bright region at its north pole (top). Observers suspect that ice deposits may survive just under the surface or in regions that never receive direct sunlight. *(Martin Slade, JPL)*

leave bright rayed craters; and the slow but intense cycle of heat and cold, which weakens the rock at the surface. The planet's crust is now thick, and although its core may be partially molten, the heat flowing outward is unable to drive plate tectonics that would actively erase craters and build folded mountain chains.

REVIEW Critical Inquiry

Why don't Earth and the moon have lobate scarps?

At first glance, we might suppose that any world with a large metallic interior should have lobate scarps. When the metallic core cools and contracts, the world should shrink, and the contraction should wrinkle and fracture the brittle crust to form lobate scarps.

Of course, Earth has a fairly large metallic core, but it has not cooled very much, and the crust is not brittle. In fact, the heat flowing out of Earth's interior keeps the crust thin, flexible, and active. If any lobate scarps ever formed on Earth, they would have been quickly destroyed by plate tectonics.

The moon, unlike Mercury, does not have a large metal component in its interior. It may be that the rocky interior did contract slightly as the moon lost its internal heat, but that smaller contraction may not have produced major lobate scarps. Also, any surface features that formed early in the moon's history would have been destroyed by the heavy bombardment.

We can, in a general way, understand lobate scarps, but not all materials contract when they cool and solidify. Water, for instance, expands when it freezes. What kinds of features would you expect to find on small, cold, icy worlds that once had interiors of liquid water? Such icy worlds exist as the moons of the outer planets, and we will explore them in Chapter 24.

Mercury, like the moon, is now an inactive world. Impact cratering, internal heat, and possible giant impacts have left their marks on these worlds, but they are now nearly dead. To find planets where the heat flowing outward from the interior still drives geological activity, we must look at larger worlds such as Venus and Mars, the subjects of the next chapter.

Summary

The moon is, at first glance, dramatically different from Earth. We see two distinct kinds of terrain on the moon. The maria, named after seas, are lowland plains with few craters. They are great lava flows and are younger than the highlands. The highlands are lighter in color and heavily cratered.

Samples returned by the Apollo missions show that the maria are dark basalts, whereas the highlands are light anorthosite. Many samples are breccias, which show how severely the moon's crust has been fractured.

The ages of the mare samples range from about 3 to 3.8 billion years, but the highland samples are about 4.0 to 4.3 billion years old. Few samples are older than 4.4 billion years. This suggests that the moon's surface was molten until about 4.4 billion years ago.

As the anorthosite crust solidified, cratering battered it and excavated great basins. Later, molten rock rose through fractures in the crust and flooded the basins to produce the maria. By the time the maria formed, cratering had declined, so the maria are marked by few craters.

The lunar surface now is evolving very slowly. Because the moon is small, it has cooled rapidly, its crust has grown thick, and it has never divided into moving plates. Now only meteorites alter the surface.

The large-impact hypothesis suggests that the Earth–moon system formed in the collision of two large planetesimals. The differentiated bodies formed a single large object that became the higher-density, iron-rich Earth. Ejected mantle material, poor in iron, formed a disk and eventually condensed to form the lower-density, iron-poor moon.

Mercury is a small world, only about 40 percent larger than the moon, and, like the moon, it has lost any permanent atmosphere it might have had, and its interior has cooled rapidly. Debris from the formation of the solar system cratered Mercury's surface as it did the moon's, and lava flows buried parts of that terrain under the intercrater plains. The Caloris Basin, formed near the end of cratering, is the largest crater basin on Mercury. Soon after the Caloris impact, lava flows created the smooth plains. These lava plains have about the same color as the intercrater plains, so the contrast between plains and highlands is not as obvious on Mercury as it is on the moon.

The interior of Mercury must contain a large metallic core to account for the planet's high density. A slight shrinkage in the diameter of the planet at the time of the cooling of the core may have led to wrinkling of the crust. This would account for the lobate scarps that mark all of the photographed parts of Mercury.

New Terms

tidal coupling	ray
terminator	secondary crater
limb	vesicular basalt
mare	anorthosite
sinuous rille	breccia
relative age	regolith
absolute age	micrometeorite
multiringed basin	jumbled terrain
ejecta	fission hypothesis

co-accretion hypothesis
capture hypothesis
large-impact hypothesis
resonance

lobate scarp
intercrater plain
smooth plain

Review Questions

1. How do we find the relative ages of the moon's maria and highlands?

2. How do we know that Copernicus is a young crater?

3. Why did the first Apollo missions land on the maria? Why were the other areas of more scientific interest?

4. How does the range of ages of lunar samples suggest that the moon formed with a molten surface?

5. Why are so many lunar samples breccias?

6. What do the vesicular basalts tell us about the evolution of the lunar surface?

7. What evidence would we expect to find on the moon if it had been subjected to plate tectonics? Do we find such evidence?

8. Cite objections to the fission, co-accretion, and capture hypotheses.

9. How does the large-impact hypothesis explain the moon's lack of iron?

10. How does the tidal coupling between Mercury and the sun differ from that between the moon and Earth?

11. What is the difference between the intercrater plains and the smooth plains in terms of time of formation?

12. What evidence do we have that Mercury has a partially molten, metallic core?

Discussion Questions

1. Old science-fiction paintings and drawings of colonies on the moon often showed very steep, jagged mountains. Why did the artists assume that the mountains would be more rugged than mountains on Earth? Why are lunar mountains actually less rugged than mountains on Earth?

2. From your knowledge of comparative planetology, propose a description of the view that astronauts would have if they landed on the surface of Mercury.

Problems

1. Calculate the escape velocity of the moon from its mass and diameter. (Hint: See Chapter 5.)

2. Why do small planets cool faster than large planets? Compare surface area to volume.

3. The smallest detail visible through Earth-based telescopes is about 1 second of arc in diameter. What size is this on the moon? (Hint: See Chapter 3.)

4. Midocean rifts and the trenches where Earth's seafloor slips downward are 1 km or less wide. Could Earth-based telescopes resolve such features on the moon? Why are we sure that such features are not present on the moon?

5. The Apollo command module orbited the moon about 200 km above the surface. What was its orbital period? (Hint: See Chapter 5.)

6. From a distance of 200 km above the surface of the moon, what is the angular diameter of an astronaut in a space suit? Could we have seen the astronauts from the command module? (Hint: See Chapter 3.)

7. If we transmitted radio signals to Mercury when it was closest to Earth and waited to hear the radar echo, how long would we wait?

8. Suppose we sent a spacecraft to land on Mercury and it transmitted radio signals to us at a wavelength of 10 cm. If we saw Mercury at its greatest angular distance west of the sun, to what wavelength would we have to tune our radio telescope to detect the signals? (Hints: See Data File Four and the Doppler shift in Chapter 7.)

9. What would the wavelength of maximum be for infrared radiation from the surface of Mercury? How would that differ for the moon?

10. Calculate the escape velocity from Mercury. How does that compare with the escape velocity from the moon and from Earth?

Critical Inquiries for the Web

1. What did the Apollo astronauts do while they were on the moon? Search for Web sites that describe the experiments performed during the astronauts' extravehicular activities (EVAs). What types of experiments/activities were performed? What did we learn from them?

2. Find the latest news about the search for water on the moon and on Mercury. What dramatic experiment was conducted to test for lunar water?

3. Search for maps of the moon that show geological information such as chemical composition, elevation, age, and so on. Do the maps confirm your understanding of the lunar surface?

4. What would it be like to walk on the lunar surface? Apollo astronauts visited six different locations on the moon exploring the variations in lunar terrain. Describe the horizons and general relief of the landing locations of the different missions by exploring Web sites that provide lunar surface photography from the missions. What differences do you see between images from landings in highlands and in maria?

5. What plans are being made to send a spacecraft to Mercury? What kind of observations will be made and what theories will be tested?

Exploring *The Sky*

1. Aside from the sun, the moon is the most prominent object in our sky.

   a. Let *The Sky* show you the moon as it appears at the present time. What kind of features can you identify? What is the phase of the moon?

   How to proceed: Use **Find** to center the moon on the screen. Click on the magnification button to give you a large image of the moon.

   b. Find the following information about the moon at the present time:

   the constellation in which it appears,
   the phase as a percentage of full,
   the angular size,
   its distance from Earth.

   How to proceed: To find the phase, angular size, and distance from Earth, click on the moon to obtain **Object Information.**

   c. Find the date of the next full moon, and find the same information as in part (a).

   How to proceed: To find the date of the next full moon, click on **Tools,** then click on **Moon Phase Calendar.** Then set the date by clicking on **Data,** and then **Site Information.**

Go to the Brooks/Cole Astronomy Resource Center (www.brookscole.com/astronomy) for critical thinking exercises, articles, and additional readings from InfoTrac College Edition, Brooks/Cole's online student library.

CHAPTER TWENTY-THREE

Venus and Mars

There wasn't a breath in that land

of death . . .

Robert Service
The Cremation of Sam McGee

Guidepost

The previous chapter grouped Earth's moon and Mercury together because they are similar worlds. This chapter groups Venus and Mars together because we might expect them to be similar. They are Earthlike in their size and location in the solar system, so it is astonishing to see how different they actually are. Much of this chapter is aimed at understanding how Venus and Mars evolved to their present states.

Neither Venus nor Mars can tell us much about the formation of the planets. Both planets have evolved since they formed. Nevertheless, we find further hints that the solar system was a dangerous place, with major impacts smashing the surfaces of the planets, a process we first suspected when we studied the moon and Mercury.

This chapter concludes our exploration of the Earthlike worlds. In the next two chapters, we will visit planets that give new meaning to the word "unearthly."

Venus Data File Five

Venus is only 5 percent smaller than Earth, but its surface is a blazing hot desert and its atmosphere is perpetually cloudy. *(NASA)*

Average distance from the sun	0.7233 AU (1.082 × 10⁸ km)
Eccentricity of orbit	0.0068
Maximum distance from the sun	0.7282 AU (1.089 × 10⁸ km)
Minimum distance from the sun	0.7184 AU (1.075 × 10⁸ km)
Inclination of orbit to ecliptic	3°23′40″
Average orbital velocity	35.03 km/s
Orbital period	0.61515 y (224.68 days)
Period of rotation	243.01 days (retrograde)
Inclination of equator to orbit	177°
Equatorial diameter	12,104 km (0.95 $D_\oplus$)
Mass	4.870 × 10²⁴ kg (0.815 $M_\oplus$)
Average density	5.24 g/cm³ (4.2 g/cm³ uncompressed)
Surface gravity	0.903 Earth gravities
Escape velocity	10.3 km/s (0.92 $V_\oplus$)
Surface temperature	745 K (472°C, or 882°F)
Albedo (cloud tops)	0.76
Oblateness	0

Without a space suit, you could live only about 30 seconds on the surface of Mars. The temperature might not be too bad. On a hot summer day at noon, the temperature might reach 20°C (68°F). But the air is 96 percent carbon dioxide, is deadly dry, and contains almost no oxygen. Also, the air pressure at the surface of Mars is only 0.01 that at the surface of Earth. Stepping out onto the surface of Mars without a space suit would be about the same as opening the door of an airplane flying at 30,000 m (100,000 ft) above Earth. (Commercial jets do not generally fly above 12,000 m, or 38,000 ft.)

No space suit would save you on Venus. The air pressure is 90 times Earth's, and the air is almost entirely carbon dioxide, with traces of various acids. Worse yet, the surface is hot enough to melt lead.

Why are Venus and Mars such rotten places for picnics? In many ways, they resemble Earth. They are made of silicates and metals, have atmospheres, and lie in the inner solar system. We might expect them to be similar. How did they get to be so different from each other and from Earth? Comparative planetology will give us some clues.

When we used comparative planetology to study the moon and Mercury in the previous chapter, we saw the importance of impact cratering and the outward flow of heat. As we study Venus and Mars, we will see the same effects, but they will be given new significance because of the larger size of Venus and Mars. In this chapter, we will discover the important principle that the size of a world determines how fast it loses its internal heat. Venus, roughly the size of Earth, has lost internal heat slowly, but Mars, intermediate in size between Earth and the moon, is nearly as cold and dead as the moon. Furthermore, these larger worlds have enough gravity to hold onto an atmosphere, and that alters the evolution of surface features.

Venus and Mars are more than just two more planets to add to our gallery of weird worlds. They give us an entirely new perspective on comparative planetology and thus a new perspective on Earth.

23-1 Venus

An astronomer once became annoyed when someone referred to Venus as a planet gone wrong. "No," she argued, "Venus is probably a fairly normal planet. It is Earth that is peculiar. The universe probably contains more planets like Venus than like Earth." To understand how unusual, how special Earth is, we have only to compare it with Venus.

In many ways, Venus is a twin of Earth, and we might expect the two planets to be quite similar. Venus is nearly 95 percent the diameter of Earth (Data File Five) and has a similar average density. It formed in the same general part of the solar nebula, so we might expect it to have a composition similar to Earth's. Planets of Earth's size cool slowly, so we might expect Venus to have a molten metallic interior and an active crust.

Unfortunately, we cannot see the surface of Venus because of perpetual thick, white clouds that completely envelop the planet. From the time of Galileo to the early 1960s, astronomers could only speculate about Earth's twin, and science fiction writers supposed that Venus was a steamy swamp planet inhabited by strange creatures.

During the last few decades, astronomers have peered below the clouds using radar, 18 spacecraft have flown past or orbited Venus, and 17 probes have landed on its surface. The resulting picture of Venus is dramatically different from the murky swamps of fiction. In fact, the surface of Venus is drier than any desert on Earth and two times hotter than a hot kitchen oven.

If Venus is not a planet gone wrong, it is certainly a planet gone down a different evolutionary path. How did Earth's twin become so different?

The Rotation of Venus

Nearly all of the planets in our solar system rotate counterclockwise as seen from the north. Uranus and Pluto are exceptions, and so is Venus.

In 1962, radio astronomers were able to transmit a radio pulse of precise wavelength toward Venus and detect the echo returning some minutes later. That is, they detected Venus by radar. The echo was not a precise wavelength, however. Part of the reflected signal had a longer wavelength, and part had a shorter wavelength. Evidently the planet was rotating; radio energy reflected from the receding edge was red shifted, and radio energy reflected from the approaching edge was blue shifted. From this Doppler effect, the radio astronomers could tell that Venus was rotating once every 243.01 days. Because the west edge of Venus produced the blue shift, it had to be rotating in the backward direction.

This rotation period turns nearly the same face toward Earth each time the planets come closest together, and it is tempting to imagine that there is a tidal resonance locking the rotation of Venus to the orbital motion of Earth. However, a true resonance would require that Venus rotate 0.06 percent slower. Thus the apparent connection between Venus and Earth may be only a coincidence.

The slow retrograde rotation of Venus probably arose during final accretion when a very large planetesimal struck proto-Venus off center and altered its rotation. The idea of such large impacts is becoming more acceptable among astronomers. In the previous chapter, we saw that large impacts may be implicated in the birth of the moon, the loss of low-density material from Mercury, and the formation of giant crater basins. We will find more evidence of these catastrophic impacts as we explore further in our solar system.

The Atmosphere of Venus

Although Venus is Earth's twin in size, the composition, temperature, and density of its atmosphere make it the most inhospitable of planets. About 96 percent of its atmosphere is carbon dioxide, and 3.5 percent is nitrogen. The remaining 0.5 percent is water vapor, sulfuric acid (H_2SO_4), hydrochloric acid (HCl), and hydrofluoric acid (HF). In fact, the thick clouds that hide the surface are believed to be composed of sulfuric acid droplets and microscopic sulfur crystals.

Soviet and American spacecraft have dropped probes into the atmosphere of Venus, and those probes have radioed data back to Earth as they fell toward the surface. These studies show that the cloud layers are much higher and much more stable than those on Earth. The highest layer of clouds, the layer we see from Earth, extends from 68 to 58 km above the surface (Figure 23-1). For comparison, the highest clouds on Earth do not extend higher than about 16 km. Another cloud layer lies from 56 to 52 km, and a third, the densest, from 52 to 48 km. Thin haze extends down to about 33 km, below which the air is surprisingly clear.

These cloud layers are highly stable because the atmospheric circulation on Venus is much more regular than that on Earth. The heated atmosphere at the **subsolar point,** the point on the planet where the sun is directly overhead, rises and spreads out in the upper atmosphere. Convection circulates this gas toward the dark side of the planet and the poles, where it cools and sinks. This circulation produces 300-km/h jet streams in the upper atmosphere, which carry the atmosphere from east to west (the same direction the planet rotates) so rapidly that the entire atmosphere seems to rotate with a period of only four days. As evidence of these circulations, we can see a distinctive Y pattern in ultraviolet images of Venus such as that in Data File Five and in Figure 23-1a.

The high-speed winds in the upper atmosphere were studied in June 1985 when the Vega 1 and Vega 2 spacecraft, on their way to rendezvous with Comet Halley, flew past Venus and dropped instrumented balloons into the upper atmosphere. Floating at an altitude of 55 km, the 3.5-m-diameter balloons found winds as high as 240 km/h. In the 46 hours the balloons survived, they were carried one-third of the way around Venus by the high-altitude winds.

The details of this atmospheric circulation are not well understood, but it seems that the slow rotation of the planet is an important factor. On Earth, large-scale circulation patterns are broken into smaller cyclonic disturbances by Earth's rapid rotation. Because Venus rotates more slowly, its atmospheric circulation is not broken up into small cyclonic storms but instead is organized as a planetwide wind pattern.

Although the upper atmosphere is cool, the lower atmosphere is quite hot (Figure 23-1b). Instrumented

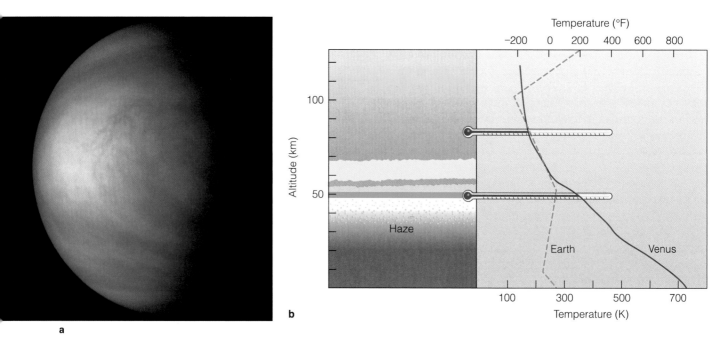

Figure 23-1

(a) This ultraviolet image of Venus was made by the Galileo spacecraft. *(NASA)* (b) The four principal cloud layers in the atmosphere of Venus are shown at left. Thermometers inserted into the atmosphere at different levels would register the temperatures indicated by the red line in the graph at the right. The blue dashed line shows the temperatures in Earth's atmosphere for comparison. Note the difference in temperatures at the surfaces of the planets.

probes that have reached the surface report that the temperature is 745 K (880°F), and the atmospheric pressure is 90 times that of Earth. Earth's atmosphere is 1000 times less dense than water, but on Venus the air is only 10 times less dense than water. If we could survive the unpleasant composition, intense heat, and high pressure, we could strap wings to our arms and fly.

The present atmosphere of Venus is extremely dry, but there is evidence that it once had significant amounts of water. As a Pioneer Venus spacecraft descended through the atmosphere of Venus in 1978, it detected about 150 times more deuterium per hydrogen atom than on Earth. This abundance of deuterium, the heavy isotope of hydrogen, could develop if ultraviolet radiation in sunlight broke water molecules into hydrogen and oxygen. The hydrogen would escape into space, but the heavier deuterium would escape more slowly than normal hydrogen. Thus, the excess of deuterium is evidence that Venus once had more water. Calculations suggest that it may have once had enough water to make a planet-wide ocean up to 25 meters deep. (For comparison, the water on Earth would make a uniform planetwide ocean 3000 meters deep.) Without an ozone layer to protect the atmosphere from ultraviolet radiation, the water was broken up and much of the hydrogen was lost to space. The oxygen presumably formed oxides in the soil. Thus, Venus is now a deadly dry world with only enough water vapor to make an ocean 0.3 meter deep.

Even if Venus once had lots of water, it was probably too warm for that water to fall from the atmosphere as rain and form oceans, and that lack of oceans is the biggest difference between Earth and Venus. Earth's oceans have made it a living paradise. The lack of oceans on Venus, through the greenhouse effect, has made it a hell.

The Venusian Greenhouse

We saw in Chapter 21 how the greenhouse effect warms Earth. Carbon dioxide (CO_2) is transparent to light but opaque to infrared (heat) radiation. Thus, energy can enter the atmosphere as light and warm the surface, but the surface cannot radiate an equal amount of energy back to space because of the CO_2. Like Earth, Venus has a greenhouse effect, but on Venus the effect is fearsome. Whereas Earth's atmosphere contains only about 0.03 percent CO_2, the atmosphere of Venus contains 96 percent CO_2, and the surface of Venus is nearly twice as hot as a hot kitchen oven.

When Venus was young, it may have been cooler than it is now, but because it formed 30 percent closer to the sun than did Earth, it was warmer than Earth, and that unleased processes that made it even hotter. As we saw in the previous section, even if Venus once had water, it probably never had large liquid-water oceans. CO_2 is highly soluble in water, but without large oceans to dissolve CO_2 and remove it from the atmosphere, Venus was trapped with a CO_2-rich atmosphere, and the greenhouse effect made the planet warmer. As the planet warmed, any small oceans that did exist evaporated, and Venus lost the ability to cleanse its atmosphere of CO_2. Volcanoes on Earth vent gases that are mostly CO_2 and water vapor, and presumably the volcanoes on Venus did the same. The water vapor was broken up by ultraviolet radiation, and the CO_2 accumulated in the atmosphere. The high temperature baked even more CO_2 out of the surface, and the atmosphere became even less transparent to infrared radiation. Consequently, the temperature rose further. This runaway greenhouse effect has made the surface so hot that even sulfur, chlorine, and fluorine have baked out of the rock and formed sulfuric, hydrochloric, and hydrofluoric acid vapors.

Earth avoided this runaway greenhouse effect because it was farther from the sun and cooler. Thus, it could form and preserve liquid-water oceans to absorb the carbon dioxide, which left a nitrogen atmosphere that was relatively transparent in some parts of the infrared. If all of the carbon dioxide in Earth's sediments was put back into the atmosphere, our air would be as dense as that of Venus, and we would suffer from a severe greenhouse effect. Recall from Chapter 21 how our use of fossil fuels and our destruction of forests is increasing the CO_2 concentration in our atmosphere and warming the planet. Venus warns us of what a greenhouse effect can do.

The Surface of Venus

Given that the surface of Venus is hidden perpetually by clouds, is hot enough to melt lead, and suffers under crushing atmospheric pressure, it is surprising how much planetary astronomers know about the geology of Venus. Early radar maps made from Earth penetrated the clouds and showed that it had mountains, plains, and some craters. The Soviet Union launched a number of spacecraft that landed on Venus, and although the spacecraft failed within an hour or so of landing, they did analyze the rock and snap a few photos. The rock seems to be basalt, typical of volcanism, and the photographs revealed dark-gray rocky plains bathed in a deep-orange glow caused by sunlight filtering down through the thick atmosphere (Figure 23-2).

Real geological studies of Venus became possible when spacecraft went into orbit round Venus and made radar maps of the rocky surface below the clouds. The Pioneer Venus probe orbited Venus in 1978 and made radar maps showing features as small as 25 km. Later two Soviet spacecraft mapped the north polar region with a resolution of 2 km. From 1992 to 1994, the NASA Magellan spacecraft orbited Venus and created radar maps showing details as small as 100 m.

The color of Venus radar maps is mostly arbitrary. Our eyes can't see radio waves, so the radar maps must be given some false colors to distinguish height or roughness or composition. Some maps use blues and greens for lowlands and yellows and reds for highlands. Coloring the lowlands blue like oceans on Earth maps might be deceptive, however, because there is no liquid water on Venus. Rather, Magellan scientists chose to use yellows and oranges for their radar maps in an effort to

a

b

Figure 23-2
The color of the surface of Venus appears orange in the original Venera 13 photograph (a), but analysis shows that the color is produced by the thick atmosphere. The corrected image (b) reveals slabs of gray rock. *(C. M. Pieters, through the Brown/Vernadsky Institute to Institute Agreement and the U.S.S.R. Academy of Sciences)*

mimic the orange color of daylight caused by the thick atmosphere (Figure 23-3). When we look at these orange images we need to remind ourselves that the true color of the rock is dark gray.

Radar maps of Venus can tell us a number of things about the surface. If we transmit a radio signal down through the clouds and measure the time until we hear the echo coming back up, we can measure the altitude of the surface. Part of the Magellan data is a detailed altitude map of the surface. We can also measure the amount of the signal that is reflected from each spot on the surface. Rough surfaces such as lava flows contain billions of tiny crevices (Figure 23-4) that bounce the radar signal around and shoot it back the way it came. Thus lava flows look bright in radar maps. Certain mineral deposits are also bright, but smooth surfaces tend to look dark. Thus, part of the Magellan data are maps of roughness.

Figure 23-5 shows a projected map of all of Venus except for the polar regions. Notice that about 65 percent of the surface is low **rolling plains** (blue and green in this map). The rest is highland regions with a few percent high mountains (red). Notice also the names on Venus. By international agreement, features on Venus are named for women, with only a few exceptions. The mountain, Maxwell Montes, 50 percent higher than Mt. Everest, was discovered in early Earth-based radar maps and is named for James Clerk Maxwell, the 19th-century physicist who first described electromagnetic

Figure 23-3
Venus without its clouds: This mosaic of Magellan radar maps has been given an orange color to mimic the sunset coloration of daylight at the surface of the planet. The image shows scattered impact craters and volcanic regions such as Beta Regio and Atla Regio. *(NASA)*

Figure 23-4
Although it is nearly 1000 years old, this lava flow near Flagstaff, Arizona, is still such a rough jumble of sharp rock that it is dangerous to venture onto its surface. Rough surfaces are very good reflectors of radio waves and look bright in radar maps. Lava flows on Venus show up as bright regions in the radar maps because they are rough. *(M. A. Seeds)*

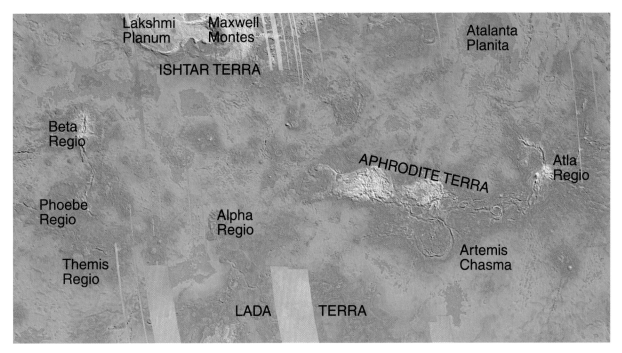

Figure 23-5
Magellan radar images have been combined and color coded to form this map of Venus. North and south polar regions are not shown. The highlands (yellow and red) consist mainly of Ishtar Terra, Aphrodite Terra, and Lada Terra. The dry lowlands are colored green and blue in this map. *(NASA)*

radiation. Alpha Regio and Beta Regio were also discovered before the naming convention was adopted. Other names on Venus are female. The highland regions Ishtar Terra and Aphrodite Terra are named for the Babylonian and Greek goddesses of love.

The low rolling plains appear to be smooth lava flows, but they are marked by about 1000 craters. Unlike the moon, where most craters are in the old highlands, the craters on Venus are uniformly scattered over the surface. Furthermore, all of the craters look sharp and fresh, and there are no old craters (Figure 23-6). There is no water on Venus, and the atmosphere moves slowly at the surface, so there is little erosion. None of the craters appear to be older than about 500 million years. Furthermore, there are no small craters. The thick atmosphere prevents small meteorites from striking the surface. Thus meteorite impacts on Venus have accumulated at the rate of one every few million years over the last half billion years. All trace of older surface has been erased by lava flows.

Volcanism is critically important on Venus. Volcanic cones, some of them very high, are common. They tend to look bright in radar images because of the rough lava flows. The volcanoes of Beta Regio and Atla Regio are visible in Figure 23-3. The volcano Sapas Mons is surrounded by a radial pattern of lava flows (Figure 23-7). Volcanoes on Venus are not located in long chains as are volcanoes on Earth. Thus the volcanism on Venus is not related to plate motion, seafloor spreading, or subduction zones but is caused by hot spots where rising currents of molten rock push up through the crust.

Figure 23-6
Impact crater Howe in the foreground of this Magellan radar image is 37 km in diameter. Craters in the background are 47 km and 63 km in diameter. This radar map has been digitally modified to represent the view from a spacecraft flying over the craters. *(NASA)*

Notice that many images created from the Magellan data are reproduced with their vertical scales stretched. This is easy to do with computer-based data (Window

Graphical Manipulation of Scientific Data

Because so much modern scientific data is stored inside computers, it is common to see data manipulated to exaggerate certain characteristics and suppress others. In some fields, this would be dishonest, but the goals of science make such manipulation valuable. By purposely distorting data, we may reveal things we would not have seen otherwise.

Computers make it easy to manipulate the colors in an image and produce false-color images. Astronomers commonly change and exaggerate the colors in images and maps to reveal details invisible to our eyes. Radio maps of galaxies and X-ray maps of supernova remnants are only two examples. As long as we remember what real galaxies and real nebulae look like, we can learn a great deal from false-color images.

Geologists, too, manipulate data. Geological maps have been color-coded for over a century, and now computers allow faster and more precise use of false color in maps and images. But geological data can also be manipulated to reveal other features. A relief map of the United States would look as smooth as glass if the map maker did not exaggerate the height of mountain ranges, and having such data in a computer makes it easy to "stretch" the heights to make them more visible. Astronomers often do the same with radar maps of Venus and surface maps of Mars.

Such data distortion would be inadmissible in a court of law. In fact, photographic evidence is generally inadmissible in court because it is so easy to fake a photograph and show the defendant in almost any incriminating situation. Faking a photograph in journalism is at least disreputable, and in some cases it is libelous. In fact, modern computer graphics make it possible to put Humphrey Bogart into a modern TV commercial with no trace of fakery, so photographic manipulation is unlikely ever to become acceptable in court or in journalism.

In science, however, our quest is not for a first impression but for detailed analysis. Data can be manipulated and distorted any way that is useful as long as the form of the manipulation is specified. Then viewers are free to form their own opinions as to the meaning of the data. Of course, this assumes that the viewer is alert and takes into consideration the nature of any data manipulation. Thus, when we look at a relief map of the United States or a radar image of a mountain on Venus, it is our responsibility to take into account any data manipulation. if we don't, we might think that the Rocky Mountains are 100 miles high and that mountains on Venus are 20 times steeper than mountains on Earth.

a

b

Figure 23-7
(a) Volcano Sapas Mons, lying along a major fracture zone, is topped by two lava-filled calderas and flanked by rough lava flows. The orange color of this radar map mimics the orange light that filters through the thick atmosphere. *(NASA)* (b) Seen by light typical of Earth's surface, Sapas Mons might look more like this computer-generated landscape. Volcano Maat Mons rises in the background. The vertical scale has been exaggerated by a factor of 15 to reveal the shape of the volcanoes and lava flows. *(Copyright © 1992, David P. Anderson, Southern Methodist University)*

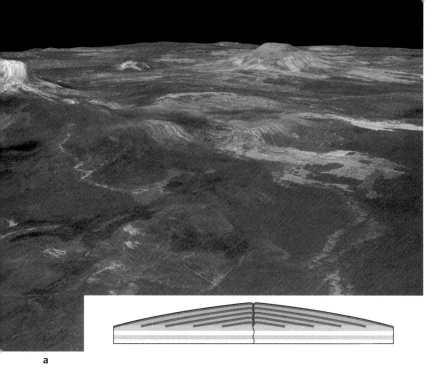

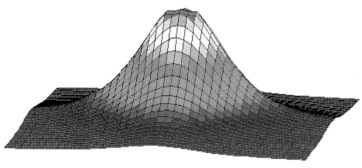

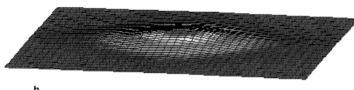

a

b

Figure 23-8
(a) A computer-generated image based on Magellan radar data shows volcanoes Gula Mons (left) and Sif Mons (right) with their associated lava flows. The vertical scale has been stretched to make surface details stand out. The true profile of the volcanoes resembles that of normal shield volcanoes (inset). *(NASA)* (b) In this computer model, details of the true shape of the mountain are dramatically revealed by a vertical magnification of 10 times (top) but are invisible when the model is reproduced with no vertical stretching (bottom). *(Model by M. A. Seeds)*

on Science 23-1), and it makes subtle details easier to study. But the distortion also makes slopes look much steeper than they really are (Figure 23-8). Volcanic peaks on Venus are **shield volcanoes,** which form from highly fluid lavas and have extremely shallow slopes. Shield volcanoes, such as Mauna Loa in Hawaii, are common on Earth.

Volcanism is evident on Venus in circular features called **coronae,** circular bulges up to 2100 km in diameter containing volcanic peaks, lava flows, and bordered by fractures. The coronae appear to be caused by rising currents of molten magma below the crust that create an uplifted dome and then withdraw to allow the surface to subside. Coronae are sometimes accompanied by features called pancake domes, circular outpourings of viscous lava. Both of these features are visible in Figure 23-9.

Lava channels are common on Venus, and they appear similar to the sinuous rills visible on Earth's moon. The longest channel on Venus is also the longest known lava channel in the solar system. It stretches 6800 km (4200 miles), roughly twice the distance from Chicago to Los Angeles. These channels are 1 to 2 km wide and can sometimes be traced back to collapsed areas where lava appears to have drained from beneath the crust (Figure 23-10).

Ishtar Terra may hold secrets to the history of Venus. Western Ishtar consists of a high volcanic plateau called Lakshmi Planum. It rises 4 km above the rolling plains and appears to have formed from lava flows from vents such as Collette and Sacajawea (Figure 23-11). Although folded mountain chains do not occur on Venus as they do on Earth, the areas north and west of Lakshmi Planum appear to be ranges of wrinkled mountains where horizontal motion of the crust has pushed up against Ishtar Terra. Faults and deep chasms are widespread over the surface of Venus. Thus there is reason to suspect that although we see no direct evidence of plate motion, there has been compression of the crust in some areas and stretching and faulting in other areas.

The history of Venus has been dominated by volcanism. While Earth's crust has broken into rigid moving plates, the surface of Venus seems more pliable. In any case, to understand Venus we will have to understand its volcanic history.

A History of Venus

The four-stage history of Venus is not well understood. We are not sure how the planet formed and differentiated, how it was cratered and flooded, or how it evolved a surface.

Venus formed only slightly closer to the sun than did Earth, so we might expect it to be a similar planet differentiated into a silicate mantle and a molten iron core. The density and size of Venus tell us that it must have a dense interior much like Earth's, but if the core is molten, then we might expect the dynamo effect to generate a magnetic field. No spacecraft has ever detected

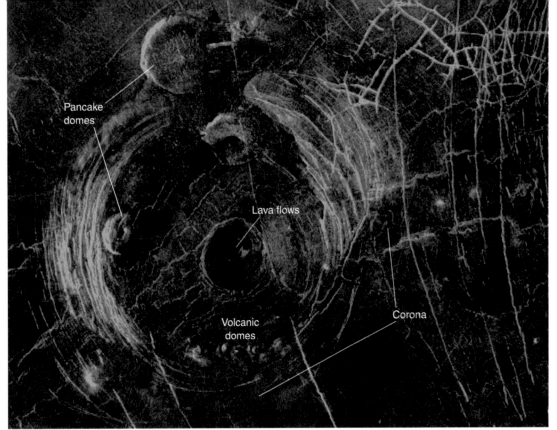

Figure 23-9
Aine Corona is about 200 km (120 miles) in diameter. It is ringed by concentric faults, which suggest that it was pushed up by magma below the crust. Two pancake domes are visible as well as a line of small volcanic domes and a recent lava flow at the center of the corona. Faults run through the area, and a network of fractures lies at upper right. *(NASA)*

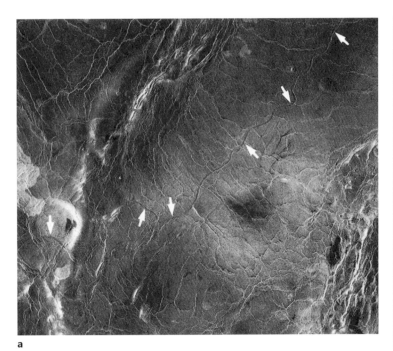

a

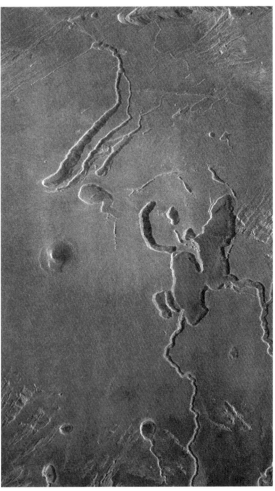

b

Figure 23-10
(a) Arrows point to a 600 km segment of Baltis Vallis, the longest lava flow channel in the solar system. It is a few hundred miles longer than the Nile river on Earth. The ends of the channel are covered by later lava flows, so it may have been longer than its visible 6800 km. (b) The Lo Chen Valles system of channels leads from a collapsed region of the crust from which subsurface lava may have drained. *(NASA)*

Figure 23-11
This radar map of Lakshmi Planum and Maxwell Montes is based on radar data from both Soviet and American spacecraft with ground-based radar data added (color). Violet and purple regions are smoothest, and orange is roughest. Note the volcanic calderas Collette and Sacajawea and the wrinkled mountains around Lakshmi Planum. *(USGS)*

The third stage of planetary evolution is flooding by lava and water, and volcanism has certainly flooded the surface of Venus with lava. Even the chemical composition of the surface rocks resembles the basalts of Earthly lava flows. But flooding by water may never have occurred on Venus. As we saw earlier, Venus may have been too warm for liquid water to exist on its surface, and the ultraviolet radiation in sunlight could easily break up any water and allow the hydrogen to escape to space. Thus, Venus suffered from lava floods but not from water floods.

Because Venus never had liquid water on its surface, it could not cleanse its atmosphere of CO_2, and that gas accumulated to form a thick blanket that traps solar energy in a dramatic example of the greenhouse effect. Had Earth been just 5 percent closer to the sun, it might never have formed oceans to dissolve CO_2 from the air and thus might have suffered a Venusian greenhouse death.

The last stage of planetary evolution is slow surface evolution. On Earth, that means erosion and plate tectonics, but, with no water, erosion is not important on Venus. Because we see no long, folded mountains, no rift valleys outlining plates, and no long features like Earth's midocean ridges and folded mountain

such a field around Venus. The magnetic field must be at least 25,000 times weaker than Earth's. Some theorists wonder if the core of the planet is solid. If it is solid, then we must wonder how Venus got rid of its internal heat so much faster than Earth did.

Because the planet lacks a magnetic field, the solar wind is deflected by the uppermost layers of the planet's atmosphere, thus forming a bow shock where the solar wind is slowed and deflected (Figure 23-12). The magnetic field carried by the solar wind drapes over Venus like seaweed over a fish hook, forming a long tail within which ions flow away from the planet. We will see in Chapter 26 that comets, which also lack magnetic fields, interact with the solar wind in the same way.

The interaction of Venus with the solar wind is clear evidence that the planet has little or no magnetic field. Thus, we know little about the differentiation of the planet into core and mantle. The size of the core shown in Figure 23-12 is estimated by analogy with Earth's.

We know a bit more about the cratering history of Venus. The Magellan radar maps reveal no small craters, apparently because the dense atmosphere breaks up and vaporizes smaller meteorites. Also, the larger craters are about 15 percent more common than craters on the lunar maria. This means that the surface of Venus is, on the average, a bit older than Earth's active surface but not nearly as old as the surface of the moon. Planetary astronomers are now trying to understand how Venus managed to erase all traces of its older crust, and some experts now discuss an age of violent volcanism that may have resurfaced Venus roughly half a billion years ago.

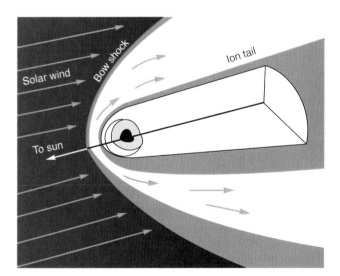

Figure 23-12
By analogy with Earth, the interior of Venus should contain a molten core (estimated here), but no spacecraft has detected a planetary magnetic field. Thus, Venus is unprotected from the solar wind, which strikes the planet's upper atmosphere and is deflected into an ion tail.

chains, it also seems very unlikely that plate tectonics is important on Venus.

Fully 70 percent of the heat from Earth's interior flows outward through the midocean rifts. But Venus lacks crustal rifts, and its numerous volcanoes cannot carry much of the heat out of the interior. Rather, Venus seems to get rid of its interior heat through large currents of hot magma that rise beneath the crust. We see coronae above such currents. Planetary scientists now speculate that cooler magma may sink back into the interior and cause minor horizontal motions in the crust that could account for the folded mountains near Ishtar Terra.

True plate tectonics is not important on Venus. For one thing, the crust is very dry and is consequently about 12 percent less dense than Earth's crust. This low-density crust is more buoyant than Earth's crust and thus resists being pushed into the interior. Also, the crust is so hot it is halfway to the melting point of rock. Such hot rock is not very stiff, so it cannot form the rigid plates typical of plate tectonics on Earth.

Some planetary astronomers speculate that we might be seeing an entirely different tectonic style on Venus. Rather than experiencing gradual plate motion, as does Earth, Venus may suffer periodic overturning as large fractions of its hot, pliable crust sink rapidly into the mantle and new crust forms from lava flows and volcanism. The resurfacing that seems to have occurred half a billion years ago may be simply the most recent of these overturnings. Such a tectonic style could allow Venus to lose large amounts of its internal heat and might explain how the planet could have a solid core.

Such overturnings could have dramatic connections with the climate. Mathematical models show that a major outbreak of volcanism on Venus could increase the greenhouse effect and drive the surface temperature up by as much as 100° C (180° F). This could soften the crust and increase the extent of the volcanism and might lead to a planetwide outburst of lava flows. These resurfacing episodes may occur periodically on Venus, or the planet may have had a geology more like Earth's plate tectonics until the surface was flooded by lava 500 million years ago. In either case, Earth's neighboring planet, its apparent twin, is in fact astonishingly unearthly.

by the Magellan spacecraft. There are no rift valleys or subduction zones outlining plates, and there are no major chains of folded mountains. Rather, we see evidence of volcanism and lava flows, and the small regions of folded mountains suggest only limited horizontal motion in the crust. Rather than being dominated by the motion of rigid crustal plates, Venus may have a crust made more flexible by the heat, a crust dominated by rising plumes of molten rock that strain the crust to produce coronae and that break through to form volcanoes and lava flows.

At first glance, comparative planetology suggested that Earth and Venus should be sister worlds, but it seems they can be no more than distant cousins. We can blame the thick atmosphere of Venus for altering its geology, but that raises another question: Why doesn't Earth have a similar atmosphere?

Venus is certainly an unearthly world, but the familiar principles of comparative planetology help us understand it. As we turn our attention to Mars, we will see the same principles in action, but we will discover that Mars is a world that lacks what Venus has in abundance—active volcanism and air.

23-2 Mars

Mercury and the moon are small. Venus and Earth are, for terrestrial planets, large. But Mars occupies an intermediate position (Figure 23-13). It is twice the diameter of the moon but only 53 percent Earth's diameter. Its small size has allowed it to cool faster than Earth,

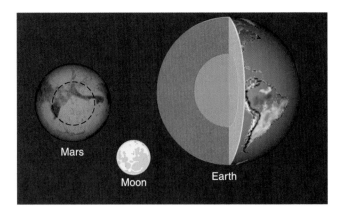

Figure 23-13
Mars is a medium-size world smaller than Earth but larger than the moon. Earth's core is hot and molten, while the moon is cold and may not have a molten core at all. The nature of the interior of Mars is unknown. Unlike the moon, Mars has an atmosphere, but it is much thinner than Earth's atmosphere.

and much of its atmosphere has leaked away. Its present carbon-dioxide atmosphere is only 1 percent as dense as Earth's.

In some ways, Mars is much like Earth. A day on Mars is nearly the same length as an Earth day—24 hours and 40 minutes—and its year lasts 1.88 Earth years. Also, just as Earth's axis is tipped 23.5°, that of Mars is tipped 25°. As the northern and southern hemispheres turn alternately toward the sun, seasonal changes are visible even through a small telescope. As spring comes to the southern hemisphere, the white polar cap shrinks and the grayish surface markings grow darker and, according to some observers, greener. At one time, these seasonal changes led some to believe that plant life flourished on Mars when spring thawed the polar cap. We will see later that this is not so.

In other ways, Mars is very different from Earth. Its surface has never broken into moving plates like those that create the topography on Earth. Instead, Mars is a one-plate planet, with its crust frozen into a solid layer but bearing signs of geological activity that include some of the largest volcanoes in the solar system.

The Canals of Mars

Even before the space age brought us photographs of the Martian surface, the planet Mars was a mysterious landscape in the public mind. In the century following Galileo's first astronomical use of the telescope, astronomers discovered dark markings on Mars as well as bright polar caps (Data File Six). Timing the motions of the markings, they concluded that a Martian day was about 24 hours long, and the similarity with Earth's day further supported the belief that Mars was another Earth.

Apparent proof that Mars was peopled by intelligent beings appeared late in the summer of 1877, when Italian astronomer Giovanni Virginio Schiaparelli, using a telescope only 8.75 in. in diameter, thought he glimpsed fine, straight lines visible when Earth's atmosphere was unusually still. When he reported his discovery, he used the Italian word *canali,* which means "channel." But the word was translated into English as "canal," an artificially dug channel, and the "canals of Mars" were born. Many astronomers could not see the canals, whereas others drew maps showing hundreds (Figure 23-14).

In the decades that followed Schiaparelli's discovery, many people assumed that the canals were water courses built by an intelligent race on Mars to carry water from the polar caps to the lower latitudes. Much of this excitement was generated by Percival Lowell, a wealthy Bostonian who founded Lowell Observatory in 1894, principally for the study of Mars. He not only mapped hundreds of canals but also publicized his results. Although some astronomers claimed the canals were merely illusions, by 1907 the general public was so sure that life existed on Mars that the *Wall Street*

Mars Data File Six

This Hubble Space Telescope view of Mars shows the northern polar cap at the top, impact craters, and dark surface features. The spot near the left edge is the volcano Ascraeus Mons, 25 km (16 mi) high, poking up through morning clouds of water-ice crystals in the thin CO_2 atmosphere. *(Philip James, University of Toledo; Steven Lee, University of Colorado, Boulder; and NASA)*

Average distance from the sun	1.5237 AU (2.279×10^8 km)
Eccentricity of orbit	0.0934
Maximum distance from the sun	1.6660 AU (2.492×10^8 km)
Minimum distance from the sun	1.3814 AU (2.066×10^8 km)
Inclination of orbit to ecliptic	1°51′09″
Average orbital velocity	24.13 km/s
Orbital period	1.8808 y (686.95 days)
Period of rotation	24^{h}37^{m}22.6^s
Inclination of equator to orbit	25°19′
Equatorial diameter	6796 km (0.53 $D_\oplus$)
Mass	0.6424×10^{24} kg (0.1075 $M_\oplus$)
Average density	3.94 g/cm^3 (3.3 g/cm^3 uncompressed)
Surface gravity	0.379 Earth gravities
Escape velocity	5.0 km/s (0.45 $V_\oplus$)
Surface temperature	−140° to 20°C (−220° to 68°F)
Average albedo	0.16
Oblateness	0.009

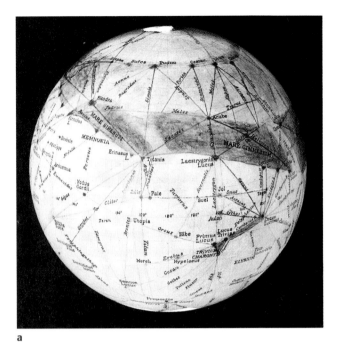
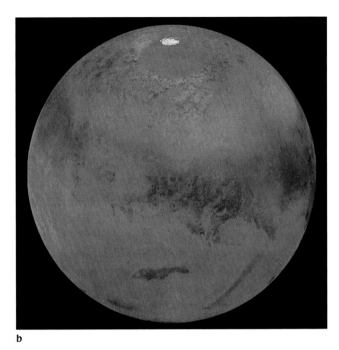

a

b

Figure 23-14

(a) Early in the 20th century, Percival Lowell mapped canals over the face of Mars and concluded that intelligent life resided there. (b) Modern images recorded by spacecraft reveal a globe of Mars with no canals. Instead, the planet is marked by craters and, in some places, volcanoes. Both of these images are reproduced with south at the top, as they appear in telescopes. Lowell's globe is inclined more nearly vertically and is rotated slightly to the right compared with the modern globe. *(a, Lowell Observatory; b, U.S. Geological Survey)*

Journal could suggest that the most extraordinary event of the previous year had been "the proof by astronomical observations . . . that conscious, intelligent human life exists upon the planet Mars." Further sightings of bright clouds and flashes of light on Mars strengthened this belief, and some urged that gigantic geometrical diagrams be traced in the Sahara Desert to signal to the Martians that Earth, too, is inhabited. All seemed to agree that the Martians were older and wiser than humans.

This fascination with men from Mars was not a passing fancy. Beginning in 1912, Edgar Rice Burroughs wrote a series of 11 novels about the adventures of the Earthman John Carter lost on Mars. Burroughs made the geography of Mars, named by Schiaparelli after Mediterranean lands both real and mythical, into household words. He also gave his Martians green skin.

By Halloween night of 1938, people were so familiar with life on Mars that they were ready to believe that Earth had been invaded. When a radio announcer repeatedly interrupted dance music to report the landing of a spaceship in New Jersey, the emergence of monstrous beings, and their destruction of whole cities, thousands of otherwise sensible people rushed to their cars and fled in panic, not knowing that Orson Welles and other actors were dramatizing H. G. Wells's book *The War of the Worlds.*

The fascination with Mars, its canals, and its little green men lasted right up to July 15, 1965, when Mari-

ner 4, the first spacecraft to fly past Mars, radioed back photos of the surface and proved that there are no canals and no Martians. Apparently, the canals are optical illusions produced by the human brain's astounding ability to assemble a field of disconnected marks into a coherent image. If your brain could not do this, the photos on these pages would be nothing but swarms of dots, and the images on the screen of a television set would never make sense. The brain of an astronomer looking for something at the edge of visibility is capable of connecting random markings on Mars into the straight lines of canals.

Even today, Mars holds a fascination for the general public. The grocery store tabloids regularly run stories about a giant face carved on Mars by an ancient race. Although planetary scientists recognize it as nothing more than chance shadows in a photograph and dismiss the issue as a silly hoax, the stories persist. A hundred years of speculation have given us high expectations for Mars. If there really was intelligent life on Mars and its representatives came to Earth, they would probably be a big disappointment to the readers of the tabloids.

The Atmosphere of Mars

If we visited Mars, our first concern, even before we opened the door of our spaceship, would be the atmosphere. Is it breathable? Even for the astronomer ob-

Figure 23-15
Mars is the red planet because of iron oxides. Here the rover, Sojourner, has left tracks in the red soil as it explores rocks at the Pathfinder landing site. The large boulder, Yogi (red arrow), and the two smaller rocks, Cradle and Barnacle Bill, may have reddish surfaces. White traces in the soil suggest deposits left long ago by water. *(NASA)*

serving safely from Earth, the atmosphere of Mars is a major concern. The gases that cloak Mars are critical to our understanding of the history of the red planet.

The air on Mars is 95 percent carbon dioxide, with a few percent each of nitrogen and argon. It contains almost no water vapor or oxygen, and its density at the surface of the planet—only 1 percent that of Earth's atmosphere—does not provide enough pressure to prevent liquid water from boiling into vapor. If you stepped outdoors on Mars without a space suit, your own body heat would make your blood boil.

Although the air is thin, it is dense enough to have weather patterns, and the Martian wind often picks up dust from the surface in great dust storms of red dust. The reddish soil is evidently caused by iron oxides (rusts), and that tells us that the oxygen we humans would prefer in the atmosphere is actually locked in chemical compounds with iron in the soil (Figure 23-15).

To understand Mars, we must ask why its atmosphere is so thin and dry and why the surface is rich in oxides. The answers to these questions lie in the origin and evolution of the Martian atmosphere.

Presumably, the gases in the Martian atmosphere were mostly outgassed from its interior. Volcanism on terrestrial planets typically releases carbon dioxide and water vapor plus other gases. Because Mars formed farther from the sun, we might expect that it incorporated more volatiles when it formed. But Mars is smaller than Earth, so it has had less internal heat to drive geological activity, and we might suspect that it has not outgassed as much as Earth. In any case, the outgassing occurred early in the planet's history, and Mars, being small, cooled rapidly and now releases little gas.

How much atmosphere a planet has depends on how rapidly it releases internal gas and how rapidly it loses gas from its atmosphere, and the rate at which a planet loses gas depends on its mass and temperature. The more massive the planet, the higher its escape velocity (Chapter 5) and the more difficult it is for gas atoms to leak into space. Mars has a mass less than 11 percent that of Earth, and its escape velocity is only 5 km/s, less than half Earth's. Thus, gas atoms can escape from it much more easily than they can escape from Earth.

The temperature of a planet's atmosphere is also important. If the gas is hot, its molecules have a higher average velocity and are more likely to exceed escape velocity. Thus, a planet that is near the sun and very hot is less likely to retain an atmosphere than a more distant, cooler planet. The velocity of a gas molecule, however, also depends on the mass of the molecule. On average, a low-mass molecule travels faster than a massive molecule. Thus, a planet loses its lowest-mass gases more easily because those molecules travel fastest.

We can see this principle of comparative planetology if we plot a diagram such as that in Figure 23-16. The points show the escape velocity versus temperature of the larger objects in our solar system. The temperature we use is the temperature of the gas that is in a position to escape. For the moon, which has essen-

tially no atmosphere, this is the temperature of the sun-lit surface. For Mars, we use the temperature at the top of the atmosphere. The lines in Figure 23-16 show the typical velocity of the fastest-traveling examples of various molecules. At any given temperature, some water molecules, for example, travel faster than others, and it is the highest-velocity molecules that escape from a planet. The diagram shows that the Jovian planets have escape velocities so high that very few molecules can escape. Earth and Venus can't hold hydrogen, and Mars can hold only the more massive molecules. Earth's moon is too small to keep any of the common gases. We will refer to this diagram again as we study the atmospheres of other worlds.

Over the 4.5 billion years since Mars formed, it has lost some of the lower-mass gases. Water molecules are massive enough for Mars to keep, but ultraviolet radiation can break up the water molecules. On Earth, the ozone layer protects water vapor from ultraviolet radiation, but Mars never had an oxygen-rich atmosphere so it never had an ozone layer. Ultraviolet photons from

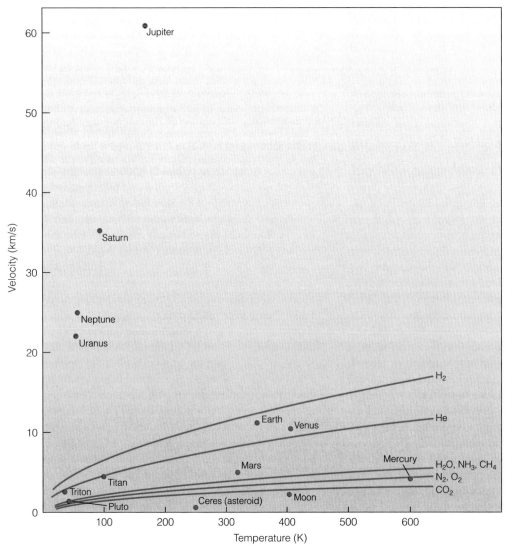

Figure 23-16
The loss of planetary gases. Dots represent the escape velocity and temperature of various solar-system bodies. The lines represent the typical highest velocities of molecules of various masses. The Jovian planets have high escape velocities and can hold on to even the lowest-mass molecules. Mars can hold only the more massive molecules, and the moon has such a low escape velocity that even the most massive molecules can escape.

the sun can penetrate deep into the atmosphere and break up molecules such as water. The hydrogen escapes, and the oxygen, a very reactive element, forms oxides in the soil—the oxides that make Mars the red planet. Thus, molecules too massive to leak into space can be lost if they break into lower-mass fragments.

The argon in the Martian atmosphere is evidence of a denser blanket of air in the past. Argon atoms are massive, almost as massive as a CO_2 molecule, and would not be lost easily. In addition, argon is inert and would not form compounds in the soil. The 1.6 percent argon in the atmosphere of Mars is evidently left over from an ancient atmosphere that was 10 to 100 times denser than the present Martian air.

Finally, we must consider the interaction of the solar wind with the atmosphere of Mars. This is not an important process for Earth because Earth has a magnetic field that deflects the solar wind. Because Mars has no magnetic field, the solar wind interacts directly with the Martian atmosphere, and detailed calculations show that significant amounts of CO_2 could have been carried away by the solar wind over the history of the planet. This process would have been most efficient long ago, however, when the sun was more active and the solar wind was stronger. We must also wonder if Mars had a magnetic field when it was younger and still retained significant internal heat. If it had a molten, conducting core, it may have had a magnetic field to protect it from the solar wind.

Although we remain uncertain as to how much of an atmosphere Mars has had in its past and how much it has lost, it is a good example for our study of comparative planetology. When we look at Mars, we see what can happen to the atmosphere of a medium-size world. Like its atmosphere, the geology of Mars is typical of intermediate-size worlds.

The Geology of Mars

Mars has been visited by a number of spacecraft beginning with Mariner 4 in 1965. Some have orbited the planet, and a few have landed. The photographs and measurements transmitted back to Earth tell us that Mars is a world of contrasts, an ancient cratered world with deep canyons and giant volcanoes (Figure 23-17).

A few generations ago, astronomers watched seasonal changes on Mars and wondered if they were seeing a cycle of growing plants. Evidently, these changes are caused by vast seasonal dust storms that cover dark rock and, as spring begins, strip the dust away. The markings grow darker and, because the entire planet has a reddish tint, the dark marks look greenish in contrast. Thus Mars is not a world of green fields, but a world of deadly dry deserts.

Photographs from the surface of Mars such as Figure 23-15 show reddish deserts of broken rock. These appear to be lava plains fractured by meteorite impacts, but they don't look much like the surface of Earth's moon. The atmosphere of Mars, thin though it is, protects the surface from the blast of micrometeorites that grinds moon rocks to dust. Also, the Martian dust storms may sweep fine dust away from some areas and leave larger rocks exposed.

Surveys by spacecraft show that Mars is divided into two parts. The southern highlands are heavily cratered, making them moonlike, while the northern lowlands are smooth and have few craters (Figure 23-18). The cratered terrain in the south is probably over 3.5 billion years old and must have formed as the heavy bombardment ended. The northern hemisphere must have been resurfaced roughly a billion years ago. This division of the planet into two fundamentally different

Figure 23-17
The red surface of Mars is marked by three giant volcanoes at left and a canyon, Valles Marineris, 4000 km long stretching across the bottom of this image made by the orbiting Viking spacecraft. Small round features are impact craters. *(NASA/USGS, courtesy Alfred S. McEwen)*

Geologists are fond of saying "The present is the key to the past." By that they mean that we can learn about Earth's history by looking at the present condition of Earth's surface. The position and composition of various rock layers in the Grand Canyon, for example, tell us that the western United States was once at the floor of an ocean. This principle of geology was astonishing when it was first formulated in the late 1700s, and it continues to be relevant today as we try to understand other worlds, such as Venus and Mars.

In the late 18th century, naturalists recognized that the present gave them clues to Earth's history. This was astonishing because most people assumed that Earth had no history. That is, they assumed either that Earth had been created in its present state as described in the Old Testament or that Earth was eternal. In either case, people commonly assumed that the hills and mountains they saw around them had always existed more or less as they were. The 18th-century naturalists began to see evidence that the hills and mountains were not eternal but were the result of past processes and were slowly changing. That gave birth to the idea that Earth had a history.

As the naturalists of the 18th century made the first attempts to thoughtfully and logically explain Earth's nature by looking at the evidence, they were inventing modern geology as a way of understanding Earth. What Copernicus, Kepler, and Newton did for the heavens in the 1600s, the first geologists did for Earth beginning in the late 1700s. Of course, the invention of geology as the study of Earth led directly to our modern attempts to understand the geology of other worlds.

Geologists and astronomers share a common goal: they are attempting to reconstruct the past (Window on Science 20-3). Whether we study Earth, Venus, or Mars, we are looking at the present evidence and trying to reconstruct the past history of the planet by drawing on observations and logic to test each step in the story. How did Venus get to be covered with lava, and how did Mars lose its atmosphere? The final goal of planetary astronomy is to draw together all of the available evidence (the present) to tell the story (the past) of how the planet got to be the way it is. Those first geologists of the late 1700s would be fascinated by the stories planetary astronomers tell today.

---

kinds of terrain is trying to tell us something about the history of Mars (Window on Science 23-2), but that message is not clear. Volcanism and lava flows have been suggested as the smoothing agent in the north, and volcanism is certainly more important in the northern hemisphere than in the south. Some astronomers argue that a great ocean filled the northern basin; we will discuss the uncertain history of water on Mars later in this

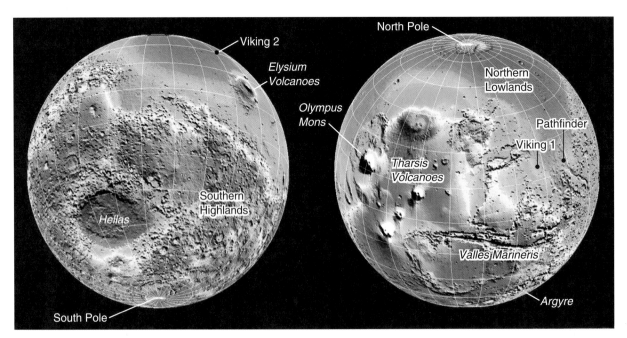

Figure 23-18
These globes of Mars are color coded to show elevation. The northern lowlands lie about 4 km below the southern highlands. Volcanoes are very high (white) and the giant impact basins, Hellas and Argyre, are low. Note the depth of the canyon Valles Marineris. *(NASA)*

chapter. It is not clear why there is an asymmetry between northern and southern hemispheres at all, but some astronomers wonder if a giant impact shattered the crust of the northern hemisphere when Mars was young.

The cratering and volcanism on Mars confirms our understanding of Mars. It is larger than Earth's moon, so it cooled more slowly and its volcanism has continued longer than on the moon. But Mars is smaller than Earth and less geologically active, so some of its ancient cratered terrain has survived undamaged by volcanism.

Martian volcanoes are shield volcanoes, which have shallow slopes showing that the lava flowed easily. Nevertheless, the largest volcano, Olympus Mons, is a vast structure (Figure 23-19). Its base is 600 km (370 mi) in diameter, and it towers 25 km (16 mi) above the surface. In contrast, the largest volcano on Earth, Mauna Loa in Hawaii, rises only 10 km (6 mi) above its base on the Pacific Ocean floor, and its base is only 225 km (140 mi) in diameter. The weight of Mauna Loa is too great to be supported by the seafloor. It has pushed the seafloor down, just as a bowling ball resting on a bed depresses the surface. This depression is visible as an undersea moat ringing Mauna Loa, but no depression is visible ringing Olympus Mons, even though it is 2.5 times higher than Mauna Loa. Evidently the crust of Mars is thicker than Earth's.

Other evidence shows that the Martian crust has been thinner and more active than the moon's. Valles Marineris is a network of canyons that stretches 4000 km (2500 mi) long and up to 600 km (400 mi) wide (Figure 23-17). At its deepest, it is four times deeper than the Grand Canyon on Earth. It is long enough to stretch from New York to Los Angeles. The canyon has been produced by faults in the crust that allowed great blocks to sink. Landslides and some erosion have further modified the canyon. Although Valles Marineris is an old feature, it does show that the crust of Mars has been more active than the crusts of the moon or Mercury, worlds that lack such dramatic canyons.

The faults that created Valles Marineris seem linked at their west end to a great volcanic rise in the crust of Mars called the Tharsis bulge. Nearly as large as the United States, the Tharsis bulge rises to 10 km above the mean radius of Mars. Tharsis is home to many smaller volcanoes, but on its summit lie three giants, and just off of its northwest edge lies the volcano Olympus Mons (Figure 23-20). The origin of the Tharsis bulge is not understood. One theory holds that it is an accumulation of lava flows that have built up over the history of the planet. A more popular theory holds that a large plume of rising magma in the mantle has pushed the crust upward and broken through to form the volcanoes. We can't tell how old the Tharsis bulge is, because the exposed surface, less than a billion years old, covers older layers.

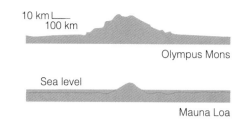

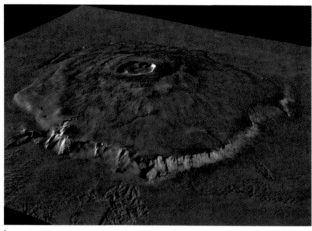

b

Figure 23-19
(a) Olympus Mons, the highest volcano in the solar system, is much larger than Mauna Loa, the largest volcano on Earth. Mauna Loa has sunk into Earth's crust, producing a moat around its base. (b) This perspective view of Olympus Mons shows the 6-km-high cliff at its base and the multiple calderas at its summit. (© *Calvin J. Hamilton, Columbia, Maryland*)

A similar uplifted volcanic bulge, the Elysium region visible in Figure 23-18, lies almost halfway around the planet. It appears to be similar to the Tharsis bulge, but it is more heavily cratered and so must be older.

Volcanoes form where the rising magma breaks through the crust, but volcanoes on Earth do not grow as large as they do on Mars, evidently because Earth's crust is in motion. On Earth, a rising current of magma, like that below the Hawaiian islands, breaks through the crust and forms a volcano. The motion of the crustal plate, however, moves the accumulated lava aside, and the next eruption produces a new volcano (Figure 23-21). Thus, the Hawaiian-Emperor island chain extends 7500 km (4700 mi) northwest across the Pacific seafloor because the crust is moving (see Figure 21-11). On Mars, the magma seems to have erupted over and over in the same place and built massive volcanoes such as Olympus Mons, the largest volcano in our solar system. The vast size of Olympus Mons is evidence that the crust of Mars is not separated into moving plates.

No spacecraft has ever photographed an erupting volcano on Mars, but it is possible that some of the volcanoes are still active. Craters in the youngest lava

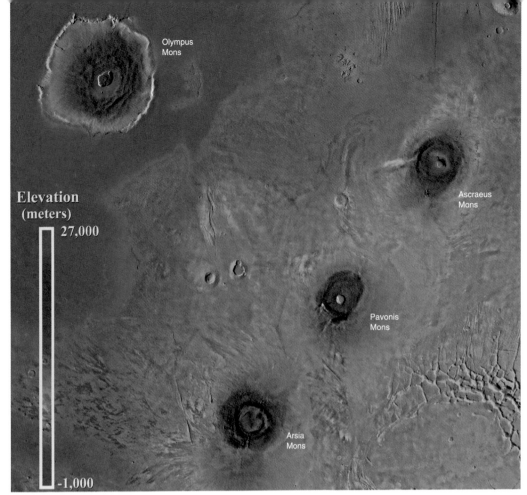

Olympus
Mons

Ascraeus
Mons

Elevation
(meters)
27,000

Pavonis
Mons

Arsia
Mons

-1,000

Figure 23-20
This color-coded image
shows the Tharsis volcanic
bulge as green rising 6 km
above the surrounding plains
(blue). The three largest
Tharsis volcanoes rise high
above the surface. The giant
volcano Olympus Mons at
upper left rests on the lower
plains but rises as high as
Ascraeus Mons. Note the
faults at lower right. Compare with Figures 23-17 and
23-18. *(© Calvin J. Hamil-*
ton, Columbia, Maryland)

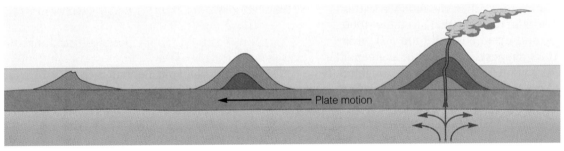

Plate motion

a

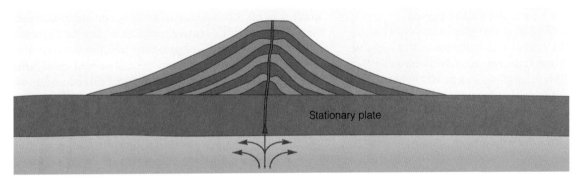

Stationary plate

b

Figure 23-21
(a) On Earth, a rising current of magma breaks through the crust to form a chain of volcanoes because the moving plate carries older volcanic peaks away. The Hawaiian-Emperor island chain formed in this way, and most
of the older peaks have now been worn down below sea level. (b) On Mars, volcanoes such as Olympus Mons
have grown to great size, suggesting that the crust is not moving. Successive eruptions occur at the same place
and add layer after layer to the bulk of the volcano.

flows around Olympus Mons show that the volcano may have been active as recently as a few hundred million years ago. The Elysium volcanoes appear to be old and inactive. Nevertheless, Mars may still retain enough heat to trigger an eruption, but the interval between eruptions may be very long.

Water on Mars

There is no trace of liquid water on Mars now, and the low atmospheric pressure would not allow liquid water to survive very long. Two kinds of geological features, however, suggest that they were carved by moving water, and that is exciting for two reasons. Water is the stuff of life, and if Mars had liquid water on its surface long ago, life may have begun there. Also, water is an important component in planet building, and we will never understand the history of Mars until we understand the history of its water.

One type of erosional feature on Mars resembles the traces left behind by violent floods that scour the surface in a matter of hours or days (Figure 23-22a). Some of these **outflow channels** are so large they must have been formed by floods 10,000 times greater than the flow of the Mississippi. One way to account for these features is to suppose that the water was frozen in the soil as permafrost. Melted by volcanic heat and under the weight of the overlying rock, water could have burst out to produce sudden floods. Photos of Mars show jumbled valleys typical of the withdrawal of subsurface water and craters surrounded by "splashes" of ejecta that suggest the soil was water-laden at the time of impact (Figure 23-22b).

In contrast to the flood features, branching **runoff channels** look like dry river beds, and photos made by the Mars Global Surveyor spacecraft show that the water flowed through the riverbeds for long periods of time, perhaps for millions of years (Figure 23-23). These channels are located in the old, cratered region of the southern hemisphere, and they have about the same age. Evidently, the atmosphere was denser then, and liquid water could survive on the surface.

The water that formed the runoff channels does not seem to have fallen as rain or snow. There are very few small tributary streambeds, as there would be if precipitation drained from a large area. Rather the water

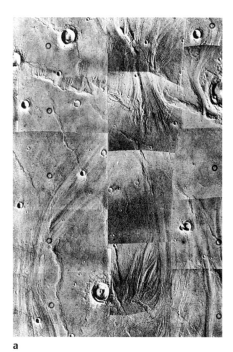 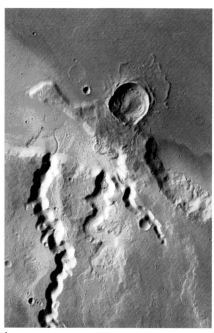

a   b

Figure 23-22

(a) Outflow channels on Mars are dramatically evident in this mosaic of images. Flowing water eroded a wide area and was diverted around obstructions. (b) The largest crater in this image is surrounded by a "splash" of ejected material, suggesting that the crust was rich in water at the time of impact. Note the meandering stream channels leading into the area of the crater. *(NASA)*

appears to have been frozen in the soil as permafrost that melted and seeped out through springs to drain away down the runoff channels.

By counting craters in the flow features, astronomers can estimate that they formed between 1 and 3 billion years ago. Thus whatever water Mars once had may have been lost or has become hidden in the crust as ice.

Some astronomers have found features on Mars that they interpret to be shorelines and related features produced by ancient oceans that lay in the northern lowlands. By counting craters, they estimate that the oceans might have existed 2.5 to 3.5 billion years ago. To explain how the water could remain a liquid, they propose that volcanic eruptions released enough gases to produce a strong greenhouse effect that raised the temperature and vaporized CO_2 and water to increase the atmospheric pressure.

If liquid water was stable on Mars, then its climate must have been different in the past. Evidence of climatic variation is visible near the polar caps (Figure 23-24). The polar caps consist of frozen carbon dioxide (dry ice) with frozen water beneath. When spring begins in a hemisphere, the polar cap there begins to shrink as the carbon dioxide turns directly into gas. The water, however, remains behind in a smaller, permanent polar cap. The region around this permanent cap is marked by layered terrain, evidently deposited by wind-borne dust that accumulates on the ice cap. As

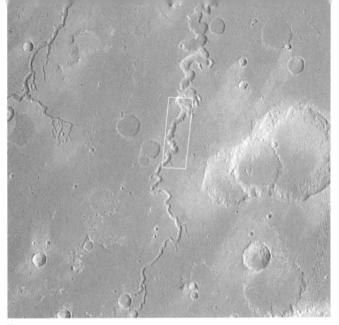

a

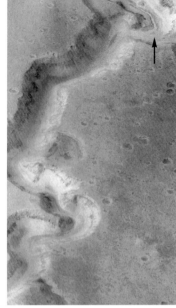

b

Figure 23-23
(a) Meandering across the plains of Mars, runoff channels suggest a steady flow of water. (b) A detailed image of the central canyon recorded by the Mars Global Surveyor spacecraft reveals a narrow streambed at the bottom of the canyon (arrow), further evidence of a sustained flow. *(Malin Space Science Systems/NASA)*

the cap shrinks, the material accumulates in layers. Each layer is a few dozen meters thick and may take about 10,000 years to accumulate. The variation in the layering suggests that some periodic change in climate, perhaps due to orbital changes, may affect the frequency and intensity of planetwide dust storms and thus alter the rate at which material is deposited. Recall that

Earth's climate may go through cycles caused by variations in its orbital motion. (See Chapter 2.)

One astonishing bit of evidence of water on Mars is the analysis of rock samples from the planet. Of course, no astronaut has ever visited Mars and brought back a rock, but over the history of the solar system occasional impacts by asteroids have blasted giant craters

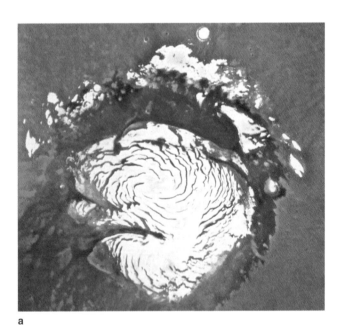

a

b

Figure 23-24
(a) The north polar cap of Mars is made of many regions of ice separated by narrow valleys free of ice. *(NASA)* (b) A summer photo taken near the north pole shows frozen water left behind when the dry ice vaporized. In the ice-free valley, layers as thin as 10 m are visible (arrows). *(NASA)* (c) In places, different sets of layers are superimposed, suggesting periodic changes in the Martian climate. *(Adapted from a diagram by J. A. Cutts, K. R. Balasius, G. A. Briggs, M. H. Carr, R. Greeley, and H. Masursky)*

c

a

b

Figure 23-25
(a) The impact crater Galle at right, known for obvious reasons as the Happy Face Crater, is 210 km in diameter. Such large impacts may occasionally eject fragments of the crust into space. A few fragments from Mars have fallen to Earth as meteorites. *(Malin Space Science System/NASA)* (b) Meteorite ALH84001, found in Antarctica, has been identified as an ancient rock from Mars. Minerals found in the meteorite were deposited in water and thus suggest that the Martian crust was once richer in water than it is now. *(NASA)*

on Mars and ejected bits of rock into space. A few of those bits of rock have fallen to Earth as meteorites, and nearly a dozen have been found and identified (Figure 23-25). These rocks are basalts, as we might expect from a planet so heavily covered by lava flows, and chemical analysis shows that the magma from which the rocks solidified must have contained about 1.4 percent water. If all of the lava flows on Mars contained that much water and it was all outgassed, it could create a planet-wide ocean 200 m deep. That isn't quite enough to explain all the flood features on Mars, but it does assure us that the planet once had more water.

Further evidence of water on Mars comes from precision measurements of color using Hubble Space Telescope images of Mars. This technique can locate minerals that formed in the presence of water (Figure 23-26). These minerals tell us that water was once present and may still be locked in the crust as water-bearing minerals or as ice.

It is dramatically clear that liquid water has been present on the surface of Mars and that it is no longer stable there. Understanding how Mars changed so dramatically will help astronomers better understand planets in general and Earth in particular.

A History of Mars

The four-stage history of Mars is a case of arrested development. Because of its small size, its interior cooled, and geological activity did not last very long. Also because of its small size, it gradually lost some of its atmosphere, and its surface fell into a deep freeze.

In the first stage of planetary evolution, the planet seems to have differentiated into a core, mantle, and crust. There are no obvious traces of plate tectonics such as folded mountain chains, but the Mars Global Surveyor spacecraft measured residual magnetic fields in the crust of Mars and found them in alternating parallel bands. The existence of any residual field means that Mars must have had a magnetic field long ago. Otherwise there would be no residual field at all trapped in the rocks of the crust. The presence of the field in parallel bands resembles the alternating bands formed in Earth's crust by plate tectonics. (See Figure 21-10.) This is far from conclusive, but it does tell us that Mars once had a molten conducting core.

During the second stage of planetary evolution, cratering, the crust of Mars was battered during the heavy bombardment as the last of the debris in the young solar system was swept up. The old southern hemisphere survives from this age about 4 billion years ago. The largest impacts blasted out the great basins like Hellas and Argyre. A few astronomers have suggested that a single major impact caused the formation of the northern lowlands.

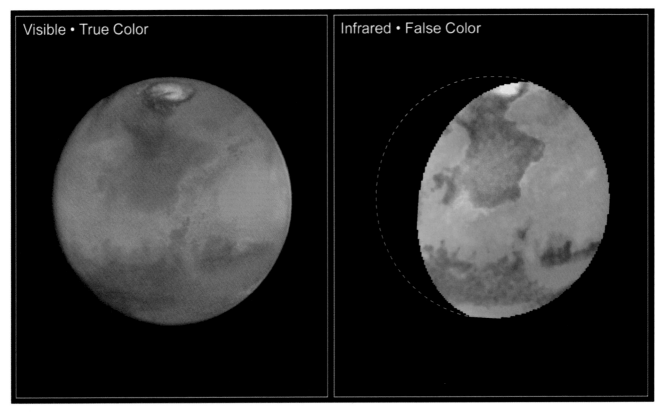

Figure 23-26
Water-related minerals on Mars can be identified through precision measurements of color. On the left, a Hubble Space Telescope image of Mars has been carefully calibrated to reproduce the colors we would see with the unaided eye. The blue shade along the edges of the disk is atmospheric haze and wispy clouds. On the right an infrared image of the same side of Mars is reproduced in false colors to show the location of reddish water-bearing minerals. The large red region, Mare Acidalium, was the site of massive flooding early in Martian history. *(Jim Bell, Cornell, Justin Maki, JPL, Mike Wolff, Space Sciences Institute, and NASA)*

The third stage of planet formation, flooding, included flooding by great lava flows that smoothed some regions. Volcanism in the Tharsis and Elysium regions was very active. Whether flooding included liquid water is controversial, because the present atmosphere is too thin to allow liquid water to exist on the surface.

The early atmosphere was evidently thicker, and liquid water carved the erosional features we see today. Some astronomers believe there was once a liquid-water ocean that filled the northern hemisphere lowlands. They argue that the region is smooth because it is old seafloor. Other astronomers argue that the northern hemisphere is smooth because it was flooded by lava flows, and that liquid-water oceans never existed on Mars. In any case, as atmospheric gases leaked into space, the surface temperature and pressure fell, and water became trapped in the soil as permafrost. Mars became a frozen world.

The history of Mars may hinge on climate variations. Recently calculated models suggest that Mars once may have rotated at a much steeper angle to its orbit, as much as 45°. This could have made the climate warmer and kept more of the CO_2 from freezing out at the poles. The rise of the Tharsis bulge could have tipped the axis to its present 25° and cooled the climate. That might account for the warmer, wetter climate implied by the flow features. Some calculations suggest that Mars goes through cycles as its rotational inclination fluctuates, and this may cause short-term variations in climate much like the ice ages on Earth. That might explain the layered terrain near the polar caps.

The fourth stage of planet formation is unremarkable on Mars. The crust of Mars is now too thick to be active. The planet has lost much of its internal heat, and now it lacks a molten core, as evidenced by the lack of a planetwide magnetic field. Although the crust was stressed by rising magma that produced the Tharsis and Elysium volcanic rises and the great canyon Valles Marineris, it is too thick for plate tectonics. Volcanism may still occur on Mars, but the crust has grown too thick for much geological activity beyond slow erosion by wind-borne dust.

Our history of Mars is far from complete, but it does illustrate how the intermediate size of Mars has influenced both its atmosphere and its geology. Neither small nor large, Mars is a medium world.

Why doesn't Mars have coronae like those on Venus?

The coronae on Venus are believed to have been caused by rising currents of molten magma in the mantle pushing upward under the crust and then withdrawing to leave the circular scars called coronae. Earth, Venus, and Mars have had significant amounts of internal heat, and there is plenty of evidence that they have had rising convection currents of magma under their crusts. Of course, we wouldn't expect to see coronae on Earth because its surface is rapidly modified by erosion and plate tectonics. Furthermore, the mantle convection on Earth seems to produce plate tectonics rather than coronae. Mars, however, is a smaller world and must have cooled faster. We see no evidence of plate tectonics, and we do see giant volcanoes that suggest rising plumes of magma erupting up through the crust at the same point over and over. Perhaps we see no coronae on Mars because the crust of Mars rapidly grew too thick to deform easily over a rising plume. On the other hand, perhaps we could think of the entire Tharsis bulge as a single giant corona.

Planetary astronomers haven't explored enough planets yet to see all the fascinating combinations nature has in store for us. Perhaps after we have visited a few hundred terrestrial planets, we will recognize the geology of Mars as typical of medium-size worlds. Of course, Mars is not medium in terms of its location. Of the terrestrial planets, Mars is the farthest from the sun. How has that affected the evolution of its atmosphere?

Our history of Mars must include a phenomenon that was not part of our histories of Mercury and Venus. Unlike those worlds, Mars has moons. These moons, and the moons we will study in the outer solar system, contain important clues to the origin of the planets.

23-3 The Moons of Mars

If we could camp overnight on Mars, we might notice its two small moons, Phobos and Deimos. Phobos, shaped like a flattened loaf of bread 20 km × 23 km × 28 km, would appear less than half as large as Earth's full moon. Deimos, only 12 km in diameter and three times farther from Mars, would look only 1/15 the diameter of Earth's moon.

Both moons are tidally locked to Mars, keeping the same side facing the planet as they orbit. Also, both moons revolve around Mars in the same direction that Mars rotates, but Phobos follows such a small orbit that it revolves faster than Mars rotates. Thus, Phobos rises in the west and sets in the east.

Origin and Evolution

Deimos and Phobos are typical of the small, rocky moons in our solar system (Figure 23-27). They are dark gray, with albedos of only about 0.06, and they have low densities, about 2 g/cm³.

Many of the properties of these moons hint that they are captured asteroids. In the outer parts of the asteroid belt, almost all asteroids are dark, low-density objects. Massive Jupiter, orbiting just outside the asteroid belt, can scatter such bodies throughout the solar system, so we should not be surprised if Mars, the next planet inside the asteroid belt, has captured a few of these as satellites.

However, capturing a passing asteroid into a closed orbit is not so easy that it happens often. The asteroid approaches the planet along a hyperbolic orbit and, if it is unimpeded, swings around the planet and disappears back into space. To convert the hyperbolic orbit into a closed orbit, the planet must slow the asteroid as it passes. Tidal forces might do this, but they would be rather weak. Interactions with other moons or grazing collisions with a thick atmosphere might also slow the asteroid.

Both satellites have been photographed by spacecraft, and those photos show that the satellites are heavily cratered. Such cratering could have occurred while the moons were either still in the asteroid belt or in orbit around Mars. In any case, the heavy battering has broken the satellites into irregular chunks of rock, and they cannot pull themselves into smooth spheres because their gravity is too weak to overcome the structural strength of the rock. We will discover that low-mass moons are typically irregular in shape, whereas more massive moons are more spherical.

Photos of Phobos reveal a unique set of narrow, parallel grooves (Figure 23-27a). Averaging 150 m wide and 25 m deep, the grooves run from Stickney, the largest crater, to an oddly featureless region on the opposite side of the satellite. One theory suggests that the grooves are deep fractures produced by the impact that formed the crater. The featureless region opposite Stickney may be similar to the jumbled terrain on Earth's moon and on Mercury. All were produced by the focusing of seismic waves from a major impact on the far side of the body. High-resolution photographs show that the grooves are lines of pits, suggesting that the pulverized rock material on the surface has drained into the fractures or that gas, liberated by the heat of impact, escaped through the fractures and blew away the dusty soil.

The Mars Global Surveyor spacecraft reveals more about the dust. Observations made with the spacecraft's infrared spectrometer show that the moon's surface cools quickly from −4°C to −112°C (from 25°F to −170°F) as it passes from sunlight into darkness. Solid rock would retain heat and cool more slowly, so the dust must be

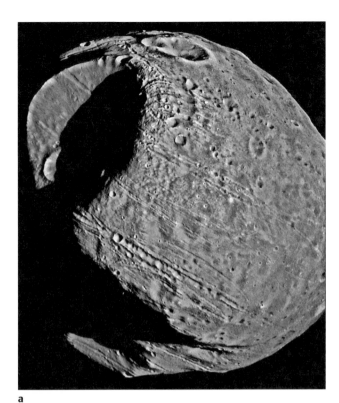

a

b

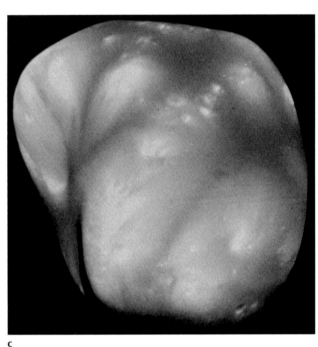

c

Figure 23-27
The moons of Mars are too small to pull themselves into spherical shape. (a) Phobos has an impact crater 10 km in diameter at one end with grooves radiating away. *(Damon Simonelli and Joseph Ververka, Cornell University/NASA)* (b) A photo from the Mars Global Surveyor spacecraft shows features as small as 8 meters in diameter. *Malin Space Science Systems/NASA)* (c) Deimos, smaller than Phobos, looks more uniform because of dusty soil covering the smaller features. *(NASA)*

at least a meter deep and very fine. In photos made by the spacecraft camera, the dust blankets the terrain, but the photos also show boulders a few meters in diameter thought to be ejected by impacts (Figure 23-27b).

Deimos not only has no grooves, it also looks smoother because of a thicker layer of dust on its surface (Figure 23-27c). This material partially fills craters and covers minor surface irregularities. It seems cer-

tain that Deimos experienced collisions in its past, so fractures may be hidden below the debris.

The debris on the surfaces of the moons raises an interesting question. How can the weak gravity of small bodies hold any fragments from meteorite impacts? Escape velocity on Phobos is only 12 m/s. An athletic astronaut could almost jump into space. Certainly, most fragments from impacts should escape, but some could fall back and accumulate on the surface.

Deimos, smaller than Phobos, has a smaller escape velocity. But it has more debris on its surface because it is farther from Mars. Phobos is close enough to Mars that most ejecta from impacts on Phobos will be drawn into Mars. Deimos, being farther from Mars, is able to keep a larger fraction of its ejecta. Phobos is so close to Mars that tides are making its orbit shrink, and it will fall into Mars or be ripped apart by tidal forces within about 100 million years.

Deimos and Phobos illustrate three principles. First, some satellites are probably captured asteroids. Second, small satellites tend to be irregular in shape and heavily cratered. And third, tidal forces can affect small moons and gradually change their orbits. We find even stronger tidal effects in Jupiter's satellite system (Chapter 24).

REVIEW Critical Inquiry

Why would we be surprised to find volcanism on Phobos or Deimos?

In discussing the moon, Mercury, Venus, Earth, and Mars, we have seen illustrations of the principle that the larger a world is the more slowly it loses its internal heat. It is the flow of that heat from the interior through the surface into space that drives geological activity such as volcanism and plate motion. A small world, like Earth's moon, cools quickly and remains geologically active for a shorter time than a larger world like Earth. Phobos and Deimos are not just small, they are tiny. However they formed, any interior heat would have leaked away very quickly; with no energy flowing outward, there can be no volcanism.

Some futurists suggest that our first human missions to Mars will be not to land on the surface but to build a colony on Phobos or Deimos. These plans speculate that there may be water deep inside the moons that colonists could use. What would happen to water released in the sunlight on the surface of such small worlds?

Venus is nearly the same size as Earth, but it is unlikely that humans will ever colonize its surface. The smaller and colder Mars, however, may someday be home to large colonies. In the next chapter, we turn our attention to Jupiter and Saturn, giant worlds so totally unlike Earth that we will need entirely new principles of comparative planetology to describe them.

noes and floods the surface with lava flows. The surface contains more impact craters than Earth but fewer than the moon, and astronomers suspect the entire planet has been resurfaced by volcanism within the last half billion years.

A number of spacecraft have landed on the surface of Venus, but none has survived for more than an hour or two. The surface temperature is about 745 K (880°F), and the atmosphere contains traces of sulfuric, hydrochloric, and hydrofluoric acids. Soviet spacecraft have taken photographs on the surface and transmitted them back to Earth. These photos show rocky terrain of broken and weathered rock. Analysis of samples at the landing sites suggests that the rocks are basalts.

No magnetic field has been detected around Venus, and that suggests that the core is not molten iron generating a magnetic field through the dynamo effect as in Earth's core. The occasional resurfacing of the planet by volcanism may have cooled the interior so it is now solid. Another mystery is the retrograde rotation. This may have been produced by a major impact during formation.

Mars is smaller than Earth but larger than Mercury. Although early astronomers saw canals on Mars and thought it to be inhabited, we now know that it is a dry desert world. Its southern highlands are heavily cratered and old; the northern lowlands have been resurfaced and lack craters. Much of Mars has been altered by volcanism, and some of its volcanoes are very large. Some may still be active.

Mars may once have had liquid water flowing over its surface. Photographs returned by spacecraft show extensive dry streambeds. Most of these seem to have been cut by sudden flooding rather than by gradual runoff. Whatever water Mars still has is now frozen in the soil or in the polar caps.

The Martian atmosphere is now very thin because the low gravity has been unable to retain most gases. But features such as the streambeds and the layered terrain near the polar caps suggest that the Martian climate may have been warmer and wetter in the past and that it may change cyclically, perhaps due to orbital changes.

Deimos and Phobos, the two small moons of Mars, may be captured asteroids. Like most small moons, they are irregular in shape because their gravities are too weak to deform their rock into spheres.

Summary

All terrestrial planets pass through a four-stage history: (1) differentiation; (2) cratering; (3) flooding of the crater basins by lava, water, or both; and (4) slow surface evolution. The importance of each of these processes in the evolution of a planet depends on the planet's mass and temperature.

Venus is Earth's twin in size and density, but the planet has evolved along divergent lines because it is slightly closer to the sun. The higher temperature evaporated any early oceans and prevented the absorption of carbon dioxide from the atmosphere. The accumulating carbon dioxide created a greenhouse effect that produced very high surface temperatures.

The crust of Venus is marked by low, rolling plains and higher uplands with some very high volcanoes. Coronae appear to be produced by rising currents of molten rock below the crust. Where the magma breaks through, it builds volca-

New Terms

subsolar point	corona
rolling plain	outflow channel
shield volcano	runoff channel

Review Questions

1. Why might we expect Venus and Earth to be similar?

2. What evidence do we have that Venus and Mars once had more water than at present? Where did that water come from? Where did it go?

3. What features will you look for in future radar maps of Venus to search for plate tectonics?

4. Why doesn't Mars have mountain ranges like those on Earth? Why doesn't Earth have large volcanoes like those on Mars?

5. What were the canals on Mars? How do they differ from the dry streambeds on Mars?

6. Propose an explanation for the nearly pure carbon-dioxide atmospheres of Venus and Mars. How did Earth avoid such a fate?

7. What evidence do we have that the climate on Mars has changed?

8. How can we estimate the ages of the streambeds on Mars?

9. Why are Phobos and Deimos nonspherical? Why is Earth's moon much more spherical?

Discussion Questions

1. From what you know of Earth, Venus, and Mars, do you expect the volcanoes on Venus and Mars to be active or extinct? Why?

2. If humans someday colonize Mars, the biggest problem may be finding water and oxygen. With plenty of solar energy beating down through the thin atmosphere, how might colonizers extract water and oxygen from the Martian environment?

Problems

1. How long would it take radio signals to travel from Earth to Venus and back if Venus were at its nearest point to Earth? at its farthest point from Earth?

2. The Pioneer Venus Orbiter circled Venus with a period of 24 hours. What was its average distance above the surface of Venus? (*Hint:* See Chapter 5.)

3. Calculate the velocity of Venus in its orbit around the sun.

4. What is the maximum angular diameter of Venus as seen from Earth? (*Hint:* Use the small-angle formula.)

5. If the Magellan spacecraft transmitted radio signals down through the clouds on Venus and heard an echo from a certain spot 0.000133 second before the main echo, how high is the spot above the average surface of Venus?

6. The smallest feature visible through an Earth-based telescope has an angular diameter of about 1 second of arc. If a canal on Mars was just visible when Mars was at its closest to Earth, how wide was the canal? (*Hint:* Use the small-angle formula.)

7. What is the maximum angular diameter of Phobos as seen from Earth? What surface features should we expect to see? (*Hint:* Use the small-angle formula.)

8. What is the maximum angular diameter of Phobos as seen from the surface of Mars?

9. Deimos is about 12 km in diameter and has a density of 2 g/cm^3. What is its mass? (*Hint:* The volume of a sphere is $\frac{4}{3}\pi r^3$.)

Critical Inquiries for the Web

1. Have you ever wondered how the various features on the planets get their names? Surface features on Venus are (mostly) named for female figures from history and mythology. Who decides how planetary features are named? Look for information on planetary nomenclature, and summarize the way different types of features on Venus are assigned names.

2. "Martians" have fascinated humans for the last century or more. Many online sources chronicle the representation of life on Mars by Earthlings throughout history and literature. Read about the Martians as represented by a particular literary work or nonfiction account and discuss to what extent they are (or are not) based on realistic views of the nature of Mars—in terms of our current understanding as well as the views of the period in which the work was written.

3. Spacecraft are studying Mars right now. Search for the latest discoveries and photographs.

Exploring *The Sky*

1. Venus and Mars exhibit phases similar to the moon's phases.

   a. Use *The Sky* to sketch the appearance of Venus and Mars at the present time. Also give the phase as a percentage of full.
   How to proceed: Use **Find** to give you a highly magnified view of a particular planet. Click on the planet to obtain **Object Information.**

   b. Explain the observed phase on the basis of geometrical relationship between the sun, Earth, and the planet as shown by the **3D Solar System Mode.**

 Go to the Brooks/Cole Astronomy Resource Center (www. brookscole.com/astronomy) for critical thinking exercises, articles, and additional readings from InfoTrac College Edition, Brooks/Cole's online student library.

CHAPTER TWENTY-FOUR

Jupiter and Saturn

There is something fascinating

about science. One gets such whole-

sale returns of conjecture out of

such a trifling investment of fact.

Mark Twain
Life on the Mississippi

Guidepost

As we begin this chapter, we leave behind the psychological security of planetary surfaces. We can imagine standing on the moon, on Venus, or on Mars, but Jupiter and Saturn have no surfaces. Thus, we face a new challenge—to use comparative planetology to study worlds so unearthly we cannot imagine being there.

One reason we find the moon and Mars of interest is that we might go there someday. Humans may become the first Martians. But the outer solar system seems much less useful, and that gives us a chance to think about the cultural value of science.

This chapter begins our journey into the outer solar system. In the next chapter, we will visit worlds out in the twilight at the edge of the sun's family.

The sulfuric acid fogs of Venus seem totally alien, but compared with the planets of the outer solar system, Venus is a tropical oasis. Of the five planets beyond the asteroid belt, four have no solid surface, and one, Pluto, is so far from the sun that most of its atmosphere lies frozen on its surface. As we explore these strange worlds, we will discover new principles of comparative planetology. We begin with Jupiter and Saturn because they are the largest of the Jovian planets.

These distant worlds can be studied from Earth, but much of what we know has been radioed back by space probes. Pioneer 10 and Pioneer 11 explored Jupiter and Saturn in the mid-1970s, but their instruments were not very sensitive. The Voyager 1 and Voyager 2 spacecraft were launched in 1977 on a mission to visit all four Jovian worlds, concluding when Voyager 2 flew past Neptune in 1989. In December, 1995, the Galileo spacecraft reached Jupiter. While an instrumented probe dropped into Jupiter's atmosphere, the main Galileo spacecraft went into orbit to begin a long-term study of the planet and its moons. Throughout this discussion of the Jovian worlds, you will find images and data returned to Earth by these robotic explorers.

Jupiter and Saturn are unearthly, but because they obey the same laws of nature, we can understand their peculiar personalities. We begin with the closest and largest Jovian planet, named for the ruler of the Roman gods.

24-1 Jupiter

Jupiter is the most massive of the Jovian planets, containing nearly three-fourths of all the planetary matter in our solar system. This high mass accentuates some processes that are less obvious or nearly absent on the other Jovian worlds. Just as we used Earth as the basis of comparison for our study of the terrestrial planets, we will examine Jupiter in detail so we can use it as a standard in our comparative study of the other Jovian planets.

The most striking feature of Jupiter is its varicolored cloud belts, which make it look like a child's rubber ball (Figure 24-1). We will examine these belts later and see that they may be related to liquid circulations deep inside the planet. However, first we must survey Jupiter in general.

Surveying Jupiter

Jupiter is interesting because it is big, massive, mostly liquid hydrogen, and very hot inside. The preceding facts are common knowledge among astronomers, but we should demand an explanation of how they know these facts. Often the most interesting thing about a fact isn't the fact itself, but how we know it.

We know that Jupiter is a big planet because it looks big. At their closest points to Earth, Jupiter is about

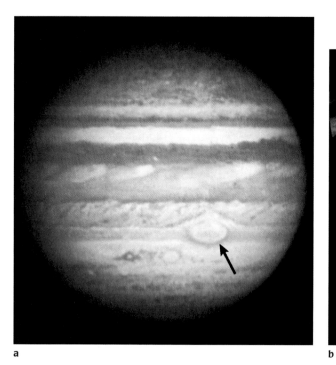

a

b

Figure 24-1

(a) The atmosphere of Jupiter is banded by turbulent belts of clouds that circulate parallel to the planet's equator. Oval spots are rotating storm systems such as the Great Red Spot (arrow). (b) The Great Red Spot, shown here in a false-color image, is roughly twice the diameter of the Earth. This infrared image from the Galileo spacecraft shows lower clouds as blue, higher thin hazes as red, and the highest clouds as white. Note the turbulent circulation of the gases. *(NASA)*

eight times farther away than Mars, but even a small telescope will reveal that the disk of Jupiter is more than two times bigger than the disk of Mars. If we use the small-angle formula, we can compute the diameter of Jupiter—a bit over 11 times Earth's diameter (Data File Seven). That's a big planet!

We know that Jupiter is massive because its moons race around it at high speed. Io is the innermost of the four Galilean moons, and its orbit is just a bit larger than the orbit of our moon around Earth. Io streaks around its orbit in less than two days, whereas Earth's moon takes a month. Jupiter has to be a very massive world to hold on to such a rapidly moving moon. In fact, we can put the radius of Io's orbit and its orbital period into Newton's version of Kepler's third law (Chapter 5) and calculate the mass of Jupiter—almost 318 times Earth's mass.

The size and mass of Jupiter are easy to find, but we might wonder how astronomers know that it is made of hydrogen. That fact is just a little bit harder to find, but the first step is to divide the mass by the volume and find Jupiter's average density, about 1.34 g/cm^3. Of course, it is more dense at the center and less dense near the surface, but this average density tells us that Jupiter can't be made totally of rock. Rock has a density of 2.5 to 4 g/cm^3, so Jupiter must contain less-dense material such as hydrogen.

Spectra recorded from Earth and from spacecraft visiting Jupiter tell us that the composition of Jupiter is much like that of the sun—it is mostly hydrogen and helium. This fact was confirmed when a probe from the Galileo spacecraft fell into the atmosphere in 1995 and radioed its results back to Earth. Jupiter is mostly hydrogen and helium with traces of heavier atoms forming molecules such as methane (CH_4), ammonia (NH_3), and water (Table 24-1).

Just as astronomers can build mathematical models of the interiors of stars, they can use the equations that describe gravity, energy, and how matter responds to pressure to build mathematical models of the interior of Jupiter. The models tell us that a planet as big and massive as Jupiter, having a low density and composed mostly of hydrogen and helium, must contain

Jupiter Data File Seven

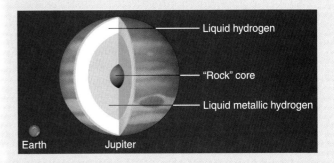

A theoretical model of the interior of Jupiter. Note the oblateness of the planet.

Average distance from the sun	5.2028 AU (7.783 × 10^8 km)
Eccentricity of orbit	0.0484
Maximum distance from the sun	5.455 AU (8.160 × 10^8 km)
Minimum distance from the sun	4.951 AU (7.406 × 10^8 km)
Inclination of orbit to ecliptic	1°18′29″
Average orbital velocity	13.06 km/s
Orbital period	11.867 y (4334.3 days)
Period of rotation	9^{h}55^{m}30^s
Inclination of equator to orbit	3°5′
Equatorial diameter	142,900 km (11.20 $D_\oplus$)
Mass	1.899 × 10^{27} kg (317.83 $M_\oplus$)
Average density	1.34 g/cm^3
Gravity at base of clouds	2.54 Earth gravities
Escape velocity	61 km/s (5.4 $V_\oplus$)
Temperature at cloud tops	−120°C (−180°F)
Albedo	0.51
Oblateness	0.0637

Table 24-1 Composition of Jupiter and Saturn (by mass)

Molecule	Jupiter (%)	Saturn (%)
H_2	78	88
He	19	11
H_2O	0.0001	—
CH_4	0.2?	0.6
NH_3	0.5?	0.2

only a small core of heavy elements at its center, a so-called "rocky core." Although the core is about the size of Earth, it must contain roughly 15 Earth masses. The models also tell us that the pressure below the atmosphere is so great that hydrogen cannot exist as a gas; most of Jupiter is liquid hydrogen.

Even the shape of the planet is revealing. By watching the cloud belts, we can tell that Jupiter rotates rapidly, in about 10 hours. Simple measurements on a photograph show that it is slightly flattened; it is 6.37 percent larger in diameter, through its equator, than it is through its poles. Astronomers refer to this by saying the **oblateness** of Jupiter is 0.0637. The amount of flattening depends on the speed of rotation and on the rigidity of the material. A planet that is mostly liquid will be more oblate than a planet that is mostly rocky. Thus, astronomers can use the observed oblateness of Jupiter to check their models of the size of its heavy-element core.

It is easy to think of the heavy-element core as a ball of rock, but astronomers are quick to point out that it is very hot, about 30,000 K. That is five times hotter than the surface of the sun, but the pressure is so great that it keeps the material confined to the core of the planet. How do we know it is so hot? Because infrared observations show that Jupiter emits almost twice as much energy as it receives from the sun. It must be very hot inside to have so much heat flowing outward into space.

The models of Jupiter's interior tell us that ordinary liquid hydrogen cannot support the pressure deep inside the planet. It is compressed into a state in which the electrons are free to move easily from atom to atom, and although the hydrogen can flow like a liquid, it can conduct electricity like a metal. This **liquid metallic hydrogen** does not occur naturally on Earth, but it has been created under tremendous pressure in laboratories.

Basic observations and the known laws of physics can tell us a great deal about Jupiter. Its vast magnetic field can tell us even more.

Jupiter's Magnetic Field

Astronomers detected a magnetic field around Jupiter as early as the 1950s, which is surprising considering that magnetic fields are invisible to the eye. In fact, Jupiter's giant magnetic field was first detected by radio.

In 1955, radio astronomers detected bursts of synchrotron radio waves coming from Jupiter. They called it **decameter radiation** because its wavelength was tens of meters. Synchrotron radiation is electromagnetic radiation from fast electrons spiraling in a magnetic field, so it was obvious that Jupiter had a magnetic field.

Because the radio signals fluctuated with a period of 9 hours 55.5 minutes, astronomers concluded that this period was the true period of rotation of Jupiter. This result is what we would expect from the dynamo theory for the generation of magnetic fields, which requires an electrically conducting fluid, convection driven by heat flowing outward, and rapid rotation. The liquid metallic hydrogen in Jupiter is a very good electrical conductor, and we have seen clear evidence that heat flows outward from inside Jupiter. That, coupled

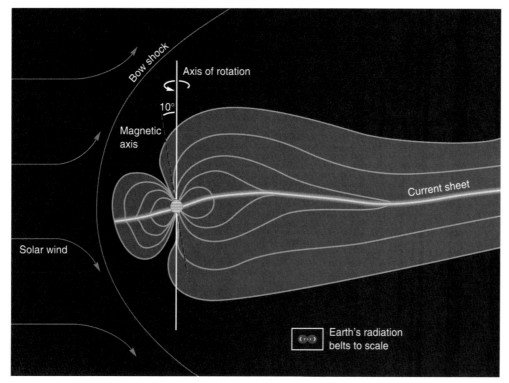

Figure 24-2
Jupiter's magnetic field traps particles from the solar wind to form powerful radiation belts. The rapid rotation of the planet forces the slightly inclined magnetic field to wobble as it rotates and confines most of the high-energy particles to a thin plane called the current sheet. Earth's magnetosphere and radiation belts are shown to scale (inset).

with Jupiter's rapid rotation, should generate a powerful dynamo effect. Thus, the fluctuation of the radio signals tells us the true rotation period of the planet's liquid interior.

By the early 1960s, radio astronomers had discovered short-wavelength **decimeter radiation,** with wavelengths a few tenths of a meter long. Although this type of radiation is also synchrotron radiation, radio telescopes revealed that it comes from a region at least three times larger than the disk of Jupiter. Detailed studies of the decimeter radiation showed that Jupiter's magnetic field was at least ten times stronger than Earth's, and, like Earth's field, was inclined about 10° to the axis of rotation.

Radio studies were helpful, but nothing is as good as a visit. In 1973 and 1974, two Pioneer spacecraft flew past Jupiter, followed in 1979 by two Voyager spacecraft. The Galileo spacecraft entered orbit around Jupiter in late 1995. These spacecraft found Jupiter surrounded by a magnetosphere over 100 times larger than Earth's. If we could see Jupiter's magnetosphere, it would appear to be more than three times larger in diameter than the full moon.

The magnetic field deflects the solar wind and traps high-energy particles in radiation belts much more intense than Earth's. The radiation is more intense because Jupiter's magnetic field is stronger and can trap and hold higher-energy particles, mostly electrons and protons. The spacecraft passing through the radiation belts received radiation doses equivalent to a billion chest X rays—at least 100 times a lethal dose for a human. Some of the electronics on the spacecraft were damaged by the radiation.

Jupiter's rapid rotation has flattened its magnetosphere, and the charged particles are concentrated in a thin layer called the **current sheet** (Figure 24-2). Because the magnetic field is inclined to the axis of rotation, the magnetosphere wobbles up and down as Jupiter rotates.

We have seen that Earth's magnetosphere produces aurorae, and the same is true on Jupiter. Charged particles in the magnetosphere leak downward along the magnetic field, and, where they enter the atmosphere, they produce aurorae 1000 times more powerful than those on Earth. The aurorae on Jupiter, like aurorae on Earth, occur in rings around the magnetic poles (Figure 24-3a).

Some of the heavier ions in the radiation belts appear to come from the inner Galilean moon, Io. As we will see later in this chapter, Io has active volcanoes that spew sulfur. Because Io orbits with a period of 1.8 days, and Jupiter's magnetic field rotates in only 10 hours, the wobbling magnetic field rushes past Io at high speed, sweeping up stray particles, accelerating them to high energy, and spreading them around Io's orbit in a doughnut of ionized gas called the **Io plasma torus.**

Jupiter's magnetic field interacts with Io to produce a powerful electric current (about a million amperes) that flows through a curving path called the **Io flux tube** from Jupiter out to Io and back to Jupiter. Spots of bright

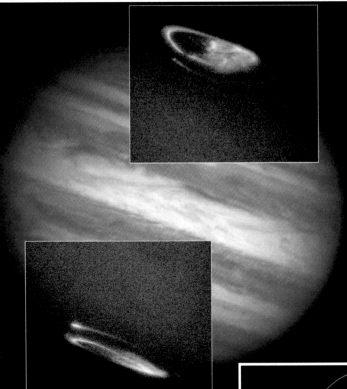

a

Figure 24-3
(a) Aurorae on Jupiter are confined to rings around the magnetic poles, as shown in these ultraviolet images superimposed on a visible-light photo of Jupiter. *(John Clarke, University of Michigan, NASA)* (b) At left a visible-light image of Jupiter and its moon Io shows the location of the Io flux tube. The UV image at right shows the location of Jupiter's auroral rings and the spots of aurora that occur at the base of the Io flux tube. Locate the same spots in part a of this figure. *(NASA)*

b

aurora lie at the two points where the Io flux tube enters Jupiter's atmosphere (Figure 24-3b).

This Io flux tube explains a mystery. Back in 1955, when radio astronomers discovered the decametric radio signals, they noticed that the signals were stronger when Io was in certain parts of its orbit. Now we can understand that the electrical current flowing through the Io flux tube is at least partially responsible for generating the decametric radio signals.

Fluctuations in the aurora tell us that the solar wind buffets Jupiter's magnetosphere, but some of the fluctuations seem to be caused by changes in the magnetic dynamo deep inside the planet. Thus, studies of the aurora on Jupiter may help astronomers learn more about its liquid depths.

Much of the power and fascination of Jupiter is invisible to our eyes, but the swirling cloud belts are beautiful in their complexity.

Jupiter's Atmosphere

If you parachuted into Jupiter with a small rubber boat, expecting to go sailing on the liquid hydrogen ocean, you would be disappointed. Mathematical models of Jupiter tell us that there is no surface. Deep below the clouds, the temperature and pressure exceed the **critical point** for hydrogen, the temperature and pressure above which liquid hydrogen and gaseous hydrogen have the same density and are indistinguishable. As you parachuted down through Jupiter's atmosphere, the temperature, pressure, and density would rise, and gradually the gas would become a fluid. You would never splash down and your rubber boat would be useless. Below the clouds of Jupiter lies the largest ocean in the solar system—and it has no surface and no waves.

When we look at Jupiter, we don't see a surface. We see those parts of the atmosphere that contain clouds. These cloud layers lie deep inside a nearly transparent atmosphere of hydrogen and helium. We can detect this atmosphere by noticing that Jupiter has limb darkening just as the sun does (Chapter 8). When we look near the limb of Jupiter (the edge of its disk), the clouds are much dimmer (Figure 24-1) because it is nearly sunset or sunrise along the limb. If we were on Jupiter at this location, we would see the sun just above the horizon, and it would be dimmed by the atmosphere. In addition, light reflected from clouds must travel out at a steep angle

through the atmosphere to reach Earth, dimming the light further. Jupiter is brighter near the center of the disk because the sunlight shines nearly straight down and the cloud layers look brighter.

Jupiter's atmosphere is not deep. The Galileo probe fell through a transparent hydrogen atmosphere and then into the clouds; it was in total darkness by the time it reached 80 km below the highest clouds. It was de-

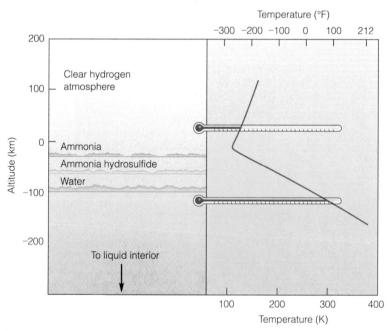

a

b

Figure 24-4

(a) The beautiful clouds we see in Jupiter's atmosphere lie deep in a clear hydrogen atmosphere. *(NASA)* (b) Thermometers inserted into the Jovian atmosphere would register temperatures as shown by the red line in the graph at right. Note the temperatures at which the three cloud layers form by extending them to the right until they touch the line on the graph. Read the temperature from the scale at the bottom and compare with Figure 21-15 and Figure 24-19.

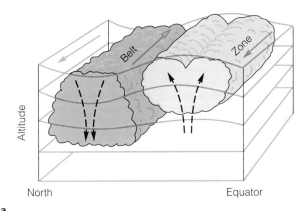

a

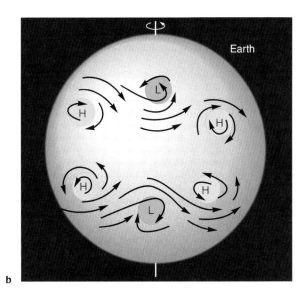

b

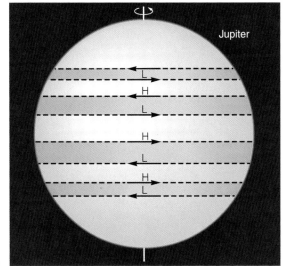

c

Figure 24-5

(a) In Jupiter's atmosphere belts are formed by low-pressure, sinking gas that is deeper in the atmosphere and not as bright as the higher-pressure rising gas in the zones. (b) In Earth's atmosphere, a wave circulation separates low- and high-pressure regions. (c) On Jupiter, low- and high-pressure regions take the form of belts of sinking gas and zones of rising gas.

stroyed by heat and pressure at a depth of 130 km, and models tell us that the gas of the atmosphere merges into the liquid interior about 1000 km below the clouds, which is not as deep as it seems when we recall that Jupiter's radius is 72,000 km. The entire atmosphere is only about 1 percent of the planet's radius. Earth's atmosphere is about 2.5 percent of Earth's radius. Furthermore, the actual clouds we see on Jupiter are confined to a layer less than 100 km deep (Figure 24-4a).

Spectroscopic observations combined with models of the atmosphere suggest that Jupiter's clouds lie in three layers (Figure 24-4b). The highest layer is composed of ammonia crystals at a temperature of about 150 K ($-180°$F). Through holes in these clouds, we can glimpse a deeper layer of ammonia hydrosulfide (NH_4SH) crystals with a temperature of about 200 K, and deeper still there may be a layer of water crystals at a temperature of about 280 K. (Make special note of the temperatures of these cloud layers. We will find similar clouds at similar temperatures when we visit Saturn.)

The clouds of Jupiter are organized into dark **belts** and bright **zones.** Observations made by spacecraft show that the zones are cooler and higher than the belts. The zones appear to be high-pressure regions of rising gas that cools as it rises and forms clouds higher in the atmosphere where they receive more sunlight and look brighter (Figure 24-5a). The belts are low-pressure regions with sinking gas and lower clouds that are not as brightly lit.

On Earth, high- and low-pressure regions are bounded by high-speed winds (Figure 24-5b), and the same is true on Jupiter. Winds blowing hundreds of kilometers per hour separate the belts and zones (Figure 24-5c). The clouds on Jupiter are twisted and swirled by the winds, but the belts and zones are highly stable.

Although their colors and brightness change sometimes, they have not changed their position since humans have been mapping them.

The colors of the clouds cannot be caused by ammonia, ammonia hydrosulfide, or water, which form colorless, white crystals. Rather, astronomers think that the colors are caused by traces of compounds containing sulfur or phosphorus modified by sunlight.

Mixed within the belt–zone circulation are light and dark spots, a few larger than Earth. Some appear and disappear in days; others last for years. Dark spots appear to be openings through which we see deeper, darker clouds, and white spots appear to be higher cloud circulations. The largest is the Great Red Spot shown in Figure 24-1, which has been in one of the southern zones for more than 330 years (Figure 24-6). In its longest dimension, it is twice the diameter of Earth. Pioneer, Voyager, and Galileo observations show that the Great Red Spot is higher than its surroundings. It is evidently

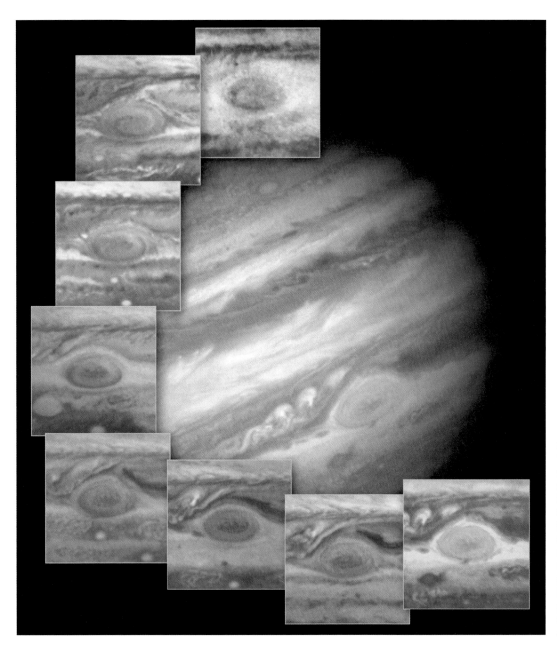

Figure 24-6
The great red spot on Jupiter changes its appearance as shown in this eight-year sequence of images, but it has been visible for over 300 years. Rotating counterclockwise with winds of 270 mi/hr, the spot is the largest storm in the solar system. *(Hubble Heritage Team STScI/AURA/NASA and Amy Simon, Cornell U)*

formed by rising gas carrying heat upward from deep below the clouds.

On Earth, rising currents of hot gas can produce thunderheads that tower 15 kilometers and spawn powerful winds and tremendous lightning. The same is true on Jupiter. Voyager photos of the night side of Jupiter revealed lightning bolts among the clouds, and Galileo images show thunderheads up to 50 km high (white clouds in Figure 24-1b upper left).

The stability of the belt–zone circulation on Jupiter has intrigued astronomers. Some models suggest that the circulation is driven by solar heat entering the atmosphere from outside (Figure 24-7a), but other models suggest that the atmosphere is dominated by circula-

tions in the deep interior of the planet. In the laboratory, scientists have imitated Jupiter by placing a fluid in a rotating plastic sphere 30 cm in diameter. Under conditions like those in Jupiter, the fluid divides into concentric cylinders, and where those cylinders touch the surface of the sphere, belt–zone circulation appears (Figure 24-7b). Models of both solar heating and internal circulation have been successful in creating belts and zones.

The controversy may have been settled by the Galileo probe. As it fell deeply into Jupiter's atmosphere, it found that the high-speed winds that bound the belts and zones are not limited to the upper cloud layers exposed to bright sunlight. Winds of 650 km/s extend at least as deep as 130 km below the cloud tops. These deep roots to the winds suggest that the atmospheric

Sending spacecraft to another world is expensive, and it may seem pointless when that world seems totally hostile to human life. What practical value is there in sending a space probe to Jupiter? To resolve that question, we need to consider the distinction between science, technology, and engineering.

Science is nothing more than the logical study of nature, and the goal of science is understanding. Although much scientific knowledge proves to have tremendous practical value, the only goal of science is a better understanding of how nature works. Technology, in contrast, is the practical application of scientific knowledge to solve a specific problem. People trying to find a faster way to paint automobiles might use all the tools and techniques of science, but if their goal has some practical outcome, we should more properly call it technology rather than science. Engineering is the most practical form of technology. An engineer is likely to use well-understood technology to find a practical solution to a problem.

Of course, there are situations in which science and technology blur together. For example, humanity has a practical and tragic need to solve the AIDS problem; we need a cure and a preventative. Unfortunately, we don't understand the AIDS virus itself or viruses in general well enough to design a simple solution, and thus much of the work involves going back to basic science and trying to better understand how viruses interact with the human body. Is this technology or science? It is hard to decide.

We might describe science that has no known practical value as basic science or basic research. Our exploration of worlds such as Jupiter would be called basic science, and it is easy to argue that basic science is not worth the effort and expense because it has no known practical use. Of course, the problem is that we have no way of knowing what knowledge will be of use until we acquire that knowledge. In the middle of the 19th century, Queen Victoria is supposed to have asked physicist Michael Faraday what good his experi-

ments with electricity and magnetism were. He answered, "Madam, what good is a baby?" Of course, Faraday's experiments were the beginning of the electronic age. Many of the practical uses of scientific knowledge that fill our world—transistors, vaccines, plastics—began as basic research. Basic scientific research provides the raw materials that technology and engineering use to solve problems.

Basic scientific research has yet one more important use that is so valuable it seems an insult to refer to it as merely practical. Science is the study of nature, and as we learn more about how nature works, we learn more about what our existence in this universe means for us. The seemingly impractical knowledge we gain from space probes to other worlds tells us about our own planet and our own role in the scheme of nature. Science tells us where we are and what we are, and that knowledge is beyond value.

circulation is dominated by the circulation of the liquid interior.

Understanding Jupiter not only helps us better understand Earth, but it also suggests that some planets are fundamentally different from our world (Window on Science 24-1). Jupiter differs from Earth in another peculiar way. It has a ring.

Jupiter's Ring

Astronomers have known for centuries that Saturn has rings, but it was not until 1979, when the Voyager 1 spacecraft sent back photos, that Jupiter's ring was discovered. Less than 100 times as bright as Saturn's rings, the ghostly ring around Jupiter is a puzzle. What is it

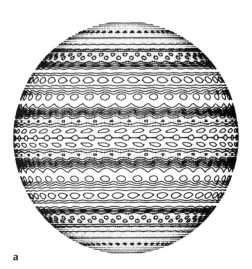

a

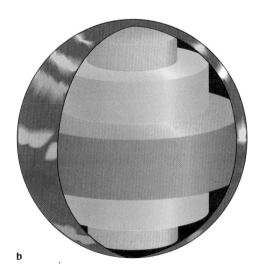

b

Figure 24-7
(a) A computer model of the Jovian atmosphere that depends only on the heat of the sun can generate belt–zone circulation *(Gareth Williams)* (b) An alternative model links the belts and zones in Jupiter's atmosphere with cylindrical circulations deep in the liquid interior. The Galileo probe traced atmospheric winds deep into the atmosphere and that supports the cylindrical circulation theory.

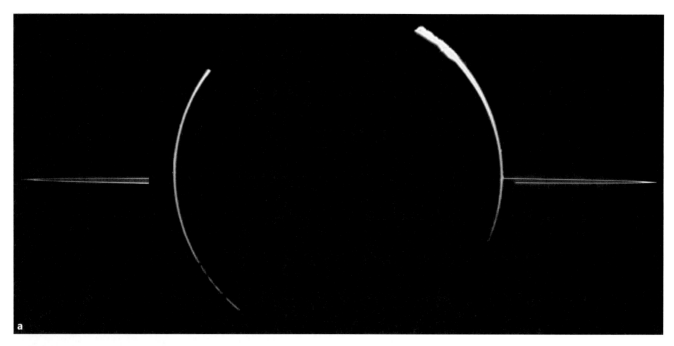

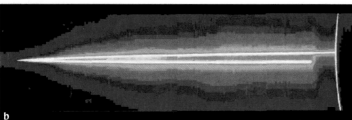

Figure 24-8
(a) The main ring of Jupiter, illuminated from behind, glows brightly in this photo made by the Galileo spacecraft from within Jupiter's shadow. (b) Digital enhancement and false color reveal the halo of ring particles that extends above and below the main ring. *(NASA)*

made of? Why is it there? A few simple observations help us understand.

Saturn's rings are made of bright ice chunks, but the particles in Jupiter's ring are very dark and reddish. We can guess that the ring is rocky rather than icy.

We can also conclude that the ring particles are mostly microscopic. Photos of the ring show that it is very bright when illuminated from behind (Figure 24-8)—it is scattering light forward. Efficient **forward scattering** occurs when particles have diameters roughly the same as the wavelength of light, a few millionths of a meter. Large particles do not scatter light forward, so a ring filled with basketball-size particles would look dark when illuminated from behind. Forward scattering tells us that the ring is made of particles about the size of those in cigarette smoke.

The location of these particles near the planet is understandable. They orbit inside the **Roche limit,** the distance from a planet within which a moon cannot hold itself together by its own gravity. If a moon stays far from its planet, then the moon's gravity will be much greater than the tidal forces caused by the planet, and the moon will be able to hold itself together. If, however, the planet's moon comes inside the Roche limit, the tidal forces overcome its gravity and pull the moon apart. The Space Shuttle can orbit inside the Roche

limit because it is held together by bolts and welds, but a moon held together by its own gravity cannot survive inside the Roche limit. If a planet and its moon have similar densities, the Roche limit is 2.44 planetary radii, and Jupiter's rings (and those of Saturn, Uranus, and Neptune) lie inside this limit.

Now we can understand the dust near Jupiter. If a dust speck gets knocked loose from a larger rock inside the Roche limit, the rock's gravity cannot hold the dust speck. And the billions of dust specks in the ring can't pull themselves together to make a new moon because of tidal forces inside the Roche limit.

We can be sure that the ring particles are not old. The pressure of sunlight and the powerful magnetic field alter the orbits of the particles, and they gradually spiral into the planet. Images show faint ring material extending down toward the cloud tops, and this is evidently dust specs spiraling into the planet. Dust is also lost from the ring as electromagnetic effects force it out of the plane of the ring to form a low-density halo above and below the ring (Figure 24-8b). Another reason the ring particles can't be old is that the intense radiation around Jupiter will grind the dust specks down to nothing in a century or so. Thus the rings we see today can't be material left over from the formation of Jupiter.

The rings of Jupiter must be continuously resupplied with new dust. Small moons that orbit near the outer edge of the rings may lose dust particles as they are hit by micrometeorite impacts. Observations made by the Galileo spacecraft show that the main ring is densest at its outer edge where the small moon Adrastea orbits and that another small moon, Metis, orbits inside the ring. Galileo images reveal much fainter rings called the **gossamer rings,** extending twice as far from the planet as the main ring. These gossamer rings are most dense at the orbits of two small moons, Amalthea and Thebe, again giving us evidence that the dust is being blasted into space by impacts on the inner moons.

Besides supplying the rings with particles, it is possible that the moons help confine the ring particles and keep them from spreading outward. This is an important process in planetary rings, and we will explore it in detail when we study the rings of Saturn later in this chapter.

Our exploration of Jupiter reveals that it is much more than just a big planet. It is the gravitational and magnetic center of an entire community of objects. Occasionally the community suffers an intruder.

Comet Impact on Jupiter

Comets are very common in the solar system, and Jupiter, because of its strong gravity, probably gets hit by comets more often than most planets. But no one had ever seen it happen until 1994, when fragments from a disrupted comet smashed into Jupiter.

The comet, named Shoemaker-Levy 9 after its discoverers, Eugene and Carolyn Shoemaker and David H. Levy, was captured into orbit around Jupiter sometime in the past. In 1992, it passed inside Jupiter's Roche limit and was pulled into at least 21 pieces that looped out away from Jupiter in long elliptical orbits. At this point, the objects were discovered on a photograph. Drawn out into a long strand of small comets, the pieces fell back and slammed into Jupiter over a period of six days in July 1994 (Figure 24-9).

The fragments were no bigger than a kilometer or so in diameter and contained a fluffy mixture of rock and ices. They hit Jupiter at speeds of about 60 km/s and released energy equivalent to a few million megatons of TNT. (At the height of the cold war, all the nuclear weapons on Earth totaled about 80,000 megatons. The Hiroshima bomb was only 0.15 megaton.)

The impacts occurred just over Jupiter's horizon as seen from Earth, but they produced fireballs almost 3000 km high, and the rapid rotation of Jupiter brought the impact points within sight of Earth only 15 minutes later. Infrared telescopes easily spotted the glowing scars where the comets hit (Figure 24-9b), and, as they cooled, the impact sites became dusty, dark spots at visible wavelengths. Visible through even small telescopes, some of these dark smudges were larger than Earth (Figure 24-9c).

The comet impact was an astonishing spectacle, but what can we learn from it? We can use the event in two different ways. First, astronomers used the impacts as probes of Jupiter's atmosphere. By making assumptions about the nature of Jupiter's atmosphere and by using the most powerful computers, astronomers created models of a high-velocity projectile penetrating into Jupiter's upper atmosphere. By comparing the observed impacts with the models, astronomers were able to fine-tune the models to better represent Jupiter's atmosphere. This method of comparing models with reality is a critical part of science; therefore, the impacts on Jupiter resulted in a better understanding of the nature of Jupiter's atmosphere.

Second, the spectacle helps us in a more general way. It reminds us that planets are hit by large objects such as asteroids and the heads of comets. Jupiter probably is hit every century or so. In 1690, the Italian astronomer Cassini observed a dark spot that appeared on Jupiter and changed in a pattern over a period of days. We now recognize that pattern as the scar of a comet impact. We have seen evidence for large impacts on Earth, the moon, Venus, Mercury, and Mars, and we will see further evidence as we continue our exploration of the solar system.

Comets hit planets all the time. The impact in 1994 was just the first such event to occur since the rise of modern astronomy. We will see in Chapter 26 that a comet or asteroid impact on Earth may have changed the climate and killed the dinosaurs. A solar system is a dangerous place to put an inhabited planet.

The History of Jupiter

Our goal in studying any planet is to be able to tell its story—to describe how it got to be the way it is. While we understand part of the story of Jupiter, we still have much to learn.

If the solar nebula theory for the origin of the solar system is correct, then Jupiter formed from the colder gases of the outer solar nebula, where ices were able to condense. Thus, Jupiter grew rapidly and eventually became massive enough to trap hydrogen and helium gas directly from the solar nebula. The denser material, such as rock and metal, sank to the center to form a core, and we now see Jupiter as a low-density world of liquid hydrogen surrounding a hot core of heavier elements.

In the deeper interior of Jupiter, the hydrogen takes the form of liquid metallic hydrogen, which is a very good electrical conductor. The planet's rapid rotation, coupled with the outward flow of heat from its hot interior, drives a dynamo effect that produces a powerful magnetic field. That vast magnetic field traps high-energy particles from the solar wind to form intense radiation belts.

The rapid rotation and large size of Jupiter cause the weather patterns to take the form of belt–zone circulation. Heat flowing upward from the interior causes rising currents in the bright zones, and cooler gas sinks

a

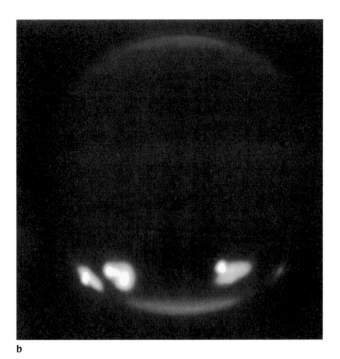

b

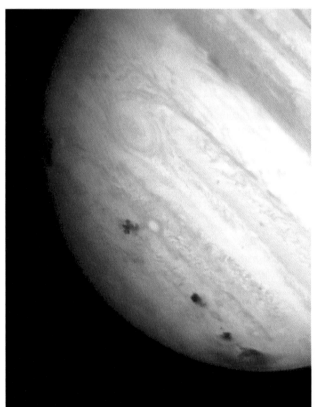

c

Figure 24-9
(a) Comet Shoemaker-Levy 9, broken into fragments by a previous encounter with Jupiter, falls toward the giant planet. Distances have been compressed in this combination of photos. Note the moon Io and its shadow on Jupiter. *(NASA)* (b) Fragments of the comet struck Jupiter's southern hemisphere over a period of six days and created tremendous fireballs, leaving hot spots visible at infrared wavelengths. *(University of Hawaii)* (c) At visible wavelengths, the impact sites show as dark smudges. The largest scar at the extreme right is the size of Earth. *(NASA)*

in the dark belts. As on Earth, winds blow at the margins of these regions, and large spots appear to be cyclonic disturbances.

Although the age of planet building is long past, Jupiter continues to be hit by meteorites and occasional comets, as do all the planets. Any debris left over from the formation of Jupiter would have been blown away long ago by the solar wind, so the dust we find trapped in Jupiter's thin ring must be young. It probably comes from meteorites hitting the innermost moons.

Our study of Jupiter has been challenging because Jupiter lacks any surface—it is difficult to imagine being there. Most of the surface features and processes we found on the terrestrial planets are missing on Jupiter, but, as the prototype of the Jovian worlds, it earns its place as the ruler of the solar system.

R E V I E W Critical Inquiry

How do we know Jupiter is hot inside?

Critical analysis hinges on testing ideas, and sometimes it is helpful to test even the most basic ideas. Doing so can give us confidence in what we know, and it can sometimes help us discover new ideas. We know that something is hot if we touch it and it burns our fingers, but we can't touch Jupiter. We also know something is hot if it is glowing bright red—it is red hot. But Jupiter is not glowing red hot. We can tell that something is hot if we can feel heat when we hold our hand near it. That is, we can detect infrared radiation with our skin. In the case of Jupiter, we need greater sensitivity than the back of our hand, but infrared telescopes tell us that Jupiter is a source of infrared. It really is glowing in the infrared. Sunlight would warm it a little bit, but it is emitting twice as much infrared as it should. That means it must be hot inside. From models of the interior, astronomers can tell us that the center must be about 30,000 K in order to heat the surface of the planet and make it glow in the infrared.

Astronomical understanding is usually based on simple observations. For example, how do we know Jupiter has a low density?

The study of Jupiter is more meteorology than geology, but when we examine its moons, we find all the excitement of impact cratering and active volcanism.

24-2 Jupiter's Family of Moons

We can divide Jupiter's satellites into two groups. The smaller moons resemble the moons of Mars and may be captured asteroids. The four larger moons, the moons seen by Galileo, are clearly not captured asteroids but complicated worlds in their own right (Figure 24-10).

Our study of the moons of Jupiter will illustrate three important principles in comparative planetology. First, a body's composition depends on the temperature of the material from which it formed. This is illustrated by the prevalence of ice as a building material in the outer solar system where sunlight is weak. The second principle is that cratering can tell us the age of a surface. Finally, we will see that internal heat has a powerful influence over the geology of these larger moons.

Callisto: The Ancient Face

The outermost of Jupiter's four large moons, Callisto is half again as large in diameter as Earth's moon (Figure 24-11). Like all of Jupiter's larger satellites, Callisto is tidally locked to its planet, keeping the same side forever facing Jupiter. From its gravitational influence on passing spacecraft, astronomers can calculate Callisto's mass, and dividing by its volume tells us that its density is 1.79 g/cm^3. Ice has a density of about 1 and rock 2.5 to 4 g/cm^3, so Callisto must be a mixture of rock and ice.

Photographs from the Voyager and Galileo spacecraft show that the surface of Callisto is dark, dirty ice heavily pocked with craters. Old, icy surfaces in the solar system become dark because of dust added by meteorites and because meteorite impacts vaporize water leaving the rock in the ice behind to form a dirty crust.

Spectra of the surface show that it is mostly a 50/50 mix of ice and rock, but some areas are ice free. Nevertheless, the slumped shape of craters suggests that the outer 10 km is mostly ice. This can be understood when we recall that the spectra tell us about the outer 1 mm of the surface, which can be quite dirty, while the shapes of craters tell us about the outermost 10 km, which could be rich in ice.

Delicate measurements of the shape of Callisto's gravitational field were made by the Galileo spacecraft, and they suggest that the moon has differentiated into a core of rock and a mantle of mixed ice and rock. Callisto has no magnetic field, so it can't have a molten metal core. However, measurements of the interaction of Callisto with Jupiter's magnetic field show that Callisto may have a layer of liquid water roughly 10 km thick somewhere below its icy crust. It isn't clear how Callisto could have grown warm enough to differentiate

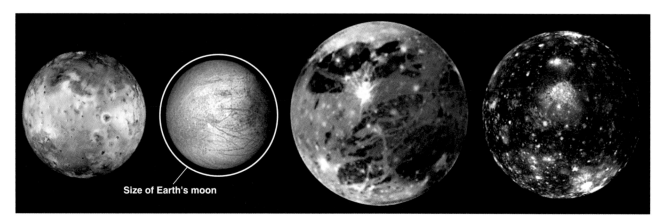

Figure 24-10
The Galilean moons of Jupiter from left to right are Io, Europa, Ganymede, and Callisto. The circle shows the size of Earth's moon. *(NASA)*

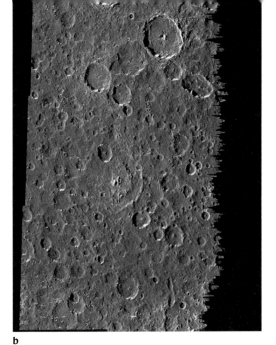

a

b

Figure 24-11

(a) Callisto's surface is heavily cratered, dirty ice, where the youngest craters look bright because they have dug into cleaner ice. The bull's-eye impact feature is called Valhalla and is roughly 3000 km in diameter. (b) A Galileo image of Callisto's surface shows the dirty icy crust and impact features. *(NASA)*

and form a liquid layer without destroying its ancient cratered surface.

Ganymede: A Hidden Past

The next Galilean moon inward is Ganymede, larger than Earth's moon, larger than Mercury, and over three-quarters the diameter of Mars. Its density is 1.9 g/cm^3, and its influence on the Galileo spacecraft reveals that it has a rocky core, an icy mantle, and a crust of ice 500 km thick. It may even have a small iron core. It is large enough for radioactive decay to have melted its interior when it formed, allowing iron to drain to its center.

The surface hints at an active past. Although a third of the surface is old, dark, and cratered, the rest is marked by lighter parallel grooves. Because this **grooved terrain** (Figure 24-12) contains fewer craters, it seems young. Highly detailed Galileo images show systems of parallel grooves lying on top of other systems of grooves.

Early theories to explain these grooves suggested that the crust broke and water flooded up through the cracks to freeze in a process that was repeated over and over to produce the multiple grooves. Actual signs of ice produced by water venting up through the surface are rare, however, and newer theories for groove

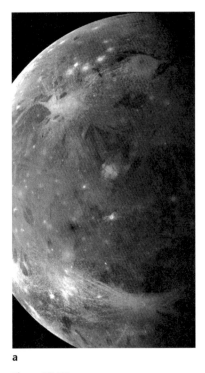

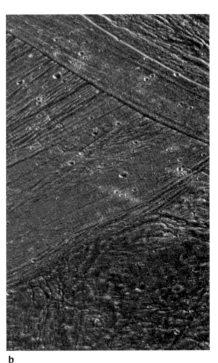

a

b

Figure 24-12

(a) The icy surface of Ganymede is divided into old, dark, cratered terrain, and bands of bright grooves (see Figure 24-10). In this image from the Galileo spacecraft, more recent impacts have broken through the dirty ice into clean ice below. (b) A close-up shows how a system of icy grooves has erased an older region, some of which is visible at the bottom, while a more recent system of grooves has erased part of the previous grooves. *(NASA)*

formation suggest that tension in the crust stretched the landscape repeatedly to form the grooves. The origin of the grooves on Ganymede is one of the most difficult puzzles in modern planetary geology.

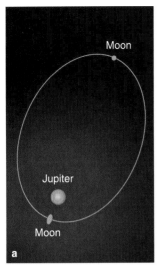

Figure 24-13
Two effects on planetary satellites. (a) Tidal heating occurs when changing tides cause friction within a moon. (b) The focusing of meteoroids exposes satellites in small orbits to more impacts than satellites in larger orbits receive.

The Galileo spacecraft found that Ganymede has a magnetic field about 10 percent as strong as the Earth's. It even has its own magnetosphere inside the larger magnetosphere of Jupiter. Theoretical calculations find it difficult to produce a magnetic field in a water-rich mantle, and there does not appear to be enough heat in Ganymede for it to have a molten metallic core. Astronomers wonder if its magnetic field is left over from a time when it was hotter and more active.

A possible heat source for Ganymede is **tidal heating,** the frictional heating of a body by changing tides (Figure 24-13a). In the past, Ganymede could have interacted with the other moons and entered a slightly elliptical orbit. Jupiter's gravity would have deformed the moon, and, as Ganymede followed its elliptical orbit, tides would have flexed it, and friction would have heated it. Such an episode of tidal heating might have driven a dynamo to produce a magnetic field and break the crust to make the grooved terrain.

The age of the grooved terrain is an important factor. If the grooves and the magnetic field were produced long ago, the magnetic field would have weakened by now. From the number of craters on the grooved terrain, astronomers think it is about 1 billion years old, but the cratering rate near Jupiter is uncertain, because Jupiter tends to focus meteoroids inward (Figure 24-13b). If the grooved terrain is 3 billion years old, it may be unrelated to the events that created the magnetic field.

Europa: A Hidden Ocean

The outer two Galilean moons are large and icy, but Europa is smaller (Figure 24-10) and has a density of 3 g/cm³. Thus, it must contain only a small percentage of water. Spectra reveal traces of water on its surface but that water seems to lie in an icy crust less than 100 km thick. As it raced past Europa, the orbital motion of the Galileo spacecraft revealed that the little moon is mostly rock and may have a small metallic core. Its bland face hides a deep secret, however. Data from Galileo suggest that a 200-km-deep ocean of water lies below the crust.

Europa is obviously youthful. Its icy surface contains almost no large craters and is bright with an albedo of 0.69. (It reflects 69 percent of the light that hits it.) As we have seen, old, icy surfaces are very dark and cratered, so we can guess that Europa's surface is recently renewed.

The long lines crossing the surface suggest a renewal process (Figure 24-14a). They look like cracks in the icy crust where water from below has welled up and frozen. Some areas look like great blocks of ice that have been shifted in a nearly frozen sea (Figure 24-14b). Other, seemingly older, regions contain numerous small craters that suggest ages of 10 million years or so. Astronomers have speculated about ice volcanoes and lava flows of liquid water, but the details remain unknown.

The youthful surface on Europa suggests that it should contain a source of heat. It is too small to have retained heat from its formation, and radioactive decay could not keep such a small world active. It contains only a very weak magnetic field, so a molten core is ruled out. Tidal heating is more effective on Europa than on Ganymede (because Europa is closer to Jupiter) and may provide enough energy to drive its activity. Thus it lacks craters because its surface is icy and cannot retain craters easily and because it is an active world and its surface is young (Figure 24-15).

Io: Bursting Energy

Geological activity is driven by heat flowing out of a planet's interior, and nothing could illustrate this principle better than Io, the innermost of Jupiter's Galilean moons. Photographs from the Voyager and Galileo spacecraft show no impact craters at all—surprising considering Jupiter's power to focus meteoroids inward (Figure 24-13b). No subtlety is needed to explain the missing craters. Dozens of active volcanoes are visible on the moon's surface blasting sulfurous gas and ash out over the surface to bury any newly formed craters (Figure 24-16a). Unlike the dead or dormant outer Galilean moons, Io is bursting with energy.

The colors of Io have been compared to those of a badly made pizza, but we can recognize the yellows, oranges, browns, blacks, and reds as typical of sulfur compounds. Spectra confirm that the surface of Io is rich in sulfur.

Spectra reveal that Io has a tenuous atmosphere of gaseous sulfur and oxygen, but those gases can't be permanent. The erupting volcanoes pour out about one ton of gases per second, but, because of Io's low escape

Figure 24-14

(a) The icy surface of Europa is shown here in natural color at left and in computer-enhanced color at right. Long lines on its surface are visible but very few craters. The impact at lower right is the crater Pwyll. (b) Icy blocks on the surface of Europa appear to have moved as do icebergs floating in pack ice on Earth. The surface is naturally blue ice. Brown stains may be caused by mineral-rich water venting from below the crust, and white areas are debris from the impact crater Pwyll. *(NASA)*

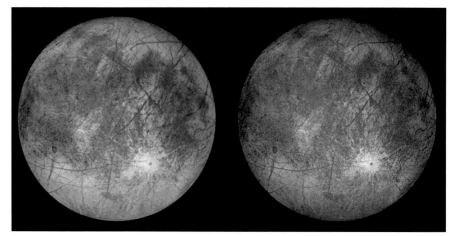

a

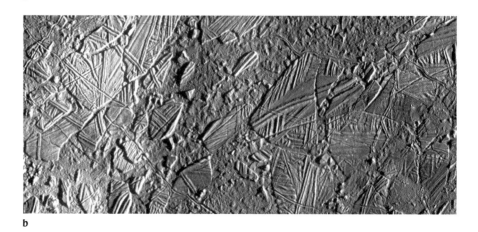

b

velocity, the gases leak into space easily. Also, any gas atoms that become ionized are swept away by Jupiter's rapidly rotating magnetic field. The ions produce a cloud of sulfur and sodium ions in a doughnut-shaped cloud enclosing Io's orbit (Figure 24-16b). Io's neighborhood is dirty (and probably smelly) with sulfur. In fact, the reddish color of the small inner moon Amalthea may be caused by sulfur pollution from Io.

We can use basic observations to deduce the nature of Io's interior. From the density, 3.55 g/cm³, we can conclude that it is mostly rock. Spectra reveal no trace of water at all, so there is no ice on Io. It is the driest world in our solar system. The oblateness of Io caused by its rotation and by the slight distortion produced by Jupiter's gravity gives astronomers clues to its interior. Model calculations suggest it contains a modest core of metals and a deep rocky mantle. The Galileo spacecraft has detected a magnetic field presumably generated by a dynamo in its molten metallic core.

The colors on the surface and the sulfur evident in spectra suggested at first that Io's crust was a mixture

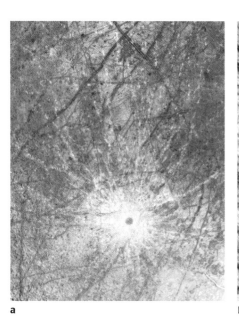

a

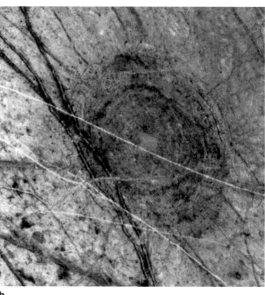

b

Figure 24-15

(a) The impact crater Pwyll, 26 km in diameter, is a young feature on the icy surface of Europa. (b) This circular bull's-eye is 140 km in diameter. It is the remains of an impact by an object about the size of a mountain. Note the linear features that cross the older impact structure. *(NASA)*

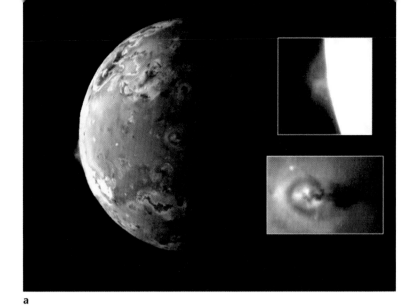

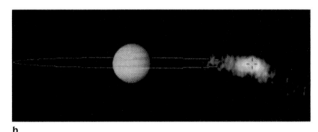

a

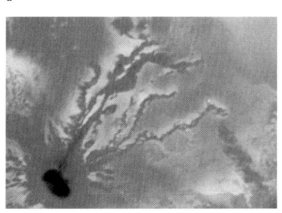

b

Figure 24-16
(a) At the left edge of the disk of Io, a volcanic plume (upper inset) rises above the caldera Pillan Patera. Near the center of the disk, a second plume rises 75 km above the volcano Prometheus and casts a long shadow to the right (lower inset). *(NASA)* (b) Sulfur vapor from Io surrounds the moon and its orbit in a glowing cloud. *(JPL, Bruce A. Goldberg)*

of sulfur-rich compounds, but later evidence argues that the crust is rock. Infrared measurements show that some volcanic eruptions are hotter than 1400 K and must be molten rock and not molten sulfur or sulfur compounds, which would melt at lower temperatures. Also, mountains up to 16 km high have been found on Io, and sulfur compounds are not strong enough to support such high mountains. The crust of Io must be mostly silicates.

Volcanism is continuous on Io. Plumes come and go over periods of months, but some, such as Pele, have been active since the Voyager spacecraft first visited Io in 1979 (Figure 24-17a). Earth's volcanoes eject lava and ash because of water dissolved in the lava. As rising lava reaches Earth's surface, the sudden decrease in pressure allows the water to come out of solution in the lava. It is like popping the cork on a bottle of champagne. The water flashes into vapor and blasts material out of the volcano, a process that was responsible for the Mount St. Helens explosion in 1987. But Io is dry. Its volcanoes appear to be powered by sulfur dioxide dissolved in the magma. When the pressure is released, the sulfur dioxide boils out of solution and blasts gas and ash high above the surface in plumes up to 300 km high. Ash falling back to the surface produces debris layers around the volcanoes, such as that around Pele in Figure 24-17a. Whitish

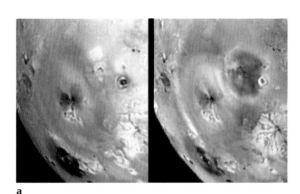

a

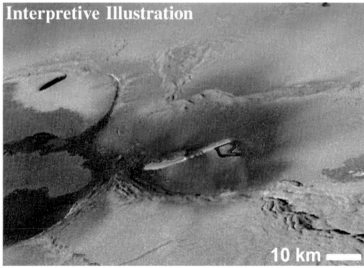

Interpretive Illustration

10 km

c

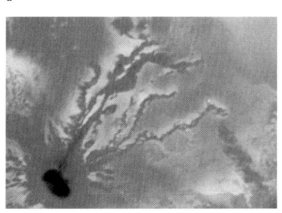

b

Figure 24-17
(a) Recorded a few months apart, these false-color images show a new volcanic feature near the reddish ash fall around the larger volcano Pele. The new area is dark ash covering an area about as large as Arizona around the vent Pillan Patera (see Figure 24-16a). (b) The volcano Ra has produced molten lava that has flowed downhill. The longest flows in this false-color image are about 320 km (200 mi) long. (c) The Galileo spacecraft imaged this curtain of hot lava fountaining out of a crack over 25 km long. *(NASA)*

areas on the surface are thought to be frosts of sulfur dioxide.

Great lava flows are visible carrying molten material downhill where it buries the surface (Figure 24-17b). In late 1999, the Galileo spacecraft photographed very hot molten lava bursting a kilometer high out of a long fissure. The lava was so bright it overloaded the camera, so the image shown in Figure 24-17c has been reconstructed to show how the curtain of glowing lava would look to our eyes. Such continuous eruptions generate lava flows and ash falls that continually resurface the crust of Io, quickly burying any newly formed impact craters.

What powers Io? It is bursting with energy, but it is only a few percent bigger than Earth's moon, which is cold and dead. Io is too small to have retained heat from its formation or to remain hot from radioactive decay. In fact, the energy blasting out of its volcanoes is about three times more energy than it could make by radioactive decay in its interior.

Io is heated by the same kind of tidal heating that is suspected in Ganymede's past and in Europa. Because Io is so close to Jupiter, tides are powerful, and gravitational interactions with Europa and Ganymede force its orbit to be slightly elliptical. As its distance from Jupiter varies, tides squeeze it, and its surface rises and falls by about 100 m. The resulting friction is enough to at least partially melt the interior and drive volcanism. In fact, the energy flowing outward is continually recycling Io's crust. Deep layers melt, are spewed out through the volcanoes, fall back to cover the surface, and are later covered themselves until they are buried so deeply that they are again melted.

Io is the most active world in our solar system because it lives so close to Jupiter. The other Galilean moons are more distant and thus have had less dramatic histories of tidal heating and geological activity.

The History of the Galilean Moons

The goal of astronomy is to tell the story of the universe. Now we must test our understanding of the Galilean moons by trying to tell their story. To do that we not only draw on what we have learned about the moons, but we also draw on what we have learned about Jupiter and about the origin of the solar system (Chapter 20).

The minor moons of Jupiter are probably captured asteroids, but the Galilean moons seem to be primordial. That is, they formed with Jupiter. Also, they seem to be a set in that they share common features and their densities are related to their distance from Jupiter (Table 24-2). This information suggests that the four moons formed in a disk-shaped nebula around Jupiter—a mini-solar nebula. We know that Jupiter was very hot when it formed, so, just as the terrestrial and Jovian planets formed in the solar nebula, the inner moons, Io

Table 24-2 The Galilean Satellites*

Name	Radius (km)	Density (g/cm³)	Orbital Period (days)
Io	1816	3.55	1.769
Europa	1563	3.04	3.551
Ganymede	2638	1.93	7.155
Callisto	2410	1.81	16.689

*For comparison, the radius of Earth's moon is 1738 km, and its density is 3.36 g/cm³.

and Europa, formed from rocky material, and the outer moons incorporated more ices.

But we must also consider the trend in geological activity, a trend that suggests tidal heating. Io is so severely heated that any water it might have had has been driven off. In contrast, the outermost moon, Callisto, has never suffered severely from tidal heating. Ganymede and Europa have less ice and more geological activity in their past, so we must suspect that gravitational interactions among the inner three moons have produced periods of tidal heating and geological upheaval.

Indeed, the three inner moons are now locked in an orbital resonance. Io orbits Jupiter twice in the time Europa takes to orbit once. Furthermore, Io orbits four times in the time Ganymede takes to orbit once. The three moons move in resonance. Callisto does not fit this pattern, and Callisto is the only one of the four moons whose surface is undisturbed. Io is bursting with energy, but the Galilean satellite system is bursting with history. Something interesting has happened there.

REVIEW Critical Inquiry

What produces Io's internal heat?

Part of critical thinking is recognizing and using basic principles. In this case, we understand that small worlds lose their internal heat quickly and become geologically inactive. Io is a bit smaller than Earth's moon, which is cold and dead, but Io is bursting with energy flowing outward.

Clearly, Io must have a powerful source of heat inside, and that heat source is tides. Io's orbit is slightly elliptical, so it is sometimes closer to Jupiter and sometimes farther away. That means that Jupiter's powerful gravity sometimes squeezes Io more than at other times, and the flexing of the little moon's interior produces heat through friction. Such tides would rapidly force Io's orbit to become circular—and then tidal heating would end and the planet would become inactive—except that the gravitation of the other moons keeps Io's orbit elliptical. Thus, it is the influence of its companions that keeps Io in such an active state.

Io has almost no impact craters, but Callisto has many. What does the distribution of craters on the Galilean satellites tell us about their history?

We have now applied the principles of comparative planetology not only to planets but also to moons. With the insight we have gained from Jupiter and its satellite system, we are ready to visit the most beautiful of the Jovian worlds.

24-3 Saturn

Saturn has played second fiddle to its own rings since Galileo first saw it in 1610. He didn't recognize the rings for what they are, but we know them now as one of the great sights in the solar system. Nevertheless, Saturn itself, only slightly smaller than Jupiter (Data File 8), is a fascinating planet with a few mysteries of its own.

Our discussion of Saturn can make use of the principles we have learned from Jupiter, but we have less information about Saturn than we did about Jupiter. Pioneer 2 and both Voyager spacecraft flew past Saturn, but no orbiter circles Saturn as Galileo circles Jupiter. Thus, our discussion of Saturn is based on less evidence and lower-quality data than we have for Jupiter.

Planet Saturn

The basic characteristics of Saturn tell us about its composition and interior. Only about a third of the mass of Jupiter and 16 percent smaller in radius, Saturn has an average density of 0.69 g/cm³. It is less dense than water—it would float! Although Saturn rotates about as fast as Jupiter, it is almost twice as oblate, so we can conclude that it must not contain a large core of heavy elements. Spectra show that its atmosphere is rich in hydrogen and helium (see Table 24-1), and models predict that it is mostly liquid hydrogen.

Infrared observations show that Saturn is radiating almost twice as much energy as it receives from the sun, telling us that heat is flowing out of its interior. It must be hot inside. In fact, it is hotter than we would expect. It should have lost more heat since it formed. Astronomers suspect that helium in the liquid hydrogen interior is condensing into droplets and falling inward. The falling droplets, releasing energy as they pick up speed, heat the planet. This heating is similar to the heating produced when a star contracts.

We can learn more about Saturn's interior from its magnetic field. The Voyager spacecraft found that Saturn's magnetic field is about 20 times weaker than Jupiter's. It also has correspondingly weak radiation belts. Models predict that the pressure inside Saturn, being

Saturn Data File Eight

Saturn's rings of ice particles are bright and beautiful, but its belt–zone circulation is subdued. A moon and its shadow lie just above center. *(NASA)*

Average distance from the sun	9.5388 AU (14.27 × 10⁸ km)
Eccentricity of orbit	0.0560
Maximum distance from the sun	10.07 AU (15.07 × 10⁸ km)
Minimum distance from the sun	9.005 AU (13.47 × 10⁸ km)
Inclination of orbit to ecliptic	2°29′17″
Average orbital velocity	9.64 km/s
Orbital period	29.461 y (10,760 days)
Period of rotation	10ʰ39ᵐ25ˢ
Inclination of equator to orbit	26°24′
Equatorial diameter	120,660 km (9.42 $D_\oplus$)
Mass	5.69 × 10²⁶ kg (95.147 $M_\oplus$)
Average density	0.69 g/cm³
Gravity at base of clouds	1.16 Earth gravities
Escape velocity	35.6 km/s (3.2 $V_\oplus$)
Temperature at cloud tops	−180°C (−292°F)
Albedo	0.61
Oblateness	0.102

Science is an expensive enterprise, and that raises the question of payment. Some science has direct technological applications, but some basic science is of no immediate practical value. Who pays the bill?

In Window on Science 24-1, we saw that technology is of immediate practical use; thus, business funds much of this kind of scientific research. Auto manufacturers need inexpensive, durable, quick-drying paint for their cars, and they find it worth the cost to hire chemists to study the way paint dries. Many industries have large research budgets, and some industries, such as pharmaceutical manufacturers, depend exclusively on scientific research to discover and develop new products.

If a field of research has immediate potential to help society, it is likely that government will supply funds. Much of the research on public health is funded by government institutions such as the National Institutes of Health and the National Science Foundation. The practical benefit of finding new ways to prevent disease, for example, is well worth the tax dollars.

Basic science, however, has no immediate practical use. That doesn't mean it is useless, but it does mean that the practical-minded stockholders of a company will not approve major investments in such research. Digging up dinosaur bones, for instance, is very poorly funded because no industry can make a profit from the discovery of a new dinosaur. Astronomy is another field of science that has few direct applications, and thus very little astronomical research is funded by industry.

The value of basic research is twofold. Discoveries that have no known practical use today may be critically important years from now. Thus, society needs to continue basic research to protect its own future. But basic research, such as studying Saturn's rings or digging up dinosaur bones, is also of cultural value because it tells us what we are. Each of us benefits in intangible ways from such research, and thus society needs to fund basic research for the same reason it funds art galleries and national parks—to make our lives richer and more fulfilling.

Because there is no immediate financial return from this kind of research, it falls to government institutions and private foundations to pay the bill. The Keck Foundation has built two giant telescopes with no expectation of financial return, and the National Science Foundation has funded thousands of astronomy research projects for the benefit of society. Debates rage as to how much money is enough and how much is too much, but, ultimately, funding basic scientific research is a public responsibility that society must balance against other needs. There isn't anyone else to pick up the tab.

---

weaker than that inside Jupiter, produces a smaller mass of liquid metallic hydrogen. Heat flowing outward must cause convection in this conducting core, and the rapid rotation drives a dynamo effect to produce the magnetic field. Unlike most magnetic fields, Saturn's field is not inclined to its axis of rotation, something we notice in ultraviolet images that show a ring of aurora around Saturn's north pole (Figure 24-18). This alignment between the magnetic axis and the axis of rotation is peculiar and isn't understood.

Saturn's atmosphere is rich in hydrogen and contains belt–zone circulation, which appears to arise in the same way as that on Jupiter. Zones are higher clouds formed by rising gas, and belts are lower clouds formed by sinking gas.

But the clouds are not very distinct on Saturn (Figure 24-19a). Measurements from Voyager spacecraft tell us that Saturn's atmosphere is much colder than Jupiter's—something we would expect because Saturn is twice as far from the sun and receives only one-fourth as much solar energy per square meter. The clouds on Saturn form at about the same temperature as the clouds on Jupiter, but those temperatures are deeper in Saturn's cold atmosphere. Compare the cloud layers in Figure 24-19b with those in Figure 24-4b. The atmo-

Figure 24-18
At ultraviolet wavelengths, the Hubble Space Telescope can photograph rings of aurora surrounding Saturn's north and south magnetic poles. (Compare with Figure 24-3a.) *(J. T. Trauger, JPL, and NASA)*

a

Figure 24-19

(a) Saturn, like Jupiter, has belt–zone circulation in its atmosphere, but it is less distinct. The white storm clouds visible here at the center of the disk are rare on Saturn. *(NASA)* (b) Because Saturn is colder than Jupiter, the clouds form deeper in the hazy atmosphere. Notice that the three cloud layers on Saturn form at about the same temperatures as do the three cloud layers on Jupiter.

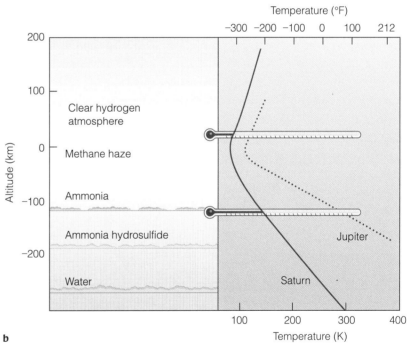

b

spheres of Jupiter and Saturn are very similar once we remember that Saturn is colder.

One dramatic difference between Jupiter and Saturn concerns the winds. On Jupiter, winds bound each of the belts and zones, but on Saturn the pattern is not the same. Saturn has fewer such winds, but they are much stronger. The eastward wind at the equator of Saturn, for example, blows at 500 m/s (1100 mph), roughly five times faster than the eastward wind at Jupiter's equator. The reason for this difference is not clear.

Saturn harbors a number of mysteries, including great storms that burst up through the clouds when it is summer in Saturn's northern hemisphere (see Figure 24-19a). Also, Saturn does not have a great spot corresponding to Jupiter's Great Red Spot. In fact, Saturn has few spots. We will learn more about these mysteries when the Cassini spacecraft goes into orbit around Saturn in 2004 (Window on Science 24-2). It will give us more and better information, not only about Saturn, but also about its rings and moons.

Saturn's Rings

Looking at the beauty and complexity of Saturn's rings, an astronomer once said, "The rings are made of beautiful physics." We might add that the physics is actually rather simple, but the result is one of the most astonishing sights in our solar system.

In 1609, Galileo was the first to see the rings of Saturn, but perhaps because of the poor optics in his telescopes, he did not recognize the rings as a disk. He drew Saturn as three objects—a central body and two smaller ones on either side. In 1659, Christian Huygens realized that the rings were a disk surrounding but not touching the planet.

The rings are labeled with letters of the alphabet (Figure 24-20a), with the A and B rings being easily visible even through a small telescope. The A ring lies outside a gap called Cassini's division (named after its discoverer), and the B ring lies inside that gap. The C ring, or Crepe ring (so named after the thin black cloth tradi-tionally associated with mourning), is a faint ring lying inside the B ring. Fainter rings are also identified by letters. From edge to edge, the A ring has a diameter about 21 times that of Earth.

The discovery that the rings are made of particles is a story of both theory and observation. Astronomers knew that the rings were transparent because they could see stars through the rings. In 1859, James Clerk Maxwell (for whom a large mountain on Venus is named) proved theoretically that the rings could not be solid. Because of Kepler's third law, the inner parts of the rings have to orbit faster than the outer parts, so solid rings cannot remain in orbit without ripping themselves apart. Much later, the American astronomer James Keeler measured Doppler shifts in the rings and proved that they did indeed obey Kepler's third law. Thus, they had to be made of particles. Astronomers knew that the rings had to be very thin because they van-

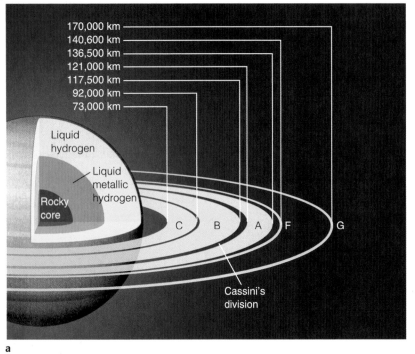

170,000 km
140,600 km
136,500 km
121,000 km
117,500 km
92,000 km
73,000 km

Liquid hydrogen

Liquid metallic hydrogen

Rocky core

C B A F G

Cassini's division

a

Figure 24-20
(a) The A and B rings of Saturn are easily visible from Earth even with a small telescope, but the C ring is more difficult to observe. The F and G ring and a few fainter rings were photographed by the Voyager spacecraft. (b) Composed of billions of ice particles, the rings of Saturn are quite transparent. Note where the rings cross in front of the planet at the lower left in the Voyager photograph. The rings are generally pale and colorless. They are marked by many gaps and ringlets produced by dynamic effects among the orbiting ring particles and Saturn's moons. *(NASA)*

b

ish when we view them edge-on as Earth passes through the ring plane. They could not be more than 2 km thick. With the development of modern infrared spectrographs, astronomers were able to detect the presence of solid water in the rings.

Cassini's division and a few other gaps in the rings that are visible from Earth were understood by an elegant application of Newtonian physics. In 1867, Daniel Kirkwood demonstrated that resonances between the orbital motion of asteroids and Jupiter could produce gaps in the asteroid belt (Chapter 26). The same theory predicted that a gap could occur in the rings of Saturn where a ring particle orbited in resonance with a moon. A ring particle in Cassini's division is in a double resonance. It orbits twice in the time the moon Mimas takes to orbit once, and the particle orbits three times in the time it takes the moon Enceladus to orbit once. Every other orbit a ring particle in Cassini's division overtakes Mimas in the same place and feels a small tug. Every third orbit it gets a tug at the same spot in its orbit from Enceladus. Gradually the tugs add up and pull the ring particle out of the region. Thus Cassini's division contains few particles because it is a double resonance. A few other gaps in the rings are related to resonances with moons.

By 1979, theory and observation accounted for the known properties of Saturn's Rings, and as the Voyager 1 spacecraft approached Saturn, astronomers expected to see thin, uniform rings of ice particles with a few well-understood gaps. What the Voyager photos actually revealed were entirely new phenomena in the rings. They were made up of roughly 1000 ringlets no wider than 2 km (Figure 24-20b), some were elliptical, and some contained kinks and knots. New phenomena called for new explanations.

For one thing, astronomers knew that the rings were not ruled by gravitation alone. Since 1896, observers have reported seeing faint radial features in Saturn's rings that are hard to understand. If the rings obeyed Kepler's laws, differential rotation should erase radial features. The Voyager images revealed that the **spokes** were real (Figure 24-21a).

Two clues are critical: The spokes are bright in forward scattering, and they sweep around Saturn at the

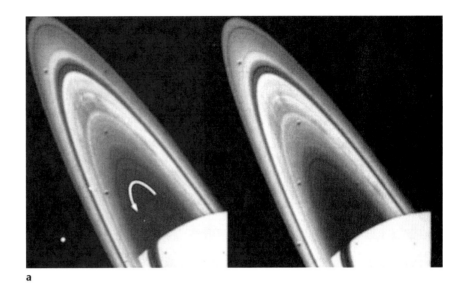

a

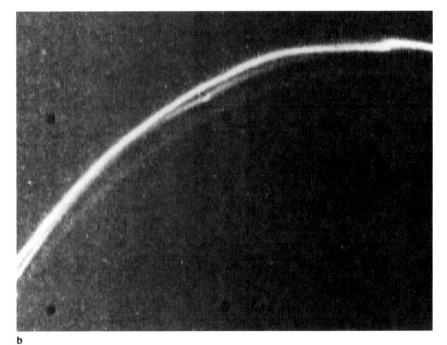

b

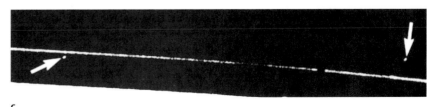

c

Figure 24-21
(a) Radial spokes in the bright B ring are visible in these Voyager images. Note the motion of the spokes (arrow) between these two frames. (b) A Voyager 1 photo of the F ring shows a number of strands with twists and kinks. (c) Another image reveals small shepherd satellites (arrows) orbiting just outside and just inside the F ring. *(NASA)*

same rate that the magnetic field rotates. Their brightness in forward scattering tells us they are made of microscopic specks. Evidently, some small specks of matter in the rings become electrically charged, and the magnetic field sweeps the particles up and out of the rings to form temporary clouds that we see as spokes.

Photographs and measurements reveal the existence of other rings around Saturn. The E ring, detected from Earth, was traced from 3 to 8 Saturn radii and shown to be densest near the orbit of the satellite Enceladus. This moon may be a source of new ring particles as existing particles slam into the moon and blast new particles free. Voyager images also revealed a D ring inside the C ring and two narrow rings labeled F and G.

The F ring posed a difficult problem. Voyager 1 photos showed it was narrow and irregularly twisted (Figure 24-21b). Planetary rings should gradually spread out due to collisions between particles and should not develop irregular twists if the particles move in simple orbits. The mystery was eventually resolved when photos showed two small satellites orbiting, one inside and one outside the F ring (Figure 24-21c).

These **shepherd satellites** interact gravitationally with wandering particles and return them to the ring, thus keeping it confined. The distortions in the F ring appear to be caused by gravitational interactions when the inner shepherd moon overtakes and passes the outer moon. Another satellite at the edge of the A ring ushers straying particles back and thus keeps the edge of the A ring sharp.

The ringlets and grooves in Saturn's rings arise from more than one cause. Some are probably produced by small moonlets imbedded in the rings. A small moon has been found in Encke's Gap located in the A ring. Some grooves are produced by spiral density waves (the same phenomena seen in spiral galaxies) that produce sets of grooves that sweep slowly outward. Other grooves are produced by another kind of waves called bending waves, that cause corrugations in the surface of the ring. Observations from Earth give a ring thickness of 1 to 2 km, but some Voyager observations suggest the rings are only 10 m thick. Evidently the bending waves make the rings look thicker, just as corrugated cardboard is thicker than the paper in the corrugations.

Individual ring particles have not been detected, but forward scattering of light shows that microscopic particles are common, especially in the bright B ring. Particles a few centimeters in diameter are more common in the C ring. Larger particles are rarer, and boulders the size of a house are quite rare.

An astronaut in a space suit could probably swim through the densest parts of the ring by pushing against the larger particles and drifting from particle to particle (Figure 24-22). This would not be dangerous because the orbiting particles in any given region travel at nearly the same speed. When particles do collide, they merely nudge each other. Occasionally, more vio-

Figure 24-22
An astronaut could probably swim through Saturn's rings. These computer models of the A ring (left) and the C ring (right) show how particle size and number differ in the rings. Particles smaller than 2 cm in diameter are not shown. Actual ring particles are probably irregular ice chunks and not the perfect spheres shown. Astronaut included to scale. *(NASA/JPL)*

lent collisions may produce a few bits of dust that slowly drift inward through the rings and finally fall into the planet. In the thinner parts of the rings, the particles are so widely spaced that ring swimming would be difficult.

The different distributions of particle sizes through the rings are responsible for differences in the appearance of the A, B, and C rings. But no one presently understands what forces sort the particles by size in the different regions. Nor do we understand the processes that must break up particles and re-form them.

Like a beautiful flower, the rings of Saturn are controlled by many different natural processes. When the Cassini spacecraft reaches orbit around Saturn in 2004, we will learn even more about the rings.

The History of Saturn

The farther we journey from the sun, the more difficult it is to understand the history of the planets. Any fully successful history of Saturn should explain its low density, its peculiar magnetic field, and its beautiful rings. We can't tell that fully successful story yet, but we can understand a few of the principles.

Certainly, we can understand that Saturn formed in the outer solar nebula, where ice particles were stable. It grew rapidly and became massive enough to capture hydrogen and helium gas directly from the nebula. The denser elements sank to the middle to form a denser core, and the hydrogen formed a liquid mantle containing liquid metallic hydrogen. The outward flow of heat from the core is believed to drive convection currents in this mantle that, coupled with the rapid rotation of the planet, produce its magnetic field. Because Saturn is smaller than Jupiter, it has less liquid metallic hydrogen, and its magnetic field is weaker.

We might be tempted at first to suppose that the rings of Saturn are primordial, made of material left over from the formation of the planet. That seems unlikely, however. Saturn, like Jupiter, should have been very hot when it formed, and that heat would have vaporized and driven off ring material. Also, such a hot Saturn would have had a very distended atmosphere, which would have slowed ring particles by friction and caused the particles to fall into the planet.

Planetary rings do not seem to be stable over 4.6 billion years, so we must suppose that the ring material is more recent. Saturn's beautiful rings may have been produced within the last 100,000 years. One suggestion is that an icy planetesimal came within Saturn's Roche limit, and tides pulled it apart. At least some of the debris would settle into the ring plane. We will see in the next two chapters that these icy planetesimals are common in the outer solar system. Another possibility is that a comet struck one of Saturn's moons. Because both comets and moons in the outer solar system are icy, such a collision would produce icy debris. Bright planetary rings such as Saturn's may

be temporary phenomena, produced when fresh ice is scattered into the rings and then wasting away as the ice is gradually lost.

The history of Saturn is complex. The Voyager spacecraft data are a treasure trove, but many mysteries remain. We may not understand the planet's story well until the Cassini spacecraft reaches Saturn in 2004.

R E V I E W Critical Inquiry

Why do the belts and zones on Saturn look so bland?

One of the most powerful tools of critical thought is simple comparison and contrast. We can understand Saturn by comparing and contrasting it with Jupiter. In the atmosphere of Jupiter, the dark belts form in regions where gas sinks, and zones form where gas rises. The rising gas cools and condenses to form icy crystals of ammonia, which we see as bright clouds. Clouds of ammonia hydrosulfide and water form deeper, below the ammonia clouds, and are not as visible. Saturn is twice as far from the sun as Jupiter, so sunlight is four times dimmer. (Remember the inverse square law from Chapter 5.) The atmosphere is colder, and rising gas currents reach colder levels and form clouds deeper in Saturn's atmosphere than in Jupiter's atmosphere. Because the clouds are deeper, they are not as brightly illuminated by sunlight and look dimmer. A layer of methane-ice-crystal haze high above the ammonia clouds makes the clouds even less distinct.

We can compare Saturn's atmosphere to Jupiter's atmosphere, but we can also learn by comparing the ring systems. How does our understanding of Jupiter's rings help us understand Saturn's?

In any contest, Saturn should win the ribbon for most beautiful planet. Its subtly marked disk and its delicate rings are both massive and graceful. Its extensive system of moons only adds to the mystery of this beautiful and curious world.

24-4 Saturn's Moons

Saturn has 18 known satellites (Figure 24-23)—far too many for us to examine individually—but these moons share characteristics common to icy worlds. Most of them are small and dead, but one is big enough to have an atmosphere and perhaps even oceans.

Titan

Saturn's largest satellite is a giant ice moon with a thick atmosphere and a mysterious surface. From Earth it is only a dot of light, and the cameras aboard the Voyager

Figure 24-23

Hubble Space Telescope images of Saturn show a few of its many moons. The upper image was recorded as the rings were edge-on to Earth. The reddish moon at the upper left is smoggy Titan casting its shadow on Saturn. At the right are the icy moons Mimas, Tethys, Janus, and Enceladus (left to right). The lower image was recorded as the rings were illuminated nearly edge-on to the sun. Tethys is at upper left and Dione at the right. Note that Dione casts a long shadow across the rings. *(E. Karkoschka, LPL, and NASA)*

spacecraft were unable to penetrate its hazy atmosphere. Nevertheless, a few basic observations tell us a great deal about this strange world.

Titan is an icy world a bit larger than the planet Mercury and almost as large as Jupiter's moon Ganymede. Unlike these worlds, Titan has a thick atmosphere. Its escape velocity is low, but it is so far from the sun it is very cold. Thus, most gas atoms don't move fast enough to escape. (See Figure 23-16.) Its density is 1.9 g/cm^3, but its uncompressed density (Chapter 20) is only 1.2 g/cm^3. Although it must have a rocky core, it must also contain a large amount of ices.

The surface of Titan is not visible in photographs because its atmosphere is filled with haze (Figure 24-24a). Methane was detected spectroscopically in 1944, as were various hydrocarbons in 1970, but most of Titan's air is nitrogen with only a few percent meth-

ane. Sunlight converts methane (CH_4) into the gas ethane (C_2H_6) plus a collection of organic molecules* (Table 24-3). Some of these molecules produce the smoglike haze, and as the smog particles have gradually settled to the surface, they may have deposited a layer of smelly, organic goo 10 m to 1 km thick.

This goo is exciting because such organic molecules may have been the precursors of life on Earth. Thus, similar chemical reactions on Titan might have created life there. Although some scientists argue that the temperature on Titan is too low to create such molecules, future explorers will want to visit Titan to look for life. We will consider this further in Chapter 27.

*Organic molecules are common in living things on Earth but do not have to be derived from living things. One chemist defines an organic molecule as "any molecule with a carbon backbone."

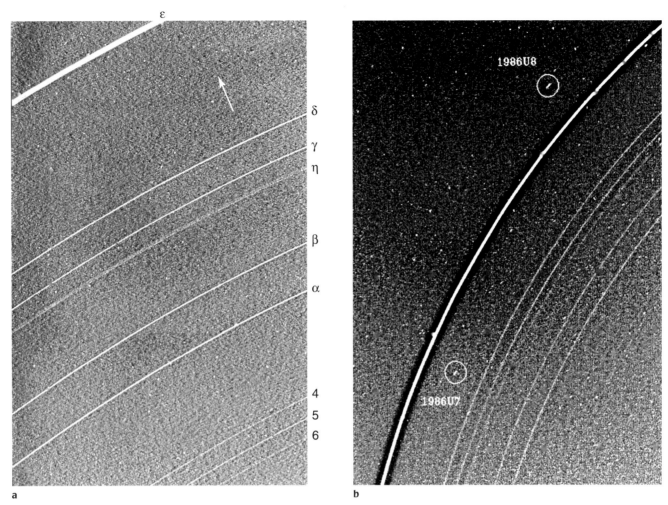

ε

δ
γ
η

β

α

4
5
6

a

b

Figure 25-8
(a) The narrow rings of Uranus are identified by Greek letter or by number. Voyager images reveal another faint ring (arrow) halfway from the ε ring to the next innermost (δ) ring. Tip this page and view the faint ring along its length. Try to locate the η ring in the occultation data (Figure 25-7a). (b) Two shepherd satellites confine the ε ring into a narrow band. Because the satellites have slightly elliptical orbits, the ring is elliptical and slightly wider where it is farther from the planet. *(NASA)*

The ring particles reflect only about 4 percent of the light that falls on them, making them about as dark as lumps of coal. The dark color may arise because the ring particles are composed of a methane-rich mixture of ices trapped in rings within the radiation belts. Radiation could decompose some of the methane to leave behind carbon, which would make the ring particles very unreflective. We will see that this same process may darken the surfaces of the Uranian moons.

The rings of Uranus are not very bright in forward scattering. That is, when the Voyager 2 camera looked back at the rings illuminated by sunlight from behind, the rings were not very bright. That must mean the rings do not contain large amounts of small dust particles. Such dust must be produced by occasional impacts between ring particles and by the radiation grinding away at particles, but small dust particles would feel strong resistance from a tenuous atmosphere of hydrogen extending high above the clouds of Uranus. The friction would make the dust spiral into Uranus. Thus small dust particles may be continuously removed from the rings.

The narrowness of the rings tells us that small moons must be present to confine the ring particles. Collisions between ring particles would cause rings to spread outward to become a uniform disk of particles too tenuous to detect. If rings have sharp edges or are very narrow, then shepherd moons must be herding wandering particles back into the rings. Voyager 2 detected two shepherd moons no more than 25 km in diameter inside and outside the ε ring (Figure 25-8b). The eccentricity of the ε ring and its variation in width around the planet are caused by the eccentricity of the two shepherd moons. Other moons are believed to shepherd the other rings, but these moons are too small to detect. Because these moons orbit inside the Roche limit of Uranus, they must be structurally strong chunks of rock in order to withstand the tidal forces that would otherwise pull them apart.

The rings of Uranus illustrate the importance of satellites in confining planetary rings. The narrow rings are anchored to the small shepherd satellites, and the natural tendency of the rings to spread pushes the satellites outward. Of course, the satellites are much more massive than the tenuous rings and respond slowly. The entire ring system of Uranus, for example, would make an icy satellite only 15 km in radius. Nevertheless, over long periods the small satellites would be pushed outward by the spreading of the rings were it not for the presence of the larger satellites. Many of the small satellites are believed to be locked into resonances with larger satellites, and this anchors the small moons. The spreading pressure of the rings is transferred through resonances to the larger and more massive satellites, and thus the rings can be confined for longer periods. Without satellites, no planet could keep a ring system for long.

It may seem fortuitous that small moons happen to take up orbits at the edges of the rings just where they are needed, but this puts the cosmic cart before the horse. Ring particles tend to become trapped in stable locations within the system of small satellites. Thus, rings exist between closely orbiting satellites where the orbits of the ring particles are most stable.

Clearly, the rings of Uranus are related to its satellite system. Not only are the rings confined by satellites, but the origin of the rings may be linked to the origin of the moons.

The Moons of Uranus

The five largest moons of Uranus (Figure 25-9), from the outermost inward, are Oberon, Titania, Umbriel, Ariel, and Miranda. Umbriel and Ariel are names from Alexander Pope's *The Rape of the Lock,* and the rest are from Shakespeare's *A Midsummer Night's Dream* and *The Tempest* (an Ariel also appears in *The Tempest*). Spectra show that they contain frozen water, and their surfaces are very dark, so astronomers assumed they were made of dirty ices. Little more was known of the moons before Voyager 2 flew through the system.

The Voyager 2 cameras discovered ten moons too small to have been seen from Earth (Figure 25-10). Also, with the construction of new-generation telescopes and

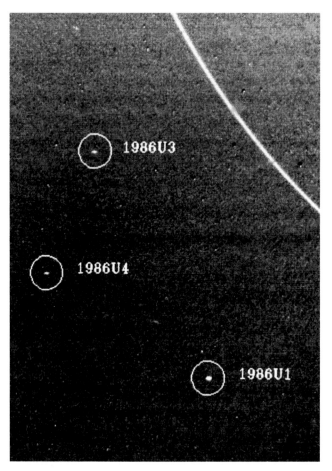

Figure 25-10
Three of the ten small moons of Uranus discovered by Voyager 2 orbit just outside the ε ring. 1986U1, known officially as Puck, is the largest, with a diameter of 170 km, and is roughly spherical. The other new satellites are smaller. The images here are elongated slightly due to motion during the time exposure needed to record the black moons. *(NASA)*

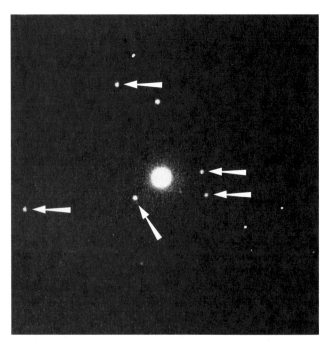

Figure 25-9
Through a telescope, Uranus is a featureless disk never more than 3.7 seconds of arc in diameter. This CCD image, made with the 4-m telescope at Cerro Tololo Inter-American Observatory in Chile, shows the disk of Uranus and its five major satellites (arrows). The other images are faint stars. *(Cerro Tololo Inter-American Observatory)*

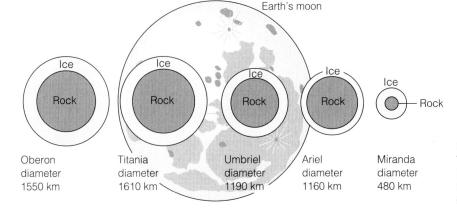

Figure 25-11
The densities of the five major satellites of Uranus suggest that they contain relatively large rocky cores with mantles of dirty ice. The size of Earth's moon is shown for comparison.

Oberon
diameter
1550 km

Titania
diameter
1610 km

Umbriel
diameter
1190 km

Ariel
diameter
1160 km

Miranda
diameter
480 km

Ice
Rock

new imaging techniques, astronomers have found even more moons orbiting Uranus. The planet appears to have at least 20 moons, and there are almost certainly more to be found.

The smaller moons are all as dark as coal. They are probably icy worlds with surfaces that have been darkened by impacts. Because they orbit near Uranus, they are inside the radiation belts, and the radiation may convert methane trapped in their ices into dark carbon deposits and further darken their surfaces.

The five large satellites are all tidally locked to Uranus, and thus their south poles were pointed toward the sun in 1986; it was these hemispheres that were photographed by Voyager 2. Their densities suggest that they contain relatively large rock cores surrounded by icy mantles (Figure 25-11).

Oberon, the outermost of the large moons, has a dark, cratered surface, but it was once an active moon (Figure 25-12a). A large fault crosses the sunlit hemisphere, and dark material, perhaps dirty water "lava," appears to have flooded the floors of some craters.

Titania is the largest of the five moons and has a heavily cratered surface, but it lacks the largest craters (Figure 25-12b). This suggests that soon after it formed, after most of the large debris had been swept up, Titania underwent an active phase in which internal melting flooded the surface with water and erased the early craters. Since then, more craters have formed, but few are as large as those that were erased. Another clue to Titania's past activity is the network of faults that crosses its surface.

Umbriel, the third moon inward, is a very dark, cratered world with no sign of faults or surface activity (Figure 25-12c). It is the darkest of the moons, with an albedo of only 0.16 compared with 0.25 to 0.45 for the other moons. Planetary astronomers suspect that Umbriel's crust is a dark mixture of rock and ice. A bright crater floor in one region suggests that some clean ice may lie at shallow depths in some regions.

Ariel has the brightest surface of the five major satellites and shows clear signs of geological activity. It is crossed by faults over 10 km deep, and some regions appear to have been smoothed by resurfacing (Figure 25-12d). Crater counts show that the smoothed regions are younger than the other regions. Apparently, past melting in the interior has caused geological activity on the moon's surface. Ariel may have been subject to tidal heating caused by an orbital resonance with Miranda and Umbriel.

Miranda, the smallest of the five satellites discovered from Earth, is the most unusual. In fact, its surface is so peculiar that one astronomer referred to it as a "moon designed by a committee." Miranda is marked by oval patterns of grooves, which at first were dubbed *circi maximi* (singular, *circus maximus*) after the ancient Roman chariot-racing tracks (Figure 25-13). Now known as **ovoids,** these features were originally thought to have been produced when Miranda was shattered by a major impact and the mutual gravity of the fragments pulled them back together to reassemble the little moon. More recent studies of the ovoids show that they are associated with faults, ice-lava flows, and rotated blocks of crust. These features suggest that a major impact was not involved. Rather, large convection currents in Miranda's icy mantle may have forced material upward to create the ovoids.

Certainly, Miranda has had a violent past. Near the equator, a huge cliff rises 20 km. An astronaut who slipped from the top of the cliff would fall for 10 minutes before hitting the surface. The cliff and the ovoids are old, however. Miranda is no longer geologically active, and we can read the history of its past on its disturbed surface.

The peculiar geology of Miranda illustrates the problem astronomers face in understanding the geology of the Jovian satellites. Has it been dominated by interior or exterior factors? If the ovoids were caused by convection in Miranda's icy mantle, then heat rising from its interior must be the dominant factor, and we must suppose that its interior was heated by tidal stress as in Jupiter's moon Io. Miranda is too small to have kept its

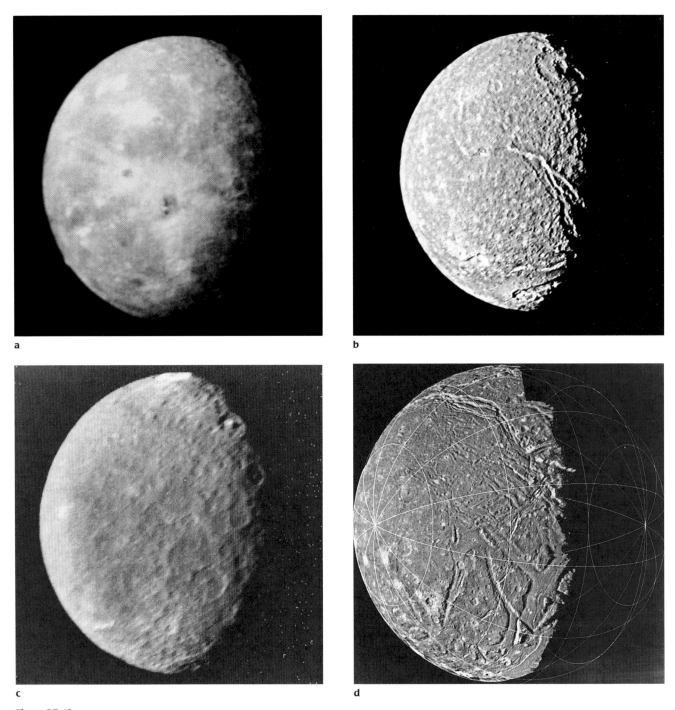

Figure 25-12
(a) Oberon appears to have an old, inactive, heavily cratered surface. (b) Titania has a cratered surface crossed by deep valleys up to 100 km across—signs of an active past. (c) Umbriel is a dark, cratered world with no sign of faults or valleys. (d) This mosaic of images of Ariel shows it crossed by faults and broad valleys that appear to have been smoothed of earlier craters. *(a, b, and c, NASA; d, U.S. Geological Survey, Flagstaff, Arizona)*

heat of formation or to have been severely heated by radioactive decay. Tidal heating could have heated Miranda if its orbit was disturbed by a resonance with other satellites. Thus, we can understand how internal factors could have driven slow convection currents rising through Miranda's icy mantle to create its ovoids.

But external factors are also a possibility. Comets sweep through the solar system, and they must occasionally strike the planets and moons. The Jovian planets would retain no permanent scars of such impacts, as Earth's astronomers saw dramatically in the summer of 1994 when a comet struck Jupiter. The moons in the outer solar system, however, could be battered or even

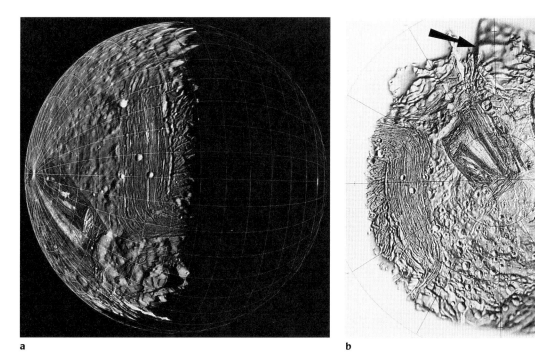

a   b

Figure 25-13
Miranda, the smallest of the five major moons of Uranus, is only 480 km in diameter, but its surface shows
signs of activity. This photomosaic of Voyager 2 images (a) reveals that it is marked by great oval systems of
grooves. In the map (b), the smallest features detected are about 1.5 km in diameter. The cliff marked by an
arrow near the top of the map is 20 km high. *(U.S. Geological Survey, Flagstaff, Arizona; map by P. Bridges)*

disrupted by such impacts. Although astronomers now
prefer internal heat to explain Miranda's geology, we
must accept that external factors such as major impacts
do occur. Thus, Miranda reminds us that modern astronomy recognizes the importance of catastrophic events
in the history of the solar system. (See Window on Science 20-1.) We will see more suggestions of catastrophic
events when we discuss the moons of Neptune.

A History of Uranus

The challenge of comparative astronomy is to tell the
story of a world, and Uranus may present the biggest
challenge of all the objects in our solar system. Not
only is it so distant that it is difficult to study, but it is
also peculiar in many ways.

Presumably, Uranus formed from the solar nebula,
as did the other Jovian planets, but it must have grown
more slowly, perhaps because the solar nebula was less
dense beyond Saturn. Also, objects orbit slowly in the
outer solar system, so protoplanets must have grown
slowly there. By the time the proto-Uranus had grown
massive enough to begin trapping hydrogen, the sun
had turned on, and the solar nebula was being blown
away. As one astronomer said, "By the time Uranus got
all dressed up, the party was over."

In fact, some theorists believe Uranus and Neptune
could not have grown to their present size in the slow-

moving orbits they now occupy. Highly complex mathematical models of planet formation in the solar nebula suggest that Uranus and Neptune might have formed
closer to the sun, in the neighborhood of Jupiter and
Saturn. Gravitational interactions among the growing
planets could have moved Uranus and Neptune outward to their present locations. This is a controversial
hypothesis, but it illustrates how uncertain the histories of the outer planets really are.

The highly inclined axis of Uranus may have originated late in its formation when it was struck by a planetesimal as large as Earth. That impact may have disturbed the interior of the world and caused it to lose
much of its heat. Uranus now radiates less than 10 percent more heat than it receives from the sun. There is
now just enough heat flowing outward to drive circulation in its slushy mantle and generate its magnetic field.

Impacts have been important in the history of the
moons and rings of Uranus. All of the satellites are
cratered, and some show signs of large impacts. Planetesimals and the nuclei of comets striking the moons
may create dust and debris that becomes trapped among
the orbits of the smaller moons to produce the narrow
rings. It seems very unlikely that the rings could have
lasted since the formation of the planet. When the final
history of Uranus is written, it will surely include some
dramatic impact events.

When Voyager 2 left Uranus, Earth's astronomers were left grappling with new planetary puzzles. Over the following years, some of the puzzles were explained, but in August 1989 Voyager 2 flew past Neptune, and the game began all over again.

25-2 Neptune

Uranus and Neptune are often discussed together. They are about the same size and density and are probably very similar planets in some ways. They do differ, however, in certain respects. Neptune has a peculiar ring and a peculiar satellite system. Even the discovery of Neptune was different from the discovery of Uranus.

The Discovery of Neptune

In 1843, the young English astronomer John Couch Adams (1819–1892) completed his degree in astronomy and immediately began the analysis of one of the great problems of 19th-century astronomy. William Herschel had discovered Uranus in 1781, but earlier astronomers had seen the planet as early as 1690 and had plotted it on charts as a star. When 19th-century astronomers tried to combine this data, they found that no planet obeying Newton's laws of motion and gravity could follow such an orbit.

Some astronomers had suggested that the gravitational attraction of an undiscovered planet was causing the discrepancies. Adams began with the variation in the positions of Uranus, never more than 2 minutes of arc, and by October 1845 had calculated the orbit and position of the undiscovered planet. He sent his prediction to the Astronomer Royal, Sir George Airy, who passed it on to an observer. Neither Airy nor his observer took the prediction very seriously, and they were very dilatory in making the search.

Meanwhile, the French astronomer Urbain Jean Leverrier (1811–1877) made the same calculations and sent his prediction to Johann Galle at the Berlin Observatory. Galle received Leverrier's prediction on the afternoon of September 23, 1846, and, after searching for 30 minutes that evening, found Neptune. It was only 2° from Adams's predicted position.

When the English announced Adams's work, the French suspected that he had plagiarized the calculations, and the controversy was bitter. In fact, both men had performed the calculations correctly. But they could have been beaten to the discovery had Galileo paid less attention to Jupiter and more attention to what he saw in the background. Recent studies of Galileo's notebooks show that he saw Neptune on December 24, 1612, and again on January 28, 1613, but he plotted it as a star in the background of drawings of Jupiter. It is interesting to speculate about the response of the Inquisition had Galileo proposed that a planet existed beyond Saturn. Unfortunately for history (perhaps fortunately for Galileo), he did not recognize Neptune as a planet, and its discovery had to wait another 234 years.

The discovery of Neptune was a triumph for Newtonian physics: The three laws of motion and the law of gravity had proved sufficient to predict the position and orbit of an unseen planet. Thus, the discovery of Neptune was fundamentally different from the discovery of Uranus—the existence of Neptune was predicted from basic laws. Discovery by prediction is important in science because confirmation of the prediction serves as confirmation of the basic laws.

The Atmosphere and Interior of Neptune

Little was known about Neptune before the Voyager 2 spacecraft swept past it in August 1989. Seen from Earth, Neptune is a tiny, blue-green dot never more than 2.3 seconds of arc in diameter. Astronomers knew it was almost four times the diameter of Earth and about 4 percent smaller in diameter than Uranus (Data File Ten). Spectra revealed that its blue-green color was caused by methane in its hydrogen-rich atmosphere. Methane absorbs red light, making Neptune look blue-green.

Neptune's density showed that it was a Jovian planet rich in hydrogen, but almost no detail is visible from Earth, so even its period of rotation was poorly known.

Voyager 2 passed only 4905 km above Neptune's cloud tops, closer than any spacecraft had ever come to one of the Jovian planets. The Voyager photos revealed that Neptune is marked by dramatic belt–zone circulation parallel to the planet's equator. Voyager 2 saw at least four cyclonic disturbances. The largest, dubbed the Great Dark Spot, looked similar to the Great Red Spot on Jupiter (Figure 25-14a). The Great Dark Spot was located in the southern hemisphere and rotated counterclockwise, with a period of about 16 days. It appeared to be caused by gas rising from the planet's interior. When the Hubble Space Telescope began imaging Neptune in 1994, the Great Dark Spot was gone, and new cloud features were seen appearing and disappearing in Neptune's atmosphere (Figure 25-14b). Thus, the cyclonic disturbances on Neptune are not nearly as long-lived as Jupiter's Great Red Spot.

The Voyager 2 images reveal other cloud features standing out against the deep blue of the methane-rich atmosphere. The white clouds are made up of crystals of frozen methane and range up to 50 km above the deeper layers. At such high altitudes, the clouds catch the sunlight and look bright (Figure 25-15a). Besides the Great Dark Spot, other, smaller spots were visible in the Voyager 2 photographs, and they, too, were presumably related to convection in the atmosphere (Figure 25-15b).

The Hubble Space Telescope has been able to monitor weather on Neptune, although the telescope's resolution is not as good as that of Voyager 2. Hubble photos show changing weather patterns on Neptune and a hint that this activity varies with solar activity. Further studies of clouds and spots on Neptune will help us understand this complex atmosphere and the way it is influenced by solar activity.

As on the other Jovian worlds, high winds circle Neptune parallel to its equator, but the winds blow at very high velocities and tend to blow backward—against the rotation of the planet. Why Neptune should have such high winds is not understood, and it is part of the larger problem of the belt–zone circulation. Now that we have seen at least traces of belts and zones on all four of the Jovian planets (if we may take the faint clouds on Uranus as belt–zone features), we can ask what drives this circulation. Because the belts and zones remain parallel to the planet's equator even when the planet rotates at a high inclination, as in the case of Uranus, it seems reasonable to believe that the atmospheric circulation is dominated by the rotation of the planet and not by solar heating.

The same observations that helped define the interior of Uranus can be applied to Neptune. The interior is believed to contain a heavy-element core, smaller than

Neptune **Data File Ten**

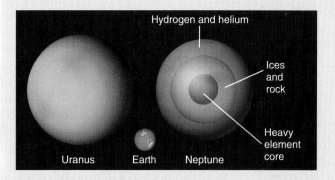

Neptune is a bit smaller than Uranus and four times larger in diameter than Earth. Its interior contains a rocky core with a deep mantle of slushy ices mixed with rock.

Average distance from the sun	30.0611 AU (44.971 × 10⁸ km)
Eccentricity of orbit	0.0100
Maximum distance from the sun	30.4 AU (45.4 × 10⁸ km)
Minimum distance from the sun	29.8 AU (44.52 × 10⁸ km)
Inclination of orbit to ecliptic	1°46′27″
Average orbital velocity	5.43 km/s
Orbital period	164.793 y (60,189 days)
Period of rotation	16ʰ3ᵐ
Inclination of equator to orbit	28°48′
Equatorial diameter	49,500 km (3.93 $D_\oplus$)
Mass	1.030 × 10²⁶ kg (17.23 $M_\oplus$)
Average density	1.66 g/cm³
Gravity	1.19 Earth gravities
Escape velocity	25 km/s (2.2 $V_\oplus$)
Temperature at cloud tops	−216°C (−357°F)
Albedo	0.35
Oblateness	0.027

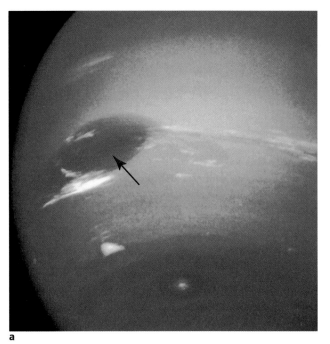 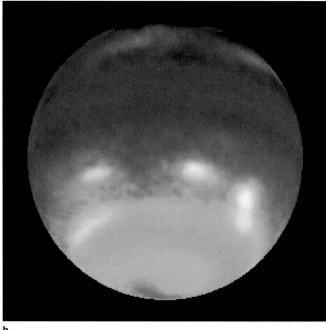

a   b

Figure 25-14

(a) A 1989 Voyager 2 image of Neptune revealed an anticyclonic disturbance about the size of Earth, which was dubbed the Great Dark Spot (arrow). High-altitude winds were deflected around the spot and formed high clouds of methane-ice crystals. *(NASA)* (b) A 1998 Hubble Space Telescope image of Neptune shows how cloud patterns have changed since Voyager 2 flew past in 1989. The Great Dark Spot has disappeared, but new spots have appeared. *(Lawrence Sromovsky, University of Wisconsin-Madison; and NASA)*

that in Jupiter, surrounded by a deep mantle of slushy ice mixed with heavier material such as rock. Neptune's magnetic field is a bit less than half as strong as Earth's and is tipped 47° from the axis of rotation. It is offset 55 percent of the way to the surface (Figure 25-5). As in the case of Uranus, the field is probably generated by the dynamo effect acting in the conducting fluid mantle.

Laboratory experiments on Earth have compressed methane and heated it to simulate conditions about 7000 km below the clouds of Neptune. The experiments show that the methane can decompose and release pure carbon, which then forms crystals of diamond. The sinking of these diamond crystals through the interior might release enough energy to help heat the planet. The same process may occur on Uranus. Needless to say, it seems very unlikely that humans will every find a way to mine the diamonds in Neptune, but the formation of this mineral warns us that we do not understand Neptune and Uranus well at all.

Neptune is a tantalizing world just big enough to be imaged by the Hubble Space Telescope but far enough away to make it difficult to study. The data from Voyager 2 revealed one detail invisible from Earth—rings.

The Rings of Neptune

No rings orbiting Neptune are directly visible from Earth, but every few years as it moves along its orbit

Neptune crosses in front of (occults) a distant star. Sometimes astronomers noticed the light of the star dim for a few seconds before or after the main occultation. This suggested that the planet had rings, but the stars did not always dim as expected. Astronomers proposed that Neptune had ring arcs rather than complete rings. The truth was revealed when Voyager 2 flew past in August 1989.

The nature of the rings and ring particles can be deduced from the Voyager 2 images. All of the rings lie within Neptune's Roche limit. Two narrow rings have widths of only a few dozen kilometers, with a fainter, broader ring in between. Another broad, faint ring with a width of 2500 km lies closer to the planet (Figure 25-16a). The rings are bright in forward-scattered light, so we can conclude that they contain significant amounts of dust. Because the rings are not very reflective, we can guess that the material is rocky rather than icy or that it is darkened by exposure to radiation in Neptune's magnetosphere. The narrow widths of the two brightest rings suggest they are confined by shepherd satellites, and observations have revealed that the small moons Despina and Galatea shepherd the innermost narrow ring. There are probably other, even smaller moons helping to confine the other rings.

Voyager 2's photographs revealed that Neptune's rings are continuous loops but that the outermost ring

a

a

b

b

Figure 25-15

(a) Cloud features seen on Neptune are caused by high, white clouds about 50 km above the darker layers. High-speed winds stretch the clouds into streaks parallel to the equator. (b) Spots on Neptune appear to be caused by circulations in the atmosphere. The spot shown at bottom center is D2, for Dark Spot 2. It was photographed by the Voyager 2 spacecraft in 1989. However, observations by the Hubble Space Telescope show that the spots seen by Voyager 2 were not permanent features like the Great Red Spot on Jupiter but rather are more like weather disturbances that come and go. *(NASA)*

Figure 25-16

(a) The brilliant disk of Neptune is hidden behind the central black bar in this Voyager 2 photo of the rings. Although the rings look uniform here, they actually contain short arcs where the ring particles are denser. *(NASA)* (b) This computer-enhanced image shows the arcs in Neptune's rings. A mathematical model reveals that the arcs are created by the gravitational influence of Neptune's moon Galatea. Arrows point to fainter arcs whose existence was predicted by the model. *(NASA image courtesy of Carolyn C. Porco)*

includes short arcs containing more particles than average (Figure 25-16b). These are the arcs observed from Earth. Theoretical calculations showed that the gravitational influence of the small moon Galatea could cause the bunching of particles to create arcs in the observed positions. In fact, the calculations predicted even smaller arcs in specific places along the ring, which were then located in photos. Nevertheless, detailed models show that Galatea can't keep the material in the ring from leaking away. There must be smaller moons near the ring that shepherd it and keep it narrow.

Compared to the brilliant rings of Saturn, the dark, tenuous rings of Neptune and Uranus are hardly impressive. Yet their origin may depend on a spectacular process. Because Neptune's rings, like those of Uranus, are too tenuous to have survived since the formation of the planet, we must suppose that some process replenishes them with fresh particles. Comets are quite common, and collisions of the nuclei of comets with small moons may occasionally supply fresh particles to the rings. Certainly, the rings and moons of Neptune are related.

The Moons of Neptune

Neptune has only two moons that are visible from Earth, but Voyager 2 discovered six more small moons. The two largest moons, Triton and Nereid, have been a puzzle for years because of their peculiar orbits.

Triton has a nearly circular orbit, but it travels backward—clockwise as seen from the north. This makes Triton the only large satellite in the solar system with a retrograde (backward) orbit. The other satellite visible from Earth, Nereid, goes in the right direction, but its orbit is highly elliptical and very large (Figure 25-17). Nereid takes 359.4 days to orbit Neptune once. Many astronomers have speculated that the orbits of the two moons are evidence of a violent event. An encounter with a massive planetesimal may have disturbed the moons, or Triton itself may have been captured into orbit during a close encounter with Neptune.

Voyager 2 found six more moons orbiting Neptune, but they all lie close to the planet among the rings. No new moons were found beyond the orbit of Triton. Some astronomers have suggested that Triton, in its retrograde orbit, would have consumed any moons near it.

The Voyager 2 photographs show that Triton is a highly complex moon. Although it is only 1360 km in radius (78 percent the radius of Earth's moon), it is so cold (34.5 K, or −398°F) that it can hold an atmosphere of nitrogen with some methane. Its air is 10^5 times less dense than Earth's atmosphere. Although a few wisps of haze are visible in the photographs, the atmosphere is transparent, and the surface is easily visible (Figure 25-18).

The surface of Triton is evidently composed of ices. Early analysis suggested that the ices were mostly water ice with methane and ammonia ice mixed in. However,

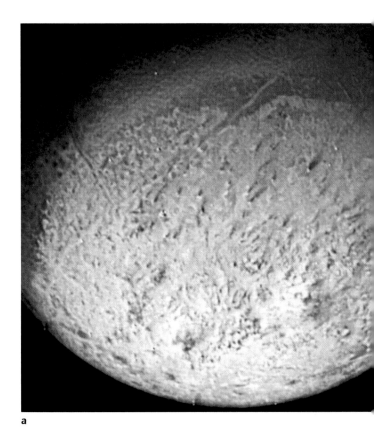

a

b

Figure 25-18
(a) Triton's south pole (bottom) had been in sunlight for 30 years when Voyager 2 flew past in 1989. The frozen nitrogen in the polar cap appears to be vaporizing, perhaps to refreeze in the darkness at the north pole. The dark smudges are produced when liquid nitrogen in the crust vaporizes and drives nitrogen geysers. Methane carried along in the nitrogen is turned to black compounds by sunlight and makes the black smudges. (b) Roughly round basins on Triton may be old impact basins flooded repeatedly by liquids from the interior. Notice the small number of craters on Triton, a clue that it is a partially active world. *(NASA)*

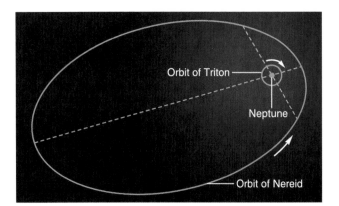

Figure 25-17
Nereid has a large, elliptical orbit around Neptune, but Triton follows a small, circular, retrograde orbit. At a distance of 30 AU from Earth, Triton is never more than 16 seconds of arc from the center of Neptune.

more recent studies reveal that the surface ice is dominated by frozen nitrogen with some methane, carbon monoxide, and carbon dioxide. This is consistent with the nitrogen-rich atmosphere. Some regions of the surface look dark and may be older terrain that has grown dark as sunlight has converted methane into dark organic compounds. The south pole of Triton had been turned toward the sun for 30 years when Voyager 2 flew past, and deposits of nitrogen frost in a large polar cap appeared to be vaporizing there and refreezing in the darkness of the north pole (Figure 25-18a). Thus, the cycle of nitrogen on Triton resembles in some ways the cycle of carbon dioxide on Mars.

The surface of Triton suggests that it has been active recently and may still be active. Few craters mark the surface of the icy world, and comparison with the rate at which we might expect meteorites to hit Triton suggests the average age of the surface is no more than 100 million years. Some process has erased older craters. Long linear features appear to be fractures in an icy crust, and some roughly round basins appear to have been flooded time after time by liquids from the interior (Figure 25-18b). Triton is much too cold for the liquid to be molten rock or even liquid water. Rather, the flooding may have been caused by water containing agents such as ammonia, which would lower the freezing point of the liquid. It isn't possible to say at present whether Triton is still active, but 100 million years isn't very long in the history of a world. It may still suffer periodic eruptions and flooding.

Another form of activity on Triton are the dark smudges visible in the bright nitrogen ices near the south pole. These may be caused by nitrogen ice beneath the crust. Warmed slightly by the sun, the nitrogen ice can change from one form of nitrogen ice to another and release heat that could vaporize some of the nitrogen. This nitrogen vapor could vent through the crust to form nitrogen gas geysers. Sunlight could convert some of the methane carried along with the nitrogen into dark organic material, which would fall to the surface and account for the dark smudges on the ice.

Active worlds must have a source of energy, so we must ask where Triton gets its energy. Triton is big enough to retain some heat from low-level radioactivity in its interior. That may be enough to melt exotic ices and cause flooding on the surface. The nitrogen geysers appear to be powered not from the interior but from sunlight. In fact, Triton is very efficient at absorbing sunlight. Its thin atmosphere allows sunlight to fall on the crust, which is composed of ices that are partially transparent to light. As the light penetrates into the ice and is gradually absorbed, it is converted to heat. The ices, however, are not very transparent to infrared radiation, and the heat is trapped in the crust. Thus, Triton may be heated, in part, by an icy form of the greenhouse effect.

The History of Neptune

Can we tell the story of Neptune? In some ways it seems to be a simple world, but its magnetic field is peculiar, and its moons and rings demand careful attention.

Neptune is much like Uranus, and we can assume that it formed from the solar nebula much as did Uranus. It is a mostly liquid world with a dense core. Its hazy atmosphere, marked by changing cloud patterns, hides a mantle of partially frozen water where astronomers suspect the dynamo effect generates its off-center magnetic field.

Neptune's satellite system suggests a peculiar history. Triton, the largest moon, revolves around Neptune in a retrograde orbit, while Nereid's long-period orbit is highly elliptical. These orbital oddities suggest that the satellite system may have been disturbed by an encounter with some other object, such as a large planetesimal or the capture of Triton. We have seen evidence in other satellite systems of impacts with large objects, so such an event is not unreasonable.

Six smaller moons orbit Neptune, but they lie near its ring system. Because the rings are bright in forward scattering, we can conclude that they are rich in dust, and the shepherding of small satellites could confine their width and produce the observed arcs. Such delicate rings could not have survived since the origin of the planets, so we must suppose that the rings are occasionally supplied with fresh dust particles generated by collisions between comets and Neptune's moons. Once again, the evidence we see of major impacts in the solar system's history assures us that such impacts do occur.

R E V I E W Critical Inquiry

Why is Neptune blue but its clouds white?

When we look *at* something, we really turn our eyes toward it and receive light *from* the object. The light we receive from Neptune is sunlight that is reflected from various layers of Neptune and journeys to our eyes. Because sunlight contains a distribution of photons of all visible wavelengths, it looks white to us, but sunlight entering Neptune's atmosphere must pass through hydrogen gas that contains a small amount of methane. While hydrogen is transparent, methane is a good absorber of longer wavelengths, so red photons are more likely to be absorbed than blue photons. Once the light is scattered from deeper layers, it must run this methane gauntlet again to emerge from the atmosphere, and again red photons are more likely to be absorbed. The light that finally emerges from Neptune and eventually reaches our telescopes is poor in longer wavelengths and thus looks blue.

The methane-ice-crystal clouds lie at high altitudes, so sunlight does not have to penetrate very far into

Neptune's atmosphere to reflect off the clouds, and thus it loses many fewer of its red photons. The clouds look white.

This discussion shows how a careful, step-by-step analysis of a natural process can help us better understand how nature works. For example, explain where the energy comes from to keep Triton active.

Neptune was the last planet visited by Voyager 2, which is now journeying out of our solar system. It will not pass near any other worlds unless, millions of years from now, it enters some distant solar system. Nevertheless, we have one more planet to analyze.

25-3 Pluto

Pluto, the farthest outpost of our solar system, is a small, icy world. Discovered through a combination of ingenuity, perseverance, and luck, it is now known to have a satellite.

The Discovery of Pluto

Percival Lowell (1855–1916) was fascinated with the idea that an intelligent race built the canals he thought he could see on Mars (Chapter 23). Lowell founded Lowell Observatory primarily for the study of Mars. Later, to improve the reputation of his observatory, he began to search for a planet beyond Neptune.

Lowell used the same method that Adams and Leverrier had used to predict the position of Neptune. Working from the observed irregularities in the motion of Neptune, Lowell predicted the location of an undiscovered planet of about 7 Earth masses, which, he concluded, would look like a 13th-magnitude stellar image in eastern Taurus. Lowell searched for the planet photographically until his death in 1916.

In the late 1920s, a 22-year-old amateur astronomer, Clyde Tombaugh, began using a homemade 9-in. telescope to sketch Jupiter and Mars from his family's wheat farm in western Kansas. He sent his drawings to Lowell Observatory, and the observatory director hired him (without an interview) to resume Lowell's search for the undiscovered planet. The young Tombaugh bought a one-way train ticket for Flagstaff not knowing what his new job would be like.

The observatory director set Tombaugh to work photographing the sky along the ecliptic around the predicted position of the planet. The search technique was a classic method in astronomy. Tombaugh obtained pairs of 14×17-inch glass plates exposed two to three days apart. To search a pair of plates, he mounted them in a blink comparator, a machine that allowed him to look through a microscope at a small spot on one plate and then at the flip of a lever see the same spot on the other plate. As he blinked back and forth, the star im-

a

b

Figure 25-19
Pluto is small and far away, so its image is indistinguishable from that of a star on most photographs. Clyde Tombaugh discovered the planet in 1930 by looking for an object that moved relative to the stars on a pair of photographs taken a few days apart. *(Lowell Observatory photographs)*

ages did not move, but a planet would have moved along its orbit during the two to three days that elapsed before the second plate was exposed. So Tombaugh searched the giant plates, star image by star image, looking for an image that moved. A single pair of plates could contain 400,000 star images. He searched pair after pair and found nothing.

The observatory director turned to other projects, and Tombaugh, working alone, expanded his search to cover the entire ecliptic. For almost a year, Tombaugh exposed plates by night and blinked plates by day. Then on February 18, 1930, nearly a year after he had left Kansas, a quarter of the way through a pair of plates, he found a 15th magnitude image that moved (Figure 25-19). He later remembered, "Oh, I thought, I had better look at my watch; this could be a historic moment. It was within about 2 minutes of 4 PM [MST]."

The discovery was announced on March 13, the 149th anniversary of the discovery of Uranus and the 75th anniversary of the birth of Percival Lowell. The planet was named Pluto after the god of the underworld and, in a way, after Lowell, because the first two letters in Pluto are the initials of Percival Lowell.

The discovery of Pluto seemed a triumph of discovery by prediction, but Tombaugh sensed something was wrong from the first moment he saw the image. It was moving in the right direction by the right amount, but it was 2.5 magnitudes too faint. Clearly, Pluto was not the 7-Earth-mass planet that Lowell had predicted. The faint image implied that Pluto was a small planet with a mass too low to seriously alter the motion of Neptune.

Later analysis has shown that the variations in the motion of Neptune, which Lowell used to predict the location of Pluto, were random errors of observation and could not have led to a trustworthy prediction. The discovery of the new planet only 6° from Lowell's predicted position was apparently an accident, which proves that if you search long enough and know what to expect, you are likely to find something (Window on Science 25-1).

Pluto as a Planet

Pluto is very difficult to observe from Earth. Only a bit larger than 0.1 second of arc in diameter, the little planet is only 65 percent the diameter of Earth's moon and shows little surface detail. The best photos by the Hubble Space Telescope reveal large areas of light and dark terrain (Data File Eleven), promising that Pluto will reveal itself to be an interesting world when spacecraft finally visit it.

Most planetary orbits in our solar system are nearly circular; but Pluto's is quite elliptical (see Figure 1-7). On the average, it is the most distant planet, but it can

Pluto Data File Eleven

The Hubble Space Telescope is just able to reveal bright and dark regions on Pluto. Some of the bright regions may be deposits of bright methane ice. *(Alan Stern, Southwest Research Institute, Marc Buie, Lowell Observatory, NASA, ESA)*

Average distance from the sun	39.44 AU (59.00×10^8 km)
Eccentricity of orbit	0.2484
Maximum distance from the sun	49.24 AU (73.66×10^8 km)
Minimum distance from the sun	29.64 AU (44.34×10^8 km)
Inclination of orbit to ecliptic	17°9′3″
Average orbital velocity	4.73 km/s
Orbital period	247.7 y (90,465 days)
Period of rotation	$6^{d}9^{h}21^{m}$
Inclination of equator to orbit	119.6°
Equatorial diameter	2370 km (0.19 $D_{\oplus}$)
Mass	1.2×10^{22} kg (0.002 $M_{\oplus}$)
Average density	2.0 g/cm^3
Surface gravity	0.06 Earth gravities
Escape velocity	1.2 km/s (0.11 $V_{\oplus}$)
Surface temperature	−230°C (−382°F)
Albedo	0.5

In 1928, Alexander Fleming noticed that bacteria in a culture dish were avoiding a spot of mold. He went on to discover penicillin. In 1895, Conrad Roentgen noticed a fluorescent screen glowing in his laboratory when he experimented with other equipment. He discovered X rays. In 1896, Henri Becquerel happened to place a mineral containing uranium on a photographic plate safely wrapped in black paper. The plate was fogged, and Becquerel discovered natural radioactivity. Like many discoveries in science, these seem to be accidental, but as we have seen in the case of the discovery of Uranus, Neptune, and Pluto, "accidental" doesn't quite describe what happened.

The most important discoveries in science are those that totally change the way we think about nature, and it is very unlikely that anyone would predict such discoveries. For the most part, scientists work within a paradigm (Window on Science 4-1), a set of models and expectations about nature, and it is very difficult to imagine natural events that lie beyond that paradigm. Ptolemy, for example, could not have imagined a universe filled with galaxies because they lie totally beyond his geocentric paradigm. That means that the most important discoveries in science are almost bound to arise unexpectedly.

An unexpected discovery, however, is not the same as an accidental discovery. Fleming discovered penicillin in his culture dish not because he was the first to see it, but because he had studied bacterial growth for many years, so when he saw what many others must have seen, he recognized it as important. Roentgen recognized the importance of the glowing screen in his lab, and Becquerel didn't simply discard the fogged photographic plate. Long years of experience prepared them to recognize the significance of what they saw.

A study of discoveries made with new telescopes from Galileo to the present re-vealed that each time astronomers built a new telescope that significantly surpassed existing telescopes, they discovered new and wonderful things, such as the moons of Jupiter, binary stars, spiral galaxies, and quasars. But nearly all the discoveries made with these telescopes were unpredicted.

While it is possible to predict the presence of Neptune from the orbital motion of Uranus, this is not typical of scientific discoveries. Most important scientific discoveries lie outside the current paradigm and so must be made by scientists who are searching for something else or studying some other phenomenon. That means that scientists pursuing true basic research are rarely able to explain the potential value of their work, just as astronomers can't predict what they will discover with a new telescope. That's why science is so exciting.

a

b

Figure 25-20
(a) A high-quality ground-based photo shows Pluto and its moon, Charon, badly blurred by seeing. *(NASA)*
(b) The Hubble Space Telescope image clearly separates the planet and its moon and allows more accurate measurements of the position of the moon. *(R. Albrecht, ESA/ESO Space Telescope European Coordinating Facility, NASA)*

come closer to the sun than Neptune. In fact, from January 21, 1979, to March 14, 1999, Pluto was closer to the sun than Neptune. The planets will never collide, however, because Pluto's orbit is inclined 17°.

Orbiting so far from the sun, Pluto is cold enough to freeze most compounds we think of as gases, and spectroscopic observations have found evidence of solid nitrogen ice on its surface with traces of frozen methane, carbon monoxide, nitrogen, and ethane. The daytime temperature of about 50 K is enough to vaporize some of the nitrogen and carbon monoxide and a little of the methane to form a thin atmosphere around Pluto. This atmosphere was detected in 1988 when Pluto passed in front of (occulted) a distant star, and the starlight faded gradually rather than winking out suddenly.

Pluto's only known moon was discovered on photographs in 1978. It is very faint, and even in photographs made with the Hubble Space Telescope it is impossible to measure the diameter of the moon directly (Figure 25-20). The moon was named Charon after the mythological ferryman who transports souls across the river Styx into the underworld. Charon follows a nearly circular but highly inclined orbit around Pluto (Figure 25-21), and the planet and the moon are tidally locked to each other, forever facing each other as they circle their common center of mass.

The discovery of Charon was important because astronomers could use it to find the mass of Pluto. Charon orbits 19,640 km from Pluto with an orbital period of 6.387 days. Kepler's third law tells us the mass of the system, about 6.5×10^{-9} solar masses or about 0.2 percent Earth's mass. Most of this mass is in Pluto, which seems to be about 12 times the mass of Charon.

From 1985 to 1990, Earth passed through the plane of Charon's orbit, something that only happens every 124 years, and astronomers were able to observe **transits** as Charon moved across the disk of Pluto and occultations as it disappeared behind Pluto (Figure 25-22). Telescopes on Earth could not resolve Pluto and Charon during these events, but astronomers could observe variations in the total brightness as if the planet and its moon were an eclipsing binary star (Chapter 10). From the changes in light intensity, astronomers were able to calculate the diameter of Pluto and Charon. Pluto is about 2370 km in diameter (68 percent the diameter of Earth's moon), and Charon is about 1152 km in diameter, roughly half the size of Pluto. The accurate diameters of Pluto and Charon were critical data for astronomers, enabling them to calculate the densities. The density of Pluto is about 2 g/cm³, and the density of Charon is just a bit less. Pluto and Charon must contain about 35 percent ice and 65 percent rock.

Spectra of Pluto reveal that its surface is composed of frozen compounds such as methane, carbon monoxide, and nitrogen. These compounds are volatile in that they vaporize easily. Spectra of Charon show that the

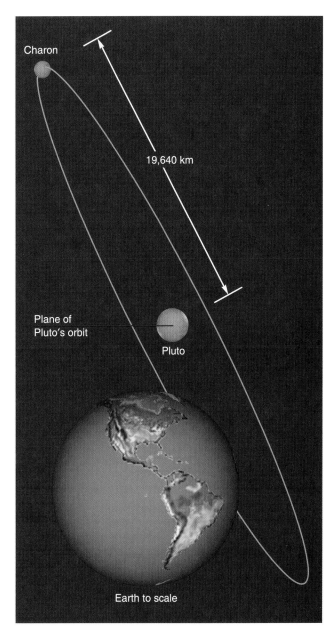

Figure 25-21
The nearly circular orbit of Charon is only a few times larger than the diameter of Earth. It is seen here nearly edge-on, and consequently it looks elliptical in this diagram. The orbit and the equator of Pluto are inclined 119.6° to the plane of Pluto's orbit around the sun.

small moon lacks these ices and has a surface that is mostly frozen water. Perhaps it has lost its more volatile compounds because of its lower escape velocity. Water ice at Charon's surface temperature is no more volatile than is a piece of rock on Earth.

Both Pluto and Charon go through dramatic seasons as they circle the sun in their highly inclined orbit. Seasons on Pluto and Charon are much like those on Uranus. This may cause changes in Pluto's atmosphere

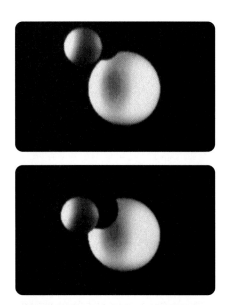

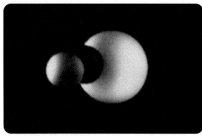

Figure 25-22
This computer simulation shows a partial transit as Charon, traveling from north to south, crosses its disk of Pluto. From Earth, this kind of detail was invisible, but the timing of changes in total brightness allowed astronomers to calculate the diameters of Pluto and Charon. *(Marc Buie)*

as it grows warmer when it is closest to the sun as it was in the late 1980s, and then freezes as it draws away.

We strain our imaginations when we try to visualize the surface of Pluto. Old surfaces should be cratered, so we might imagine an icy, cratered world. The tidal lock between Pluto and Charon suggests an age of tidal heating that may have driven geological activity and left traces on the surfaces of these worlds. However, the seasonal vaporization and refreezing of ices may have erased or covered much of the older surface with ices. Thus Pluto and Charon may be worlds unlike any we have seen before. Yet the biggest puzzle may be not their geology, but their origin.

The Origin of Pluto and Charon

Is Pluto really a planet? It is so dramatically different from its Jovian neighbors that it doesn't seem to fit into the outer solar system, and a few people have argued that it isn't a planet at all. In fact, this depends almost entirely on our arbitrary definition of the word "planet." Although the news media find this debate interesting, astronomers are much more interested in the history of Pluto than in the definition of a word. A better question is: "How did Pluto form?"

An old theory proposed that Pluto was an escaped moon of Neptune. Pluto is much like Neptune's moon Triton. It is roughly the same density, and it is only 15 percent smaller; it could easily fit in with the icy moons of the Jovian planets. The disturbed orbits of Triton and Nereid could be the result of a complex interaction with a passing massive planetesimal in which Pluto might have been ejected from an orbit around Neptune. This theory has been mostly abandoned, because such an interaction seems unlikely. A newer hypothesis seems more probable.

Modern astronomers have begun thinking of Pluto as the largest of a family of small icy bodies that were once common in the outer solar system. Searches for other icy worlds beyond the orbit of Neptune have turned up over 100 with hints that there are a vast number of more distant bodies. (We will discuss this belt of icy planetesimals in detail in Chapter 26.) Pluto is locked in a 3:2 orbital resonance with Neptune; that is, Neptune orbits the sun three times while Pluto orbits twice. This may give Pluto's orbital motion added stability. Indeed, many of the smaller icy bodies found beyond the orbit of Neptune are caught in resonances with Neptune. Thus, we can think of Pluto as the largest of the icy planetesimals that never became part of a larger planet.

The moon Charon is presumably another of these icy worlds. A collision between Charon and Pluto could have tilted the axis of rotation of Pluto and trapped Charon in an elliptical orbit around Pluto. Tidal forces between Pluto and Charon could then have forced the orbit of Charon to become nearly circular as it is today. If that happened, then we might hope to see dramatic evidence of tidal heating in Pluto and in Charon when spacecraft finally reach Pluto.

At present no spacecraft is heading for Pluto. Plans have been proposed, but no funds have been set aside. It may be a long time before we see the surface of these cold worlds, but our study of comparative planetology helps our imagination go where our cameras cannot.

REVIEW Critical Inquiry

What evidence do we have that cataclysmic impacts have occurred in our solar system?

We would expect objects that formed together from the solar nebula to rotate and revolve in the plane of the solar system, so the high inclination of Pluto and the orbit of Charon suggest a past collision or, at least, a close gravitational interaction with a passing body. And the Pluto–Charon system is not the only evidence for major impacts in the past. The peculiar orbits of Neptune's moons Triton and Nereid, the fractured and cratered state of the Jovian satellites, and the peculiar rotation of Uranus all hint that impacts and encounters with large planetesimals or the nuclei of comets have been important in the history of these worlds. Furthermore, the existence of planetary rings suggests that impacts have scattered small particles and replenished the ring systems. Even in the inner solar system, the backward rotation of Venus, the density of Mercury, and the formation of Earth's moon are seen as possible consequences of major impacts when the solar system was young.

Certainly, we do not imagine that planets bounce around like billiard balls, but we must recognize that the planets are not totally isolated now and may have been victims (or beneficiaries) of large impacts as they formed long ago. For example, how do the size and composition of Pluto and Charon, compared with the moons of the outer Jovian worlds, suggest an unexpected population of icy worlds in the outer solar system?

Like clues at the scene of a crime, the evidence we find in the outer solar system alerts us to the reality that all of the planets in our solar system have been and still are sitting ducks for impacts by the smaller objects that hurtle along their orbits around the sun. What are these objects? What can they tell us about the origin and evolution of our solar system? We explore these questions in the next chapter.

Summary

Although Uranus was discovered unexpectedly in 1781, we should probably not refer to its discovery as accidental. In 1846, the position of Neptune was predicted from the orbital motion of Uranus, and Pluto was discovered in 1930 based on predictions derived from the motion of Neptune.

Uranus and Neptune are commonly classified as Jovian planets, but their interiors are different from the interiors of Jupiter and Saturn. The density, rotation periods, and oblateness of Uranus and Neptune tell us that they contain small cores of heavy materials, such as rock and metal, and deep mantles of partly frozen ices mixed with denser material, such as rock. Their atmospheres are rich in hydrogen and helium, and absorption of red light by small amounts of methane causes both planets to have a blue color.

The axis of rotation of Uranus is tipped $97.9°$ from the perpendicular to its orbit, and its south pole was facing the sun when Voyager 2 flew past. The spacecraft discovered that belts and zones lie deep in the hazy atmosphere. Thus, the rotation of the planet, rather than solar energy, dominates atmospheric circulation.

The rings of Uranus are very narrow and dark. Two shepherd satellites were discovered orbiting on either side of the ε ring, and other shepherd satellites are believed to bound the other rings. These satellites and the larger satellites appear to anchor the rings and prevent them from spreading and dissipating. The dark color of the ring particles appears to come from the methane-rich ices darkened by the radiation belts around the planet. The surfaces of the moons are also very dark.

The rings of Uranus are not very bright in forward scattering and thus must be very poor in small particles. Some dust was detected, but it seems to be falling into the planet due to the resistance of low-density hydrogen gas leaking out of the upper atmosphere.

Voyager 2 discovered ten small satellites orbiting Uranus, and more have been found from Earth-based telescopes. All of these moons are small and dark. The five largest moons show signs of impact cratering and some surface activity in the past. The innermost moon, Miranda, is marked by large ovals called ovoids. Astronomers believe that Miranda was heated by tides when it was young and that great convection currents stirred its partially frozen mantle and formed the ovoids.

Weather features on Uranus and Neptune do change. Voyager 2 found Uranus almost cloudless, and on Neptune it found the Great Dark Spot, an anticyclonic storm much like the Great Red Spot on Jupiter. In the years since, the Hubble Space Telescope has observed the appearance of methane clouds on Uranus and the disappearance of the Great Dark Spot and the appearance of new weather patterns on Neptune.

Like Uranus, Neptune has a highly tilted off-center magnetic field presumably generated in a circulating mantle of slushy ices. Also like Uranus, Neptune has narrow, dim rings shepherded by small moons.

Triton, Neptune's largest moon, is an icy world with a thin atmosphere of nitrogen and methane. A polar cap of methane ice was visible as Voyager 2 flew past, but few impact craters mark the surface, hinting that the moon has been active. Nitrogen geysers detected in the south polar cap appear to be powered by heat from the sun.

Pluto and its moon Charon are rocky worlds with deep ice mantles. Pluto is large enough to keep a thin atmosphere of nitrogen, carbon monoxide, and methane. These two icy worlds may be among the largest of the icy planetesimals that filled the outer solar system during planet formation.

New Terms

occultation

ovoid

transit

Review Questions

1. Describe the location of the equinoxes and solstices in the Uranian sky. What are seasons like on Uranus?

2. Why is belt–zone circulation difficult to detect on Uranus?

3. Discuss the origin of the rings of Uranus and Neptune. Cite evidence to support hypotheses.

4. How do the magnetic fields of Uranus and Neptune suggest that the mantles inside those planets are fluid?

5. If Neptune had no satellites at all, would you expect it to have rings? Why or why not?

6. Why might the surface brightness of ring particles and small moons orbiting Uranus and Neptune depend on whether those planets have magnetic fields?

7. Both Uranus and Neptune have a blue-green tint when observed through a telescope. What does that tell us about their composition?

8. How can small worlds like Triton and Pluto have atmospheres when a larger world such as Ganymede has none?

9. Why do we suspect that Triton has had an active past? What source of energy could power such activity?

10. If you visited Pluto and found Charon a full moon directly overhead, where would Charon be in the sky when it was new? when it was first quarter?

11. What evidence do we have that Pluto and Charon are made of mixtures of rock and ice?

Discussion Questions

1. Why might it be unfair to describe William Herschel's discovery of Uranus as accidental? Why might it be unfair to describe the discovery of the rings of Uranus as accidental?

2. Suggest a single phenomenon that could explain the inclination of the rotation axis of Uranus, the orbits of Neptune's satellites, and the existence of Pluto's moon.

Problems

1. What is the maximum angular diameter of Uranus as seen from Earth? of Neptune? of Pluto? (*Hint:* Use the small-angle formula.)

2. One way to recognize a distant planet is by its motion along its orbit. If Uranus circles the sun in 84 years, how many seconds of arc will it move in 24 hours? (This does not include the motion of Earth. Assume a circular orbit for Uranus.)

3. What is the orbital velocity of Miranda around Uranus?

4. What is the escape velocity from the surface of Miranda? (*Hints:* Radius = 242 km. Assume that the density is 2 g/cm^3. See Chapter 5.)

5. The magnetosphere of Uranus rotates with the planetary interior in 17.24 hours. What is the velocity of the outer portion of the magnetic field just beyond the orbit of Oberon? (*Hint:* The circumference of a circle is $2\pi r$.)

6. If the ε ring is 60 km wide and the orbital velocity of Uranus is 6.81 km/s, how long a blink should we expect to see when the ring crosses in front of a star? Is this consistent with Figure 25-7a?

7. What is the escape velocity from the surface of an icy moon with a diameter of 20 km? (*Hints:* The density of ice is 1 g/cm^3. The volume of a sphere is $\frac{4}{3}\pi r^3$. See Chapter 5.)

8. What is the difference in the orbital velocities of the two shepherd satellites Cordelia and Ophelia? (*Hints:* Orbital radii = 49,800 km and 53,800 km. See Chapter 5.)

9. Repeat Problem 2 for Pluto. That is, ignoring the motion of Earth, how far across the sky would Pluto move in 24 hours? (Assume a circular orbit for Pluto.)

10. Given the size of Triton's orbit (R = 355,000 km) and its orbital period (P = 5.877 days), calculate the mass of Neptune. (*Hint:* See Chapter 5.)

Critical Inquiries for the Web

1. Who was William Herschel? What can you find out about his life after he discovered Uranus?

2. What is weather on Uranus like now? Can you locate recent photos of cloud patterns there?

3. How many moons does Uranus have? More are discovered every year, it seems, so search for the latest discoveries.

4. What happened to John Couch Adams and Urbain Jean Leverrier after the discovery of Neptune? Did they make other important discoveries? What is the Adams Society? Can you find out what Leverrier did in the revolution of 1848?

4. Repeat Inquiries 2 and 3 for Neptune. Remember that finding new moons orbiting Neptune is difficult because it is further away.

5. Who was Clyde Tombaugh? What can you find out about his life after he discovered Pluto?

 Go to the Brooks/Cole Astronomy Resource Center (www. brookscole.com/astronomy) for critical thinking exercises, articles, and additional readings from InfoTrac College Edition, Brooks/Cole's online student library.

CHAPTER TWENTY-SIX

Meteorites, Asteroids, and Comets

When they shall cry

"PEACE, PEACE" then cometh

sudden destruction!

COMET'S CHAOS?—What

Terrible events will the Comet bring?

From a religious pamphlet predicting the end of the world because of the appearance of Comet Kohoutek, 1973

Guidepost

In Chapter 20, we began our study of planetary astronomy by asking how our solar system formed. In the five chapters that followed, we surveyed the planets, but we gained only limited insight into the origin of the solar system. The planets are big, and they have evolved as heat has flowed out of their interiors. In this chapter, we have our best look at unevolved matter left over from the solar nebula. These small bodies are, in fact, the last re-mains of the nebula that gave birth to the planets.

This chapter is unique in that it covers small bodies. In past chapters, we have used the principles of com-parative planetology to study large objects—the planets. In this chapter, we see that some of these principles apply to smaller bodies, but we also see that we need some new tools in order to think about the tiniest worlds in the solar system.

In 1910, Comet Halley was spectacular. On the night of May 19, Earth actually passed through the tail of the comet—and millions of people panicked. The spectrographic discovery of cyanide gas in the tails of comets led many to believe that life on Earth would end. Householders in Chicago stuffed rags around doors and windows to keep out the gas, and bottled oxygen was sold out. Con artists in Texas sold comet pills and inhalers to ward off the noxious fumes. An Oklahoma newspaper reported (in what was apparently a hoax) that a religious sect tried to sacrifice a virgin to the comet.

Throughout history, bright comets have been seen as portents of doom. Even in our own time, the appearance of bright comets has generated predictions of the end of the world. Comet Kohoutek in 1973 and Comet Halley in 1986 caused concern among the superstitious. More recently Comet Hyakutake in 1996 (Figure 26-1) and Comet Hale-Bopp produced spectacular sights in the sky and superstitious concern on Earth. A bright comet moving slowly through the night sky is so out of the ordinary that we should not be surprised if it generates some instinctive alarm.

In fact, comets are graceful and beautiful visitors to our skies. Astronomers think of comets as messengers from the age of planet building. By studying comets, we learn about the conditions in the solar nebula from which planets formed. But comets are only the icy remains of the solar nebula; the asteroids are the rocky debris left over from planet building. Because we cannot easily visit comets and asteroids, we begin our discussion with the fragments of those bodies that fall into our atmosphere—the meteorites.

26-1 Meteorites

In the afternoon of November 30, 1954, Mrs. E. Hulitt Hodges of Sylacauga, Alabama, lay napping on her living room couch. An explosion and a sharp pain jolted her awake, and she found that a meteorite had smashed through the ceiling and bruised her left leg. Mrs. Hodges is the only person known to have been injured by a meteorite. By the way, Mrs. Hodges lived right across the street from the Comet Drive-In Theater.

Figure 26-1
Comet Hyakutake swept through the inner solar system in 1996 and was dramatic in the northern sky. Seen here from Kitt Peak National Observatory, the comet passed close to the north celestial pole (behind the observatory dome in this photo). Notice the Big Dipper below and to the left of the head of the comet. *(Courtesy Tod Lauer)*

Meteorite impacts on homes are not common, but statistical calculations show that a meteorite should damage a building somewhere in the world about once every 16 months. About two meteorites large enough to produce visible impacts strike Earth each day, but most meteors are small particles ranging from a few centimeters down to microscopic dust. Earth gains about 40,000 tons of mass per year from meteorites of all sizes.

Recall from Chapter 20 that astronomers distinguish between the words *meteoroid, meteor,* and *meteorite.* A small body in space is a meteoroid, but once it begins to vaporize in Earth's atmosphere it is called a meteor. If it survives to strike the ground, it is called a meteorite.

We have two questions to consider about meteorites: Where in the solar system do these objects originate? What can meteorites tell us about the origin of the solar system? To answer these questions, we must consider the orbits of meteoroids and the minerals found in meteorites.

Meteoroid Orbits

Meteoroids are much too small to be visible through even the largest telescope. We see them only when they fall into Earth's atmosphere at 10 to 30 km/s, roughly 30 times faster than a rifle bullet, and are heated to glowing by friction with the air. The average meteoroid is about the mass of a paper clip and vaporizes at an altitude of about 80 km above Earth's surface, but it produces a bright streak of fire that we see as a meteor. The trail of a meteor points back along the path of the meteoroid, so if we could study the direction and speed of meteors, we could get clues to the orbits of the meteoroids.

One way to backtrack meteor trails is to observe meteor showers. On any clear night, you can see 3 to 15 meteors an hour, but on some nights you can see a shower of dozens of meteors an hour that are obviously related to each other. To confirm this, try observing a

a   b

Figure 26-2
(a) Meteors in a meteor shower enter Earth's atmosphere along parallel paths, but perspective makes them appear to diverge from a radiant point. (b) Similarly, parallel railroad tracks appear to diverge from a point on the horizon.

meteor shower. Pick a shower from Table 26-1, and on the appropriate night stretch out in a lawn chair and watch a large area of the sky. When you see a meteor, sketch its path on the appropriate sky chart from the back of this book. In just an hour or so you will discover that most of the meteors you see seem to come from a single area of the sky, the **radiant** of the shower (Figure 26-2a). In fact, meteor showers are named after the constellation from which they seem to radiate. Thus the Perseids are a meteor shower that radiates from the constellation Perseus.

Observing a meteor shower is relaxing, and the show is natural fireworks, but it is even more exciting when we understand what a meteor shower tells us (Window on Science 26-1). The significant fact is that the meteors in a shower appear to come from a single point in the sky, the radiant. That must mean the meteoroids were traveling through space along parallel paths. When they encounter Earth and are vaporized in the upper atmosphere, we see their fiery tracks in perspective; they appear to come from a single radiant point, just as railroad tracks seem to come from a single point on the horizon (Figure 26-2b).

Studies of meteor-shower radiants reveal that these meteors are produced by bits of matter orbiting the sun along the paths of comets. The vaporizing head of the comet releases bits of debris that become spread along the entire orbit (Figure 26-3). When Earth passes through this stream of material, we see a meteor shower. In some cases the comet has wasted away and is no longer visible, but in other cases the comet is still prominent though somewhere else along its orbit. For example,

Table 26-1 Meteor Showers

Shower	Dates	Hourly Rate	Radiant* R. A.	Radiant* Dec.	Associated Comet
Quadrantids	Jan. 2–4	30	15^h24^m	50°	
Lyrids	April 20–22	8	18^h4^m	33°	1861 I
η Aquarids	May 2–7	10	22^h24^m	0°	Halley?
δ Aquarids	July 26–31	15	22^h36^m	−10°	
Perseids	Aug. 10–14	40	3^h4^m	58°	1982 III
Orionids	Oct. 18–23	15	6^h20^m	15°	Halley?
Taurids	Nov. 1–7	8	3^h40^m	17°	Encke
Leonids	Nov. 14–19	6	10^h12^m	22°	1866 I Temp
Geminids	Dec. 10–13	50	7^h28^m	32°	

*R. A. and Dec. give the celestial coordinates (right ascension and declination) of the radiant of each shower.

We can enjoy a meteor shower as Mother Nature's fireworks, but our enjoyment is much greater once we begin to understand what causes meteors and why meteors in a shower follow a pattern. When we know that meteor showers help us understand the origin of our world, the evening display of shooting stars is even more exciting. Science typically increases our enjoyment of the natural world by revealing the significance of things we might otherwise enjoy only in a casual way.

Everyone likes flowers, for example. We enjoy them in a casual way, admiring their colors and fragrances, their graceful petals and massed blooms. But botanists know that evolution has carefully designed flowers to attract insects and spread pollen. The bright colors attract insects, and the shapes of the flowers provide little runways so the insect will find it easy to land. Some color patterns even guide the insect in like landing lights at an airport, and many flowers, such as orchids and snapdragons, force the insect to crawl inside in just the right way to exchange pollen and fertilize the flower. The nectar is nothing more than bug bait. Once we begin to understand what flowers are for, a visit to a garden becomes not only an adventure of color and fragrance, but also an adventure in understanding the beauty of design and function, an adventure in meaning as well as an adventure of the senses.

And our understanding of a natural phenomenon helps us understand and enjoy related phenomena. For example, some flowers attract flies for pollination, and such blossoms smell like rotting meat. A few flowers depend on bats, and they open their blossoms at night. Many plants, such as pine trees, depend on the wind to spread their pollen, and those plants do not have colorful flowers at all; but flowers that depend on hummingbirds have long trumpet-shaped blossoms that just fit the hummingbirds' beaks.

The more science tells us about nature, the more enjoyable the natural world is. We can enjoy a meteor shower as nothing more than a visual treat, but the more we know about it, the more interesting and exciting it becomes. The natural world is filled with meaning, and science, as a way of discovering and understanding that meaning, gives us new opportunities to enjoy the world around us.

in May Earth comes near the orbit of Comet Halley, and we see the Eta Aquarids shower. Around October 20 Earth passes near the other side of the orbit of Comet Halley, and we see the Orionids shower.

Our pleasant observations of meteor showers tell us that at least some meteoroids come from comets. Even when there is no shower, we still see meteors, which are called **sporadic meteors** because they are not part of specific showers. Many of these are produced by stray bits of matter that were released long ago by comets. Such comet debris gets spread through the inner solar system, and bits fall into Earth's atmosphere continuously.

Another way to backtrack meteor trails is to photograph a meteor from two locations on Earth a few miles apart. Then astronomers can use triangulation to find the altitude, speed, and direction of the meteor as it moves through the atmosphere. Then we could work backward to find out what its orbit looked like before it entered Earth's atmosphere. These studies reveal that meteors belonging to showers and some sporadic meteors have orbits that are similar to the orbits of comets. In many cases, the comets can be identified, so the origin of these meteors is well known. But some sporadic meteors have orbits that lead back to the asteroid belt. Thus, we can conclude that meteors have a dual source: Some come from comets, and some come from the asteroid belt. To learn more about these objects, we must examine those that make it to Earth's surface, the meteorites.

Figure 26-3
As a comet's ices evaporate, it releases rocky and metallic bits of material that spread along its orbit. If Earth passes through such material, it experiences a meteor shower.

Figure 26-4
The Barringer Meteorite Crater, a popular tourist attraction about an hour east of Flagstaff, Arizona, is one of the youngest impact craters on Earth. It was formed only 50,000 years ago and is only slightly modified by erosion. *(USGS)*

An Analysis of Meteorites

When a meteor is massive enough and strong enough to survive its plunge through Earth's atmosphere, it finally reaches Earth's surface, and we refer to it as a meteorite. A large meteorite hitting Earth might dig an impact crater much like those on the moon (see Figure 22-5). Most meteorites are small, and, because their energy of motion is spent in the atmosphere, they finally fall to Earth with no more energy than a falling hailstone.

Over 150 impact craters have been found on Earth. The Barringer Meteorite Crater near Flagstaff, Arizona is a good example (Figure 26-4). It was created about 50,000 years ago by a meteorite about as large as a big building (80 to 100 m). Debris at the site shows that the meteorite was composed of iron, and astronomers can calculate from the size of the crater that the meteorite hit at a speed of about 11 km/s and released as much energy as a large thermonuclear bomb.

The Barringer crater is 1.2 km in diameter and 200 m deep. It seems large when you stand on the edge, and the hike around it, though beautiful, is long and dry (Figure 26-5). Nevertheless, the crater is actually small compared with some impact features on Earth. About 65 million years ago, an impact in the northern Yucatan of Central America produced a crater, now covered by sediment, that was between 180 and 300 km in diameter,

Figure 26-5
Nearly a mile in diameter, the Barringer Meteorite Crater was formed when a building-size iron meteorite struck the horizontal limestone layers of the Arizona desert. Some rock was vaporized, some powdered, and some broken to fragments and ejected from the crater. Notice the raised and deformed rock strata all around the crater. *(M. A. Seeds)*

a   b

Figure 26-6
(a) W. A. Cassidy, leader of several expeditions to Antarctica, holds a numbering device near an achondritic meteorite exposed on the icy surface. (b) An iron meteorite shown as it was found. *(NASA photographs, Ursula B. Marvin)*

almost as big as the state of Ohio. This impact has been blamed for changing Earth's climate and causing the extinction of the dinosaurs, something we will discuss later. The 1994 comet impacts on Jupiter (shown in Figure 24-9) were even larger, but they formed no craters because Jupiter has no surface. During the Leonid meteor shower in 1999, astronomers watching the dark portion of the moon's disk saw a number of bright flashes apparently caused by meteorite impacts. No new craters on the moon have been identified, but it is clear that meteorite impacts and the formation of impact craters are a continuing process in our solar system.

Large meteorites can produce craters, but most meteorites are small and don't create significant craters. Also, craters erode rapidly on Earth, so most meteorites are found without associated craters. The best place to look for meteorites turns out to be certain areas of Antarctica—not because more meteorites fall there but because the meteorites are easy to recognize on the icy terrain. Most meteorites look like Earth rocks, but on the ice of Antarctica there are no natural rocks. Also, the slow creep of the ice toward the sea concentrates the meteorites in certain areas where the ice runs into mountain ranges, slows down, and evaporates. Thus, teams of scientists can easily recover meteorites from the ice (Figure 26-6). Such unbiased collections of meteorites are important in helping define the frequency of different meteorite types.

Meteorites that are seen to fall are called **falls**. A fall is known to have occurred at a given time and place, and thus the meteorite is well-documented. A meteorite that is discovered but was not seen to fall is called

a **find.** Such a meteorite could have fallen thousands of years ago.

Studies of meteorites reveal three major classes: irons, stones, and stony-irons. **Iron meteorites** make up 66 percent of finds (Table 26-2) but only 6 percent of falls. Why? Because an iron meteorite does not look like a rock. If you trip over one on a hike, you are more likely to carry it home and show it to your local museum. Also, some stone meteorites deteriorate rapidly when exposed to weather; irons survive longer. Thus, there is a **selection effect** that makes it more likely that iron meteorites will be found (Window on Science 26-2). That only 6 percent of falls are irons shows that iron meteorites are fairly rare.

When iron meteorites are sliced open, polished, and etched with nitric acid, they reveal regular bands called **Widmanstätten patterns** (Figure 26-7). These patterns are caused by alloys of iron and nickel called kamacite and taenite that formed as the molten iron

Table 26-2 Proportions of Meteorites

Type	Falls (%)	Finds (%)
Stony	92	26
Iron	6	66
Stony-iron	2	8

Many different kinds of science depend on selecting objects to study. Scientists studying insects in the rain forest, for example, must choose which ones to catch. They can't catch every insect they see, so they might decide to catch and study any insect that is red. If they are not careful, a selection effect could bias their data and lead them to incorrect conclusions without their ever knowing it. The reason selection effects are dangerous in science is that they can have powerful influences without being evident. Only by carefully designing a research project can scientists avoid selection effects.

For example, suppose we decide to measure the speed of cars on a highway. There are too many cars to measure every one, so we reduce the work load and measure only the speed of red cars. It is quite possible that this selection criterion will mislead us because people who buy red cars may be more likely to be younger and drive faster. Should we measure only brown cars? No, because we might suspect that only older, more sedate people would buy a brown car. Should we measure the speed of any car in front of a truck? Perhaps we should pick any car following a truck? Again, we may be selecting cars that are traveling a bit faster or a bit slower than normal. Only by very carefully designing our experiment can we be certain that the cars we measure are traveling at representative speeds.

Public-opinion companies that survey voters in shopping malls must think carefully about selection effects. If they are seeking opinions about property taxes and decide to interview only men with gray hair, they will get a biased result. Interviewing men wearing T-shirts would give them quite a different result. Thus, they must be very careful to avoid unnoticed selection effects.

Astronomers face the danger of selection effects quite often. Iron meteorites are easier to find than stone meteorites. Very luminous stars are easier to see at great distances than faint stars. Spiral galaxies are brighter, bluer, and more noticeable than elliptical galaxies. What we see through a telescope depends on what we notice, and that is powerfully influenced by selection effects.

Scientists engaged in observation must spend considerable time designing their experiments. They must be careful to observe an unbiased sample if they expect to make logical deductions from their results. The scientists in the rain forest, for example, should not catch and study only the red insects. Often, the most brightly colored insects are poisonous (or at least taste bad) to predators. Catching only brightly colored insects could produce a highly biased sample of the insect population. Careful scientists must plan their work with great care and avoid any possible selection effects.

---

cooled and solidified. The size and shape of the bands indicate that the molten metal cooled very slowly, no faster than 20 K per million years.

A lump of molten metal floating in space would cool very quickly, so the presence of the Widmanstätten pattern tells us that the molten metal must have been well insulated to have cooled so slowly. Such slow cooling is typical of the interiors of bodies at least 30 to 50 km in diameter. On the other hand, the iron meteorites show no evidence of the very high pressures that would develop in larger bodies. Thus, the iron meteorites appear to have formed from the interiors of molten objects of planetesimal size.

In contrast to irons, **stony meteorites** are common. Among falls, 92 percent are stones. Although there are many different types of stony meteorites, we will classify them into two subgroups depending on their physical and chemical content—chondrites and achondrites.

Roughly 80 percent of all meteorite falls are stony meteorites called **chondrites.** These are dark gray, granular rocks containing **chondrules,** round bits of glassy rock no larger than 5 mm across (Figure 26-8a). Chondrules must have cooled very quickly (in minutes), but their origin is not clear. They may be bits of molten rock splashed into space by impacts between planetesimals. Another possibility is that they are bits of rock that condensed from the solar nebula and were later melted by a sudden flare-up of the young sun or by electrical discharges (lightning). Whatever their origin, they are very old, and, because melting of the meteoroid would have destroyed them, their presence in

Figure 26-7
An iron meteorite sliced open, polished, and etched with acid reveals a Widmanstätten pattern. Such patterns are evidence that the iron cooled from a molten state no faster than a few degrees per million years. *(Jordan Marché, North Museum, Franklin and Marshall College)*

a

b

Figure 26-8
(a) A stony meteorite sliced open and polished to show the small, spherical inclusions called chondrules. *(Jordan Marché, North Museum, Franklin and Marshall College)* (b) A sliced portion of the Allende meteorite showing round chondrules and irregularly shaped white inclusions called CAIs. *(NASA)*

hot, the carbonaceous chondrites have never even been warm. They are the least modified of all the meteorites.

One of the most important meteorites ever recovered was a carbonaceous chondrite that was seen falling on the night of February 8, 1969, near the Mexican village of Pueblito de Allende. The brilliant fireball was accompanied by tremendous sonic booms and showered an area about 50 km by 10 km with over 4 tons of fragments. About 2 tons were recovered.

Studies of the Allende meteorite disclosed that it contained—besides volatiles and chondrites—small, irregular inclusions rich in calcium, aluminum, and titanium. Now called **CAIs,** for calcium–aluminum-rich inclusions, these bits of matter are highly refractory (Figure 26-8b); that is, they can survive very high temperatures. If we could scoop out a ton of the sun's surface matter and cool it, the CAIs would be the first particles to form. As the temperature fell, other materials would condense in accord with the condensation sequence described in Chapter 20. When the material finally reached room temperature, we would find that all of the hydrogen, helium, and a few other gases like argon and neon had escaped and that the remaining lump, weighing about 18 kg (40 lb), had almost the same composition as the Allende meteorite, including CAIs. Thus, the Allende meteorite seems to be a very old sample of the solar nebula.

Unlike the carbonaceous chondrites, stony meteorites called **achondrites** (7.1 percent of falls) are highly modified. They contain no chondrules and no volatiles. This suggests that they have been hot enough to melt chondrules and drive off volatiles, leaving behind rock with compositions similar to Earth's lavas.

Iron meteorites and stony meteorites make up most falls, but 2 percent of falls are meteorites that are made up of mixed iron and stone. These **stony-iron meteorites** appear to have solidified from a region of molten iron and rock—the kind of environment we might expect deep inside a planetesimal with a molten iron core and a rock mantle.

The Origins of Meteorites

The properties of meteorites suggest that they formed in the solar nebula. Their radioactive ages are about 4.6 billion years (Chapter 20), and some appear not to have been modified since they condensed from the solar nebula. Others have been heated or melted sometime after formation.

Meteorites almost certainly do not come from comets. Most cometary particles are very small specks of low-density, almost fluffy, material. When these enter the atmosphere, we see them incinerated as meteors, and most meteors are produced by this cometary debris. But such meteors are small and weak and do not survive to reach the ground. Although most meteors come from comets, most meteorites are stronger chunks of matter—more like fragments of asteroids.

chondrites indicates that these meteorites have never been melted. Chondrites can never have been hotter than 1000 to 1300 K.

On the other hand, the chondrites have been heated to moderate temperatures for long periods sometime in their past, as shown by studies of the granulation and chemical composition of the rock. Also, the chondrites contain no volatiles. These easily vaporized compounds evidently either were baked out of the chondrites or were never present because the chondrites formed at too high a temperature.

In contrast, **carbonaceous chondrites,** a subclass of chondrites, are rare, only about 5.7 percent of falls, but they are important because they contain volatiles. They are rich in water and compounds of carbon that would have been driven out if they had ever been heated above about 500 K. Thus, if the chondrites have never been

Thus, when we talk of the origin of the meteorites, we are concerned with the history of asteroidlike planetesimals rich in metals and silicates, not the cometlike planetesimals that were rich in ices. Indeed, many of the properties of the meteorites suggest that they originated in stony-iron planetesimals, which evolved in complicated ways and were eventually broken up by collisions. Here we will concentrate on the answers to two questions: How did these planetesimals evolve? When did these planetesimals break up?

The iron and achondritic meteorites show that at least some of the planetesimals melted, but what produced this heat? Planets the size of Earth can accumulate a great deal of heat from the decay of radioactive atoms such as uranium, thorium, and the radioactive isotope of potassium. The heat is trapped deep underground by many kilometers of insulating rock. But in a small planetesimal, the insulating layers are not as thick, and the heat leaks out into space as fast as the slowly decaying atoms can produce it. If a small planetesimal is to melt, it must have a more rapidly decaying heat source.

Recent studies have shown that some meteorites may have contained the radioactive element aluminum-26. These atoms are now gone, decayed to magnesium-26, but if they were once present their rapid decay could have melted the center of a planetesimal as small as 20 km in diameter. This is a sufficient heat source to melt the parent bodies of the meteorites.

Where did this aluminum-26 come from? Supernova explosions can manufacture aluminum-26, but it decays very rapidly with a half-life of only 715,000 years. If the solar nebula was enriched by aluminum-26 from a nearby supernova explosion, the explosion must have occurred just before the formation of our solar system. In fact, some astronomers wonder if the shock waves from the supernova explosion compressed gas clouds and triggered the formation of the sun and planets.

Once the planetesimals were melted and the aluminum-26 had decayed, they cooled and slowly solidified. In the meantime, however, the planetesimals would have differentiated. The heavy metals would have sunk to the center to form a molten metal core, and the less dense silicates would have floated upward to form a stony mantle. In some cases, the surface may not have been melted. If such a planetesimal were broken up (Figure 26-9), fragments from the center would look much like iron meteorites with their Widmanstätten patterns, and fragments from the outer portions might look like stony-irons and achondrites.

The chondrites and carbonaceous chondrites might have formed as smaller bodies farther from the sun. Planetesimals in the outer asteroid belt would have been cooler and could have retained volatiles more easily. They could not have been large, or heating and differentiation would have altered them.

The evidence seems strong that the meteorites are the result of the breaking up of larger bodies. For example, some meteorites are breccias—collections of fragments cemented together. Breccias are found on Earth and are very common on the moon, but studies of the meteoric breccias show that they were produced

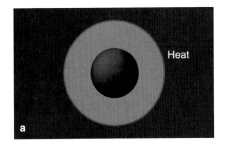

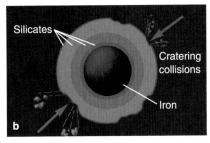

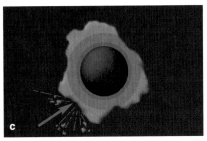

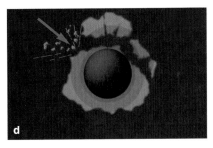

Figure 26-9
Production of various kinds of meteoroids from asteroids. If it melted, an asteroid (a) could differentiate into an iron core surrounded by layers of different silicate compositions (b). Impacts (c, d, e) could break up these layers and produce various kinds of meteorites. *(Adapted from a diagram by C. R. Chapman)*

by impacts. A collision between planetesimals produces fragments, and the slower-moving particles fall back to the surface of the planetesimal to form a regolith, a soil of broken rock fragments. Later impacts may add to this regolith and stir it. Still later, an impact may be so violent that the fragments are pressed together and momentarily melt where they touch. Almost instantly, the material cools, and the fragments weld themselves together to form a breccia. Most of the breccias were formed as the planetesimals formed. Much later, as the planetesimals were broken up by collisions, the layers of breccia were exposed, shattered, and became brecciated meteoroids.

The collisions that broke up the planetesimals could not have happened long ago, when the solar system was young. Small meteoroids would have been swept up by the planets in only a billion years or less. Also, cosmic rays striking meteoroids in space produce isotopes such as helium-3, neon-20, and argon-38. Studies of these atoms in meteorites show that most meteorites have not been exposed to cosmic rays for more than a few tens of millions of years. The meteorites must have been buried under protective layers until recently. Thus, the thousands of meteorites now in museums around the world must have broken off of planetesimals somewhere in our solar system within the last billion years. Where are those planetesimals? To answer that question, me must recall that many meteoroid orbits lead back to the asteroid belt. The asteroids are evidently the planetesimals from which we receive meteorites.

REVIEW Critical Inquiry

How can we say that meteors come from comets, but meteorites come from asteroids?

A selection effect can determine what we notice when we observe nature, and a very strong selection effect prevents us from finding meteorites that originated in comets. Cometary particles are physically weak, and they vaporize in Earth's atmosphere easily. Thus, very few ever reach the ground, and we are unlikely to find them. Furthermore, even if a cometary particle reached the ground, it would be so fragile that it would weather away rapidly, and, again, we would be unlikely to find it. Asteroidal particles, however, are made from rock and metal and so are stronger. They are more likely to survive their plunge through the atmosphere and more likely to survive erosion on the ground. Meteors from the asteroid belt are rare. Almost all of the meteors we see come from comets, but not a single meteorite is known to be cometary.

A similar selection effect makes it more likely that we will find iron meteorites than stony meteorites. Irons

are easier to notice and survive longer. Thus, most *finds* are iron, but most *falls* are stone.

The meteorites are valuable because they give us hints about the process of planet building in the solar nebula. What evidence do we find in the meteorites that tells us they were once part of larger bodies broken up by impacts?

The vast majority of meteorites must have originated in the asteroids. Our next goal is to try to understand the nature and origin of these tiny rocky worlds.

26-2 Asteroids

Until recently, few astronomers knew or cared much about asteroids. They were small chunks of rock drifting between the orbits of Mars and Jupiter that occasionally marred long-exposure photographs by drifting past more interesting objects. Asteroids were more irritation than fascination.

Now we know differently. The evidence of meteorites tells us that the asteroids are the last remains of the planetesimals that built the planets 4.6 billion years ago. Spacecraft have photographed a few asteroids and revealed them to be small, complex worlds with irregular shapes and heavily cratered surfaces (Figure 26-10). The study of the asteroids gives us a way to explore the ancient past of our planetary system.

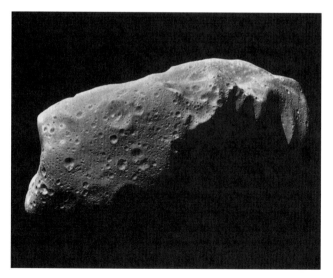

Figure 26-10
The asteroid Ida is shown in this image recorded by the Galileo spacecraft as it flew past the asteroid. Ida is irregular in shape and stretches 52 km (32 miles) across its longest diameter. Our eyes would see a gray color much like that in this black-and-white image. *(NASA)*

The Asteroid Belt

The first asteroid was discovered on January 1, 1801 (the first night of the 19th century), by the Sicilian monk Giuseppe Piazzi. It was later named Ceres after the Roman goddess of the harvest (thus our word *cereal*). We know now that it is a nearly spherical ball of rock about 900 km in diameter.

Astronomers were excited by Piazzi's discovery. There seemed to be a gap where a planet might exist between Mars and Jupiter at an average distance from the sun of 2.8 AU. Ceres fit. Its average distance from the sun is 2.766 AU. It was a bit small to be a planet, and because three more objects—Pallas, Juno, and Vesta—were discovered in the following years, astronomers realized that Ceres and the other asteroids were not true planets.

Today thousands of asteroids have accurately known orbits, and there are many more as yet undiscovered; but they are all small bodies. All of the larger asteroids in the asteroid belt have been found.

When an asteroid is discovered, the discoverer is allowed to choose a name for it, and asteroids have been named for spouses, lovers, dogs, Greek gods, politicians, and others.* Once an orbit has been calculated, the asteroid is assigned a number listing its position in the catalog known as the *Ephemerides of Minor Planets*. Thus, Ceres is known as 1 Ceres, Pallas as 2 Pallas, and so on. Only three are larger than 400 km in diameter, and most are much smaller (Figure 26-11).

Most of these objects orbit the sun in the asteroid belt between Mars and Jupiter, and we might suspect that massive Jupiter was responsible for their origin. Certainly the distribution of asteroids in the belt shows the importance of Jupiter's gravitation. Certain regions of the belt, called **Kirkwood's gaps** (Chapter 24) after their discoverer, are almost free of asteroids (Figure 26-12). These gaps lie at certain distances from the sun where an asteroid would find itself in resonance with Jupiter. For example, if an asteroid lay 3.28 AU from the sun, it would revolve twice around its orbit in the time it took Jupiter to revolve once around the sun. Thus, on alternate orbits, the asteroid would find Jupiter

*Some sample asteroid names: Olga, Chicago, Vaticana, Noel, Ohio, Tea, Gaby, Fidelio, Hagar, Geisha, Dudu, Tata, Mimi, Dulu, Tito, Zulu, Beer, and Zappafrank (after the late musician Frank Zappa).

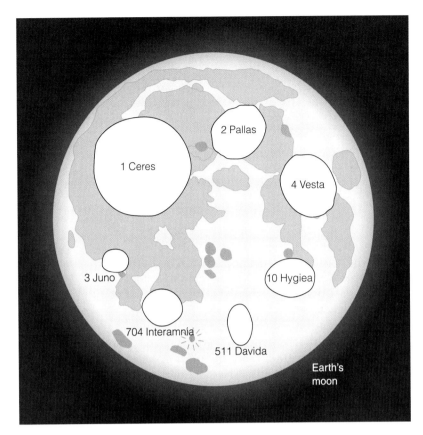

Figure 26-11
The relative size and approximate shape of the larger asteroids are shown here compared with the size of Earth's moon. Smaller asteroids can be highly irregular in shape.

at the same place in space tugging outward. The cumulative perturbations would rapidly change the asteroid's orbit until it was no longer in resonance with Jupiter. Our example is in a 2:1 resonance, but gaps occur in the asteroid belt at many resonances, including 3:1, 5:2, and 7:3.

Modern research shows that the motion of asteroids in Kirkwood's gaps is described by a theory in mathematics that deals with chaotic behavior. The smooth motion of water sliding over the edge of a waterfall decays rapidly into a chaotic jumble. The same mathematical theory of chaos that describes the motion of the water shows how the slowly changing orbit of an asteroid within one of Kirkwood's gaps can suddenly change into a long, elliptical orbit that carries the asteroid into the inner solar system, where it is likely to be removed by a collision with Mars, Earth, or Venus. This ejection process also explains how Jupiter's gravity can throw so many meteoroids from the asteroid belt into the inner solar system.

Nonbelt Asteroids

Not all asteroids orbit within the asteroid belt. Some of the most interesting follow orbits that cross the orbits

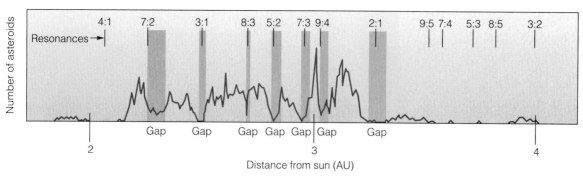

Figure 26-12
Here the red curve shows the number of asteroids at different distances from the sun. Yellow bars mark Kirkwood's gaps where there are few asteroids. Note that these gaps match resonances with the orbital motion of Jupiter.

of the terrestrial planets or wander among the Jovian worlds. In fact, some asteroids even share the orbits of the larger planets.

The **Apollo–Amor objects** are asteroids whose orbits carry them into the inner solar system. The Amor objects follow orbits that cross the orbit of Mars but don't reach the orbit of Earth. The Apollo objects have Earth-crossing orbits. These Apollo–Amor objects are dangerous. Jupiter's influence makes their orbits precess, so they are bound to interact with a planet eventually. About one-third will be thrown into the sun, and a few will be ejected from the solar system, but many of these objects are doomed to collide with a planet—perhaps ours. Earth is hit by an Apollo object once every 250,000 years, on average. With a diameter of up to 2 km, they hit with the power of a 100,000-megaton bomb and can dig craters 20 km in diameter.

Dozens of Apollo objects have known orbits, and none of those will hit Earth in the foreseeable future. The bad news is that objects as small as a kilometer in diameter could cause planetwide catastrophes, and most of the Apollo asteroids have not been found. A number of teams are now searching for these near-Earth asteroids, and NEAT, the Near-Earth Asteroid Tracking team, now searches for such objects with an automated telescope. They estimate that there are about 700 near-Earth asteroids larger than 1 km in diameter and that automated searches will be able to locate and determine orbits for all of these objects by about 2010 AD.

This is a serious issue because the inner solar system contains large numbers of small asteroids. For example, in January 1999, an asteroid about 9 m in diameter was discovered as it raced past Earth at a distance only half that of the moon. Had that asteroid struck Earth, it would have hit with an energy ten times greater than that of the bomb that destroyed Hiroshima. Objects of this size, a few tens of meters or less, fragment and explode in the atmosphere, but the shock waves from their explosion could cause serious damage on Earth's surface. Declassified data from military satel-

lites show that Earth is hit about once a week by these house-size asteroids. The largest impact ever detected was the atmospheric explosion of a rocky meteorite over the South Pacific on February 1, 1994. It was seen by fishermen as a blast as bright as the sun. It produced about five times the energy of the Hiroshima bomb and was probably a rocky body about 15 m in diameter—probably a small Apollo object. We will see later in this chapter what a slightly larger impactor can do.

It is easy to assume that the Apollo–Amor objects are rocky asteroids that have been thrown into their extreme orbits by events in the main asteroid belt. At least some of these objects, however, may be comets that have exhausted their ices and become trapped in short orbits that keep them in the inner solar system. We will see later that the distinction between comets and asteroids is not totally clear.

While the Apollo–Amor objects rush through the inner solar system, there are also nonbelt asteroids in the outer solar system. These objects, being farther from the sun, move more slowly. The object Chiron, found in 1977, appears to be a body about 170 km in diameter. Its orbit carries it from just inside the orbit of Uranus to just inside the orbit of Saturn. Although it was first classified as an asteroid, its status is now less certain. Ten years after its discovery, Chiron surprised astronomers by suddenly brightening with a release of vapor and dust. Like a comet, Chiron was releasing a cloud of gas and dust in jets. Studies of older photographs showed that Chiron had been brighter even when it was farther from the sun, and astronomers now suspect that it may have a rocky crust covering deposits of ices such as solid nitrogen, methane, and carbon monoxide. Thus, Chiron may be more comet than asteroid, and it warns us that the distinction is not clear-cut.

Chiron's orbit is not stable. It cannot have remained in its present orbit for the entire history of the solar system but must have been perturbed into its present orbit within the last million years or so. Other, smaller bodies have been found in the outer solar system, and astronomers are now trying to understand how their origin

Composition and Origin

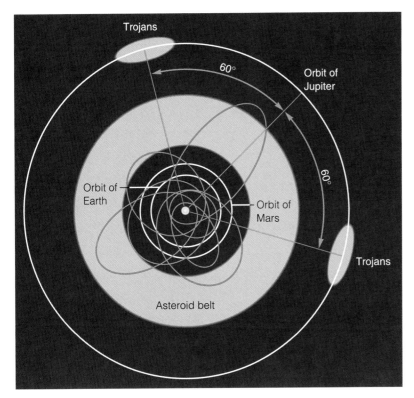

Figure 26-13
The Trojan asteroids are trapped in the Lagrangian points 60° ahead of and 60° behind Jupiter. The Apollo asteroids move along orbits that cross Earth's orbit. The orbits of only 7 of the 31 known Apollo asteroids are shown here.

An old theory held that the asteroids are the remains of a planet that broke up, but modern astronomers discount that idea. Once formed, a planet is very difficult to tear apart. Rather, astronomers believe the asteroids are the remains of planetesimals left over from the formation of the planets. Jupiter's gravity could have stirred the planetesimals just inside its orbit and caused collisions at unusually high velocities. These impacts would have tended to break up the planetesimals rather than assemble them into a planet. Resonances and chaos would have helped eject material, and most of the planetesimal objects have now been lost—swept up by planets, captured as satellites, or ejected from the solar system. The asteroid belt today contains hardly a fifth the mass of Earth's moon.

Collisions are surely common among the asteroids, and we can see evidence of catastrophic impacts powerful enough to break up an asteroid. Early in the 20th century, Japanese astronomer Kiyotsugu Hirayama discovered that some groups of asteroids share similar orbits. They have the same average distance from the sun, the same eccentricity, and the same inclination. Up to 19 of these **Hirayama families** are known, and modern observations show that the asteroids in a family typically share similar spectroscopic characteristics. Evidently, a family is produced by a catastrophic collision that breaks a single asteroid into a family of fragments that continue traveling along similar orbits around the sun.

In 1983, the Infrared Astronomy Satellite detected the infrared glow of sun-warmed dust scattered in bands throughout the asteroid belt. These dust bands appear to be the products of past collisions. The dust will eventually be blown away, but collisions occur constantly in the asteroid belt, so new dust bands will presumably be produced as the present bands dissipate.

Beginning in the 1990s, astronomers have gotten close-up looks at a few asteroids. One way to study asteroids is to bounce radio signals off of asteroids passing near Earth and construct images from the timing and Doppler shifts in the echoes. These images revealed that some asteroids appear to be peanut shaped. They are either two objects that have collided and fused or two objects orbiting close to each other. Such binary asteroids remind us that some asteroid collisions can be gentle.

Spacecraft images of asteroids show that they are irregular in shape and heavily cratered. The asteroid

and history are linked to the formation of the planets. We saw in the previous chapter that Pluto, Charon, and Triton might be large members of this family of icy planetesimals.

The icy asteroids of the outer solar system seem to be widely scattered, but Jupiter ushers two groups of asteroids within its own orbit. These nonbelt asteroids have become trapped in the Lagrangian points along Jupiter's orbit. These points lie 60° ahead of and 60° behind the planet and are regions where the gravitation of the sun and Jupiter combine to trap small bodies (Figure 26-13). Like cosmic sinkholes, the Lagrangian points have trapped chunks of debris now called **Trojan asteroids** and named after the heroes of the Trojan War (588 Achilles, 624 Hektor, 659 Nestor, and 1143 Odysseus, for example). Although about 15 Trojans have been found and named, astronomers estimate that 700 may be trapped in the Lagrangian points.

In 1990, observers found an asteroid at the trailing (L_5) Lagrangian point in the orbit of Mars, and growing evidence suggests that other planets may have Trojan asteroids trapped in their orbits. As we search for smaller objects, we are finding that our solar system still contains large numbers of these small bodies. The challenge is to explain their physical nature and their origin.

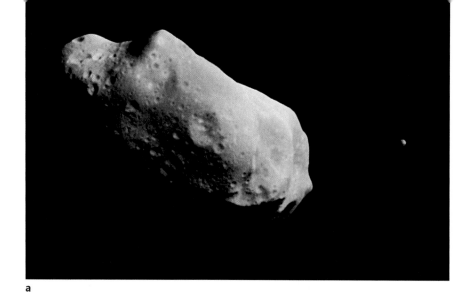

a

the comet Shoemaker-Levy 9. The spacecraft has transmitted back to Earth detailed photos and spectroscopic data (Figure 26-15). Eros is an Apollo asteroid and is highly elongated. It is probably a fragment of a larger body that has itself been battered by many impacts.

In the 1930s, astronomers discovered that some asteroids are redder than others, but only in the past few decades have ultraviolet and infrared observations made it possible to study the color of the light

b

Figure 26-14
(a) Asteroid Ida, imaged by the passing Galileo spacecraft, is shown here with the small moon Dactyl. The moon is named for the Dactyli, the companions and protectors of the nymph Ida in Greek mythology. *(NASA)* (b) Asteroid Mathilde is scarred by craters of all sizes including a giant impact feature nearly as large as the asteroid. *(Johns Hopkins University, Applied Physics Laboratory)*

a

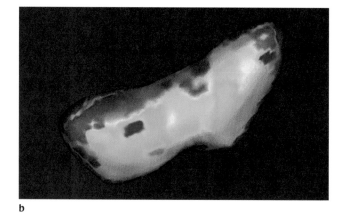

b

Figure 26-15
(a) Two views of asteroid Eros show its highly elongated shape and heavily cratered surface. (b) This infrared map of Eros from the NEAR-Shoemaker spacecraft shows the distribution of the mineral pyroxene. Such spectrographic maps will allow astronomers to study the mineralogy of Eros in detail. *(Johns Hopkins University, Applied Physics Laboratory)*

Gaspra is shown in Figure 20-15, and Ida appears in Figure 26-10. One image of Ida revealed that it has a moon of its own, 1.5-km diameter Dactyl (Figure 26-14a). The moon is probably a fragment from an impact. Other asteroids with moons are known or suspected, and double-impact craters on planetary surfaces may tell us that such double asteroids are common.

The asteroid Mathilde looks like other asteroids (Figure 26-14b), but measurements of its gravitational field by the passing NEAR spacecraft revealed that the asteroid has a very low density. Mathilde may not be solid rock but rather may be a rubble pile with large empty spaces. Observations of other asteroids suggest that many may be low-density rubble piles produced by repeated collisions rather than chunks of solid rock.

The NEAR spacecraft went into orbit around asteroid Eros in early 2000 and was immediately renamed in honor of the late Gene Shoemaker, a codiscoverer of

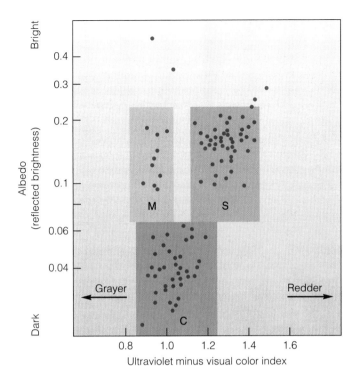

Figure 26-16
A graph of albedo or reflected brightness versus color separates asteroids into three types. The C types are dark, while the M and S types are brighter. S types are redder than M types. These differences in color and brightness are related to the chemical compositions at the surfaces of the asteroids. *(Adapted from a diagram by B. Zellner)*

reflected from asteroids and compare it with the colors reflected from various kinds of minerals and meteorites. This has allowed astronomers to plot asteroids in a diagram and separate out at least three different kinds of asteroids, the C, S, and M asteroids (Figure 26-16).

The C-type asteroids are very dark, with albedos of less than 0.06. That is, they reflect less than 6 percent of the light that hits them and are thus darker than a lump of sooty coal. Their properties match those of carbonaceous chondrite meteorites.

The S-type asteroids are brighter, with albedos of about 0.1 to 0.2, and they tend to be reddish in color. In Figure 26-16 they lie above and to the right of the C-type asteroids. Astronomers now think that the S-type asteroids are similar to C-type asteroids but have had their surfaces brightened and reddened by an unknown chemical or physical process.

The M-type asteroids are moderately bright, with albedos about the same as those of the S types, but the M types are not as red. Thus, we can separate the two types in the albedo–color diagram (Figure 26-16). Their optical properties suggest that M types are composed of iron and nickel alloys and contain few silicates. Thus the M-type asteroids may be the unbroken iron cores of once larger planetesimals.

A few other types of asteroids are known, and a number of individual asteroids have been found that are unique, but these three classes contain a majority of the known asteroids.

How did these three types originate? A clue lies in their distribution in the asteroid belt. The S-type asteroids are much more common in the inner belt. In fact, there are almost none beyond a distance of about 3.45 AU. The C types are rather rare in the inner belt but are very common in the outer belt. Many astronomers believe that this distribution reflects the differences in temperature during the formation of the planetesimals. It was cooler in the outer belt, and the planetesimals there tended toward the composition of carbonaceous chondrites. In the warmer regions of the inner belt, the composition of the growing planetesimals was more like that of the chondrites.

The large asteroid Vesta is reminding astronomers that these are complex worlds. With a diameter of 525 km, Vesta is too small to image from Earth's surface, but the Hubble Space Telescope reveals that it has a complex surface with light and dark patches. The spectra of these regions are like those of basalts typical of solidified lavas. Furthermore, a study of the best photos shows a giant crater 460 km in diameter (87 percent of Vesta's diameter) near Vesta's south pole (Figure 26-17). A class of basaltic meteorites called eucrites have infrared spectra nearly identical to that of Vesta, and the 2-km-long asteroid Braille has the same kind of infrared spectrum. Planetary astronomers believe that the impact that formed the large crater on Vesta could have blasted loose about 1 percent of the asteroid's mass. A tiny fraction of that debris fell to Earth as the eucrite meteorites, and a large piece became the asteroid Braille. Thus the eucrite meteorites give us a sample from the surface of Vesta, and because those meteorites are basalts that resemble solidified lava flows, we can cite them as strong evidence that Vesta has at some time in the past been at least partly resurfaced by lava flowing up from its interior.

How could a small asteroid be heated enough to produce lava flows? One source of heat in newly formed asteroids would be the decay of short-lived radioactive elements such as aluminum-26. Such radioactive elements could have been produced in a supernova explosion occurring shortly before the formation of our solar system. The smallest asteroids lose their heat too fast to melt, but the decay of aluminum-26 occurs fast enough to melt the interiors of bodies larger than a few dozen kilometers in diameter. Thus, it is reasonable to suppose that Vesta and some other asteroids larger than about 100 km in diameter might have been strongly heated and modified by lava flows.

In contrast, the largest asteroid, Ceres, about 900 km in diameter, does not seem to have been modified by internal heating. The light reflected from its surface

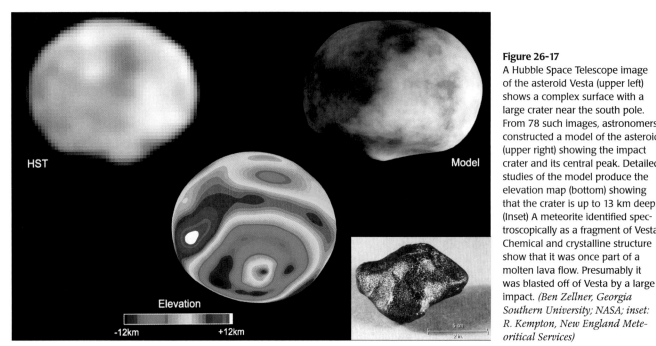

Figure 26-17
A Hubble Space Telescope image of the asteroid Vesta (upper left) shows a complex surface with a large crater near the south pole. From 78 such images, astronomers constructed a model of the asteroid (upper right) showing the impact crater and its central peak. Detailed studies of the model produce the elevation map (bottom) showing that the crater is up to 13 km deep. (Inset) A meteorite identified spectroscopically as a fragment of Vesta. Chemical and crystalline structure show that it was once part of a molten lava flow. Presumably it was blasted off of Vesta by a large impact. *(Ben Zellner, Georgia Southern University; NASA; inset: R. Kempton, New England Meteoritical Services)*

suggests a claylike surface related to carbonaceous chondrites. This is surprising, because clays form when minerals are exposed to water. In fact, spectroscopic observations reveal what seem to be traces of water ice on Ceres. Thus, we must suspect that some asteroids may have had significant amounts of water bound into their crusts when they were young.

All of the evidence suggests that the asteroids are the broken remains of planetesimals that formed in the solar nebula as planet building began. The largest remaining asteroids, such as Ceres and Vesta, may be largely unbroken planetesimals, and some of these may have evolved due to internal heat or the presence of water. Nevertheless, the vast majority of the asteroids are fragments of larger bodies, and many may consist of bodies that have been shattered and then re-formed as gravity pulled the fragments back together. Some compositional differences seem to have originated when the asteroids formed in that the objects in the outer asteroid belt contain more carbon-rich compounds. The presence of massive Jupiter orbiting just beyond this region prevented the original planetesimals from accreting to build a planet. Instead, collisions fragmented the planetesimals, and nearly all of the material is now lost. The entire mass of the asteroid belt today adds up to only 0.08 percent of Earth's mass.

Some experts discuss the mining of the asteroid belt for its mineral riches, but the asteroid belt is nearly empty. The small, fragmented asteroids drift through space far from one another and only rarely collide with each other or with planets. A voyage to explore or exploit the asteroid belt would find the distance between bodies forbiddingly large.

R E V I E W Critical Inquiry

What evidence do we have that the asteroids have been fragmented?

First, we might note that the solar nebula theory of the formation of the solar system predicts that planetesimals collided and either stuck together or fragmented. This is suggestive, but it is not evidence. A theory can never be used as evidence to support some other theory or hypothesis. Evidence means observations or the results of experiments, so we must turn to observations of asteroids. Spacecraft photographs of Ida, Gaspra, and Eros show irregularly shaped little worlds heavily scarred by impact craters, and the radar images obtained so far of asteroids also show irregularly shaped bodies that appear to be the fragments of larger bodies. In fact, radar images of some asteroids show what may be pairs of bodies in contact, and the Galileo image of Ida reveals its small satellite, Dactyl. Furthermore, some meteorites appear to come from the asteroid belt, and a few have been linked to specific asteroids such as Vesta. All of this evidence suggests that the asteroids have suffered violent collisions in their history.

The impact fragmentation of asteroids has been important, but it has not erased all traces of the original planetesimals from which the asteroids formed. What evidence do we have to tell us what those planetesimals were like?

The asteroids are only the last crumbs left behind by failed planet building near Jupiter. Farther out in the solar system are other remains—ice shards from the age of planet formation.

26-3 Comets

Few things in astronomy are more beautiful than a bright comet hanging in the night sky (Figure 26-18). Whereas meteors shoot across the sky like demented fireflies, a comet moves with the stately grace of a great ship at sea, its motion hardly apparent. Night by night it shifts against the stars and may remain visible for months. Faint comets are common; a number are discovered every year. But a truly bright comet appears about once a decade. Comet Hyakutake in 1996 (Figure 26-1) and Comet Hale-Bopp in 1997 (Figure 26-18b) were both dramatic, but the later comet was so bright that we might class it with the great comets such as Halley's comet in 1910. A patient person might see half a dozen or more bright comets in a lifetime.

While everyone enjoys the beauty of comets, astronomers study them for their cargo of clues to the origin of our solar system.

Observational Properties of Comets

We saw in Chapter 20 that the overall properties of comets could be explained by the dirty snowball hypothesis. It proposes that a comet is a lump of dirty ices, mostly water ice, orbiting the sun in a long elliptical orbit. Only when that nucleus of ice comes within a few astronomical units of the sun does the ice vaporize to release gases and imbedded dust specks, which the sun blows away to form a tail pointing away from the sun.

Now we are ready to refine this hypothesis by comparison with evidence. Observations of recent bright comets provide lots of data. Comets Hyakutake and Hale-Bopp were studied from Earth in the late 1990s with modern telescopes and instruments, and Comet Halley was visited in 1986 by five different spacecraft that flew past the nucleus of the comet.

We can think of a comet as comprising a head and a tail. The visible head, or **coma**, of a comet is a vast cloud

a

b

Figure 26-18
(a) Comet Ikeya-Seki was beautiful in the dawn sky in 1965. The long, graceful tail sprang from a nucleus of ice only a dozen kilometers in diameter. *(Stephen M. Larson)* (b) Comet Hale-Bopp passed through the inner solar system in 1997. Its large nucleus, estimated to be 40 km in diameter, produced large amounts of gas and dust, making it quite bright in the dawn and later in the evening sky. *(Dean Ketelsen)*

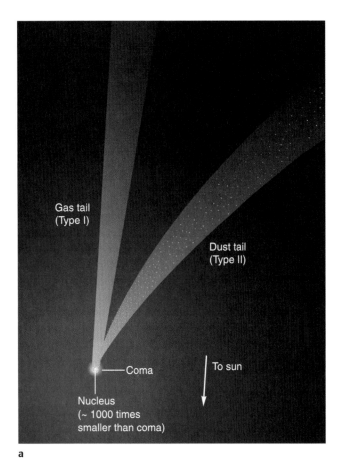

a

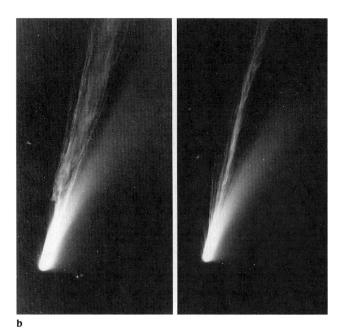

b

Figure 26-19
(a) The parts of a comet include an extended coma, or head, of gas and dust surrounding an icy nucleus only a few kilometers in diameter. Gas is carried away by the solar wind in a wispy gas tail, but dust moves more slowly and forms a curving dust tail. (b) Comet Mrkos in 1957 had pronounced gas and dust tails, which changed dramatically from night to night. *(Palomar Observatory/Caltech)*

of gas and dust up to a million kilometers in diameter—that is, roughly 80 times Earth's diameter. Spacecraft flying toward Comet Halley began running into dust particles 47 Earth radii from the nucleus. The gas in the coma contains H_2O, CO_2, CO, H, OH, O, S, C, and so on. These atoms and ions are what we would expect to find if ultraviolet radiation from the sun broke down vaporized ices consisting of water (H_2O) and carbon dioxide (CO_2) containing traces of methane (CH_4), ammonia (NH_3), and similar common gases.

Imbedded in the gas of the coma is the solid debris from the dirty snowball. Most of the particles are silicate dust specs smaller than those in cigarette smoke.

The tail of a comet springs from the coma and typically extends 10 to 100 million km through space. The longest tails can have a length of 1 AU, about 150 million kilometers. Seen from Earth, the tail of a major comet can span 30° in the sky. This immense structure contains nothing more than very tenuous gas and scattered bits of microscopic dust. Comets have been described as "as close to nothing as something can be and still be something."

A comet can have two kinds of tails, one of gas and one of dust (Figure 26-19a). Both point generally away from the sun. A **type I tail** is a gas tail; it is straight and wispy and looks like a narrow trail of smoke. Its spectrum contains emission lines of ionized gases excited by the ultraviolet radiation from the sun (Table 26-3). The gas is blown outward by the solar wind with its embedded magnetic field.

Occasionally, a comet will disconnect its gas tail, allow it to blow away, and begin developing a new tail. This seems to happen when the comet moves across a boundary in the solar wind into a region where the magnetic field is reversed. This illustrates the importance of the magnetic field in the solar wind in controlling the structure of gas tails. In September 1985, a spacecraft called ICE (International Cometary Explorer) flew through the tail of the comet Giacobini-Zinner and confirmed the magnetic nature of the gas tail. The spacecraft found the magnetic field of the solar wind draped

Table 26-3 Elements and Ions Detected in Comets

Coma	Gas Tail
H, OH, O	CO^+, CO_2^+
C, C_2, C_3, CH	H_2O^+, OH^-
CN, CO, CS, S	CH^+, CN^+, N_2^+
NH, NH_2, HCN, CH_3CN	C^+, Ca^+
Na, Fe, K, Ca, V	
Cr, Mn, Co, Ni, Cu	

over the head of the comet like seaweed draped over a fishhook. The gas tail was composed of ionized gas trapped in this magnetic tail, streaming outward with the solar wind.

A **type II tail** is a dust tail. It contains dust but not ionized gas, so it merely reflects the solar absorption spectrum. A dust tail is unaffected by the magnetic field in the solar wind, and the dust is pushed gently outward by the pressure of sunlight. Dust tails generally have a curve because the dust particles, once released by the comet, are pushed outward by sunlight, and, obeying Kepler's laws, their orbital velocity decreases. As the dust particles fall behind the head of the comet, the dust tail develops a curve. A dust tail lacks the wisps and twists of a gas tail because it contains no ionized gas and is not affected by kinks and shifts in the magnetic field of the solar wind. Thus type II tails can often be recognized by their smooth appearance.

Comet dust poses a threat to spacecraft that try to approach the head of a comet. Spacecraft flying past Comet Halley's nucleus at high speed were damaged by impacts with dust particles, and one was knocked tumbling. Astronauts may someday visit the nucleus of a comet, but they will have to approach slowly.

Dust released by comets fills the inner solar system, and the Infrared Astronomy Satellite found that comets and their orbits are rich in dust warmed by sunlight. Spacecraft analysis of the dust grains around Comet Halley revealed that the dust is partly silicate material with a high content of carbon. Thus it resembles the meteorites known as carbonaceous chondrites. Comet dust falls into Earth's atmosphere continuously and can be collected by high-flying aircraft (Figure 26-20). It settles unnoticed on roofs and parking lots all over the world, but one team of geologists found a novel way to collect this material. They vacuumed up about 10 kg of comet dust from the bottom of an ice lake in Greenland. Such dust must have been collecting on the ice and washing into the lake over thousands of years.

The observed properties of comets are determined by the icy nuclei at their hearts. We are now ready to explore the nature of these peculiar little worlds.

The Geology of Comet Nuclei

The nucleus of a comet is quite small and cannot be resolved by any telescopes now in use. Nevertheless, data from spacecraft that flew past Halley's comet and modern infrared observations of comet nuclei give astronomers a way to estimate the size of the nuclei. The nucleus of Comet Halley was observed to be 16 by 8 by 7 km. The nucleus of Comet Hale-Bopp is estimated to be significantly larger, about 40 km in diameter.

The nuclei of comets are commonly thought of as balls of ice, but water has been difficult to detect. Observations made in the ultraviolet from Earth orbit can detect a vast cloud of hydrogen surrounding the nu-

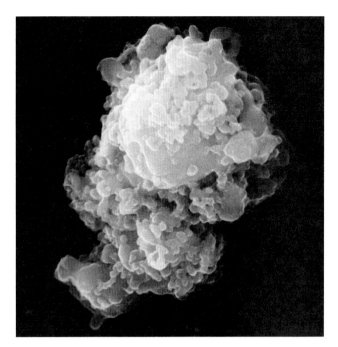

Figure 26-20
NASA aircraft flying at an altitude of 18 km (59,000 feet) can collect dust falling into Earth's atmosphere from space. This dust speck is only 6 millionths of a meter in diameter. Most of this dust is released by comets as they pass through the inner solar system. *(NASA)*

cleus of bright comets. This sphere is larger than the sun and must be continuously resupplied with fresh hydrogen coming from the head of the comet at the prodigious rate of about 10^{29} atoms per second (Figure 26-21a). Apparently, these atoms come from disrupted water molecules.

The spacecraft that flew past the nucleus of Comet Halley in 1986 sent back photos showing a dark, irregularly shaped body venting gas and dust from active regions on its sunward side (Figure 26-21b). The surface is so dark it reflects only 4 percent of the light that hits it; a sooty lump of coal reflects 6 percent. The dark color of this dust suggests a composition similar to that of carbonaceous chondrite meteorites. Spacecraft measurements show that the temperature of this dark surface is 300 to 400 K, far above the temperature of vaporizing ices. Evidently, the dark crust insulates the interior from the sun's heat.

From the gravitational effects the nucleus had on the passing spacecraft, astronomers could find the mass of the nucleus and thus its density, 0.1 to 0.25 g/cm³. This is much less than the density of ice. From these observations we can conclude that the nucleus is not solid ices but must be a fluffy mixture of ices with empty voids covered by a black, dusty crust of carbonaceous material that may vary from a few centimeters to a kilometer in depth.

Photographs of the coma of a comet often show jets springing from the nucleus and being swept back by

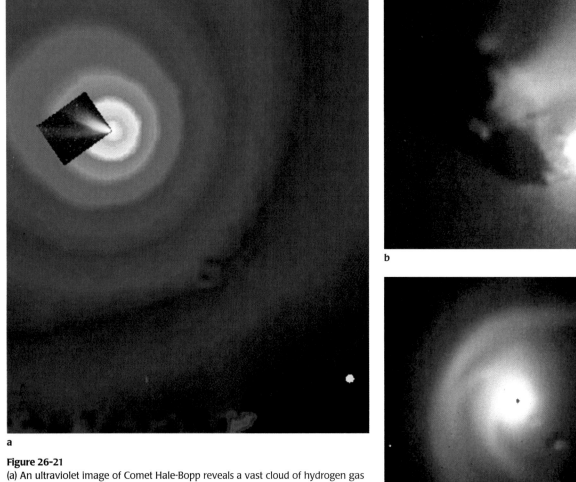

Figure 26-21
(a) An ultraviolet image of Comet Hale-Bopp reveals a vast cloud of hydrogen gas surrounding the visible comet (inset). Concentric bands in the cloud are produced by the digitizing process. At lower right is the sun to scale. *(SOHO/SWAN Consortium, ESA, and NASA)* (b) This image of the nucleus of Comet Halley taken by the Giotto spacecraft shows jets of dust and gas venting from the dark nucleus. *(©1986 Max-Planck Institute)* (c) In the inner coma of Comet Halley, jets from the nucleus are bent back to form the tail. The black dot is produced by the imaging process. The actual nucleus is too small to detect. *(Steven Larson)*

the pressure of sunlight and by the solar wind to form the tail (Figure 26-21c). Studies of the motions of these jets as the nucleus rotates compared with the photographs of the nucleus of Comet Halley reveal that the jets originate from active regions that may be faults extending through the crust into the ices below. As the rotation of a cometary nucleus carries an active region into sunlight, it rapidly begins venting gas and dust, and, as it rotates into darkness, it quickly shuts down. These jets are strong enough to act as small rocket motors slowly altering the orbital motion of the comet.

The irregular shape of the nucleus of Comet Halley suggests that the crust is a fluffy layer of rock and dust left behind by the vaporizing ices. As regions of ice sublime from solid to vapor, the crust collapses into the resulting voids. Some regions of cometary nuclei seem extra rich in highly volatile ices, so the nuclei are evidently not uniform mixtures.

The nuclei of comets are not strong. A nucleus can be ripped apart by the violence of gases bursting through the crust or by tidal forces produced if the comet passes near a massive body. A number of comets have been observed to break into two or more parts while near Jupiter or near the sun. On July 8, 1992, Comet Shoemaker-Levy 9 passed within 1.29 planetary radii of Jupiter's center, well within its Roche limit, and tidal forces ripped the nucleus into at least 21 pieces (Figure 26-22a). The fragmented pieces were as large as a few kilometers in diameter, and they released gas and dust and spread into a long string of objects that looped away from Jupiter and then fell back to strike the planet and produce massive impacts over a period of six days in July 1994 (see Figure 24-9). The fragmentation of comet nuclei may explain long chains of craters found on the Earth's moon and on some other planetary surfaces in the solar system. A comet breaking into pieces could produce a chain of impact features such as the ones shown in Figure 26-22b and c.

The nucleus of a comet is doomed. Even if it does not break up and even if it does not hit a planet, a comet in the inner solar system is a snowball in hell. Each passage near the sun costs such an object many millions of tons of ices lost as venting gas, so the nuclei slowly loose their ices until there is nothing left but dust and rock, the dead heart of a comet.

The Origin of Comets

Family relationships among the comets give us clues to their origin. Most comets have long, elliptical orbits with periods greater than 200 years. These are known as long-period comets. Their orbits are randomly inclined with comets falling into the inner solar system from all directions. As many circle the sun clockwise as counterclockwise.

In contrast, about 100 or so of the 600 well-studied comets have orbits with periods less than 200 years. These short-period comets follow orbits that lie within 30° of the plane of the solar system, and most revolve around the sun counterclockwise—the same direction the planets orbit. Comet Halley, with a period of 76 years, is a short-period comet.

Comets cannot survive long before the heat of the sun vaporizes their ices and reduces them to inactive bodies of rock and dust. A comet may last only 100 to 1000 orbits around the sun. The shortest-period comet known, Encke's Comet, has a period of only 3.3 years. Because of one or more encounters with planets long ago, Encke's comet is trapped in the inner solar system and has lost much of its ice. It no longer produces a dramatic coma and tail. This means that the comets we see in our skies can't have survived 4.6 billion years since the formation of the solar system. There must be a continuous supply of new comets. Where do new comets come from?

In the 1950s, Dutch astronomer Jan Oort proposed that the long-period comets are objects that fall in from a spherical cloud of icy bodies believed to extend from 10,000 to 100,000 AU from the sun (Figure 26-23). Astronomers estimate that the cloud, now known as the **Oort cloud,** contains several trillion icy bodies. Far from the sun, they are very cold, lack comas and tails, and are invisible. The gravitational influence of occasional passing stars could perturb a few of these objects to fall into the inner solar system, where the heat of the sun warms their ices and transforms them into comets. Because the Oort cloud is spherical, these long-period comets fall inward from random directions.

It is not outlandish that stars pass close enough to affect the Oort cloud. Data from the Hipparcos satellite show that the star Gliese 710 will

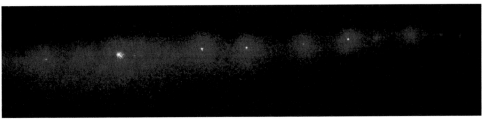
a

b

c

Figure 26-22
(a) Tides from Jupiter pulled apart the nucleus of Comet Shoemaker-Levy 9 to form a long strand of icy bodies and dust that fell back to strike Jupiter two years later. (b) A 40-km-long crater chain on the moon and (c) another 140-km-long crater chain on Jupiter's moon Callisto were apparently formed by the impact of fragmented comet nuclei. The impacts that form such chains probably occur within a span of seconds. *(NASA)*

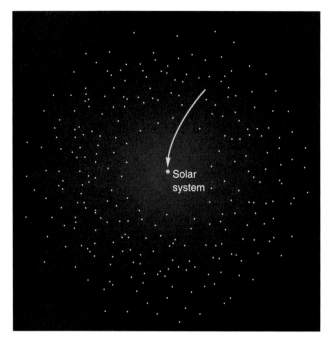

Figure 26-23
The long-period comets appear to originate in the Oort cloud. Objects that fall into the solar system from this cloud arrive from all directions.

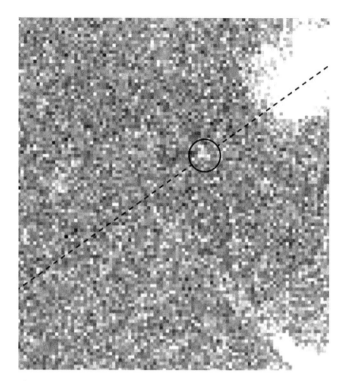

Figure 26-24
A possible Kuiper belt object lies in the circle in this exposure with the Hubble Space Telescope. The object, probably an icy body a few kilometers in diameter, was found following an orbit (dashed line) typical of a Kuiper belt object. Hubble Space Telescope and ground-based observations have located dozens of such objects. *(A. Cochran, University of Texas, and NASA)*

pass within 1 ly (about 63,000 AU) of our solar system in about a million years. That may shower the inner solar system with Oort cloud objects, which, warmed by the sun, will become comets.

Theoretical calculations show that comets falling in from the spherical Oort cloud cannot interact with the planets strongly enough to become the short-period comets. The short-period comets must come from a different reservoir of icy bodies. In 1951, Dutch-American astronomer Gerard P. Kuiper proposed that the formation of the solar system should have left behind a belt of small, icy planetesimals beyond the Jovian planets and in the plane of the solar system. Known today as the **Kuiper belt,** this disk extends from 30 AU to as much as 100 AU from the sun. The entire Kuiper belt would be hidden in Figure 26-23 behind the yellow dot symbolizing the solar system. The short-period comets appear to be visitors from this Kuiper belt.

Although the Kuiper belt was originally proposed as a theory, astronomers are now able to observe some of the Kuiper belt objects (Figure 26-24). Searches for small icy bodies in the outer solar system have so far identified over 100 objects that are members of the Kuiper belt. Also, the physical and chemical similarities among the icy moons of the outer solar system such as Triton and Charon suggest these are large Kuiper belt objects that passed too close to planets and were captured as moons. In fact, astronomers commonly think of Pluto as the largest of the remaining Kuiper belt objects. Working at the extreme limit of its capability, the

Hubble Space Telescope has produced statistical evidence of a population of Kuiper belt objects roughly the size of the nucleus of Comet Halley.

We can find more evidence that our solar system has a Kuiper belt by looking at other stars. A number of nearby stars such as Beta Pictoris (Chapter 20) are surrounded by disks of dust believed to be released by comets in the equivalent of Kuiper belts around these stars. Thus the detection of the dust implies the presence of Kuiper belts.

The sky seems abuzz with comets, but no comet has ever been identified as coming from beyond our solar system. That is, all known comet orbits are consistent with origin in the Oort cloud or in the Kuiper belt. Interstellar comets may occasionally fall through our solar system, but they are sufficiently rare that none have been recognized yet.

The Origin of Comets

When we say comets come from the Oort cloud or the Kuiper belt we only push the mystery back one step. Where did those icy bodies come from? Beginning in Chapter 20, we have studied the origin and evolution of our solar system so carefully that the answer leaps out at us. Those icy bodies are the last icy planetesimals remaining from the solar nebula.

When Gerard Kuiper first suggested the existence of the Kuiper belt, he proposed that the icy bodies there had formed from the solar nebula. Closer to the sun, massive planets had accreted or ejected the planetesimals, but beyond Neptune there are no massive planets. The icy planetesimals are safe there, especially those locked in orbital resonances with the outer planets. Those icy little worlds must have grown by condensation and accretion as the rest of the planetary system was forming.

The bodies in the Oort cloud, however, could not have formed at their present location. The solar nebula would not have been dense enough so far from the sun. Also, had they formed from the solar nebula, we would expect them to be distributed in a disk and not in a sphere. Rather, astronomers think the future comets formed as icy planetesimals in the outer solar system near the present orbits of Uranus and Neptune. As those planets grew more massive, they swept up many of these planetesimals, but they would have ejected some into long orbits leading far out of the solar system. These are the objects that became the Oort cloud.

REVIEW Critical Inquiry

How do comets help us understand the formation of the planets?

According to the solar nebula hypothesis, the planets formed from planetesimals that accreted in a disk-shaped nebula around the forming sun. The planetesimals that formed in the inner solar nebula were warm and could not incorporate much ice. The asteroids may be the last remains of such bodies. In the outer solar nebula, it was colder, and the planetesimals would have contained large amounts of ices. Many of these planetesimals were destroyed when they fell together to make the planets, but some may have survived. The icy bodies of the Oort cloud and the Kuiper belt may be the last surviving icy planetesimals in our solar system. When these bodies fall into the inner solar system, we see them as comets, and the gases they release tell us that many of them are rich in volatile materials such as water and carbon dioxide. These are the ices we would expect to find in the icy planetesimals. Furthermore, comets are rich in dust, and the planetesimals must have included large amounts of dust frozen into the ices when they formed. Thus, the nuclei of comets seem to be frozen samples of the ancient solar nebula.

Nearly all of the mass of a comet is in the nucleus, but the light we see comes from the coma and the tail. What kind of spectra do we get from comets, and what does that tell us about the process that converts a dirty iceberg into a comet?

Our discussion of comets has revealed that there is a continuous supply of these objects, that they fragment easily when they encounter planets, and that entire comets or fragments of comets can hit planets. That makes us wonder what happens when a comet hits Earth.

26-4 Impacts on Earth

Asteroids and comets fall through our solar system like runaway trucks on a busy highway. These objects must hit the planets. In fact, it is these collisions that built the planets, and most of the original planetesimals have now been either incorporated into the planets or ejected from the solar system. Nevertheless, planets still get hit, as was dramatically demonstrated in 1994 when Comet Shoemaker-Levy hit Jupiter. Like any other planet, Earth must get hit now and then, and we should look for evidence that such impacts have occurred.

Impacts and Dinosaurs

Comet nuclei and asteroids may seem small compared to planets, but they hit Earth with tremendous impact. Many scientists suspect that such impacts throw vast quantities of dust into the atmosphere and cause widespread changes in climate. The extinction of entire species, such as the dinosaurs, may have been caused by comet impacts (Figure 26-25a). One of the most important clues is an element called iridium.

In 1980, Luis and Walter Alvarez announced the discovery of an excess of the element iridium in sediments laid down at the end of the Cretaceous period—just when the dinosaurs and many other species became extinct. Iridium is rare in Earth's crust but common in meteorites, which led the Alvarezes to suggest that a major meteorite impact at the end of the Cretaceous threw vast amounts of iridium-rich dust into the atmosphere. This dust might have plunged Earth into a winter that lasted many years, killing off many species of plants and animals, including the dinosaurs.

This theory was rejected at first, but soon scientists found the iridium anomaly in sediments of the same age all over the world. Others found related elements and mineral forms typical of meteorite impacts. At the same time, theorists studying nuclear weapons predicted that a nuclear war would throw so much dust into the atmosphere that our planet would be plunged into a "nuclear winter" that could last a number of years. Within a few years, scientists generally agreed that the Cretaceous extinctions were caused by one or more major impacts.

Mathematical models and observations of the impact of the comet fragments on Jupiter in 1994 have combined to create a plausible scenario of the events

a

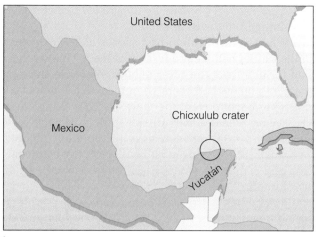

b

Figure 26-25

(a) The theory that the impact of one or more comets altered Earth's climate and drove dinosaurs to extinction has become so popular it appeared on this Hungarian stamp. The spacecraft shown (ICE) flew through the tail of Comet Giacobini-Zinner in 1985. Note the dead dinosaurs in the background. (b) The giant impact scar buried in Earth's crust near the village of Chicxulub in the northern Yucatán was formed about 65 million years ago.

following a major impact on Earth. Of course, creatures living near the site of the impact would probably die in the initial shock, and an impact at sea would create tsunamis (tidal waves) many hundreds of meters high that would devastate coastal regions for many kilometers inland even halfway around the world from the impact. But the worst effects would begin after the initial explosion. On land or sea, a major impact would excavate large amounts of pulverized rock and loft it high above the atmosphere. As this material fell back, Earth's atmosphere would be turned into a glowing oven of red-hot rock streaming through the air as a rain of meteors, and the heat would trigger massive forest fires around the world. Soot from such fires has been detected in the layers of clay laid down at the end of the Cretaceous. Once the fire storms cooled, the remaining dust in the atmosphere would block sunlight and produce deep darkness for a year or more. No matter where on Earth an impact occurred, it would almost certainly vaporize large amounts of limestone. The carbon dioxide released from the limestone would produce intense acid rain. All of these consequences make it surprising that any life could have survived such an impact.

Geologists have located a crater at least 150 km in diameter centered near the village of Chicxulub in the northern Yucatán (Figure 26-25b). Although the crater is totally covered by sediments, mineral samples show that it contains shocked quartz typical of impact sites and that it is the right age. A gigantic impact formed the crater about 65 million years ago, just when the dinosaurs and many other species died out, and many Earth scientists now believe that this is the scar of the impact that ended the Cretaceous.

At first, these climate-changing impacts were blamed on large meteorites, but the nucleus of a comet would do as much damage. In fact, we have seen in this chapter that the distinction between an asteroid and a comet is not clear-cut.

Some studies of mass extinctions have suggested that they occur every 27 to 30 million years, and a few scientists have suggested that this might be linked to an astronomical cycle. If the sun had a low-mass companion star orbiting it with a period of 27 to 30 million years, it would be too distant to detect easily and would have no significant influence on the solar system. If its orbit were elliptical, however, it would come closer to the sun once on each orbit and might stir up the Oort cloud and cause a storm of comets to fall into the inner solar system. A few of these could hit Earth and trigger extinctions. This is an entertaining idea, and the proposed companion star has even been given a name—Nemesis, after the goddess of divine retribution. Nevertheless, the evidence for the periodic nature of extinctions is very shaky, and there is no evidence at all to support the existence of Nemesis.

The controversy continues, but it is an interesting example of how scientists work with new ideas. The iridium anomaly and its meaning for the Cretaceous extinctions are now widely accepted. The periodic nature of mass extinctions and impacts is viewed with caution and, in some cases, skepticism. Suggestions that showers of comets caused periodic extinctions are highly speculative, and most scientists are waiting for better evidence before they take such hypotheses seriously.

A solar system is a dangerous place to put a planet. Comets, asteroids, and meteoroids constantly rain down on the planets, and Earth gets hit quite often. The Chicxulub Crater isn't the only large impact scar on Earth. For example, a giant crater has been found buried under sediment in Iowa, and another giant impact crater may underlie most of the Chesapeake Bay. What would it be like to live on Earth when such an impact occurs? Humanity got a hint in 1908 when something hit Siberia.

The Tunguska Event

On the morning of June 30, 1908, scattered reindeer herders and homesteaders in central Siberia were startled to see a brilliant blue-white fireball brighter than the sun streak across the sky. Still descending, it exploded with a blinding flash and an intense pulse of heat.

The whole northern part of the sky appeared to be covered with fire. . . . I felt great heat as if my shirt had caught fire . . . there was a . . . mighty crash. . . . I was thrown on the ground about [7 m] from the porch. . . . A hot wind, as from a cannon, blew past the huts from the north.

The blast was heard up to 1000 km away, and the resulting pulse of air pressure circled Earth twice. For a number of nights following the blast, European astronomers, who knew nothing of the explosion, observed a glowing reddish haze high in the atmosphere.

Travel in the wilderness of Siberia was difficult early in this century; moreover, World War I, the Bolshevik Revolution, and the Russian Civil War prevented expeditions from reaching the site before 1927. When at last an expedition arrived, it found that the blast had occurred above the Stony Tunguska River valley and had flattened trees in an irregular pattern extending out 30 km (Figure 26-26). The trees were knocked down radially away from the center of the blast, and limbs and leaves had been stripped away. The trunks of trees at the very center of the area were still standing, although they had lost all their limbs. No expedition has ever found a crater, so it seems that the explosion, equaling a 12-megaton nuclear weapon, occurred at least a few kilometers above the ground.

All kinds of speculative theories have been proposed in tabloid newspapers and occult books. One popular idea was that a flying saucer with engine trouble tried to make an emergency landing and exploded.

Figure 26-26
The aftermath of the Tunguska blast. Trees were knocked down pointing radially away from the center of the explosion. *(Sovfoto)*

More responsible ideas have proposed that Earth was hit by a piece of antimatter or a miniature black hole. The evidence does not necessarily rule out either of these last two ideas, but they violate an important principle in science. They ask us to assume more than we need to. Science always prefers the simplest adequate explanation for any given phenomenon, and a small asteroid or comet would explain the event.

The comet hypothesis proposes that the head of a small comet struck Earth and exploded in the atmosphere. We might not see such a comet if it was coming from the direction of the sun. Witnesses reported that the fireball in the atmosphere came from the east, which in the morning means from the direction of the sun. If the object was a comet, whatever tail it had would have been projected straight at Earth and would not have been visible. More likely, say some astronomers, the object was the exhausted core of a comet and had no appreciable tail.

A fragment of a partially exhausted comet should consist of silicate material resembling carbonaceous chondrites, mixed with some remaining ices. Modern studies of the area show that thousands of tons of powdered material resembling carbonaceous chondrites are scattered in the soil. A chunk of dirty ice 50 m in diameter would contain about the right mass and would be totally invisible approaching from the sun. Also, a comet fragment would be so fragile that it would explode in the atmosphere rather than dig a crater, as would a denser, stronger meteorite.

The falling-comet hypothesis has been very popular with astronomers. It does not ask us to make any fantastic assumptions, and it explains the observed phenomena, even the reddish glow over Europe. Some astronomers claim the glow was the faint traces of the comet's scattered dust, trapped high in the atmosphere and illuminated by the summer sun shining over the North Pole.

As compelling as the comet hypothesis may seem, not all astronomers agree. In the early 1980s, a detailed analysis of all the evidence suggested that the object was not cometary. Its speed and direction were not consistent with the motion of a comet. The analysis suggested instead that the object was a small Apollo asteroid, 90 to 190 m in diameter.

In 1993, astronomers studied computer models of objects entering Earth's atmosphere at high speed and concluded that the fragile head of a comet would have exploded much too high in the atmosphere. An iron-rich asteroid, in contrast, would be so strong that it would survive to reach the ground and would form a large crater. The most likely candidate seems to be a stony asteroid about 30 m in diameter. It would explode at just about the right height to produce the Tunguska blast.

One natural question is: Why should a piece of rock explode at all? To understand, we must recall that a meteor enters Earth's atmosphere at roughly 15 km/s. The air resistance is tremendous, and as the meteor descends lower and meets denser air, the air resistance increases rapidly. The resistance that the meteor experiences depends on its surface area. When the meteor breaks into smaller pieces, its surface area suddenly increases, so the air resistance suddenly increases and thus causes more fragmentation. Because air resistance depends on surface area, when the meteor begins to break up, it explodes and vaporizes almost instantly. That is most likely what created the blast that equaled a 12-megaton explosion at Tunguska.

Could such an impact happen again? Earth gets hit by small meteorites every day and by larger objects less often. Impacts by large asteroids may happen many millions of years apart, but they continue to happen. In mid-March 1998, newspaper headlines announced, "Mile-Wide Asteroid to Hit Earth in October 2028." The frightening story was true except that the news media did not mention the uncertainty in the orbit. Within days, astronomers found the asteroid on old photographs, recalculated the orbit with the new data, and concluded that the asteroid, known as 1997XF$_{11}$, will miss Earth by 600,000 miles. There will be no impact by this asteroid in 2028, but there are plenty more asteroids that haven't been discovered yet.

Some people have argued that the danger from asteroid and comet impacts is so great that we should develop massive nuclear-tipped missiles, ready to blast a meteoroid to pieces long before it can reach Earth. Some astronomers point out that lots of small fragments slamming into Earth may be even worse than one big impact. Other astronomers argue that the big objects are so rare we can ignore them. The real danger, they say, lies in the smaller, more common meteoroids, which are too small to detect with current telescopes.

Some people may prefer to believe that the Tunguska event was an alien spaceship; that is a titillating idea. But the truth about the Tunguska event is far more exciting. Earth is struck by meteorites every day. Some impacts are large enough to alter the climate, and the future of our civilization on Earth may depend on how much time we have until the next big one.

R E V I E W   Critical Inquiry

How can a large meteorite impact alter Earth's climate?

Even a meteorite no bigger than a house strikes Earth with as much energy as a nuclear bomb and can produce a crater a mile in diameter. A large impact could produce a crater 100 km or more in diameter. Rock from Earth's crust would be pulverized and lifted high above the atmosphere. As it fell back in a rain of fiery meteors, it would be so hot it would trigger massive fires over much of Earth's surface. We can detect traces of such fires at about the time of the Chicxulub impact, 65 million years ago. Even after the fires cooled, the smoke and dust in the atmosphere would prevent sunlight from reaching the ground, and plants dependent on sunlight would be starved of energy. This worldwide darkness, called a nuclear winter, might last a year or more before the dust settled out of the atmosphere. By that time, many species would have died out, and the climate could have changed dramatically.

As scary as this is, we need not ask if it will happen. The probability of such an event is 100 percent. Our question can only be: How soon? In fact, such large impacts happen occasionally on all of the terrestrial planets. Summarize the evidence we have to show that large impacts have occurred on Earth and on other planets and satellites in our solar system.

Life on Earth seems fragile and exposed. Are we just lucky to have survived so long? Does life exist on other worlds? If so, has it, too, survived long enough to develop intelligence? That is the story of the next chapter.

Summary

Most of the meteors we see, and certainly those in showers, are caused by small particles released by comets. When Earth crosses the orbit of a comet, we see a meteor shower. Most meteorites, however, are more massive, stronger fragments that probably originated in the asteroid belt.

The different kinds of meteorites could have originated in a planetesimal-size body that was heated by the decay of short-lived radioactive nuclei like aluminum-26. Differentiation would form an iron–nickel core that would cool slowly. Fragments from this core would show the Widmanstätten pattern common in iron meteorites. The outer parts of such a body might resemble the stony-iron and achondritic meteorites.

Chondrites, with their beadlike chondrules, may have formed as smaller bodies farther from the sun. The carbonaceous chondrites, being rich in volatile compounds, also probably formed farther from the sun. The presence of volatiles, chondrules, and CAIs shows that the carbonaceous chondrites are the least altered of the meteorites.

Collisions must have been common among the planetesimals, just as they are still common in the asteroid belt. We find breccias in some meteorites that must have been formed in impacts long ago. In the asteroid belt today, we can find evidence that the asteroids are fragmented. The Hirayama families appear to be fragments with similar orbits. Photographs of asteroids taken by spacecraft and radar images of other asteroids made from Earth show irregular-shaped bodies heavily battered by impact craters.

Of the three kinds of asteroids, the S types are more common in the inner belt, whereas the C types are more common in the outer belt. This may represent a difference in temperature at the time of formation or different surface weathering. The M types appear to be the exposed iron–nickel cores of broken planetesimals.

Some of the asteroids show clear evidence of geological processes. The reflected light from Vesta, for instance, looks much like the light reflected from achondritic meteorites that resemble basalts. Thus, it seems that Vesta has melted and has had lava flows on its surface.

Jupiter clearly dominates the asteroid belt and may have prevented the formation of a planet there long ago. It now causes Kirkwood's gaps and traps the Trojan asteroids at the Lagrangian points along its orbit.

According to the dirty snowball theory, a comet is a ball of dirty ices that produces a coma as it approaches within a few astronomical units of the sun. The pressure of the solar wind pushes ionized gas back into a type I, or gas, tail; the pressure of sunlight pushes dust particles back into a type II, or dust, tail.

When Comet Halley passed through the inner solar system in 1986, five spacecraft flew past it and discovered that its nucleus was larger and blacker than expected. The nucleus was about 16 km × 8 km × 7 km and was covered with a black crust having an albedo of only 0.04. The crust is probably made of material resembling carbonaceous chondrites. Gas and dust jets vent from active regions.

According to the Oort cloud theory, comets occupy a hollow spherical cloud centered on the sun and extending out about 100,000 AU. Perturbations from the motions of nearby stars may cause a few of these icy bodies to fall into the inner solar system, where they become long-period comets. These dirty snowballs originated as icy planetesimals among the Jovian planets and were ejected into the Oort cloud. Comets are almost unaltered samples of the original solar nebula.

The short-period comets have small orbits and periods of less than 200 years. Comet Halley is an example. These comets appear to originate in the Kuiper belt, a disk-shaped cloud of icy bodies believed to extend from 40 AU to at least 100 AU from the sun. Over 100 of these bodies have been found,

and astronomers speculate that Pluto, Charon, Triton, and Chiron may be icy worlds related to the Kuiper-belt objects.

Earth runs into space debris quite often. The Apollo asteroids must hit Earth with great regularity, and the heads of short-period comets must also intersect Earth now and then. One such collision, the Tunguska event, occurred as recently as 1908.

New Terms

radiant	achondrite
sporadic meteor	stony-iron meteorite
fall	Kirkwood's gaps
find	Apollo–Amor objects
iron meteorite	Trojan asteroids
selection effect	Hirayama families
Widmanstätten pattern	coma
stony meteorite	gas (type I) tail
chondrite	dust (type II) tail
chondrule	Oort cloud
carbonaceous chondrite	Kuiper belt
CAI	

Review Questions

1. If most falls are stony meteorites, why are most finds iron?

2. How do observations of meteor showers reveal one of the sources of meteoroids?

3. How can most meteors be cometary if most, perhaps all, meteorites are asteroidal?

4. Why are meteorites easier to find in Antarctica? Why are stony meteorites better represented among these finds than among finds in the United States?

5. What evidence do we have that some meteorites have originated inside large bodies?

6. Why couldn't uranium, thorium, and radioactive potassium have melted the planetesimals? What evidence do we have that some asteroids have had molten interiors?

7. How do Kirkwood's gaps resemble Cassini's division?

8. The first asteroids discovered were much larger than the typical asteroid known today. How does this illustrate a selection effect?

9. Describe three types of asteroids (S, M, and C), and explain a theory to account for their differences.

10. How might the Apollo–Amor objects have originated?

11. Why doesn't a type II comet tail ever suffer a disconnection?

12. If comets are icy planetesimals left over from the formation of the solar system, why haven't they all vaporized by now?

Discussion Questions

1. Futurists suggest that we may someday mine the asteroids for materials to build and supply space colonies. What kinds of materials could we get from asteroids? (*Hint:* What are S-, M-, and C-type asteroids made of?)

2. If cometary nuclei were heated by internal radioactive decay rather than by solar heat, how would comets differ from what we observe?

3. From what you know now, do you think the government should spend money to locate near-Earth asteroids? How serious is the risk?

Problems

1. Large meteorites are hardly slowed by Earth's atmosphere. Assuming that the atmosphere is 100 km thick and that a large meteorite falls vertically toward the ground, how long does it spend in the atmosphere? (*Hint:* How fast do meteoroids travel?)

2. If a single asteroid 1 km in diameter were fragmented, how many meteoroids 1 m in diameter could it yield? (*Hint:* The volume of a sphere is $\frac{4}{3}\pi r^3$.)

3. What is the orbital velocity of a meteoroid orbiting the sun at the distance of Earth? (*Hint:* See Chapter 5.)

4. If a half-million asteroids, each 1 km in diameter, were assembled into one body, how large would it be? (*Hint:* The volume of a sphere is $\frac{4}{3}\pi r^3$.)

5. The asteroid Vesta has a mass of 2×10^{20} kg and a radius of about 250 km. What is its escape velocity? Could you jump off the asteroid? (*Hint:* See Chapter 5.)

6. The asteroid Pallas has a mass of 2.5×10^{20} kg. What is the orbital velocity of a small satellite orbiting 500 km from the center of Pallas? (*Hint:* See Chapter 5.)

7. What is the maximum angular diameter of Ceres as seen from Earth? Could Earth-based telescopes detect surface features? Could the Hubble Space Telescope (Chapter 6)? (*Hint:* Use the small-angle formula.)

8. At what distances from the sun would you expect to find Kirkwood's gaps where the orbital period of asteroids is one-half of and one-third of the orbital period of Jupiter? Compare your results with Figure 26-12. (*Hint:* Use Kepler's third law.)

9. If the velocity of the solar wind is about 400 km/s and the visible tail of a comet is 100 million km long, how long does it take an atom to travel from the nucleus to the end of the visible tail?

10. If you saw Comet Halley when it was 0.7 AU from Earth and it had a visible tail 5° long, how long was the tail in kilometers? Suppose that the tail was not perpendicular to the line of sight. Is your answer too large or too small? (*Hint:* Use the small-angle formula.)

11. Calculate the orbital velocity of a comet while it is in the Oort cloud. (*Hint:* See Chapter 5.)

12. The mass of a comet's nucleus is about 10^{14} kg. If the Oort cloud contains 2 trillion (2×10^{12}) cometary nuclei, what is the mass of the cloud in Earth masses? (*Hint:* Earth's mass $= 6 \times 10^{24}$ kg.)

Critical Inquiries for the Web

1. Some nights are better for looking for meteors than others (see Table 26-1). We know that these showers are associated with comets, but how are these associations made? Several sites on the Internet provide information on meteor showers, including historical data and information on parent comets. Pick a shower whose parent comet is known, and summarize how we came to know that the meteors and the comet are related.

2. Chances are small that you will be killed by an asteroid impact, but if objects are out there that astronomers are not aware of whose orbits intersect Earth, we could be in for a surprise one day. Look for information on the Spacewatch project, NEAT, and other searches for near-Earth asteroids. How many such objects have been discovered? What is the record for closest known passage of an asteroid to Earth?

3. Search for the IAU Minor Planet Center Web pages, and find out what asteroids and what comets have come closest to Earth.

4. Search for information about comets in the sky right now. Are any comets bright enough for you to see? Are any bright comets expected soon?

Exploring *The Sky*

1. Many people have seen meteors and comets. However, few have seen asteroids, because they tend to be rather faint and usually require binoculars or telescopes to see. This is your opportunity to see an asteroid—at least a virtual asteroid! Use *The Sky* to locate the largest of the asteroids, Ceres, and determine:

   a. The constellation in which to find Ceres at the present time.

   b. Its present distance from the sun (between which two planetary orbits is it located?).

   c. Its present distance from Earth.

   d. Its magnitude (if it's brighter than 6th magnitude, it may be possible to see it with the unaided eye).

 Go to the Brooks/Cole Astronomy Resource Center (www. brookscole.com/astronomy) for critical thinking exercises, articles, and additional readings from InfoTrac College Edition, Brooks/Cole's online student library.

CHAPTER TWENTY-SEVEN

Life on Other Worlds

Did I solicit thee from darkness to

promote me?

John Milton
Paradise Lost

Guidepost

This chapter is either unnecessary or critical depending on our point of view. If we believe that astronomy is the study of the physical universe above the clouds, then this chapter does not belong here. But if we believe that astronomy is the study of our position in the universe, not only our physical position but also our position as living beings in the origin and evolution of the universe, then everything else in this book is just preparation for this chapter.

Astronomy is the only science that truly acts as a mirror. In studying the universe up there, we learn what we are down here. Astronomy is not really about stars, galaxies, and planets; it is about us.

Science is a way of understanding the world around us, and the heart of that understanding is the explanations that science gives us for natural phenomena. Whether we call these explanations stories, histories, theories, or hypotheses, they are attempts to describe how nature works based on fundamental rules of evidence and intellectual honesty. While we may take these explanations as factual truth, we should understand that they are not the only explanations that satisfy the rules of logic.

A separate class of explanations involves religion, and those explanations can be quite logical. The Old Testament description of the creation of the world, for instance, does not fit scientific observations, but if we accept the existence of an omnipotent being, then the biblical explanation is internally logical and acceptable. Of course, it is not a scientific explanation, but religion is a matter of faith and not subject to the rules of evidence. Religious explanations follow their own logic, and we would be wrong to demand that they follow the rules of evidence that govern scientific explanations.

If scientific explanations are not the only logical explanations, then why do we give them such weight? First, we must notice the tremendous success of scientific explanations in producing technological advances in our daily lives. Smallpox is a disease of the past thanks to the application of scientific explanations to modern medicine. The power of science to shape our world can lead us to think that its explanations are unique. Second, the process we call science depends on the use of evidence to test and perfect our explanations, and the logical rigor of this process gives us great confidence in our conclusions.

Scientific explanations have given us tremendous insight into the workings of nature, and consequently both scientists and nonscientists tend to forget that there can be other logical explanations. The so-called conflict between science and religion has been symbolized for centuries by the trial of Galileo. That conflict is easier to understand when we consider the nature of scientific explanations and the role of evidence in testing scientific understanding.

As living things, we have been promoted from darkness. We are made of heavy atoms that could not have formed at the beginning of the universe. Successive generations of stars fusing light elements into heavier elements have built the atoms so important to our existence. When a dark cloud of interstellar gas enriched in these heavy atoms fell together to form our sun, a small part of the cloud gave birth to the planet we inhabit.

Are there intelligent beings living on other planets? That is the last and perhaps the most challenging question in our study of astronomy. We will try to answer it in three steps, each dealing with a different aspect of life.

First, we must decide what we mean by life. A living thing is not so much a physical object as a process. We are not simply the matter that forms our bodies but rather a tremendously complex system that has the ability to duplicate and protect itself. Thus, life is based on information that contains the directions for the processes of duplication and preservation.

Our second step is to study the origin of life. Direct investigation is limited to Earth, but if we can understand how life began here, we can better estimate the chances that it occurred elsewhere. We will find that life on Earth probably began with simple chemical reactions that happened naturally. If these gave rise to life on Earth, similar reactions may have provided the spark on other worlds.

Our third step is to study evolution, the process by which life improves its ability to survive. The survival of stable species has transformed the simple organisms that began in Earth's oceans into a wide variety of creatures with special adaptations. The rose's thorn, the deer's quickness, and the human's intelligence are protective adaptations. Evolution is a natural, physical process. If it can work on Earth, then it surely works on any planet where life begins, and if we assume that intelligence is a valuable trait, then intelligent beings may eventually emerge.

If life is common in the universe, where might we look for it, and how might we communicate with other intelligent beings? Certainly the prospects of finding life, intelligent or otherwise, on any of the other planets in our solar system are bleak. If we are to find extraterrestrial life, we must go beyond our solar system and search among any planets that may orbit other stars.

Communication with intelligent races on other worlds may be possible, but we cannot expect to travel between solar systems. Interstellar distances are so great that only in science fiction do spaceships flit from star to star. However, it may be possible to communicate via radio. If civilizations can survive for long periods of time at a technological level at which they can build large radio telescopes, then we may be able to send and receive messages. Such messages would mark a turning point in the history of humanity. If life is common in the universe, such signals may be detected during our lifetime.

Our goal in this chapter is to use our knowledge of astronomy, combined with the rules of evidence that guide all of science, to try to understand the range of possibilities for life in our universe (Window on Science 27-1). We begin with a simple question: Is life on other worlds possible?

27-1 The Nature of Life

What is life? Philosophers have struggled with that question for thousands of years, so it is unlikely that we will answer it here. But we must agree on a working model of life before we can speculate on its occurrence on other worlds. To that end, we will identify in living things three properties: a process, a physical base, and a unit of controlling information.

The life process is aimed at survival. Living things extract energy from their environment and use that energy to modify their surroundings to make their own preservation more likely. For example, human beings obtain energy by eating and breathing, and they use that energy to build houses, cities, and stable societies to protect themselves. The same can be said of a bacterium absorbing food and reinforcing its cell structure.

This apparently selfish protection of the individual is aimed at the preservation of the race through safe reproduction. The ability to reproduce is one of the distinguishing characteristics of living organisms, and any organism that does not, in some way, ensure safe reproduction will not survive many generations. The entire life process is aimed at safe reproduction because any other target is self-destructive.

The Physical Basis of Life

On Earth, the physical basis of life is the carbon atom (Figure 27-1). Because of the way this atom bonds to other atoms, it can form long, complex, stable chains that are capable of extracting, storing, and utilizing energy. Other chemical bases of life may exist. Science-fiction stories and movies abound with silicon creatures, living things whose body chemistry is based on silicon rather than carbon. However, silicon forms weaker bonds than carbon does, and it cannot form double bonds as easily. Consequently, it cannot form the long, complex, stable chains that carbon can. Silicon is 135 times more common on Earth than carbon is, yet there are no silicon creatures among us. All Earth life is carbon based. Thus, the likelihood that distant planets are inhabited by silicon people seems small, but we cannot rule out life based on noncarbon chemistry.

a

Figure 27-1
All living things on Earth are based on carbon chemistry. (a) Katie, a complex mammal containing about 30 astronomical units of DNA. *(Michael Seeds)* (b) Tobacco mosaic virus. Each rod is a single spiral strand of RNA about 0.01 mm long surrounded by a protein coat. *(L. D. Simon)*

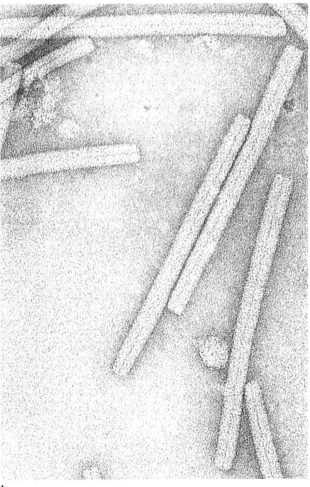

b

In fact, nonchemical life might be possible. What is required is some mechanism capable of supporting the extraction and utilization of energy that we have identified as life. One could at least imagine life based on electromagnetic fields and ionized gas. No one has ever met such a creature, but science-fiction writers conjure up all sorts.

Clearly, we could range far in space and time, theorizing about different bases for alien life, but to make progress we must discuss what we know best—carbon-based life on Earth. How can a lump of carbon-rich matter live? The answer lies in the information that guides its life processes.

Information Storage and Duplication

The key to understanding life is information—the information the organism uses to control its utilization of energy. We must discover how life stores and uses that information and how the information changes and thus preserves the species.

The unit of life on Earth is the cell, the self-contained factory capable of absorbing nourishment from its surroundings, maintaining its own existence, and, in multicellular organisms, performing its task within the larger organism (Figure 27-2). The foundation of the cell's activity is a set of patterns that describe how it is to function. This information must be stored in the cell in some safe location, yet it must be passed on easily to new

cells and be used readily to guide the cell's activity. To understand how matter can be alive, we must have an understanding of how the cell stores, reproduces, and uses this information.

The information is stored in long carbon-chain molecules called **DNA (deoxyribonucleic acid),** most of which reside in the cell nucleus. The structure of DNA resembles a long, twisted ladder. The rails of the ladder are made of alternating phosphates and sugars; the rungs are made of pairs of molecules called bases (Figure 27-3). Only four kinds of bases are present in DNA, and the order in which they appear on the DNA ladder represents the information the cell needs to function. One human cell stores about 1.5 meters of DNA, containing about 4.5 billion pairs of bases. Thus 4.5 billion pieces of information are available to run a human cell. That is enough to record all of the works of Shakespeare over 200 times. Because the human body contains about 60×10^{12} cells, the total DNA in a single human adult would stretch 9×10^{13} m, about 600 AU.

Storing all this data in each cell does the organism no good unless the data can be reproduced and passed on to new cells. The DNA molecule is specially adapted for duplicating itself by splitting its ladder down the center of the rungs, producing two rails with protruding bases (Figure 27-4). These quickly bond with the proper bases, phosphates, and sugars to reconstruct the missing part of the molecule, and presto—the cell has two complete copies of the critical information. One set goes to each of the newly forming cells. Thus, the DNA is the genetic information passed from parent to offspring.

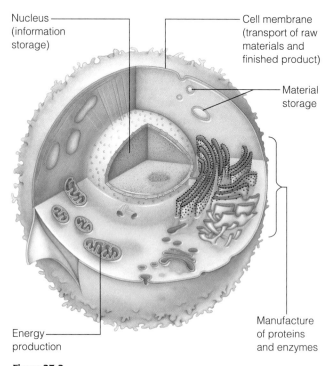

Nucleus (information storage)

Cell membrane (transport of raw materials and finished product)

Material storage

Manufacture of proteins and enzymes

Energy production

Figure 27-2
A living cell is a self-contained factory that absorbs raw materials from its surroundings and uses them to maintain itself and manufacture finished products for the use of the organism as a whole.

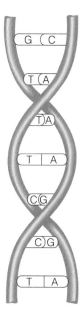

Figure 27-3
The DNA molecule consists of two vertical strands of sugars and phosphates (dark) and rungs made of bases: adenine (A), cytosine (C), guanine (G), and thymine (T).

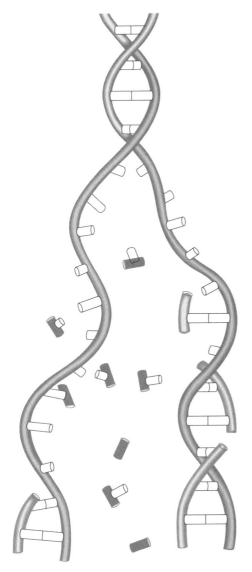

Figure 27-4
The DNA molecule can duplicate itself by splitting in half (top), assembling on each half matching bases, sugars, and phosphates (center), and thus producing two DNA molecules (bottom). The actual duplication process is significantly more complex than this schematic diagram.

Segments of the DNA molecules are patterns for the production of **proteins.** Many proteins are structural molecules—the cell might make protein to repair its cell wall, for example. **Enzymes** are special proteins that control other processes—growth, for example. Thus, the DNA molecule contains the recipes to make all of the different molecules required in an organism.

Actually, the cell does not risk its precious DNA patterns by involving them directly in the manufacture of protein. The DNA stays safely in the cell nucleus, where it produces a copy of the patterns by assembling a long carbon-chain molecule called **RNA (ribonucleic acid).** This RNA carries the information out of the nucleus and then assembles the proteins from simple molecules called **amino acids,** the basic building blocks of protein. Thus the RNA acts as a messenger, carrying copies of the necessary plans from the central office to the construction site.

Although the information coded in the DNA must be preserved for the survival of the organism, it must be changeable or the species will not survive very long at all. To see why, we must study evolution, the process that rewrites the data in the DNA.

Modifying the Information

If living things are to survive for many generations, then the information stored in their DNA must change as the environment changes. Without change in DNA, a slight warming of the climate, for example, might kill a species of plant, in turn starving the rabbits, deer, and other plant eaters, and leaving the hawks, wolves, and mountain lions with no prey. If the information stored in DNA could never change, then environmental changes would quickly drive life forms to extinction. If life is to survive in a changing world, then the information in DNA must be changeable. Living things must evolve.

Species evolve by **natural selection.** Each time an organism reproduces, its offspring receive the data stored in the DNA, but some variation is possible. For example, most of the rabbits in a litter may be normal, but it is possible for one to get a DNA recipe that gives it stronger teeth. If it has stronger teeth, it may be able to eat something other than the plant the others depend on, and if that plant is becoming scarce, the rabbit with stronger teeth has a survival advantage. It can eat other plants and so will be healthier than its littermates and have more offspring. Some of these offspring may also have stronger teeth, as the altered DNA data are handed down to the new generation. Thus nature selects and preserves those attributes that contribute to the survival of the species. Those that are unfit die. Natural selection is merciless to the individual, but it gives the species the best possible chance to survive.

The only way nature can obtain new DNA patterns from which to select the best is from DNA molecules that have changed. This can happen through chance mismatching of base pairs—errors—in the reproduction of the DNA molecule. Another way this can occur is through damage to reproductive cells from exposure to radioactivity. Cosmic rays or natural radioactivity in the soil might perform this function. In any case, an offspring born with altered DNA is called a **mutant.** Most mutations are fatal, and the individual dies long before it can have offspring of its own. But in rare cases, a mutation may give a species a new survival advantage. Then natural selection makes it likely that the new DNA message will survive and be handed down, making the species more capable of surviving.

Why can't the information in DNA be permanent?

The information stored in a creature's DNA provides the recipes that make the creature what it is. For example, the DNA in a starfish must contain all the recipes for making the various kinds of proteins needed to consume and digest food. That information must be passed on to offspring starfish, or they will be unable to survive. But the information must be changeable because our environment is changeable. Ice ages come and go, mountains rise, lakes dry up, and ocean currents shift. If the environment changes in some way, one or more of the recipes may no longer work. In our example, a change in the temperature of the ocean water may kill off the shellfish the starfish eat. If they can't digest other shellfish, the entire species will become extinct. Natural variation in DNA means that among all the infant starfish in any generation, some of the recipes are different; if the environment changes, all of the old-style starfish may die, but a few—those with the different DNA—can carry on.

The survival of life depends on this delicate balance between reliable reproduction and the introduction of small variations in DNA information. What are some of the ways these small changes in DNA can arise?

Life is based not only on information but also on the duplication of information. Today, that process seems so complex that it is hard to imagine how it could have begun.

27-2 The Origin of Life

Clearly the carbon chemistry of life on Earth is extremely complex. How could it have ever gotten started? Obviously, 4.5 billion chemical bases didn't just happen to drift together to form the DNA formula for a human. The key is evolution. Once a life form begins to reproduce itself, natural selection preserves the most advantageous traits. Over long periods spanning thousands, perhaps millions, of generations, the life form becomes more fit to survive. This nearly always means the life form becomes more complex. Thus, life could have begun as a very simple process that gradually became more sophisticated as it was modified by evolution.

We begin our study on Earth, where fossils and an intimate familiarity with carbon-based life give us a glimpse of the first living matter. Once we discover how Earthly life could have begun, we can look for signs that life began on other planets in our solar system. Finally, we can speculate on the chances that other planets, orbiting other stars, have conditions that give rise to life.

The Origin of Life on Earth

The oldest fossils hint that life began in the oceans. The oldest easily identified fossils appear in sedimentary rocks that formed 500 to 600 million years ago—the **Cambrian period.** Such Cambrian fossils were simple ocean creatures, the most complex of which were trilobites (Figure 27-5), but there are no Cambrian fossils of land plants or animals. Evidently, land surfaces were totally devoid of life until only 400 million years ago.

Precambrian deposits contain no obvious fossils, but microscopes reveal microfossils that were the ancestors of the Cambrian creatures. Fig-tree chert, a rock that resembles flint, in South Africa is 3.0 to 3.3 billion years old, and Onverwacht shale, also found in South Africa, may be as old as 3.6 billion years. These and similar deposits contain structures that appear to be microfossils of bacteria or simple algae such as those that live in water (Figure 27-6). Apparently, life was already active in Earth's oceans a billion years after the planet formed.

The key to the origin of this life may lie in an experiment performed by Stanley Miller and Harold Urey

Figure 27-5
Trilobites made their first appearance in the Cambrian oceans about 600 million years ago. This example, about the size of a human hand, lived 400 million years ago in an ocean floor that is now a limestone deposit in Pennsylvania. *(Grundy Observatory)*

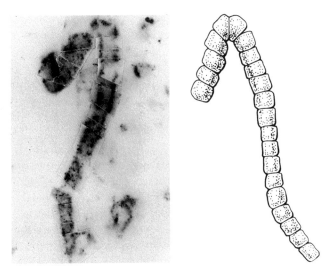

Figure 27-6
Among the oldest fossils known, this microscopic filament resembles modern bacterial forms (artist's reconstruction at right). This fossil was found in the 3.5-billion-year-old chert of the Pilbara Block in northwestern Australia. *(J. William Schopf)*

in 1952. This **Miller experiment** sought to reproduce the conditions on Earth under which life began. In a closed glass container, the experimenters placed water (to represent the oceans), the gases hydrogen, ammonia, and methane (to represent the primitive atmosphere), and an electric arc (to represent lightning bolts). The apparatus was sterilized, sealed, and set in operation (Figure 27-7).

After a week, Miller and Urey stopped the experiment and analyzed the material in the flask. Among the many compounds the experiment produced, they found four amino acids (building blocks of protein), various fatty acids, and urea, a molecule common to many life processes. Evidently, the energy from the electric arc had molded the atmospheric gases into some of the basic components of living matter. Other energy sources such as hot silica (to simulate hot lava spilling into the sea) and ultraviolet radiation (to simulate sunlight) give similar results.

Recent studies of the composition of meteorites and models of planet formation suggest that Earth's first atmosphere did not resemble the gases used in the Miller experiment. Earth's first atmosphere was probably composed of carbon dioxide, nitrogen, and water vapor. This change, however, does not invalidate the Miller experiment. The point of the experiment is to show how easily organic molecules occur in natural settings.

The Miller experiment did not create life, nor did it necessarily imitate the exact conditions on the young Earth. Rather, it is important because it shows that complex organic molecules form naturally in a wide variety of circumstances. The chemical deck is stacked to deal nature a hand of complex molecules. If we could travel back in time, we would probably find Earth's oceans filled with a rich mixture of organic compounds in what some have called the **primordial soup.**

The next step on the journey toward life is for the compounds dissolved in the oceans to link up and form

a

b

Figure 27-7
The Miller experiment (a) circulated gases through water in the presence of an electric arc. This simulation of primitive conditions on Earth produced amino acids, the building blocks of proteins. (b) Stanley Miller with a Miller apparatus. *(Courtesy Stanley Miller)*

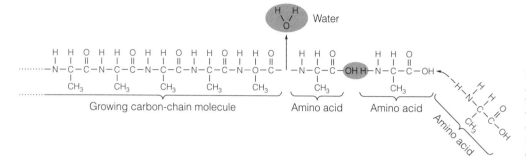

Figure 27-8
Amino acids can link through the release of a water molecule to form long carbon-chain molecules. The amino acid in this hypothetical example is alanine, one of the simplest.

larger molecules. Amino acids, for example, can link to form proteins. This linkage occurs when amino acids join end to end and release a water molecule (Figure 27-8). For many years, scientists have assumed that this process must have happened in sun-warmed tidal pools where evaporation concentrated the broth. But recent studies suggest that the young Earth was subject to extensive volcanism and large meteorite impacts that periodically modified the climate enough to destroy any life forms exposed on the surface. It seems most likely that the early growth of complex molecules took place among the hot springs along the midocean ridges. The heat from such springs could have powered the growth of long protein chains. Deep in the oceans, they would have been safe from climate changes.

Although these proteins might have contained hundreds of amino acids, they were not alive. Not yet. Such molecules did not reproduce but merely linked and broke apart at random. However, because some molecules are more stable than others and because some molecules bond more readily than others, this **chemical evolution** led to the concentration of the varied smaller molecules into the most stable larger forms. Eventually, somewhere in the oceans, a molecule took shape that could reproduce a copy of itself. At that point, the chemical evolution of molecules became the biological evolution of living things.

It is natural for us to assume that the first reproducing molecule was some form of DNA, but that is not necessary. DNA is a very complex molecule. Some experts believe that the first reproducing molecule was a primitive form of RNA, a simpler molecule. The first life may have lived in an RNA world, and DNA may have developed later. Still others argue that the first replicating molecules were some form of protein that then formed the first RNA. The details are far from clear, but there seem to be a number of ways that reproducing molecules could have formed.

An alternative theory proposes that primitive living things such as reproducing molecules did not originate on Earth but came here in meteorites or comets. Radio astronomers have found a wide variety of organic molecules in the interstellar medium, and some studies have found similar compounds inside meteorites (Figure 27-9). Such molecules form so readily

that we would be surprised if they were not present in space. A few investigators, however, have speculated that living, reproducing molecules originated in space and came to Earth as a cosmic contamination. If this is true, every planet in the universe is contaminated with the seeds of life. However entertaining this theory may be, it is presently untestable, and an untestable theory is of little use in science. Experts studying the origin of life proceed on the assumption that life began as reproducing molecules in Earth's oceans.

Figure 27-9
A sample of the Murchison meteorite, a carbonaceous chondrite that fell in 1969 near Murchison, Australia. Analysis of the interior of the meteorite revealed evidence of amino acids. Whether the first building blocks of life originated in space is unknown, but the amino acids found in meteorites illustrate how commonly amino acids and other complex molecules occur even in the absence of living things. *(Courtesy Chip Clark, National Museum of Natural History)*

Which came first, reproducing molecules or the cell? Because we think of the cell as the basic unit of life, this question seems to make no sense, but in fact the cell may have originated during chemical evolution. If a dry mixture of amino acids is heated, the acids form long, proteinlike molecules that, when poured into water, collect to form microscopic spheres that function in ways similar to cells (Figure 27-10). They have a thin membrane surface, absorb material from their surroundings, grow in size, and divide and bud just as cells do. However, they contain no large molecule that copies itself. Thus, the structure of the cell may have originated first and the reproducing molecules later.

An alternative theory supposes that the replicating molecule developed first. Such a molecule would be exposed to damage if it were bare, so the first to manufacture or attract a protective coating of protein would have a significant survival advantage. If this is the case, the protective cell membrane is a later development of biological evolution.

The first living things must have been single-celled organisms much like modern bacteria. Some of the oldest fossils known are **stromatolites,** structures produced by communities of photosynthesizing bacteria that grew in mats and, year by year, deposited layers of minerals that were later fossilized. One of the oldest such fossils known is 3.5 billion years old (Figure 27-11). If such bacteria were common when Earth was young, the early

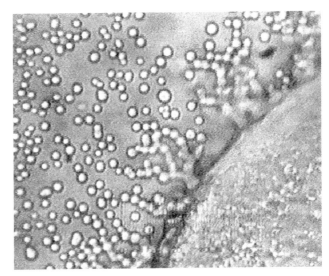

Figure 27-10
Single amino acids can be assembled into long proteinlike molecules. When such material cools in water, it can form microspheres, microscopic spheres with double-layered boundaries similar to cell membranes. Microspheres may have been an intermediate stage in the evolution of life between complex molecules and cells holding molecules reproducing genetic information. *(Courtesy Sidney Fox and Randall Grubbs)*

atmosphere may have contained a small amount of oxygen produced by the photosynthesis. Recent studies suggest that an oxygen abundance of only 0.1 percent

Figure 27-11
A 3.5-billion-year-old fossil stromatolite from western Australia is one of the oldest known fossils (below). Stromatolites were formed, layer by layer, by mats of bacteria living in shallow water. Such life may have been common in shallow seas when Earth was young (right). Stromatolites are still being formed today in similar environments. *(Mural by Peter Sawyer; photo courtesy Chip Clark, National Museum of Natural History)*

would have been sufficient to provide an ozone screen that would protect organisms from the sun's ultraviolet radiation.

How evolution shaped creatures to live in the ancient oceans, to photosynthesize and respire, to become multicellular, and to reproduce sexually is a fascinating story, but we cannot explore it in detail here. We can see that life could begin through simple chemical reactions building complex molecules and that once some DNA-like molecule formed, it protected its own survival with selfish determination. Over billions of years, the genetic information stored in living things kept those qualities that favored survival and discarded the rest. As Samuel Butler said, "The chicken is the egg's way of making another egg." In that sense, all living matter on Earth is merely the physical expression of DNA's mindless determination to continue its existence.

Perhaps this seems harsh. Human experience goes far beyond mere reproduction. *Homo sapiens* has art, poetry, music, philosophy, religion, science—all of the great, sensitive accomplishments of our intelligence. Perhaps that is more than mere reproduction of DNA, but intelligence, the ability to analyze complex situations and respond with appropriate action, must have begun as a survival mechanism. For example, a fixed escape strategy stored in DNA is a disadvantage for a creature that frequently moves from one environment to another. A rodent that always escapes from predators by automatically climbing the nearest tree would be in serious jeopardy if it met a hungry fox in a treeless clearing. Even a faint glimmer of intelligence might allow the rodent to analyze the situation and, finding no trees, to choose running over climbing. Thus, intelligence, of which *Homo sapiens* is so proud, may have developed in ancient creatures as a way of making them more versatile.

Any discussion of the evolution of life seems to involve highly improbable coincidences until we consider how many years have passed in the history of Earth. We read the words—4.6 billion years—so easily, but in truth it is hard to grasp the meaning of such a long period of time.

Geologic Time

Humanity is a very new experiment on planet Earth. We can fit all of the evolution that led from the primitive life forms in the oceans of the Cambrian period 600 million years ago to fishes, amphibians, reptiles, and mammals into a single chart such as Figure 27-12. We can take comfort in thinking that creatures like us have walked on Earth for at least 3 million years, but when we add our history to the chart, we discover that the entire history of humanity makes up no more than a thin line at the top. In fact, if we try to represent the entire 4.6-billion-year history of Earth on the chart, the portion describing the rise of life on land is an unreadably small segment.

One way to represent the evolution of life is to compress the 4.6-billion-year history of Earth into a 1-year-long videotape. In such a program, Earth forms as the video begins on January 1, and through all of January and February it cools and is cratered and the first oceans form. Search as we might, we will find no trace of life in these oceans until sometime in March or early April, when the first living things develop. The slow development of these simplest of living forms grinds on slowly through the spring and summer of our videotape. The entire 4-billion-year history of Precambrian evolution lasts until the video reaches mid-November, when the primitive ocean life begins to evolve into more complex organisms such as trilobites.

While our year-long videotape plays on and on, we might amuse ourselves by looking at the land instead of the oceans, but we would be disappointed. The land is a lifeless waste, with no plants or animals of any kind. Not until November 28 in our video does life appear on the land, but once it does, it evolves rapidly into a wide range of plants and animals. Dinosaurs, for example, appear about December 12 and vanish by Christmas evening as mammals and birds flourish.

Throughout the 1-year run of our video, there have been no humans, and even during the last days of the year, as the mammals rise to dominate the landscape, there are no people. In the early evening of December 31, vaguely human forms move through the grasslands, and by late evening they begin making stone tools. The Stone Age lasts until about 11:45 PM, and the first signs of civilization, towns and cities, do not appear until 11:54 PM. The Christian era begins only 14 seconds before the New Year, and the Declaration of Independence is signed with but 1 second to spare.

By converting the history of Earth into a year-long videotape, we have placed the rise of life in perspective. Tremendous amounts of time were needed for the first simple living things to evolve in the oceans, and even more time was needed for the evolution of complex creatures that could colonize the land. As life became more complex, it evolved and diversified faster and faster, as if evolution were drawing on a growing library of solutions that had previously been invented with great effort to solve earlier problems. The burst of diversity on land led slowly to the rise of intelligent creatures like us, a process that has taken 4.6 billion years.

The vast scope of geologic time can be illustrated in another way. The atoms in your body have a wonderful history, and we can now tell their story. The matter was born in the big bang, but it was almost entirely hydrogen and helium. The heavier atoms in your body, such as carbon, nitrogen, and oxygen, were created by nuclear fusion inside stars. At least two or three generations of stars must have lived and died to create the heavier atoms that were present in the nebula that gave

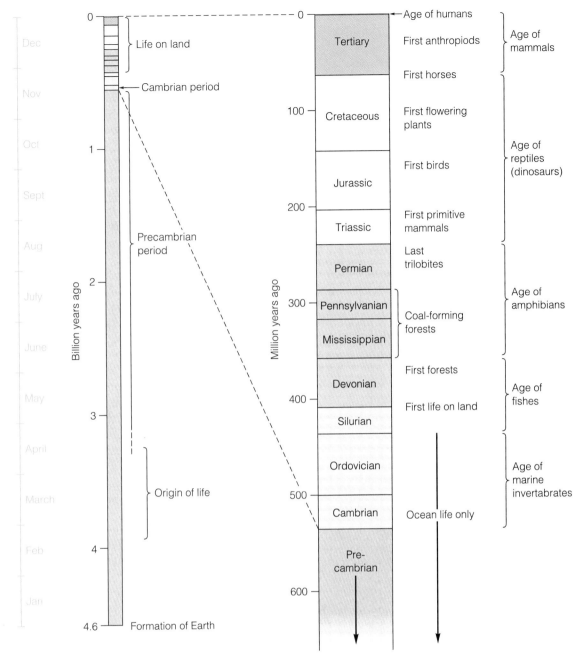

Figure 27-12
Complex life has developed on Earth only recently. If the entire history of Earth were represented in a time line (left), we would have to magnify the end of the line to see details such as life leaving the oceans and dinosaurs appearing. The age of humans would still be only a thin line at the top of our diagram. If Earth's history were a year-long videotape, humans would not appear until the last hours of December 31st.

birth to the sun and Earth. The atoms in your body were present in the first oceans and were part of the first organisms. When life emerged onto the land, some of your atoms were present. Most of the carbon atoms in your body have been inside at least one dinosaur. We only borrow our atoms from Earth. They will have further adventures in the future.

Astronomers take a cosmic view of life. The long sweep of geologic time is only part of a longer story of stars and atoms.

If life could originate on Earth and develop into intelligent creatures, perhaps the same thing could have happened on other planets. This raises three questions. First, could life originate if conditions were suitable? We have seen that the origin of life seems to be inherent in the way atoms form chemical bonds. If conditions are right on another world, the atoms must form the same kinds of molecules as did atoms on Earth, and that should lead to life.

The second question is more difficult. If life begins on a world, will it evolve toward intelligence? It is a common error to think that evolution is random. Variation is random. Some elephants are bigger than others. But natural selection is not random. It favors those who reproduce successfully. The genes for large size may help an elephant have many offspring, and thus those genes will be handed down to future generations. But genes for the intelligence to form social bonds in a protective herd can also further reproduction. Intelligence is a highly sophisticated trait, and it may take a long time to develop, but it seems likely that evolution leads to increasing levels of intelligence among some species. Our tentative answer to question two is yes; evolution should lead eventually to intelligence.

The third question gives us pause. Are suitable conditions so rare that life almost never gets started? The only way to answer that question is to search for life on other worlds. We begin, in the next section, with the other planets and moons in our solar system.

Life in Our Solar System

Although we can imagine life based on something other than carbon chemistry, we know of no examples to tell us how such life might arise and survive. We must limit our discussion to life as we know it and the conditions it requires. The most important requirement is the presence of liquid water, not only as part of the chemical reactions of life, but also as a medium to transport nutrients and wastes within the organism. Also, it seems that life on Earth began in the oceans and developed there for almost 4 billion years before it was able to emerge onto the land. Certainly, any world where we hope to find life must have liquid water, and that means it must have moderate temperatures.

The water requirement automatically eliminates many worlds in our solar system. The moon is airless, and although some data suggest ice frozen in the soil at its poles, it has never had liquid water on its surface. In the vacuum of the lunar surface, liquid water would boil away rapidly. Mercury too is airless and cannot have had liquid water on its surface for long periods of time. Venus has some traces of water vapor in its atmosphere, but it is much too hot for liquid water to survive. If there were any lakes or oceans of water on its surface when it was young, they must have evaporated quickly. Even if life began there, no traces would be left now.

The inner solar system seems too hot, and the outer solar system seems too cold. The Jovian planets have deep atmospheres, and at a certain level, they have moderate temperatures where water might condense into liquid droplets. But it seems unlikely that life could begin there. The Jovian planets have no surfaces where oceans could nurture the beginning of life, and currents in the atmosphere seem destined to circulate gas and water droplets from regions of moderate temperature to other levels that are much too hot or too cold for life to survive.

A few of the satellites of the Jovian planets might have suitable conditions for life. Jupiter's moon Europa seems to have a liquid water ocean below its icy crust (Figure 27-13), and minerals dissolved in that water would provide a rich broth of possibilities for chemical evolution. Nevertheless, Europa is not a promising site to search for life because conditions may not have remained stable for the billions of years needed for life to evolve beyond the microscopic stage. The subsurface ocean is kept from freezing by tidal heating. If Jupiter's moons interact gravitationally and modify their orbits, Europa may have been frozen solid at some points in its history. Such periods of freezing would probably prevent life from developing. Drilling through the icy crust of Europa to search for life in its ocean will be a wonderful adventure for future generations, but Europa does not seem to be a good bet to harbor any form of complex life.

Saturn's moon Titan has an atmosphere of nitrogen, argon, and methane and may have oceans of liquid methane and ethane on its surface. We saw in Chapter 24 how sunlight can convert the methane in the atmosphere to organic smog particles that settle to the surface. The chemistry of life that might crawl or swim on such a world is unknown, but life there may be unlikely because of the temperature. The surface of Titan is a deadly −179° C (−290° F). Chemical reactions occur slowly or not at all at such low temperatures, so the chemical evolution needed to begin life may never have occurred on Titan.

Mars is the most likely place for life in our solar system. The evidence, however, is not encouraging. In 1976, two robotic spacecraft, Viking 1 and Viking 2,

Figure 27-13
The icy surface of Europa, digitally enhanced and colored here, is covered by a network of cracks whose shape suggests that the crust floats on an ocean of water. Heat flowing from the interior might provide the energy to support living things in the dark ocean. *(NASA)*

landed at two different places on the Martian surface. The spacecraft scooped up soil samples, and subjected them to tests for the presence of living organisms. For example, they gave some soil samples a dose of nutrient-rich water and watched for signs the nutrients were taken up by biological processes. The results seem negative. Although some peculiar chemical processes were detected, no evidence was found that clearly indicated the presence of any living things in the soil.

The chemical experiments performed by the Viking landers were sophisticated, but they could only search for traces of currently living things. They could not search for signs of past life, such as fossils. For that, we need to send a geologist to Mars or bring rocks from Mars back to Earth. Nature has performed the latter task for us.

Meteorite ALH84001 (Figure 27-14a) was found on the Antarctic ice in 1984 and only later recognized as one of a class of meteorites that originated on Mars. How do we know they came from Mars? About a dozen meteorites are classified as SNC meteorites because they have very similar physical and chemical properties. One of those meteorites contains gases trapped in small, glassy inclusions. Analysis of these gases shows that they have the same abundance of oxygen isotopes as the gases in the atmosphere of Mars as measured by the two Viking landers. The abundances of the isotopes are like fingerprints, and they match exactly. It seems conclusive that the SNC meteorites are rocks that were blasted off of the surface of Mars at some point long ago when a comet or an asteroid struck the planet. From traces of radioactive elements in ALH84001, scientists conclude it left the surface of Mars about 16 million years ago, orbited the sun for a long time, and then fell to Earth in Antarctica about 13,000 years ago.

In 1996, NASA scientists announced that they had found three kinds of evidence that life once existed in ALH84001. Part of the evidence consisted of carbonate deposits that resemble those produced by some Earthly

a

b

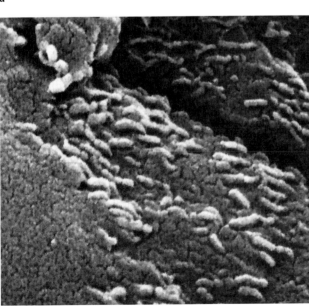

c

Figure 27-14
(a) ALH84001 is one of a dozen meteorites known to have originated on Mars. (b) Microscopic carbonate globules found inside the meteorite resemble similar deposits made by Earthly bacteria. (c) Concentrations of features that resemble Earthly bacteria have been found inside the meteorite. *(NASA)*

bacteria (Figure 27-14b). A second part of the evidence was complex organic molecules that could be the decayed remains of ancient bacteria. The third piece of evidence was revealed in electron microscope images that show what seem to be small fossils of living things (Figure 27-14c).

This discovery was received with great excitement by the press and the public. It would, indeed, be dramatic evidence if true, because it would show that life could begin on another world.

Scientists, being professionally skeptical (see Window on Science 20-2), began testing these results immediately, and in many cases the tests did not confirm the conclusion that life once existed in the meteorite. There is evidence that the water that deposited the carbonates was much too hot for living things to survive. Also, the organic molecules found in the meteorite may have formed from nonbiological processes, or they may be contamination. There is conclusive evidence that the interior of the meteorite has been contaminated by chemical compounds such as amino acids that exist in the Antarctic ice. Furthermore, the microscopic fossils may be natural mineral features of the rock samples and not the remains of living things. Although studies of ALH84001 continue, it does not give us unchallenged evidence that life once existed on Mars.

The issue is not closed. Some scientists continue to defend the evidence of life found in ALH84001, and other meteorites are being studied. Spacecraft now visiting Mars may help us understand the past history of water there and paint a more detailed picture of present conditions. Landers are planned that will eventually land on Mars, collect rocks, and return them to Earth. Nevertheless, conclusive evidence may have to wait until a geologist in a space suit can wander the dry streambeds of Mars cracking open rocks and searching for fossils.

We are left to conclude that, so far as we know, our solar system is bare of life except for Earth. Consequently our search for life in the universe takes us to other planetary systems.

Life in Other Planetary Systems

Might life exist in other solar systems? To consider this question, let us try to decide how common planets are and what conditions a planet must fulfill for life to originate and evolve to intelligence. The first question is astronomical; the second is biological. Our ability to discuss the problem of life outside our solar system is severely limited by our lack of experience.

In Chapter 20 we concluded that planets form as a natural by-product of star formation, and we learned that a number of extra-solar planets have been found circling nearby stars. From this we can conclude that planetary systems are very common.

If a planet is to become a suitable home for life, it must have a stable orbit around its sun. This is simple in a solar system like our own, but in a binary system most planetary orbits are unstable. Most planets in such systems would not last long before they were swallowed up by one of the stars or ejected from the system.

Thus, single stars are the most likely to have planets suitable for life. Because our galaxy contains about 10^{11} stars, half of which are single, there should be roughly 5×10^{10} planetary systems in which we might look for life.

A few million years of suitable conditions does not seem to be enough time to originate life. On our planet it took at least 0.5 to 1 billion years for the first cells to evolve and 4.6 billion years for intelligence to evolve. Clearly, conditions on a planet must remain acceptable over a long time. This eliminates massive stars that remain stable on the main sequence for only a few million years. If it takes a few billion years for life to originate and evolve to intelligence, no star hotter than about F5 will do. This is not really a serious restriction, because upper-main-sequence stars are rare anyway.

In previous sections, we decided that life, at least as we know it, requires liquid water. That requirement defines a **life zone** (or ecosphere) around each star, a region within which a planet has temperatures that permit the existence of liquid water.

The size of the life zone depends on the temperature of the star (Figure 27-15). Hot stars have larger life zones because the planets must be more distant to remain cool. But the short main-sequence lives of these stars make them unacceptable. M stars have small life zones because they are extremely cool—only planets very near the star receive sufficient warmth. However, planets that are close to a star would probably become tidally coupled, keeping the same side toward the star. This might cause the water and atmosphere to freeze in the perpetual darkness of the planet's night side and end all chance of life. Also, M stars are subject to sudden flares that might destroy life on a planet close to the star. Thus, the life zone restricts our search for life to main-sequence G and K stars.

Even a star on the main sequence is not perfectly stable. Main-sequence stars gradually grow more luminous as they convert their hydrogen to helium, and thus the life zone around a star gradually moves outward. A planet might form in the life zone, and life might begin and evolve for billions of years only to be destroyed as the slowly increasing luminosity of its star moves the life zone outward, evaporates the planet's oceans, and drives off its atmosphere. If a planet is to remain in the life zone for 4 to 5 billion years, it must form on the outer edge of the zone. This may be the most serious restriction we have yet discussed.

We should pause here to note that our concept of a life zone may be very self-centered. We admit that life might exist on worlds such as Europa and Titan, but

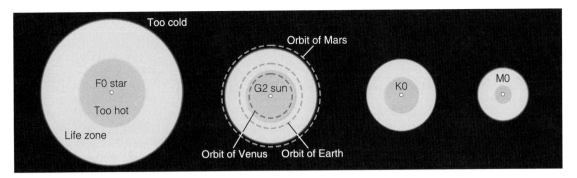

Figure 27-15
The life zone around a star (green) is the region where a planet would have a moderate temperature. M stars are not good candidates, because they are cool and their life zone is small. F stars have a large life zone but evolve too quickly to allow the rise of complex life. K and G stars like our sun seem the best candidates in a search for intelligent life.

they lie outside our conventionally defined life zone. Furthermore, scientists on Earth are discovering life in highly unlikely places such as in the Antarctic ice, in hot springs on the ocean floor, and even in solid rock deep underground. As we learn more about living things, we may have to extend our definition of the life zone to include a wider range of conditions.

The fundamental question in our discussion is simple. If conditions are right, will life begin? Early in this chapter, we decided that life could begin through simple chemical reactions, so perhaps we should change our question and ask: What could prevent life from beginning? Given what we know about life, it should arise whenever conditions permit, and our galaxy should be filled with planets that are inhabited with living creatures.

REVIEW Critical Inquiry

What evidence do we have that life is at least possible on other worlds?

Our evidence is limited almost entirely to Earth, but it is promising. The fossils we find show that life originated in Earth's oceans almost 4 billion years ago, and biologists have proposed relatively simple chemical processes that could have created these first reproducing molecules. The fossils show that life developed from a very slow beginning into more and more complex creatures that filled the oceans. The pace of evolution quickened dramatically about 600 million years ago, the beginning of the Cambrian period, when life began taking on complex forms; and later, when life emerged onto the surface of the land, it evolved rapidly to produce the tremendous diversity we see around us. Human intelligence has been a very recent development; it is only a few million years old.

If this process occurred on Earth, then it seems reasonable that it could have occurred on other worlds as

well. Thus, the evidence suggests we can expect that life might begin and evolve to intelligent forms on any world where conditions are right. What are the conditions we should expect of other worlds that host life?

It is both easy and fun to speculate about life on other worlds, but it leads to a simple question: Is there really life beyond Earth? If we can't leave Earth and visit other worlds, then the only life we can detect will be beings intelligent enough to communicate with us. Why haven't we heard from them?

27-3 Communication with Distant Civilizations

If other civilizations exist, perhaps we can communicate with them in some way. Sadly, travel between the stars appears more difficult in real life than in science fiction. It may in fact be almost impossible. If we can't physically visit, perhaps we can communicate by radio. Again, nature places restrictions on such conversations, but the restrictions are not too severe. The real problem lies with the nature of civilizations.

Travel Between the Stars

Roaming among the stars is, in practice, tremendously difficult because of three limitations: distance, speed, and fuel. The distances between stars are almost beyond comprehension. It does little good to explain that if we represent the sun by a golf ball in New York City, the nearest star would be another golf ball in Chicago. It is only slightly better to note that the fastest commercial jet would take about 4 million years to reach the nearest star.

When we discuss life on other worlds, we might be tempted to use UFO sightings and supposed visits by aliens from outer space as evidence to test our hypotheses. We don't do so for two reasons, both related to the reliability of these observations.

First, the reputation of the sources of UFO sightings and alien encounters does not give us confidence that these data are reliable. Most people hear of such events via grocery store tabloids, daytime talk shows, or sensational "specials" on viewer-hungry cable networks. We must take note of the low reputation of the media that report UFOs and space aliens. Most of these reports are simply made up for the sake of sensation, and we cannot use them as reliable evidence.

Second, the remaining UFO sightings, those not simply made up, do not survive careful examination. Most are mistakes or unconscious misinterpretations of natural events made by honest people. A number of unbiased studies have found no grounds for believing in UFOs. In short, there is no conclusive evidence that Earth has been visited by aliens from space.

That's too bad. A confirmed visit by intelligent creatures from beyond our solar system would answer many of our questions. It would be exciting, enlightening, and, like any real adventure, a bit scary. But none of the UFO sightings are dependable, and we are left with no direct evidence of intelligent life on other worlds.

The second limitation is a speed limit—we cannot travel faster than the speed of light. Although science fiction writers invent hyperspace drives so their heroes can zip from star to star, the speed of light is a natural and unavoidable limit that we cannot exceed. This, combined with the large distances between stars, makes interstellar travel very time consuming.

The third limitation says that we can't even approach the speed of light without using a fantastic amount of fuel. Even if we ignore the problem of escaping from Earth's gravity, we must still use energy stored in fuel to accelerate to high speed and to decelerate to a stop when we reach our destination. To return to Earth, assuming we wish to, we have to repeat the process. These changes in velocity require a tremendous amount of fuel. If we flew a spaceship as big as a large yacht to a star 5 ly (1.5 pc) away and wanted to get there in only 10 years, we would use 40,000 times as much energy as the United States consumes in a year.

Travel for a few individuals might be possible if we accepted very long travel times. That would require some form of suspended animation (currently unknown) or colony ships that carry a complete, though small, society in which people are born, live, and die generation after generation. Whether the occupants of such a ship would retain the social characteristics of humans over a long voyage is questionable.

These three limitations not only make it difficult for us to leave our solar system but also make it difficult for aliens to visit Earth. Reputable scientists have studied UFOs and related phenomena and have never found any evidence that Earth is being visited or has ever been visited by aliens from other worlds (Window on Science 27-2). Thus, it seems unlikely that humans will ever meet an alien face to face. The only way we can communicate with other civilizations is via radio.

Radio Communication

Nature places two restrictions on our ability to communicate with distant societies by radio. One has to do with simple physics, is well understood, and merely makes the communication difficult. The second has to do with the fate of technological civilizations, is still unresolved, and may severely limit the number of societies we can detect by radio.

Radio signals are electromagnetic waves that travel at the speed of light. Because even the nearest civilizations must be a few light-years away, this limits our ability to carry on a conversation with distant beings. If we ask a question of a creature 10 ly away, we will have to wait 20 years for a reply. Clearly, the give-and-take of normal conversation will be impossible.

Instead, we could simply broadcast a radio beacon of friendship to announce our presence. Such a beacon would have to consist of a pattern of pulses obviously designed by intelligent beings to distinguish it from natural radio signals emitted by nebulae, pulsars, and so on. For example, pulses counting off the first dozen prime numbers would do. In fact, we are already broadcasting a recognizable beacon. Short-wavelength radio signals, such as TV and FM, have been leaking into space for the last 50 years or so. Any civilization within 50 ly might already have detected us.

If we intentionally broadcast a signal, we can anticode it. A coded message is arranged to prevent strangers from reading it, but an anticoded message is arranged to make it as easy as possible to read even by strangers who do not know our language. To create an anticoded message to broadcast into space, we could transmit pulses to represent 1s and gaps to represent 0s. A message counting through the first few prime numbers would distinguish our signal from natural sources of radio noise and attract the attention of intel-

An anticoded message

1 0 1 0 0 1 1 1 1 1
0 0 1 0 1 0 0 1 0 0
0 1 0 1 0 0 1 0 1 0
0 1 0 1 0

5 rows of 7

1	0	1	0	0	1	1
1	1	1	0	0	1	0
1	0	0	1	0	0	0
1	0	1	0	0	1	0
1	0	0	1	0	1	0

7 rows of 5

1	0	1	0	0
1	1	1	1	1
0	0	1	0	1
0	0	1	0	0
0	1	0	1	0
0	1	0	1	0
0	1	0	1	0

Figure 27-16
An anticoded message is designed for easy decoding. Here a string of 35 radio pulses, represented as 1s and 0s, can be arranged in only two ways, as 5 rows of 7 or 7 rows of 5. The second way produces a friendly message. Any number of pulses can be used so long as it is the product of two prime numbers. Then the pulses can be arranged in only two ways.

ligent listeners on other worlds. We could even transmit a picture by sending a string of 1s and 0s that can be arranged in only two ways. One way produces nonsense, but the other way produces a meaningful picture (Figure 27-16).

In 1974, at the dedication of the 1000-ft radio telescope at Arecibo, radio astronomers transmitted such a signal toward the globular cluster M13, which is located in the constellation Hercules 26,000 ly from Earth. When the signal finally arrives, any aliens who detect it will be able to arrange its 1679 pulses in only two ways, as 23 rows of 73 or 73 rows of 23. The first arrangement will produce only a jumble, but the second arrangement will form a picture that describes life on Earth (Figure 27-17). Whether humans will still survive on Earth in 52,000 years when any reply to the message arrives is open to debate.

It took only minutes to transmit the Arecibo message. If more time were taken, a more detailed picture could be sent, and if we were sure our radio telescope was pointed at a listening civilization, we could send a series of pictures. With pictures we could teach aliens our language and tell them all about our life, our difficulties, and our accomplishments.

If we can think of sending such signals, aliens can think of it too. If we point our radio telescopes in the right direction and listen at the right wavelength, we might hear other intelligent races calling out to one another. This raises two questions: Which stars are the best candidates, and what wavelengths are most likely? We have already answered the first question. Main-sequence G and K stars have the most favorable characteristics. But the second question is more complex.

Only certain wavelengths are useful for communication. We cannot use wavelengths longer than about

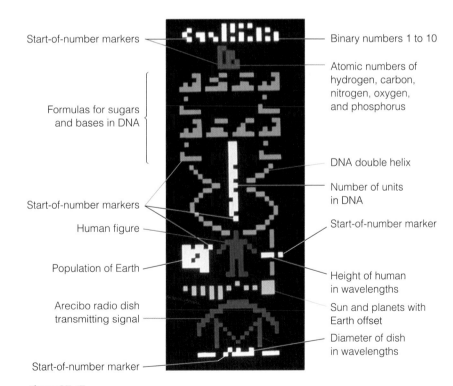

Start-of-number markers
Binary numbers 1 to 10
Atomic numbers of hydrogen, carbon, nitrogen, oxygen, and phosphorus
Formulas for sugars and bases in DNA
DNA double helix
Number of units in DNA
Start-of-number markers
Start-of-number marker
Human figure
Population of Earth
Height of human in wavelengths
Sun and planets with Earth offset
Arecibo radio dish transmitting signal
Diameter of dish in wavelengths
Start-of-number marker

Figure 27-17
The Arecibo message to M13 begins by counting from 1 to 10 and goes on to describe our solar system and the biochemistry of life on Earth (color added for clarity). Binary numbers give the height of the human figure (1110) and the diameter of the telescope dish (100101111110) in units of the wavelength of the signal, 12.3 cm. *(NASA)*

30 cm because the signal would be lost in the background radio noise from our galaxy. Nor can we go to wavelengths much shorter than 1 cm because of absorption within our atmosphere. Thus only a certain range of wavelengths, a radio window, is open for communication.

This communications window is very wide, so a radio telescope would take a long time to tune over all the wavelengths searching for intelligent signals. Nature may have given us a way to narrow the search, however. Within the communications window lie the 21-cm line of neutral hydrogen and the 18-cm line of OH. The interval between these two lines has been dubbed the **water hole,** because the combination of H and OH yields water (H_2O). Water is the fundamental solvent in our life form, so it might seem natural for similar water creatures to call out to each other at wavelengths in the water hole (Figure 27-18). But even silicon creatures would be familiar with the 21-cm line of hydrogen.

This is not idle speculation. A number of searches for extraterrestrial radio signals have been made, and some major searches are now under way. The field has become known as **SETI,** Search for Extra-Terrestrial Intelligence, and it has generated heated debate. Some scientists and philosophers argue that life on other worlds can't possibly exist, so it is a waste of money to search. Others argue that life on other worlds is common and could be detected with present technology. Congress funded a NASA search for a short time but then ended support because political leaders feared public reaction. In fact, the annual cost of a major search is only about as much as a single Air Force attack helicopter. The controversy may spring in part from the theological and philosophical controversy that would result from the discovery of intelligent life on another world.

In spite of the controversy, some searches have been made, and others are continuing. The first was Project Ozma in 1960. It listened to two stars at a wavelength of 21 cm. Since then, numerous small-scale searches have been conducted, but in order to have some assurance of success a search must scan the entire sky at millions of closely spaced frequency bands. Only recently has technology made that possible, and as technology improves, the searches can scan more sky and more frequencies.

From 1985 to 1994, META, Megachannel Extra-Terrestrial Assay, searched the entire sky at millions of frequency bands in the water hole. Funded by the Planetary Society, the search used radio antennas at Harvard and near Buenos Aires, Argentina. The project has instituted a second, even more efficient search called BETA. Since it began, the search has detected dozens of candidate signals. That is, it has found dozens of radio signals that satisfy its most basic criteria. Unfor-

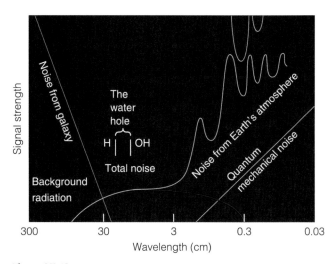

Figure 27-18
Radio noise from various sources makes it difficult to detect distant signals at wavelengths longer than 30 cm or shorter than 1 cm. In this range, radio emission from H atoms and from OH mark a small wavelength range dubbed the water hole, which may be a likely place for communication.

tunately, none of these signals has proved to be continuous, and no signal has been detected when the radio telescopes were redirected at any of the candidate stars. Many of these signals are probably unusual noise from sources on Earth, but the candidate signals are still under study.

One ingenious search is called SERENDIP, Search for Extraterrestrial Radio Emission from Nearby Developed Intelligent Populations. Rather than monopolize an entire radio telescope, SERENDIP rides piggyback on the 305-m Arecibo Telescope. Wherever the radio astronomers point the telescope, the receiver samples the signal, looking for intelligent signals over millions of frequency bands. If candidate signals are found, the receivers note the position of the radio telescope for future investigation.

The NASA search originally called SETI was renamed High Resolution Microwave Survey to reduce criticism, but Congress cut off all of its funds in the fall of 1993. The search continues, however, with private funding as Project Phoenix at the SETI Institute. It uses an 8.4-million-channel receiver on major radio telescopes and plans to survey the entire sky in 20 years. In addition, it will listen carefully to 800 nearby solar-type stars. The search officially began on October 12, 1992, 500 years after the landing of Columbus in the New World.

What should we expect if signals are found? By international agreement, those searching for signals have agreed to share candidate signals and obtain reliable confirmation before making public announcements. False alarms would be embarrassing. The chances of success depend on the number of inhabited worlds in our galaxy, and that number is difficult to estimate.

How Many Inhabited Worlds?

The technology exists, and given enough time the searches will find other inhabited worlds, assuming there are at least a few out there. If intelligence is common, then we should find the signals soon—in the next few decades—but if intelligence is rare in the universe, it may be a very long time before we confirm that we are not alone.

Simple arithmetic can give us an estimate of the number of technological civilizations with which we might communicate. The formula for the number of communicative civilizations in a galaxy, N_c, is:

$$N_c = N^* \cdot f_P \cdot n_{LZ} \cdot f_L \cdot f_I \cdot F_S$$

N^* is the number of stars in a galaxy, and f_P represents the probability that a star has planets. If all single stars have planets, f_P is about 0.5. The factor n_{LZ} is the average number of planets in a solar system suitably placed in the life zone, f_L is the probability that life will originate if conditions are suitable, and f_I is the probability that the life form will evolve to intelligence. These factors can be roughly estimated, but the remaining factor is much more uncertain.

F_S is the fraction of a star's life during which the life form is communicative. Here we assume that a star lives about 10 billion years. If a society survives at a technological level for only 100 years, our chances of communicating with it are small. But a society that stabilizes and remains technological for a long time is much more likely to be in the communicative phase at the proper time to signal to us. If we assume that technological societies destroy themselves in about 100 years, F_S is 100 divided by 10 billion, or 10^{-8}. But if societies can remain technological for a million years, then F_S is 10^{-4}. The influence of the factors in the formula is shown in Table 27-1.

If the optimistic estimates are true, there may be a communicative civilization within a few dozen light-years of us, and we could locate it by searching through only a few thousand stars. On the other hand, if the pessimistic estimates are correct, we may be the only planet in our galaxy capable of communication. We may never know until we understand how technological societies function.

Table 27-1 The Number of Technological Civilizations per Galaxy

	Variables	Estimates	
		Pessimistic	Optimistic
N^*	Number of stars per galaxy	2×10^{11}	2×10^{11}
f_P	Fraction of stars with planets	0.01	0.5
n_{LZ}	Number of planets per star that lie in life zone for longer than 4 billion years	0.01	1
f_L	Fraction of suitable planets on which life begins	0.01	1
f_I	Fraction of life forms that evolve to intelligence	0.01	1
F_S	Fraction of star's life during which a technological society survives	10^{-8}	10^{-4}
N_c	Number of communicative civilizations per galaxy	2×10^{-5}	10×10^6

R E V I E W Critical Inquiry

Why does the number of inhabited worlds we might hear from depend on how long civilizations survive at a technological level?

When we turn radio telescopes toward the sky and scan millions of frequency bands, we take a snapshot of the universe at the particular time when we are living and able to build radio telescopes. To detect other civilizations, they must be in a similar technological stage so they will be broadcasting either intentionally or accidentally. If we search for decades and detect no other signal, it may mean not that life is rare but rather that civilizations do not survive at a technological level for very long. Only a few decades ago, our civilization was threatened by nuclear war, and now we are threatened by the pollution of our environment. If nearly all civilizations in our galaxy are either on the long road up from primitive life forms in oceans or on the long road down from nuclear war or environmental collapse, there may be no one transmitting during the short interval when we are capable of building radio telescopes with which to listen.

Radio communication between inhabited worlds is limited because it requires very fast computers to search many frequency intervals. Why must we search so many frequencies when we suspect that the water hole would be a good place to listen?

Are we the only thinking race? If we are, we bear the sole responsibility to understand and admire the universe. Then we are the sole representatives of that state of matter called intelligence. The mere detection of signals from another civilization would demonstrate that we share the universe with others. Although we might never leave our solar system, such communication would end the self-centered isolation of humanity and stimulate a reevaluation of the meaning of our existence. We may never realize our full potential as humans until we communicate with nonhuman intelligent life.

Summary

To discuss life on other worlds, we must first understand something about life in general, life on Earth, and the origin of life. In general, we can identify three properties in living things—a process, a physical basis, and a controlling unit of information. The process must extract energy from the surroundings, maintain the organism, and modify the surroundings to promote the organism's survival. The physical basis is the arrangement of matter and energy that implements the life process. On Earth all life is based on carbon chemistry. The controlling information is the data necessary to maintain the organism's function. Data for Earth life are stored in long carbon-chain molecules called DNA.

The DNA molecule stores information in the form of chemical bases linked together like the rungs of a ladder. When these patterns are copied by RNA molecules, they can direct the manufacture of proteins and enzymes. Thus, DNA information is the chemical formulas the cell needs to function. When a cell divides, the DNA molecule splits lengthwise and duplicates itself so that each of the new cells has a copy of the information. Errors in the duplication or damage to the DNA molecule can produce mutants, organisms that contain new DNA information and have new properties. Natural selection determines which of these new organisms are most suited to survive, and the species evolves to fit its environment.

The Miller experiment duplicated conditions in Earth's primitive environment and suggests that energy sources such as lightning could have formed amino acids and other complex molecules. Chemical evolution would have connected these together in larger and more complex, but not yet living, molecules. When a molecule acquired the ability to produce copies of itself, natural selection perfected the organism through biological evolution. Although this may have happened in the first billion years, life did not become diverse and complex until the Cambrian period, about 600 million years ago. Life emerged from the oceans about 400 million years ago, and humanity developed only a few million years ago.

It seems unlikely that there is now life on other planets in our solar system. Most of the planets are too hot or too cold; Mars may have been suitable in the past, but it is now a cold, dry desert. In 1996, proposed evidence of ancient life was found in a meteorite known to have come from Mars, but that evidence has now been cast into serious doubt.

To find life, we must look beyond our solar system. Because we suspect that planets form from the leftover debris of star formation, we suspect that most stars have planets. The rise of intelligence may take billions of years, however, so short-lived massive stars and binary stars with unstable planetary orbits must be discounted. The best candidates are G and K main-sequence stars.

The distances between stars are too large to permit travel, but communication by radio could be possible. A certain wavelength range called a radio window is suitable, and a small range between the radio signals of H and OH, the so-called water hole, is especially likely.

We now have the technology to search for other intelligent life in the universe. Although such searches are controversial, a number of radio astronomers are now searching for radio signals from extraterrestrial civilizations. If life is common in the universe, success seems inevitable.

New Terms

DNA (deoxyribonucleic acid)	Miller experiment
protein	primordial soup
enzyme	chemical evolution
RNA (ribonucleic acid)	stromatolite
amino acid	life zone
natural selection	water hole
mutant	SETI
Cambrian period	

Review Questions

1. If life is based on information, what is that information?

2. What would happen to a life form if the information handed down to offspring was always the same? How would that endanger the future of the life form?

3. How does the DNA molecule produce a copy of itself?

4. Give an example of natural selection acting on new DNA patterns to select the most advantageous characteristics.

5. Why do we believe that life on Earth began in the sea?

6. Why do we think that liquid water is necessary for the origin of life?

7. What is the difference between chemical evolution and biological evolution?

8. What was the significance of the Miller experiment?

9. How does intelligence make a creature more likely to survive?

10. How do astronomers know that meteorite ALH84001 really came from Mars?

11. Discuss the evidence pro and con that there is or has been life on Mars.

12. Why are upper-main-sequence stars unlikely sites for intelligent civilizations?

13. Why do we suspect that travel between stars is nearly impossible?

14. How does the stability of technological civilizations affect the probability that we can communicate with them?

15. What is the water hole, and why would it be a good place to look for other civilizations?

Discussion Questions

1. What would you change in the Arecibo message if humanity lived on Mars instead of Earth?

2. What do you think it would mean if decades of careful searches for radio signals for extraterrestrial intelligence turned up nothing?

Problems

1. A single human cell encloses about 1.5 m of DNA containing 4.5 billion base pairs. What is the spacing between these base pairs in nanometers? That is, how far apart are the rungs on the DNA ladder?

2. If we represent the history of the Earth by a line 1 m long, how long a segment would represent the 400 million years since life moved onto the land? How long a segment would represent the 3-million-year history of human life?

3. If a human generation, the time from birth to childbearing, is 20 years, how many generations have passed in the last million years?

4. If a star must remain on the main sequence for at least 5 billion years for life to evolve to intelligence, how massive could a star be and still harbor intelligent life on one of its planets? (*Hint:* See Chapter 13.)

5. If there are about 1.4×10^{-4} stars like the sun per cubic light-year, how many lie within 100 light-years of Earth? (*Hint:* The volume of a sphere is $\frac{4}{3}\pi r^3$.)

6. Mathematician Karl Gauss suggested planting forests and fields in a gigantic geometric proof to signal to possible Martians that intelligent life exists on Earth. If Martians had telescopes that could resolve details no smaller than 1 second of arc, how large would the smallest element of Gauss's proof have to be? (*Hint:* Use the small-angle formula.)

7. If we detected radio signals with an average wavelength of 20 cm and suspected that they came from a civilization on a distant planet, roughly how much of a change in wavelength should we expect to see because of the orbital motion of the distant planet? (*Hint:* What is the Doppler shift?)

8. Calculate the number of communicative civilizations per galaxy from your own estimates of the factors in Table 27-1.

Critical Inquiries for the Web

1. The popular movie *Contact* focused interest in the SETI program by profiling the work of a radio astronomer dedicated to the search for extraterrestrial intelligence. Visit Web sites that give information about the movie, SETI programs, and radio astronomy. Discuss whether or not the movie realistically depicted how such research is done.

2. How many inhabited worlds are out there? Search for a Web site that allows interactive manipulation of our equation for the number of technological civilizations (sometimes called the Drake Equation) and consider a range of possibilities in the variables.

3. Where outside the solar system would you look for habitable planets? NASA has increasingly focused its interests on this question. Look for information online about programs dedicated to detect planets on which the existence of life as we know it might be possible. What criteria are used to choose targets for the planned searches? What methods will be used to carry out the searches?

 Go to the Brooks/Cole Astronomy Resource Center (www.brookscole.com/astronomy) for critical thinking exercises, articles, and additional readings from InfoTrac College Edition, Brooks/Cole's online student library.

Afterword

Supernatural is a null word.

Robert A. Heinlein
The Notebooks of Lazarus Long

Our journey is over, but before we part company, there is one last thing to discuss—the place of humanity in the universe. Astronomy gives us some comprehension of the workings of stars, galaxies, and planets, but its greatest value lies in what it teaches us about ourselves. Now that we have surveyed astronomical knowledge, we can better understand our own position in nature.

To some, the word *nature* conjures up visions of furry rabbits hopping about in a forest glade dotted with pastel wildflowers. To others, nature is the blue-green ocean depths filled with creatures swirling in a mad struggle for survival. Still others think of nature as windswept mountaintops of gray stone and glittering ice. As diverse as these images are, they are all Earth-bound. Having studied astronomy, we can view nature as a beautiful mechanism composed of matter and energy interacting according to simple rules to form galaxies, stars, planets, mountaintops, ocean depths, and forest glades.

Perhaps the most important astronomical lesson is that we are a small but important part of the universe. Most of the universe is lifeless. The vast reaches between the galaxies appear to be empty of all but the thinnest gas, and the stars, which contain most of the mass, are much too hot to preserve the chemical bonds that seem necessary for life to survive and develop. Only on the surfaces of a few planets, where temperatures are moderate, could atoms link together to form living matter.

If life is special, then intelligence is precious. The universe must contain many planets devoid of life, planets where the wind has blown unfelt for billions of years. There may also exist planets where life has developed but has not become complex, planets on which the wind stirs wide plains of grass and rustles dark forests. On some planets, insects, fish, birds, and animals may watch the passing days unaware of their own existence. It is intelligence, human or alien, that gives meaning to the landscape.

Science is the process by which intelligence tries to understand the universe. Science is not the invention of new devices or processes. It does not create home computers, cure the mumps, or manufacture plastic spoons—that is engineering and technology, the adaptation of scientific understanding for practical purposes. Science is understanding nature, and astronomy is understanding on the grandest scale. Astronomy is the science by which the universe, through its intelligent lumps of matter, tries to understand its own existence.

As the primary intelligent species on this planet, we are the custodians of a priceless gift—a planet filled with living things. This is especially true if life is rare in the universe. In fact, if Earth is the only inhabited planet, our responsibility is overwhelming. In any case, we are the only creatures who can take action to preserve the existence of life on Earth, and, ironically, our own actions are the most serious hazards.

The future of humanity is not secure. We are trapped on a tiny planet with limited resources and a population growing faster than our ability to produce food. In our efforts to survive, we have already driven some creatures to extinction and now threaten others. If our civilization collapses because of starvation, or if our race destroys itself somehow, the only bright spot is that the rest of the creatures on Earth will be better off for our absence.

But even if we control our population and conserve and recycle our resources, life on Earth is doomed. In 5 billion years, the sun will leave the main sequence and swell into a red giant, incinerating Earth. However, Earth will be lifeless long before that. Within the next few billion years, the growing luminosity of the sun will first alter Earth's climate and then boil its atmosphere and oceans. Earth, like everything else in the universe, is only temporary.

To survive, humanity must eventually leave Earth and search for other planets. Colonizing the moon and other planets of our solar system will not save us, because they will face the same fate as Earth when the sun dies. But travel to other stars is tremendously difficult and may be impossible with the limited resources we have in our small solar system. We and all of the living things on Earth may be trapped.

This is a depressing prospect, but a few factors are comforting. First, everything in the universe is temporary. Stars die, galaxies die, perhaps the entire universe will fall back in a "big crunch" and die. That our distant future is limited only assures us that we are a part of a much larger whole. Second, we have a few billion years to prepare, and a billion years is a very long time. Only a few million years ago, our ancestors were learning to walk erect and communicate. A billion years ago, our ancestors were microscopic organisms living in the primeval oceans. To suppose that a billion years hence we humans will still be human, that we will still be the dominant species on Earth, or that we will still be the sole intelligence on Earth is the ultimate conceit.

Our responsibility is not to save our race for all eternity but to behave as dependable custodians of our planet, preserving it, admiring it, and trying to understand it. That calls for drastic changes in our behavior toward other living things and a revolution in our attitude toward our planet's resources. Whether we can change our ways is debatable—humanity is far from perfect in its understanding, abilities, or intentions. However, we must not imagine that we and our civilization are less than precious. We have the gift of intelligence, and that is the finest thing this planet has ever produced.

We must not cease from exploration and the end of all our exploring will be to arrive where we began and to know the place for the first time.

—T. S. Eliot

Units and Astronomical Data

Introduction

The metric system is used worldwide as the system of units, not only in science but also in engineering, business, sports, and daily life. Developed in 18th-century France, the metric system has gained acceptance in almost every country in the world because it simplifies computations.

A system of units is based on the three fundamental units for length, mass, and time. Other quantities such as density and force are derived from these fundamental units. In the English (or British) system of units (commonly used in the United States) the fundamental unit of length is the foot, composed of 12 inches. The metric system is based on the decimal system of numbers, and the fundamental unit of length is the meter, composed of 100 centimeters.

To see the advantage of having a decimal-based system, try computing the volume of a bathtub that is 5′9″ long, 1′2″ deep, and 2′10″ wide. In the metric system the length of the tub is 1 m and 75 cm or 1.75 m. The other dimensions are 0.35 m and 0.86 m, and the volume is just $1.75 \times 0.35 \times 0.86 \text{ m}^3$. To make the computation in English units, we must first convert inches to feet by dividing by 12. We can convert centimeters to meters by the simpler process of moving the decimal point. Thus, the computation is much easier if we measure the tub in meters and centimeters instead of feet and inches.

Because the metric system is a decimal system, it is easy to express quantities in larger or smaller units as is convenient. We can express distances in centimeters, meters, kilometers, and so on. The prefixes specify the relation of the unit to the meter. Just as a cent is $\frac{1}{100}$ of a dollar, a centimeter is $\frac{1}{100}$ of a meter. A kilometer is 1000 m, and a kilogram is 1000 g. The meanings of the commonly used prefixes are given in Table A-1.

The SI Units

Any system of units based on the decimal system would be easy to use, but by international agreement, the preferred set of units, known as the *Système International d'Unités* (SI units) is based on the meter, kilogram, and second. These three fundamental units define the rest of the units as given in Table A-2.

The SI unit of force is the newton (N), named after Isaac Newton. It is the force needed to accelerate a 1 kg mass by 1 m/s^2, or the force roughly equivalent to the weight of an apple at Earth's surface. The SI unit of energy is the joule (J), the energy produced by a force of 1 N acting through a distance of 1 m. A joule is roughly the energy in the impact of an apple falling off a table.

Exceptions Units help us in two ways. They make it possible to make calculations, and they help us to conceive of certain quantities. For calculations the metric

Table A-1 Metric Prefixes

Prefix	Symbol	Factor
Mega	M	10^6
Kilo	k	10^3
Centi	c	10^{-2}
Milli	m	10^{-3}
Micro	μ	10^{-6}
Nano	n	10^{-9}

Table A-2 SI Metric Units

Quantity	SI Unit	English Unit
Length	Meter (m)	Foot
Mass	Kilogram (kg)	Slug (sl)
Time	Second (s)	Second (s)
Force	Newton (N)	Pound (lb)
Energy	Joule (J)	Foot-pound (fp)

system is far superior, and we will use it for our calculations throughout this book.

But Americans commonly use the English system of units, so for conceptual purposes we can express quantities in English units. Instead of saying the average person would weigh 133 N on the moon, we could express the weight as 30 lb. Thus, the text will commonly give quantities in metric form followed by the English form in parentheses: The radius of the moon is 1738 km (1080 miles).

In SI units, density should be expressed as kilograms per cubic meter, but no human can enclose a cubic meter in his or her hand, so this unit does not help us grasp the significance of a given density. This book will refer to density in grams per cubic centimeter. A gram is roughly the mass of a paper clip, and a cubic centimeter is the size of a small sugar cube, so we can conceive of a density of 1 g/cm^3, roughly the density of water. This is not a bothersome departure from SI units because we will not make complex calculations using density.

Conversions

To convert from one metric unit to another (from meters to kilometers, for example), we have only to look at the prefix. However, converting from metric to English or English to metric is more complicated. The conversion factors are given in Table A-3.

Example: The radius of the moon is 1738 km. What is this in miles? Table A-3 indicates that 1 mile equals 1.609 km, so:

$$1738 \text{ km} \times \frac{1 \text{ mile}}{1.609 \text{ km}} = 1080 \text{ miles}$$

Temperature Scales

In astronomy, as in most other sciences, temperatures are expressed on the Kelvin scale, although the centigrade (or Celsius) scale is also used. The Fahrenheit scale commonly used in the United States is not used in scientific work.

Temperatures on the Kelvin scale are measured from absolute zero, the temperature of an object that contains no extractable heat. In practice, no object can be as cold as absolute zero, although laboratory apparatuses have reached temperatures less than 10^{-6} K. The scale is named after the Scottish mathematical physicist William Thomson, Lord Kelvin (1824–1907).

The centigrade scale refers temperatures to the freezing point of water (0°C) and to the boiling point of water (100°C). One degree centigrade is 1/100th the temperature difference between the freezing and boiling points of water. Thus the prefix *centi.* The centigrade scale is also called the Celsius scale after its inventor, the Swedish astronomer Anders Celsius (1701–1744).

The Fahrenheit scale fixes the freezing point of water at 32°F and the boiling point at 212°F. Named after the German physicist Gabriel Daniel Fahrenheit (1686–1736), who made the first successful mercury thermometer in 1720, the Fahrenheit scale is used only in the United States.

It is easy to convert temperatures from one scale to another using the information given in Table A-6 inside the back cover of this book.

Powers of 10 Notation

Powers of 10 make writing very large numbers much simpler. For example, the nearest star is about 43,000,000,000,000 km from the sun. Writing this number as 4.3×10^{13} km is much easier.

Very small numbers can also be written with powers of 10. For example, the wavelength of visible light is about 0.0000005 m. In power of 10 this becomes 5×10^{-7} m.

The powers of 10 used in this notation appear below. The exponent tells us how to move the decimal point. If the exponent is positive, we move the decimal point to the right. If the exponent is negative, we move the decimal point to the left. Thus, 2×10^3 equals 2000.0, and 2×10^{-3} equals 0.002.

$$\vdots$$

10^5	=	100,000
10^4	=	10,000
10^3	=	1,000
10^2	=	100
10^1	=	10
10^0	=	1
10^{-1}	=	0.1
10^{-2}	=	0.01
10^{-3}	=	0.001
10^{-4}	=	0.0001

$$\vdots$$

Table A-3 Conversion Factors

1 inch = 2.54 centimeters	1 centimeter = 0.394 inch
1 foot = 0.3048 meter	1 meter = 39.36 inches = 3.28 feet
1 mile = 1.6093 kilometers	1 kilometer = 0.6214 mile
1 slug = 14.594 kilograms	1 kilogram = 0.0685 slug
1 pound = 4.4482 newtons	1 newton = 0.2248 pound
1 foot-pound = 1.35582 joules	1 joule = 0.7376 foot-pound
1 horsepower = 745.7 joules/s	1 joule/s = 1 watt

If you use scientific notation in calculations, be sure you correctly enter numbers into your calculator. Not all calculators can accept scientific notation, but those that can have a key labeled EXP, EEX, or perhaps EE that allows you to enter the exponent of ten. To enter a number such as 3×10^8, we press the keys 3 EXP 8. To enter a number with a negative exponent, we must use the change-sign key, usually labeled $+/-$ or CHS. To enter the number 5.2×10^{-3}, we press the keys 5.2 EXP $+/-$ 3. Try a few examples.

To read a number in scientific notation from a calculator we must read the exponent separately. The number 3.1×10^{25} may appear in a calculator display as 3.1 25 or on some calculators as 3.1 10^{25}. Examine your calculator to determine how such numbers are displayed.

$* * *$

Astronomy, and science in general, is a way of learning about nature and understanding the universe we live in. To test hypotheses about how nature works, scientists use observations of nature. The tables that follow—in this Appendix and inside the back cover of this book—contain some of the basic observations that support our best understanding of the astronomical universe. Of course, these data are expressed in the form of numbers, not because science reduces all understanding to mere numbers, but because the struggle to understand nature is so demanding we must use every tool available. Quantitative thinking, reasoning mathematically, is one of the most powerful tools ever invented by the human brain. Thus these tables are not nature reduced to mere numbers, but numbers supporting humanity's growing understanding of the natural world around us.

Table A-4 Constants

astronomical unit (AU)	$= 1.495979 \times 10^{11}$ m
parsec (pc)	$= 206,265$ AU
	$= 3.085678 \times 10^{16}$ m
	$= 3.261633$ ly
light-year (ly)	$= 9.46053 \times 10^{15}$ m
velocity of light (c)	$= 2.997925 \times 10^8$ m/s
gravitational constant (G)	$= 6.67 \times 10^{-11}$ N $\cdot$ m^2/kg^2
mass of Earth ($M_\oplus$)	$= 5.976 \times 10^{24}$ kg
equatorial radius of Earth ($R_\oplus$)	
	$= 6378.164$ km
mass of sun ($M_\odot$)	$= 1.989 \times 10^{30}$ kg
radius of sun ($R_\odot$)	$= 6.9599 \times 10^8$ m
solar luminosity ($L_\odot$)	$= 3.826 \times 10^{26}$ J/s
mass of moon	$= 7.350 \times 10^{22}$ kg
radius of moon	$= 1738$ km
mass of H atom	$= 1.67352 \times 10^{-27}$ kg

Table A-5 Principal Satellites of the Solar System

Planet	Satellite	Radius (km)	Distance from Planet (10^3 km)	Orbital Period (days)	Orbital Eccentricity	Orbital Inclination
Earth	Moon	1738	384.4	27.322	0.055	5°8′43″
Mars	Phobos	14 × 12 × 10	9.38	0.3189	0.018	1°.0
	Deimos	8 × 6 × 5	23.5	1.262	0.002	2°.8
Jupiter	Metis	20	126	0.29	0.0	0°.0
	Adrastea	12 × 8 × 10	128	0.294	0.0	0°.0
	Amalthea	135 × 100 × 78	182	0.4982	0.003	0°.45
	Thebe	50	223	0.674	0.0	1°.3
	Io	1820	422	1.769	0.000	0°.3
	Europa	1565	671	3.551	0.000	0°.46
	Ganymede	2640	1071	7.155	0.002	0°.18
	Callisto	2420	1884	16.689	0.008	0°.25
	Leda	~8	11,110	240	0.146	26°.7
	Himalia	~85	11,470	250.6	0.158	27°.6
	Lysithea	~20	11,710	260	0.12	29°
	Elara	~30	11,740	260.1	0.207	24°.8
	Ananke	15	21,200	631	0.169	147°
	Carme	22	22,350	692	0.207	163°
	Pasiphae	35	23,300	735	0.40	147°
	Sinope	20	23,700	758	0.275	156°

Planet	Satellite	Radius (km)	Distance from Planet (10^3 km)	Orbital Period (days)	Orbital Eccentricity	Orbital Inclination
Saturn	Pan	10	133.570	0.574	0.000	0°
	Atlas	20 × 15 × 15	137.7	0.601	0.002	0°.3
	Prometheus	70 × 40 × 50	139.4	0.613	0.003	0°.0
	Pandora	55 × 35 × 50	141.7	0.629	0.004	0°.05
	Epimetheus	70 × 50 × 50	151.42	0.694	0.009	0°.34
	Janus	110 × 80 × 100	151.47	0.695	0.007	0°.14
	Mimas	196	185.54	0.942	0.020	1°.5
	Enceladus	250	238.04	1.370	0.004	0°.0
	Tethys	530	294.67	1.888	0.000	1°.1
	Calypso	17 × 11 × 12	294.67	1.888	0.0	~1°.?
	Telesto	12	294.67	1.888	0.0	~1°?
	Dione	560	377	2.737	0.002	0°.0
	Helene	20 × 15 × 15	377	2.74	0.005	0°.15
	Rhea	765	527	4.518	0.001	0°.4
	Titan	2575	1222	15.94	0.029	0°.3
	Hyperion	205 × 130 × 110	1484	21.28	0.104	~0°.5
	Iapetus	720	3562	79.33	0.028	14°.72
	Phoebe	110	12,930	550.4	0.163	150°
Uranus	Cordelia	20	49.8	0.3333	~0	~0°
	Ophelia	15	53.8	0.375	~0	~0°
	Bianca	25	59.1	0.433	~0	~0°
	Cressida	30	61.8	0.462	~0	~0°
	Desdemona	30	62.7	0.475	~0	~0°
	Juliet	40	64.4	0.492	~0	~0°
	Portia	55	66.1	0.512	~0	~0°
	Rosalind	30	69.9	0.558	~0	~0°
	Belinda	30	75.2	0.621	~0	~0°
	S/1986 U10	20	76.2	0.638	~0	~0°
	Puck	85 ± 5	85.9	0.762	~0	~0°
	Miranda	242 ± 5	129.9	1.414	0.017	3°.4
	Ariel	580 ± 5	190.9	2.520	0.003	0°
	Umbriel	595 ± 10	266.0	4.144	0.003	0°
	Titania	805 ± 5	436.3	8.706	0.002	0°
	Oberon	775 ± 10	583.4	13.463	0.001	0°
	Caliban	40	7164	579	0.082	139.2°
	Sycorax	80	12,174	1284	0.509	152.7°
	S/1999 U2	~20	~10,000?			
	S/1999 U3	~20	~10,000?			
	S/1999 U1	~20	~25,000?			
Neptune	Naiad	30	48.2	0.296	~0	~0°
	Thalassa	40	50.0	0.312	~0	~0°
	Despina	90	52.5	0.333	~0	~0°
	Galatea	75	62.0	0.396	~0	~0°
	Larissa	95	73.6	0.554	~0	~0°
	Proteus	205	117.6	1.121	~0	~0°
	Triton	1352	354.59	5.875	0.00	160°
	Nereid	170	5588.6	360.125	0.76	27.7°
Pluto	Charon	626	19.7	6.38718	~0	122°

Observing the Sky

Observing the sky with the naked eye is of no more importance to modern astronomy than picking up pretty pebbles is to modern geology. But the sky is a natural wonder unimaginably bigger than the Grand Canyon, the Rocky Mountains, or any other natural wonder that tourists visit every year. To neglect the beauty of the sky is equivalent to geologists neglecting the beauty of the minerals they study. This supplement is meant to act as a tourist's guide to the sky. We analyzed the universe in the regular chapters, but here we will admire it.

The brighter stars in the sky are visible even from the centers of cities with their air and light pollution. But in the countryside only a few miles beyond the cities, the night sky is a velvety blackness strewn with thousands of glittering stars. From a wilderness location, far from the city's glare, and especially from high mountains, the night sky is spectacular.

Using Star Charts

The constellations are a fascinating cultural heritage of our planet, but they are sometimes a bit difficult to learn because of Earth's motion. The constellations above the horizon change with the time of night and the seasons.

Because Earth rotates eastward, the sky appears to rotate around us westward. A constellation visible in the southern sky soon after sunset will appear to move westward, and in a few hours it will disappear below the horizon. Other constellations will rise in the east, so the sky changes gradually through the night.

In addition, Earth's orbital motion makes the sun appear to move eastward among the stars. Each day the sun moves about twice its own diameter eastward along the ecliptic, and each night at sunset, the constellations are about 1° farther toward the west.

Orion, for instance, is visible in the southern sky in January, but as the days pass, the sun moves closer to Orion. By March, Orion is difficult to see in the southwest sky soon after sunset. By June, the sun is so close to Orion it sets with the sun and is invisible. Not until late July is the sun far enough past Orion for the constellation to become visible rising in the eastern sky just before dawn.

Because of the rotation and orbital motion of Earth, we must use more than one star chart to map the sky. Which chart we select depends on the month and the time of night. The charts given at the end of this book show the evening sky for each month.

Two sets of charts are included for two typical locations on Earth. The Northern Hemisphere charts show that sky as seen from a northern latitude typical of the United States and central Europe. The Southern Hemisphere star charts at the end of this book are appropriate for readers in Earth's Southern Hemisphere including Australia, southern South America, and southern Africa.

To use the charts, select the appropriate chart, and hold it overhead as shown in Figure B-1. If you face south, turn the chart until the words *southern horizon* are at the bottom of the chart. If you face other directions, turn the chart appropriately.

Figure B-1
To use the star charts in this book, select the appropriate chart for the date and time. Hold it overhead, and turn it until the direction at the bottom of the chart is the same as the direction you are facing.

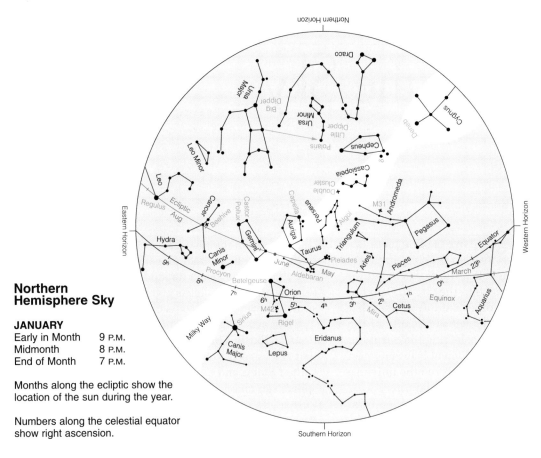

Northern Hemisphere Sky

JANUARY

Early in Month	9 P.M.
Midmonth	8 P.M.
End of Month	7 P.M.

Months along the ecliptic show the location of the sun during the year.

Numbers along the celestial equator show right ascension.

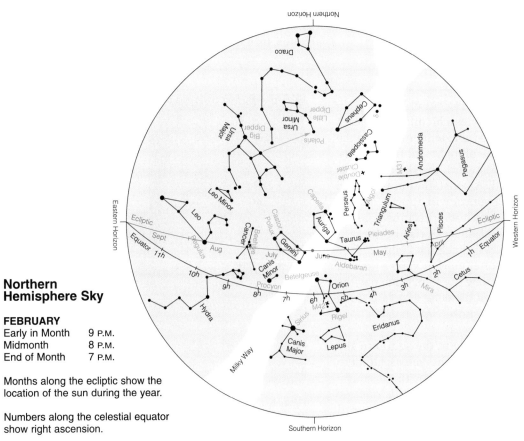

Northern Hemisphere Sky

FEBRUARY

Early in Month	9 P.M.
Midmonth	8 P.M.
End of Month	7 P.M.

Months along the ecliptic show the location of the sun during the year.

Numbers along the celestial equator show right ascension.

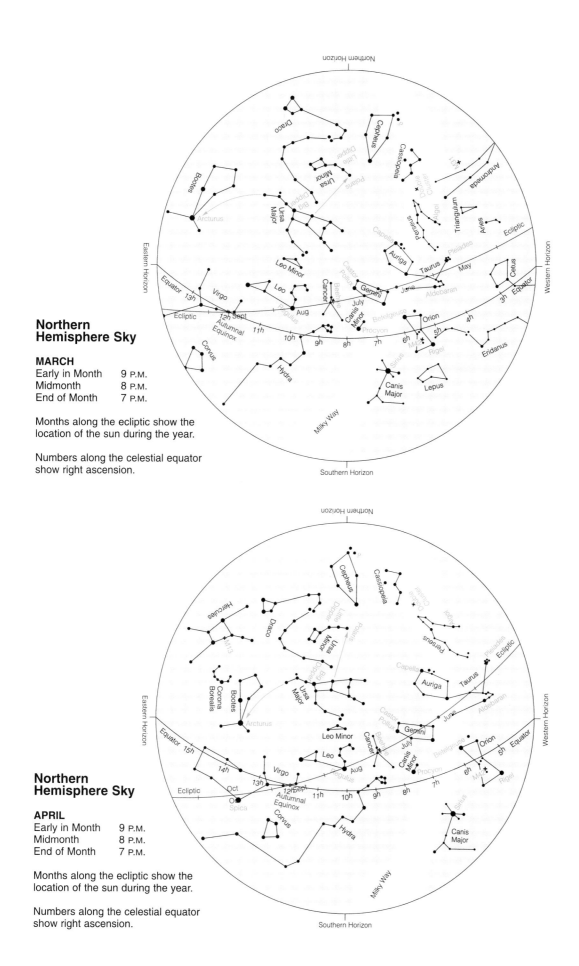

Northern Hemisphere Sky

MARCH

Early in Month	9 P.M.
Midmonth	8 P.M.
End of Month	7 P.M.

Months along the ecliptic show the location of the sun during the year.

Numbers along the celestial equator show right ascension.

Northern Hemisphere Sky

APRIL

Early in Month	9 P.M.
Midmonth	8 P.M.
End of Month	7 P.M.

Months along the ecliptic show the location of the sun during the year.

Numbers along the celestial equator show right ascension.

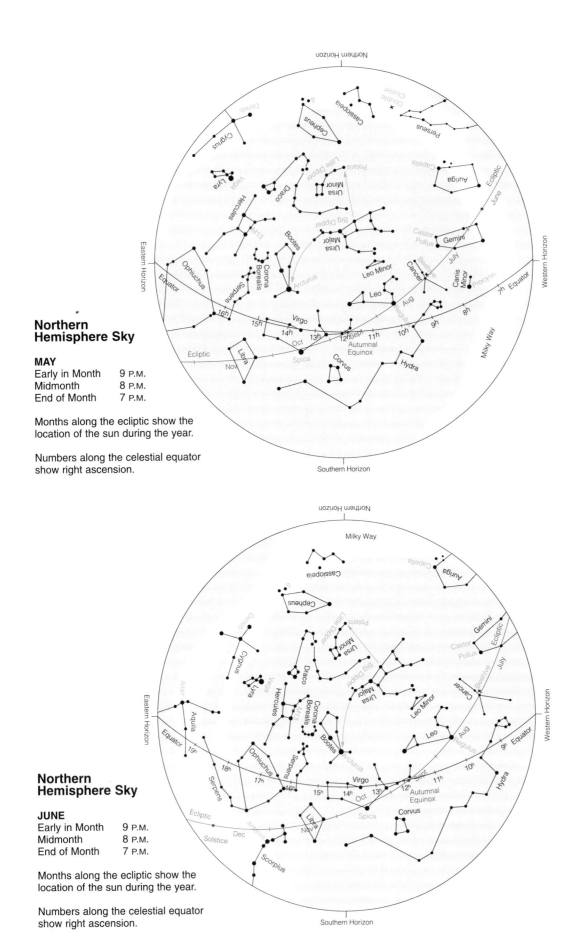

Northern Hemisphere Sky

MAY
Early in Month	9 P.M.
Midmonth	8 P.M.
End of Month	7 P.M.

Months along the ecliptic show the location of the sun during the year.

Numbers along the celestial equator show right ascension.

Northern Hemisphere Sky

JUNE
Early in Month	9 P.M.
Midmonth	8 P.M.
End of Month	7 P.M.

Months along the ecliptic show the location of the sun during the year.

Numbers along the celestial equator show right ascension.

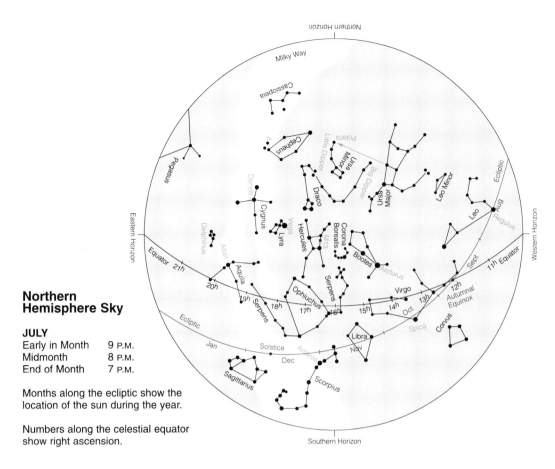

Northern Hemisphere Sky

JULY

Early in Month	9 P.M.
Midmonth	8 P.M.
End of Month	7 P.M.

Months along the ecliptic show the location of the sun during the year.

Numbers along the celestial equator show right ascension.

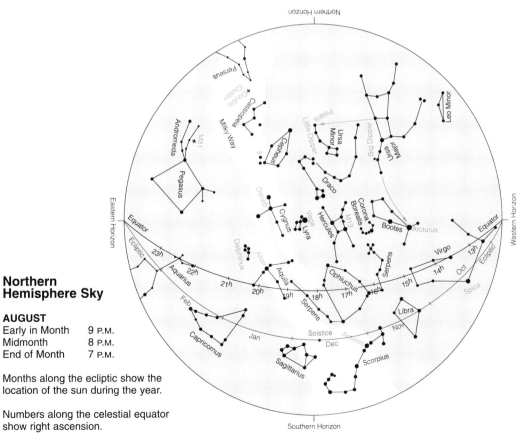

Northern Hemisphere Sky

AUGUST

Early in Month	9 P.M.
Midmonth	8 P.M.
End of Month	7 P.M.

Months along the ecliptic show the location of the sun during the year.

Numbers along the celestial equator show right ascension.

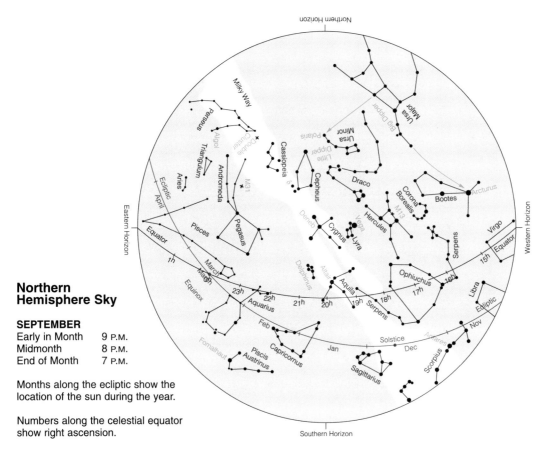

Northern
Hemisphere Sky

SEPTEMBER

Early in Month	9 P.M.
Midmonth	8 P.M.
End of Month	7 P.M.

Months along the ecliptic show the
location of the sun during the year.

Numbers along the celestial equator
show right ascension.

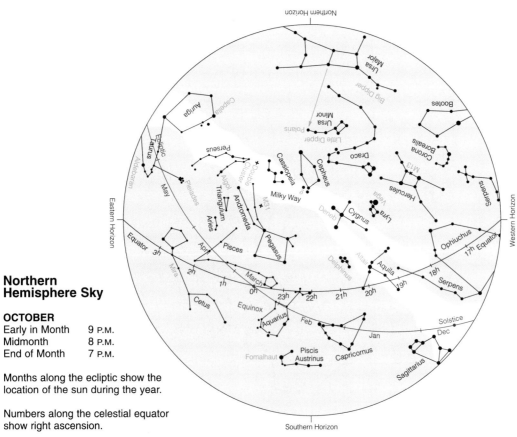

Northern
Hemisphere Sky

OCTOBER

Early in Month	9 P.M.
Midmonth	8 P.M.
End of Month	7 P.M.

Months along the ecliptic show the
location of the sun during the year.

Numbers along the celestial equator
show right ascension.

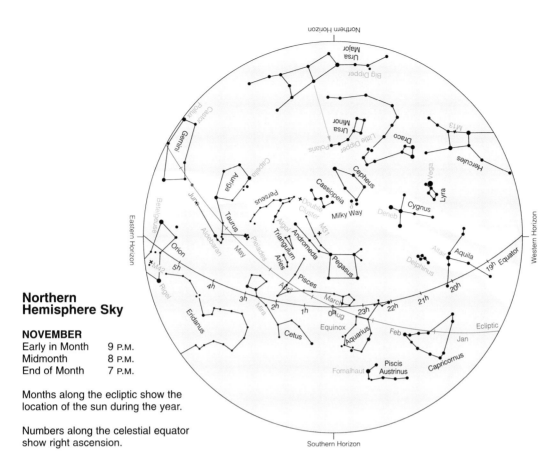

Northern Hemisphere Sky

NOVEMBER

Early in Month	9 P.M.
Midmonth	8 P.M.
End of Month	7 P.M.

Months along the ecliptic show the location of the sun during the year.

Numbers along the celestial equator show right ascension.

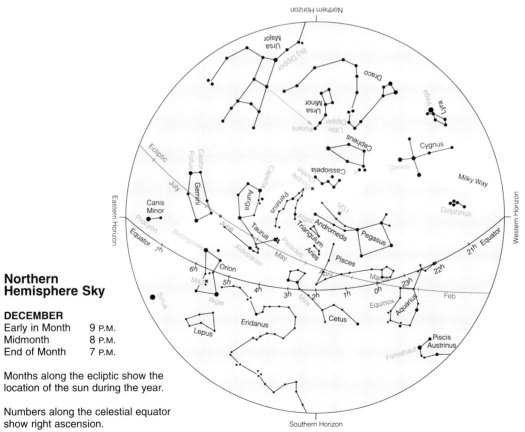

Northern Hemisphere Sky

DECEMBER

Early in Month	9 P.M.
Midmonth	8 P.M.
End of Month	7 P.M.

Months along the ecliptic show the location of the sun during the year.

Numbers along the celestial equator show right ascension.

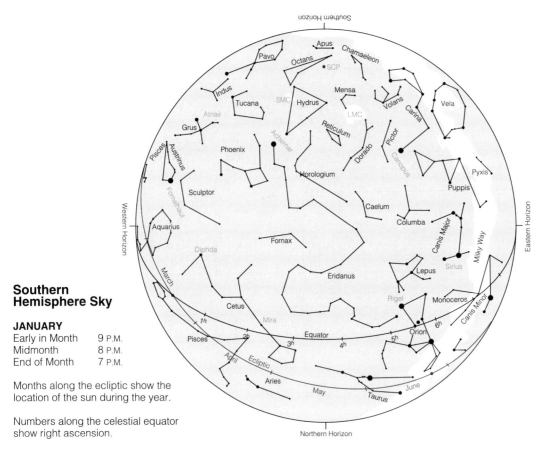

Southern Hemisphere Sky

JANUARY

Early in Month	9 P.M.
Midmonth	8 P.M.
End of Month	7 P.M.

Months along the ecliptic show the location of the sun during the year.

Numbers along the celestial equator show right ascension.

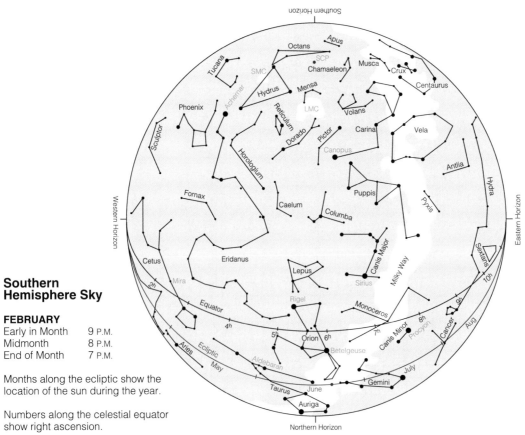

Southern Hemisphere Sky

FEBRUARY

Early in Month	9 P.M.
Midmonth	8 P.M.
End of Month	7 P.M.

Months along the ecliptic show the location of the sun during the year.

Numbers along the celestial equator show right ascension.

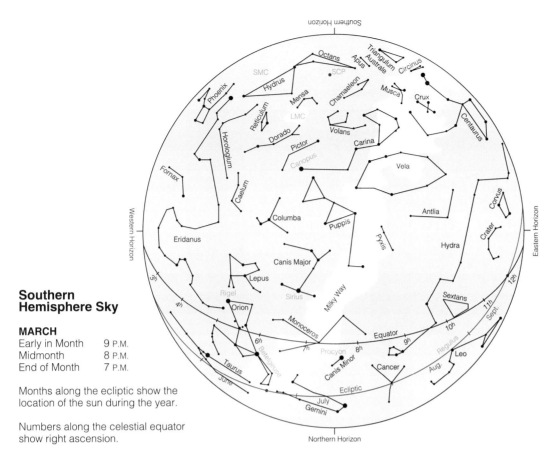

Southern Hemisphere Sky

MARCH

Early in Month	9 P.M.
Midmonth	8 P.M.
End of Month	7 P.M.

Months along the ecliptic show the location of the sun during the year.

Numbers along the celestial equator show right ascension.

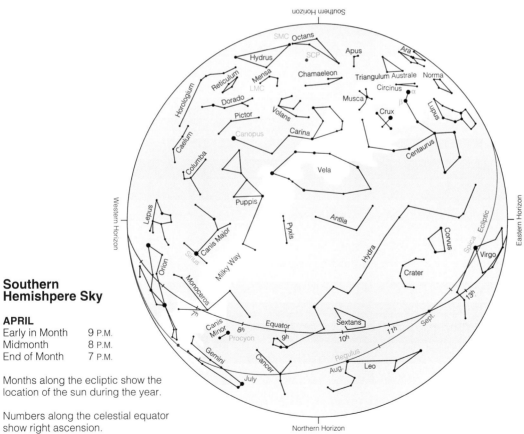

Southern Hemishpere Sky

APRIL

Early in Month	9 P.M.
Midmonth	8 P.M.
End of Month	7 P.M.

Months along the ecliptic show the location of the sun during the year.

Numbers along the celestial equator show right ascension.

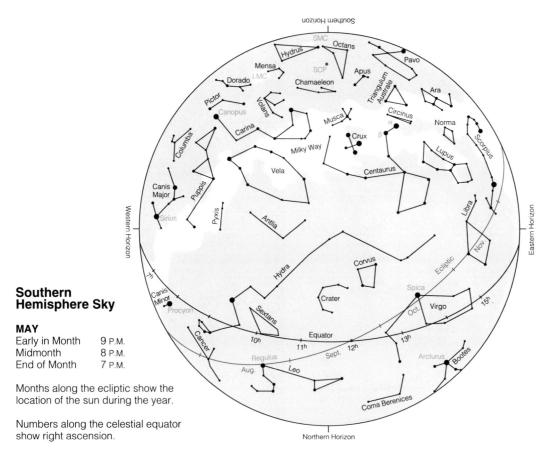

Southern
Hemisphere Sky

MAY
Early in Month	9 P.M.
Midmonth	8 P.M.
End of Month	7 P.M.

Months along the ecliptic show the location of the sun during the year.

Numbers along the celestial equator show right ascension.

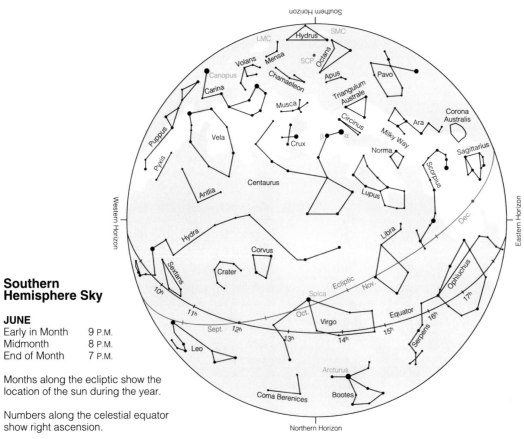

Southern
Hemisphere Sky

JUNE
Early in Month	9 P.M.
Midmonth	8 P.M.
End of Month	7 P.M.

Months along the ecliptic show the location of the sun during the year.

Numbers along the celestial equator show right ascension.

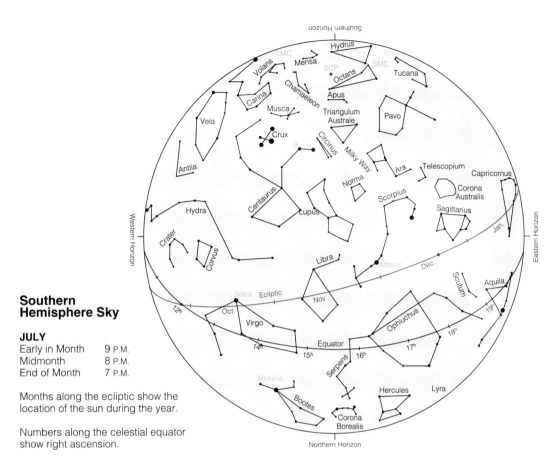

Southern Hemisphere Sky

JULY

Early in Month	9 P.M.
Midmonth	8 P.M.
End of Month	7 P.M.

Months along the ecliptic show the location of the sun during the year.

Numbers along the celestial equator show right ascension.

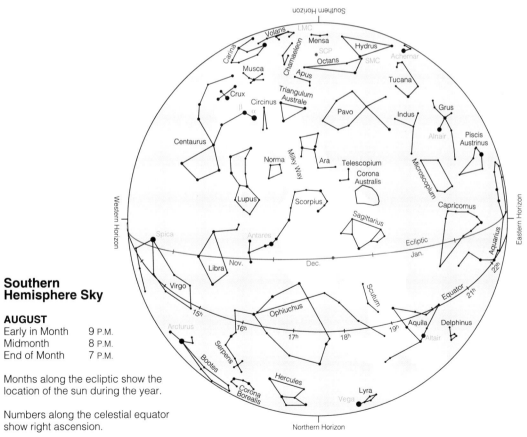

Southern Hemisphere Sky

AUGUST

Early in Month	9 P.M.
Midmonth	8 P.M.
End of Month	7 P.M.

Months along the ecliptic show the location of the sun during the year.

Numbers along the celestial equator show right ascension.

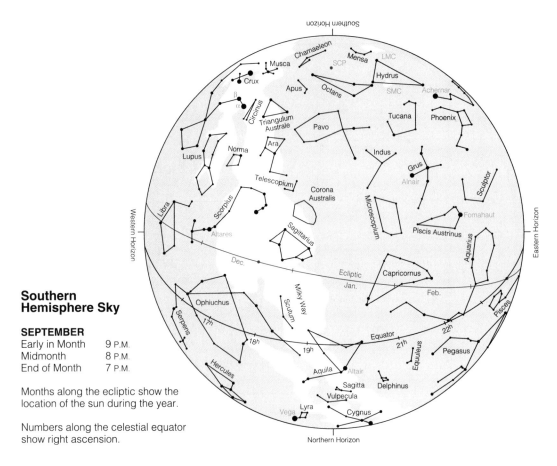

Southern Hemisphere Sky

SEPTEMBER

Early in Month	9 P.M.
Midmonth	8 P.M.
End of Month	7 P.M.

Months along the ecliptic show the location of the sun during the year.

Numbers along the celestial equator show right ascension.

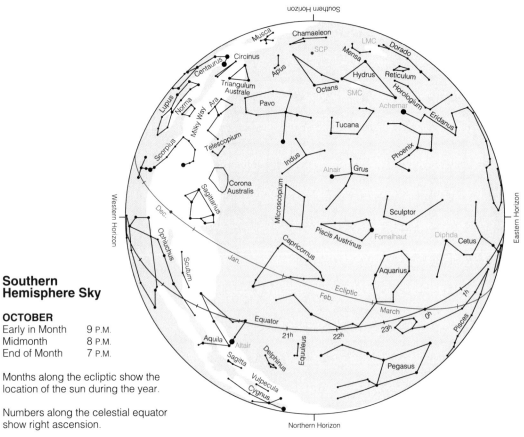

Southern Hemisphere Sky

OCTOBER

Early in Month	9 P.M.
Midmonth	8 P.M.
End of Month	7 P.M.

Months along the ecliptic show the location of the sun during the year.

Numbers along the celestial equator show right ascension.

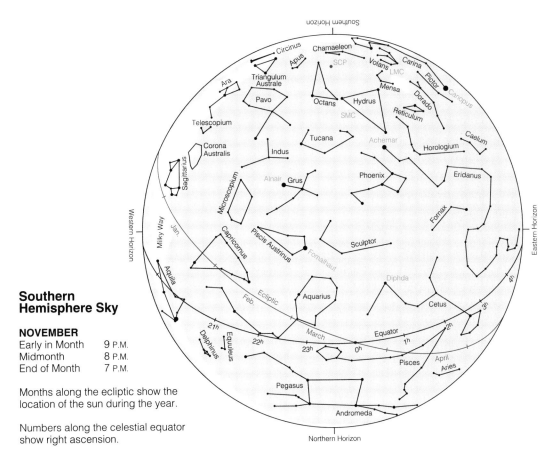

Southern Hemisphere Sky

NOVEMBER

Early in Month	9 P.M.
Midmonth	8 P.M.
End of Month	7 P.M.

Months along the ecliptic show the location of the sun during the year.

Numbers along the celestial equator show right ascension.

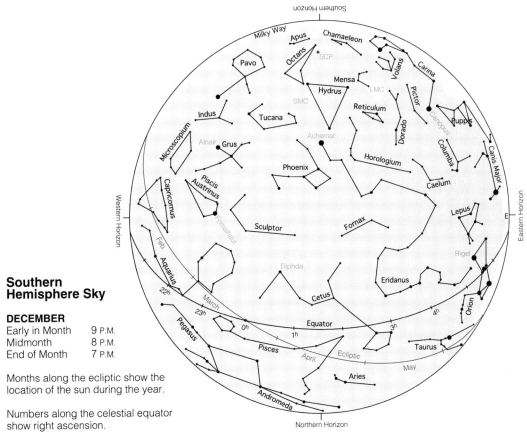

Southern Hemisphere Sky

DECEMBER

Early in Month	9 P.M.
Midmonth	8 P.M.
End of Month	7 P.M.

Months along the ecliptic show the location of the sun during the year.

Numbers along the celestial equator show right ascension.

Glossary

Numbers refer to the page where the term is first discussed in the text.

absolute age (460)   An age determined in years, as from radioactive dating (cf. relative age).

absolute bolometric magnitude (177)   The absolute magnitude we would observe if we could detect all wavelengths.

absolute visual magnitude (M_v) (176)   Intrinsic brightness of a star; the apparent visual magnitude the star would have if it were 10 pc away.

absolute zero (125)   The lowest possible temperature; the temperature at which the particles in a material, atoms or molecules, contain no energy of motion that can be extracted from the body.

absorption line (133)   A dark line in a spectrum; produced by the absence of photons absorbed by atoms or molecules.

absorption spectrum (dark-line spectrum) (133)   A spectrum that contains absorption lines.

acceleration (81)   A change in a velocity; a change in either speed or direction. (See **velocity**.)

acceleration of gravity (78)   A measure of the strength of gravity at a planet's surface.

accretion (430)   The sticking together of solid particles to produce a larger particle.

accretion disk (285)   The whirling disk of gas that forms around a compact object such as a white dwarf, neutron star, or black hole as matter is drawn in.

achondrite (572)   Stony meteorite containing no chondrules or volatiles.

achromatic lens (103)   A telescope lens composed of two lenses ground from different kinds of glass and designed to bring two selected colors to the same focus and correct for chromatic aberration.

active galactic nucleus (AGN) (365)   The central energy source of an active galaxy.

active galaxy (365)   A galaxy that is a source of excess radiation, usually radio waves, X rays, gamma rays, or some combination.

active optics (109)   Optical elements whose position or shape is continuously controlled by computers.

active region (154)   An area on the sun where sunspots, prominences, flares, etc., occur.

adaptive optics (106)   Computer-controlled telescope mirrors that can at least partially compensate for seeing.

albedo (451)   The fraction of the light that hits an object that is reflected.

alt-azimuth mounting (104)   A telescope mounting capable of motion parallel to and perpendicular to the horizon.

amino acids (597)   Carbon-chain molecules that are the building blocks of protein.

Angstrom (Å) (100)   A unit of distance; $1\ \text{Å} = 10^{-10}$ m; often used to measure the wavelength of light.

angular diameter (16)   A measure of the size of an object in the sky; numerically equal to the angle in degrees between two lines extending from the observer's eye to opposite edges of the object.

angular distance (16)   A measure of the separation between two objects in the sky; numerically equal to the angle in degrees between two lines extending from the observer's eye to the two objects.

angular momentum (87)   The tendency of a rotating body to continue rotating; mathematically, the product of mass, velocity, and radius.

angular momentum problem (415)   An objection to Laplace's nebular hypothesis that cited the slow rotation of the sun.

annular eclipse (45)   A solar eclipse in which the solar photosphere appears around the edge of the moon in a bright ring, or annulus. The corona, chromosphere, and prominences cannot be seen.

anorthosite (464)   Rock of aluminum and calcium silicates found in the lunar highlands.

antimatter (398)   Matter composed of antiparticles, which upon colliding with a matching particle of normal matter annihilate and convert the mass of both particles into energy. The antiproton is the antiparticle of the proton, and the positron is the antiparticle of the electron.

aphelion (23)   The orbital point of greatest distance from the sun.

apogee (45)   The orbital point of greatest distance from Earth.

Apollo–Amor objects (576)   Asteroids whose orbits cross that of Earth (Apollo) and Mars (Amor).

apparent relative orbit (191)   The orbit of one star in a visual binary with respect to the other star as seen from Earth. (See **true relative orbit**.)

apparent visual magnitude (m_v) (13)   The brightness of a star as seen by human eyes on Earth.

archaeoastronomy (53)   The study of the astronomy of ancient cultures.

association (319)   Group of widely scattered stars (10 to 1000) moving together through space; not gravitationally bound into a cluster.

asterism (10)   A named group of stars not identified as a constellation, e.g., the Big Dipper.

asteroids (424)   Small rocky worlds, most of which lie between Mars and Jupiter in the asteroid belt.

astrometric binary (193)   A binary star identified by its irregular proper motion.

astronomical unit (AU) (4)   Average distance from the earth to the sun; 1.5×10^8 km, or 93×10^6 miles.

atmospheric window (101)   Wavelength regions in which our atmosphere is transparent—at visual wavelengths, infrared, and at radio wavelengths.

aurora (161)   The glowing light display that results when a planet's magnetic field guides charged particles toward the north and south magnetic poles, where they strike the upper atmosphere and excite atoms to emit photons.

autumnal equinox (21)   The point on the celestial sphere where the sun crosses the celestial equator going southward. Also, the time when the sun reaches this point and autumn begins in the Northern Hemisphere—about September 22.

Babcock model (158)   A model of the sun's magnetic cycle in which the differential rotation of the sun winds up and tangles the solar magnetic field in a 22-year cycle. This is thought to be responsible for the 11-year sunspot cycle.

Balmer series (134)   Spectral lines in the visible and near-ultraviolet spectrum of hydrogen produced by transitions whose lowest orbit is the second.

barred spiral galaxy (345) A spiral galaxy with an elongated nucleus resembling a bar from which the arms originate.

basalt (445) Dark, igneous rock characteristic of solidified lava.

belts (513) Dark bands of clouds that circle Jupiter parallel to its equator; generally red, brown, or blue-green; believed to be regions of descending gas.

big bang theory (396) The theory that the universe began with a violent explosion from which the expanding universe of galaxies eventually formed.

binary stars (190) Pairs of stars that orbit around their common center of mass.

binding energy (129) The energy needed to pull an electron away from its atom.

bipolar flows (229) Oppositely directed jets of gas ejected by some protostellar objects.

birth line (228) In the H–R diagram, the line above the main sequence where protostars first become visible.

black body radiation (126) Radiation emitted by a hypothetical perfect radiator; the spectrum is continuous, and the wavelength of maximum emission depends only on the body's temperature.

black dwarf (274) The end state of a white dwarf that has cooled to low temperature.

black hole (303) A mass that has collapsed to such a small volume that its gravity prevents the escape of all radiation; also, the volume of space from which radiation may not escape.

blazar (375) See **BL Lac objects.**

BL Lac objects (375) Objects that resemble quasars; thought to be highly luminous cores of distant galaxies.

blue shift (140) The shortening of the wavelengths of light observed when the source and observer are approaching each other.

Bok globules (210) Small, dark clouds only about 1 ly in diameter that contain 10 to 1000 $M_\odot$ of gas and dust; believed related to star formation.

bow shock (444) The boundary between the undisturbed solar wind and the region being deflected around a planet or comet.

breccia (464) A rock composed of fragments of earlier rocks bonded together.

bright-line spectrum (133) See **emission spectrum.**

brown dwarf (248) A very cool, low-luminosity star whose mass is not sufficient to ignite nuclear fusion.

butterfly diagram See **Maunder butterfly diagram.**

CAI (572) Calcium–aluminum-rich inclusions found in some meteorites; believed to be very old.

Cambrian period (598) A geological period 0.6 to 0.5 billion years ago during which life

on earth became diverse and complex. Cambrian rocks contain the oldest easily identifiable fossils.

capture hypothesis (469) The theory that the moon formed elsewhere in the solar system and was later captured by Earth.

carbonaceous chondrites (572) Stony meteorites that contain both chondrules and volatiles. They may be the least altered remains of the solar nebula still present in the solar system.

carbon–nitrogen–oxygen (CNO) cycle (232) A series of nuclear reactions that use carbon as a catalyst to combine four hydrogen atoms to make one helium atom plus energy; effective in stars more massive than the sun.

Cassegrain telescope (103) A reflecting telescope in which the secondary mirror reflects light back down the tube through a hole in the center of the objective mirror.

catastrophic hypotheses (414) Explanations for natural processes that depend on dramatic and unlikely events, such as the collision of two stars to produce our solar system.

celestial equator (16) The imaginary line around the sky directly above Earth's equator.

celestial sphere (14) An imaginary sphere of very large radius surrounding Earth and to which the planets, stars, sun, and moon seem to be attached.

center of mass (84) The balance point of a body or system of bodies.

Cepheid variable star (262) Variable star with a period of 1 to 60 days; the period of variation is related to luminosity.

Chandrasekhar limit (274) The maximum mass of a white dwarf, about 1.4 $M_\odot$; a white dwarf of greater mass cannot support itself and will collapse.

charge-coupled device (CCD) (111) An electronic device consisting of a large array of light-sensitive elements used to record very faint images.

chemical evolution (600) The chemical process that led to the growth of complex molecules on the primitive Earth. This did not involve the reproduction of molecules.

chondrite (571) A stony meteorite that contains chondrules.

chondrule (571) Round, glassy body in some stony meteorites; believed to have solidified very quickly from molten drops of silicate material.

chromatic aberration (103) A distortion found in refracting telescopes because lenses focus different colors at slightly different distances. Images are consequently surrounded by color fringes.

chromosphere (44) Bright gases just above the photosphere of the sun.

circular velocity (84) The velocity required to remain in a circular orbit about a body.

circumpolar constellation (17) Any of the constellations so close to the celestial pole that they never set (or never rise) as seen from a given latitude.

closed orbit (86) An orbit that returns to its starting point; a circular or elliptical orbit. (See **open orbit.**)

closed universe (395) A model universe in which the average density is great enough to stop the expansion and make the universe contract.

cluster method (351) The method of determining the masses of galaxies based on the motions of galaxies in a cluster.

CNO cycle (232) See **carbon–nitrogen–oxygen cycle.**

co-accretion hypothesis (469) The theory that Earth and its moon formed together.

cocoon (227) The cloud of gas and dust around a contracting protostar that conceals it at visible wavelengths.

cold dark matter (402) Invisible matter in the universe composed by heavy, slow-moving particles such as WIMPs.

collapsar (310) See **hypernova.**

collisional broadening (142) The smearing out of a spectral line because of collisions among the atoms of the gas.

color index (127) A numerical measure of the color of a star.

coma (581) The glowing head of a comet.

comet (426) One of the small, icy bodies that orbit the sun and produce tails of gas and dust when they near the sun.

compact object (274) A star that has collapsed to form a white dwarf, neutron star, or black hole.

comparative planetology (439) The study of planets by comparing the characteristics of different examples.

comparison spectrum (113) A spectrum of known spectral lines used to identify unknown wavelengths in an object's spectrum.

condensation (430) The growth of a particle by addition of material from surrounding gas, atom by atom.

condensation sequence (429) The sequence in which different materials condense from the solar nebula as we move outward from the sun.

conservation of energy law (245) One of the basic laws of stellar structure. The amount of energy flowing out of the top of a shell must equal the amount coming in at the bottom plus whatever energy is generated within the shell.

conservation of mass law (245) One of the basic laws of stellar structure. The total mass of the star must equal the sum of the masses of the shells, and the mass must be distributed smoothly throughout the star.

constellation (10) One of the stellar patterns identified by name, usually of mythological gods, people, animals, or objects;

also, the region of the sky containing that star pattern.

continuous spectrum (132) A spectrum in which there are no absorption or emission lines.

convection (149) Circulation in a fluid driven by heat; hot material rises, and cool material sinks.

corona (44, 487) The faint outer atmosphere of the sun; composed of low-density, very hot, ionized gas. On Venus, round networks of fractures and ridges up to 1000 km in diameter.

coronagraph (151) A telescope designed to photograph the inner corona of the sun.

coronal gas (217) Extremely high-temperature, low-density gas in the interstellar medium.

coronal hole (162) An area of the solar surface that is dark at X-ray wavelengths; thought to be associated with divergent magnetic fields and the source of the solar wind.

cosmic microwave background radiation (398) Radiation from the hot clouds of the big bang explosion. Because of its large red shift, it appears to come from a body whose temperature is only 2.7 K.

cosmic ray (119) A subatomic particle traveling at tremendous velocity that strikes Earth's atmosphere from space.

cosmological constant (Λ) (405) Einstein's constant that represents a repulsion in space to oppose gravity.

cosmological principle (391) The assumption that any observer in any galaxy sees the same general features of the universe.

cosmology (389) The study of the nature, origin, and evolution of the universe.

Coulomb barrier (165) The electrostatic force of repulsion between bodies of like charge; commonly applied to atomic nuclei.

Coulomb force (129) The repulsive force between particles with like electrostatic charge.

critical density (400) The average density of the universe needed to make its curvature flat.

critical point (512) The temperature and pressure at which the vapor and liquid phases of a material have the same density.

current sheet (511) The thin equatorial layer of charged particles in a planet's magnetosphere.

dark-line spectrum See **absorption spectrum.**

dark matter (352) Nonluminous material that is detected only by its gravitational influence.

dark nebula (210) A nebula consisting of dust and gas blocking our view of more distant stars.

decameter radiation (510) Radio signals from Jupiter with wavelengths of about 10 m.

decimeter radiation (511) Radio signals from Jupiter with wavelengths of about 0.1 m.

deferent (59) In the Ptolemaic theory, the large circle around Earth along which the center of the epicycle moved.

degenerate matter (253) Extremely high-density matter in which pressure no longer depends on temperature, due to quantum mechanical effects.

density (143) The amount of matter per unit volume in a material; measured in grams per cubic centimeter, for example.

density wave theory (329) Theory proposed to account for spiral arms as compressions of the interstellar medium in the disk of the galaxy.

diamond ring effect (45) A momentary phenomenon seen during some total solar eclipses when the ring of the corona and a bright spot of photosphere resemble a large diamond set in a silvery ring.

differential rotation (157) The rotation of a body in which different parts of the body have different periods of rotation; this is true of the sun, the Jovian planets, and the disk of the galaxy.

differentiation (431) The separation of planetary material according to density.

diffraction fringe (106) Blurred fringe surrounding any image caused by the wave properties of light. Because of this, no image detail smaller than the fringe can be seen.

dirty snowball model (426) The model of comets as kilometer-size balls of ices with embedded impurities.

disk component (319) All material confined to the plane of the galaxy.

distance indicators (347) Objects whose luminosities or diameters are known; used to find the distance to a star cluster or galaxy.

distance modulus ($m_v - M_v$) (177) The difference between the apparent and absolute magnitude of a star; a measure of how far away the star is.

distance scale (348) The combined calibration of distance indicators used by astronomers to find the distances to remote galaxies.

DNA (deoxyribonucleic acid) (597) The long carbon-chain molecule that records information to govern the biological activity of the organism. DNA carries the genetic data passed to offspring.

Doppler broadening (142) The smearing of spectral lines because of the motion of the atoms in the gas.

Doppler effect (140) The change in the wavelength of radiation due to relative radial motion of source and observer.

double exhaust model (371) The theory that double radio lobes are produced by pairs of jets emitted in opposite directions from the centers of active galaxies.

double-line spectroscopic binary (196) A spectroscopic binary star in which spectral lines from both stars are visible in the spectrum.

double-lobed radio source (367) A galaxy that emits radio energy from two regions (lobes) located on opposite sides of the galaxy.

double stars (190) A pair of stars close together in the sky; not all double stars are necessarily in orbit around each other.

dust tail (type II) (583) The tail of a comet formed of dust blown outward by the pressure of sunlight. (See **gas tail.**)

dynamo effect (157) The process by which a rotating, convecting body of conducting matter, such as Earth's core, can generate a magnetic field.

east point (16) The point on the eastern horizon exactly halfway between the north point and the south point; exactly east.

eccentric (58) In astronomy, an off-center circular path.

eccentricity (e) (70) A measure of the flattening of an ellipse. An ellipse of $e = 1$ is circular. The smaller e becomes, the more flattened the ellipse.

eclipse season (47) That period when the sun is near a node of the moon's orbit and eclipses are possible.

eclipse year (48) The time the sun takes to circle the sky and return to a node of the moon's orbit; 346.62 days.

eclipsing binary (199) A binary star system in which the stars eclipse each other.

ecliptic (20) The apparent path of the sun around the sky.

ejecta (461) Pulverized rock scattered by meteorite impacts on a planetary surface.

electromagnetic radiation (99) Changing electric and magnetic fields that travel through space and transfer energy from one place to another—for example, light, radio waves, etc.

electron (125) Low-mass atomic particle carrying a negative charge.

ellipse (70) A closed curve enclosing two points (foci) such that the total distance from one focus to any point on the curve back to the other focus equals a constant.

elliptical galaxy (342) A galaxy that is round or elliptical in outline; it contains little gas and dust, no disk or spiral arms, and few hot, bright stars.

emission line (133) A bright line in a spectrum caused by the emission of photons from atoms.

emission nebula (208) A cloud of glowing gas excited by ultraviolet radiation from hot stars.

emission spectrum (bright-line spectrum) (133) A spectrum containing emission lines.

energy (88) The capacity of a natural system to perform work—for example, thermal energy.

energy level (131)   One of a number of states an electron may occupy in an atom, depending on its binding energy.

enzymes (597)   Special proteins that control processes in an organism.

epicycle (59)   The small circle followed by a planet in the Ptolemaic theory. The center of the epicycle follows a larger circle (deferent) around Earth.

equant (59)   The point off-center in the deferent from which the center of the epicycle appears to move uniformly.

equatorial mounting (104)   A telescope mounting that allows motion parallel to and perpendicular to the celestial equator.

ergosphere (305)   The region surrounding a rotating black hole within which one could not resist being dragged around the black hole. It is possible for a particle to escape from the ergosphere and extract energy from the black hole.

escape orbit (86)   An orbit that does not return to its starting point. (See **open orbit**.)

escape velocity (86)   The initial velocity an object needs to escape from the surface of a celestial body.

evening star (24)   Any planet visible in the sky just after sunset.

event horizon (303)   The boundary of the region of a black hole from which no radiation may escape. No event that occurs within the event horizon is visible to a distant observer.

evolutionary hypotheses (414)   Explanations for natural events that involve gradual changes as opposed to sudden catastrophic changes—for example, the formation of the planets in the gas cloud around the forming sun.

excited atom (131)   An atom in which an electron has moved from a lower to a higher orbit.

extra-solar planet (417)   A planet orbiting a star other than the sun.

eyepiece (103)   A short-focal-length lens used to enlarge the image in a telescope; the lens nearest the eye.

fall (570)   A meteorite seen to fall. (See **find**.)

false-color image (111)   A representation of graphical data in which the colors are altered or added to reveal details.

field (gravitational, electric, or magnetic) (83)   A way of explaining action at a distance; a particle produces a field of influence to which another particle in the field responds.

filtergram (151)   A photograph (usually of the sun) taken in the light of a specific region of the spectrum—e.g., an H-alpha filtergram.

find (570)   A meteorite that is found but was not seen to fall. (See **fall**.)

fission hypothesis (469)   The theory that the moon formed by breaking away from Earth.

flare (160)   A violent eruption on the sun's surface.

flatness problem (403)   In cosmology, the circumstance that the early universe must have contained almost exactly the right amount of matter to close space-time (to make space-time flat).

flat universe (395)   A model of the universe in which space-time is not curved.

flocculent (331)   Woolly, fluffy; used to refer to certain galaxies that have a woolly appearance.

focal length (102)   The distance from a lens to the point where it focuses parallel rays of light.

folded mountain chain (447)   A long range of mountains formed by the compression of a planet's crust—for example, the Andes on Earth.

forbidden line (209)   A spectral line that does not occur in the laboratory because it depends on an atomic transition that is highly unlikely.

forward scattering (516)   The optical property of finely divided particles to preferentially direct light in the original direction of the light's travel.

free-fall contraction (227)   The early contraction of a gas cloud to form a star during which internal pressure is too low to resist contraction.

frequency (99)   The number of times a given event occurs in a given time; for a wave, the number of cycles that pass the observer in 1 second.

galactic cannibalism (355)   The theory that large galaxies absorb smaller galaxies.

galactic corona (322)   The low-density extension of the halo of a galaxy; now suspected to extend many times the visible diameter of the galaxy.

galaxy (6)   A very large collection of gas, dust, and stars orbiting a common center of mass. We live in the Milky Way Galaxy.

galaxy seeds (407)   Small centers of gravitational attraction in the early universe that caused the concentration of matter into clusters of galaxies, filaments, and walls.

Galilean satellites (424)   The four largest satellites of Jupiter, named after their discoverer, Galileo.

gamma-ray burster (309)   An object that produces a sudden burst of gamma rays; thought to be associated with neutron stars and black holes.

gas tail (type I) (582)   The tail of a comet produced by gas blown outward by the solar wind. (See **dust tail**.)

general theory of relativity (92)   Einstein's more sophisticated theory of space and time, which describes gravity as a curvature of space-time.

geocentric universe (57)   A model universe with Earth at the center, such as the Ptolemaic universe.

geosynchronous orbit (85)   An eastward orbit whose period is 24 hours. A satellite in such an orbit remains above the same spot on Earth's surface.

giant molecular clouds (217)   Very large, cool clouds of dense gas in which stars form.

giant stars (181)   Large, cool, highly luminous stars in the upper right of the H-R diagram; typically 10 to 100 times the diameter of the sun.

glitch (296)   A sudden change in the period of a pulsar.

globular cluster (260)   A star cluster containing 50,000 to 1 million stars in a sphere about 75 ly in diameter; generally old, metal-poor, and found in the spherical component of the galaxy.

gossamer ring (517)   The dimmest part of Jupiter's ring produced by dust particles orbiting near small moons.

grand unified theories (GUTs) (403)   Theories that attempt to unify (describe in a similar way) the electromagnetic, weak, and strong forces of nature.

granulation (149)   The fine structure visible on the solar surface caused by rising currents of hot gas and sinking currents of cool gas below the surface.

grating (112)   A piece of material in which numerous microscopic parallel lines are scribed; light encountering a grating is dispersed to form a spectrum.

gravitational collapse (429)   The stage in the formation of a massive planet when it grows massive enough to begin capturing gas directly from the nebula around it.

gravitational lens (380)   The effect of focusing of light from a distant galaxy or quasar by an intervening galaxy to produce multiple images of the distant body.

gravitational red shift (305)   The lengthening of the wavelength of a photon due to its escape from a gravitational field.

gravitational wave (299)   The transport of energy by the motion of waves in a gravitational field; predicted by general relativity.

greenhouse effect (451)   The process by which a carbon dioxide atmosphere traps heat and raises the temperature of a planetary surface.

grooved terrain (520)   Regions of the surface of Ganymede consisting of parallel grooves; believed to have formed by repeated fracture of the icy crust.

ground state (131)   The lowest permitted electron orbit in an atom.

half-life (427)   The time required for half of the atoms in a radioactive sample to decay.

halo (316)   The spherical region of a spiral galaxy containing a thin scattering of stars, star clusters, and small amounts of gas.

heat of formation (432)   In planetology, the heat released by the infall of matter during the formation of a planetary body.

heavy bombardment (434)   The period of intense meteorite impacts early in the formation of the planets when the solar system was filled with debris.

heliocentric universe (60)   A model of the universe with the sun at the center, such as the Copernican universe.

helioseismology (152)   The study of the interior of the sun by the analysis of its modes of vibration.

helium flash (254)   The explosive ignition of helium burning that takes place in some giant stars.

Herbig-Haro objects (229)   Small nebulae that vary irregularly in brightness; believed to be associated with star formation.

Hertzsprung-Russell diagram (179)   A plot of the intrinsic brightness versus the surface temperature of stars; it separates the effects of temperature and surface area on stellar luminosity; commonly absolute magnitude versus spectral type, but also luminosity versus surface temperature or color.

high-velocity star (321)   A star with a large space velocity. Such stars are halo stars passing through the disk of the galaxy at steep angles.

Hirayama families (577)   Families of asteroids with orbits of similar size, shape, and orientation; believed to be fragments of larger bodies.

homogeneity (391)   The assumption that, on the large scale, matter is uniformly spread through the universe.

horizon (14)   The line that marks the apparent intersection of Earth and the sky.

horizon problem (403)   In cosmology, the circumstance that the primordial background radiation seems much more isotropic than could be explained by the standard big bang theory.

horizontal branch (260)   In the H-R diagram of a globular cluster, the sequence of stars extending from the red giants toward the blue side of the diagram; includes RR Lyrae stars.

horoscope (25)   A chart showing the positions of the sun, moon, planets, and constellations at the time of a person's birth; used in astrology to attempt to read character or foretell the future.

hot dark matter (402)   Invisible matter in the universe composed of low-mass, high-velocity particles such as neutrinos.

hot spot (368)   In geology, a place on Earth's crust where volcanism is caused by a rising convection cell in the mantle below. In radio astronomy, a bright spot in a radio lobe.

H-R diagram (179)   (See **Hertzsprung-Russell diagram**.)

HI cloud (213)   An interstellar cloud of neutral hydrogen.

HII region (209)   A region of ionized hydrogen around a hot star.

Hubble constant (H) (349)   A measure of the rate of expansion of the universe; the average value of velocity of recession divided by distance; presently believed to be about 70 km/s/megaparsec.

Hubble law (349)   The linear relation between the distance to a galaxy and its radial velocity.

Hubble time (405)   An upper limit on the age of the universe derived from the Hubble constant.

hydrostatic equilibrium (235)   The balance between the weight of the material pressing downward on a layer in a star and the pressure in that layer.

hypernova (310)   The explosion produced as a very massive star collapses into a black hole; thought to be responsible for at least some gamma-ray bursts.

hypothesis (71)   A conjecture, subject to further tests, that accounts for a set of facts.

inflationary universe (403)   A version of the big bang theory that includes a rapid expansion when the universe was very young.

infrared cirrus (217)   A fine network of filaments covering the sky detected in the far infrared by the IRAS satellite; believed associated with dust in the interstellar medium.

infrared radiation (100)   Electromagnetic radiation with wavelengths intermediate between visible light and radio waves.

instability strip (264)   The region of the H-R diagram in which stars are unstable to pulsation; a star passing through this strip becomes a variable star.

intercloud medium (213)   The hot, low-density gas between cooler clouds in the interstellar medium.

intercrater plains (473)   The relatively smooth terrain on Mercury.

interstellar absorption lines (212)   Dark lines in some stellar spectra that are formed by interstellar gas.

interstellar extinction (211)   The dimming of starlight by gas and dust in the interstellar medium.

interstellar medium (208)   The gas and dust distributed between the stars.

interstellar reddening (211)   The process in which dust scatters blue light out of starlight and makes the stars look redder.

intrinsic variable (262)   A variable driven to pulsate by processes in its interior.

inverse square law (82)   The rule that the strength of an effect (such as gravity) decreases in proportion as the distance squared increases.

Io flux tube (511)   A tube of magnetic lines and electric currents connecting Io and Jupiter.

ion (129)   An atom that has lost or gained one or more electrons.

ionization (129)   The process in which atoms lose or gain electrons.

Io plasma torus (511)   The doughnut-shaped cloud of ionized gas that encloses the orbit of Jupiter's moon Io.

iron meteorite (570)   A meteorite composed mainly of iron–nickel alloy.

irregular galaxy (346)   A galaxy with a chaotic appearance, large clouds of gas and dust, and both population I and population II stars, but without spiral arms.

island universe (341)   An older term for a galaxy.

isotopes (128)   Atoms that have the same number of protons but a different number of neutrons.

isotropy (391)   The assumption that in its general properties the universe looks the same in every direction.

joule (J) (88)   A unit of energy equivalent to a force of 1 newton acting over a distance of 1 meter; 1 joule per second equals 1 watt of power.

Jovian planet (422)   Jupiter-like planet with large diameter and low density.

jumbled terrain (467)   Strangely disturbed regions of the moon opposite the locations of the Imbrium Basin and Mare Orientale.

Kelvin temperature scale (125)   The temperature, in Celsius (Centigrade) degrees, measured above absolute zero.

Keplerian motion (321)   Orbital motion in accord with Kepler's laws of planetary motion.

Kerr black hole (305)   A solution to the equations of general relativity that describes the properties of a rotating black hole.

kiloparsec (kpc) (315)   A unit of distance equal to 1000 pc, or 3260 ly.

Kirchhoff's laws (133)   A set of laws that describes the absorption and emission of light by matter.

Kirkwood's gaps (575)   Regions in the asteroid belt in which there are very few asteroids; caused by resonances with Jupiter.

Kuiper belt (586)   The collection of icy planetesimals believed to orbit in a region from just beyond Neptune out to 100 AU or more.

Lagrangian points (284)   Points of stability in the orbital plane of a binary star system, planet, or moon. One is located 60° ahead and one 60° behind the orbiting bodies; another is located between the orbiting bodies.

large-impact hypothesis (469)   The theory that the moon formed from debris ejected during a collision between Earth and a large planetesimal.

L dwarf (139)   A type of star that is even cooler than the M stars.

life zone (606)   A region around a star within which a planet can have temperatures that permit the existence of liquid water.

light curve (200) A graph of brightness versus time commonly used in analyzing variable stars and eclipsing binaries.

light-gathering power (105) The ability of a telescope to collect light; proportional to the area of the telescope objective lens or mirror.

lighthouse theory (294) The theory that a neutron star produces pulses of radiation by sweeping radio beams around the sky as it rotates.

light pollution (106) The illumination of the night sky by waste light from cities and outdoor lighting, which prevents the observation of faint objects.

light-year (ly) (5) The distance light travels in 1 year.

limb (149, 458) The edge of the apparent disk of a body, as in "the limb of the moon."

limb darkening (149) The decrease in brightness of the sun or other body from its center to its limb.

line of nodes (47) The line across an orbit connecting the nodes; commonly applied to the orbit of the moon.

line profile (142) A graph of light intensity versus wavelength showing the shape of an absorption line.

liquid metallic hydrogen (425, 510) A form of hydrogen under high pressure that is a good electrical conductor.

lithosphere (445) The outer layers of Earth, including the crust and the upper mantle.

lobate scarp (473) A curved cliff such as those found on Mercury.

local bubble (221) A region of high-temperature, low-density gas in the interstellar medium in which the sun happens to be located.

look-back time (349) The amount by which we look into the past when we look at a distant galaxy; a time equal to the distance to the galaxy in light-years.

luminosity (L) (177) The total amount of energy a star radiates in 1 second.

luminosity class (182) A category of stars of similar luminosity; determined by the widths of lines in their spectra.

lunar eclipse (39) The darkening of the moon when it moves through Earth's shadow.

Lyman series (134) Spectral lines in the ultraviolet spectrum of hydrogen produced by transitions whose lowest orbit is the ground state.

MACHOs (402) Massive Compact Halo Objects; low-luminosity objects such as planets and brown dwarfs that contribute to the mass of the halo.

Magellanic Clouds (314) Small, irregular galaxies that are companions to the Milky Way; visible in the southern sky.

magnetar (310) A class of neutron stars that have exceedingly strong magnetic fields; thought to be responsible for soft gamma-ray repeaters.

magnetosphere (444) The volume of space around a planet within which the motion of charged particles is dominated by the planetary magnetic field rather than the solar wind.

magnifying power (106) The ability of a telescope to make an image larger.

magnitude-distance formula (176) The mathematical formula that relates the apparent magnitude and absolute magnitude of a star to its distance.

magnitude scale (12) The astronomical brightness scale; the larger the number, the fainter the star.

main sequence (179) The region of the H-R diagram running from upper left to lower right, which includes roughly 90 percent of all stars.

mantle (443) The layer of dense rock and metal oxides that lies between the molten core and Earth's surface; also, similar layers in other planets.

mare (sea) (458) One of the lunar lowlands filled by successive flows of dark lava.

mass (81) A measure of the amount of matter making up an object.

mass–luminosity relation (204) The more massive a star is, the more luminous it is.

Maunder butterfly diagram (157) A graph showing the latitude of sunspots versus time; first plotted by W. W. Maunder in 1904.

Maunder minimum (163) A period of less numerous sunspots and other solar activity from 1645 to 1715.

megaparsec (Mpc) (347) A unit of distance equal to 1 million pc.

metals (322) In astronomical usage, all atoms heavier than helium.

metastable level (209) An atomic energy level from which an electron takes a long time to decay; responsible for producing forbidden lines.

meteor (426) A small bit of matter heated by friction to incandescent vapor as it falls into Earth's atmosphere.

meteorite (427) A meteor that has survived its passage through the atmosphere and strikes the ground.

meteoroid (426) A meteor in space before it enters Earth's atmosphere.

micrometeorite (464) Meteorite of microscopic size.

midocean rift (445) Chasms that split the midocean rises where crustal plates move apart.

midocean rise (445) One of the undersea mountain ranges that push up from the seafloor in the center of the oceans.

Milankovitch hyposthesis (26) The hypothesis that small changes in Earth's orbital and rotational motions cause the ice ages.

Milky Way (6) The hazy band of light that circles the sky, produced by the combined light of billions of stars in our Milky Way Galaxy.

Milky Way Galaxy (6) The spiral galaxy containing our sun; visible at night as the Milky Way.

Miller experiment (599) An experiment that reproduced the conditions under which life began on earth and manufactured amino acids and other organic compounds.

millisecond pulsar (298) A pulsar with a period of approximately a millisecond, a thousandth of a second.

minute of arc (16) An angular measure; each degree is divided into 60 minutes of arc.

model (14) In science a mental conception of how a specific aspect of nature works; could be expressed mathematically.

molecular cloud (217) An interstellar gas cloud that is dense enough for the formation of molecules; discovered and studied through the radio emissions of such molecules.

molecule (129) Two or more atoms bonded together.

momentum (80) The tendency of a moving object to continue moving; mathematically, the product of mass and velocity.

morning star (24) Any planet visible in the sky just before sunrise.

multiringed basin (461) Very large impact basins in which there are concentric rings of mountains.

mutant (597) Offspring born with altered DNA.

nanometer (100) A unit of length equal to 10^{-9} m.

natural law (71) A conjecture about how nature works in which scientists have overwhelming confidence.

natural motion (78) In Aristotelian physics, the motion of objects toward their natural places—fire and air upward and earth and water downward.

natural selection (597) The process by which the best traits are passed on, allowing the most able to survive.

neap tides (37) Ocean tides of low amplitude occurring at first- and third-quarter moon.

nebula (208) A cloud of gas and dust in space.

nebular hypothesis (415) The proposal that the solar system formed from a rotating cloud of gas

neutrino (166) A neutral, massless atomic particle that travels at or nearly at the speed of light.

neutron (128) An atomic particle with no charge and about the same mass as a proton.

neutron star (291) A small, highly dense star composed almost entirely of tightly packed neutrons; radius about 10 km.

Newtonian focus (103)   The focal arrangement of a reflecting telescope in which a diagonal mirror reflects light out the side of the telescope tube for easier access.

node (47)   A point where an object's orbit passes through the plane of Earth's orbit.

north celestial pole (14)   The point on the celestial sphere directly above Earth's North Pole.

north circumpolar constellation (17)   One of the constellations near the north celestial pole; such constellations never set as seen from moderate northern latitudes.

north point (15)   The point on the horizon directly below the north celestial pole; exactly north.

nova (269)   From the Latin "new," a sudden brightening of a star, making it appear as a "new" star in the sky; believed associated with eruptions on white dwarfs in binary systems.

nuclear bulge (316)   The spherical cloud of stars that lies at the center of spiral galaxies.

nuclear fission (164)   Reaction that splits nuclei into less massive fragments.

nuclear fusion (164)   Reaction that joins the nuclei of atoms to form more massive nuclei.

nucleosynthesis (323)   The production of elements heavier than helium by the fusion of atomic nuclei in stars and during supernovae explosions.

nucleus (of an atom) (128)   The central core of an atom containing protons and neutrons; carries a net positive charge.

O association (229)   A large, loosely bound cluster of very young stars.

objective lens (103)   In a refracting telescope, the long-focal-length lens that forms an image of the object viewed; the lens closest to the object.

objective mirror (103)   In a reflecting telescope, the principal mirror (reflecting surface) that forms an image of the object viewed.

oblateness (510)   The flattening of a spherical body; usually caused by rotation.

observable universe (391)   The part of the universe that we can see from our location in space and time.

occultation (545)   The passage of a larger body in front of a smaller body.

Olbers' paradox (390)   The conflict between observation and theory as to why the night sky should or should not be dark.

Oort cloud theory (585)   The hypothesis that the source of comets is a swarm of icy bodies believed to lie in a spherical shell extending to 100,000 AU from the sun.

opacity (234)   The resistance of a gas to the passage of radiation.

open cluster (260)   A cluster of 10 to 10,000 stars with an open, transparent appearance and stars not tightly grouped; usually relatively young and located in the disk of the galaxy.

open orbit (86)   An orbit that does not return to its starting point; an escape orbit. (See **closed orbit**.)

open universe (395)   A model universe in which the average density is less than the critical density needed to halt the expansion.

optical double (190)   A pair of stars that lie at different distances from Earth and only appear to be near each other.

oscillating universe theory (400)   The theory that the universe begins with a big bang, expands, is slowed by its own gravity, and then falls back to create another big bang.

outflow channel (499)   Geological feature on Mars that appears to have been caused by sudden flooding.

outgassing (431)   The release of gases from a planet's interior.

ovoid (549)   Geological features on Uranus's moon Miranda thought to be produced by circulation in the solid icy mantle and crust.

ozone layer (451)   In earth's atmosphere, a layer of oxygen ions (O_3) lying 15 to 30 km high that protects the surface by absorbing ultraviolet radiation.

paradigm (63)   A commonly accepted set of scientific ideas and assumptions.

parallax (*p*) (51)   The apparent change in the position of an object due to a change in the location of the observer. Astronomical parallax is measured in seconds of arc.

parsec (pc) (173)   The distance to a hypothetical star whose parallax is one second of arc; 1 pc = 206,265 AU = 3.26 ly.

partial eclipse (41)   A lunar eclipse in which the moon does not completely enter Earth's shadow; a solar eclipse in which the moon does not completely cover the sun.

Paschen series (134)   Spectral lines in the infrared spectrum of hydrogen produced by transitions whose lowest orbit is the third.

passing star hypothesis (413)   The proposal that our solar system formed when two stars passed near each other and material was pulled out of one to form the planets.

path of totality (43)   The track of the moon's umbral shadow over Earth's surface. The sun is totally eclipsed as seen from within this path.

penumbra (39)   The portion of a shadow that is only partially shaded.

penumbral eclipse (41)   A lunar eclipse in which the moon enters the penumbra of Earth's shadow but does not reach the umbra.

perigee (45)   The orbital point of closest approach to Earth.

perihelion (23)   The orbital point of closest approach to the sun.

period–luminosity relation (263)   The relation between period of pulsation and intrinsic brightness among Cepheid variable stars.

permitted orbit (129)   One of the energy levels in an atom that an electron may occupy.

photon (100)   A quantum of electromagnetic energy; carries an amount of energy that depends inversely on its wavelength.

photosphere (44)   The bright visible surface of the sun.

planet (4)   A nonluminous object larger than a comet or asteroid that orbits a star.

planetary nebula (271)   An expanding shell of gas ejected from a star during the latter stages of its evolution.

planetesimal (430)   One of the small bodies that formed from the solar nebula and eventually grew into protoplanets.

plastic (443)   A material with the properties of a solid but capable of flowing under pressure.

plate tectonics (445)   The constant destruction and renewal of Earth's surface by the motion of sections of crust.

polar axis (104)   The axis around which a celestial body rotates.

poor galaxy cluster (357)   An irregularly shaped cluster that contains fewer than 1000 galaxies, many spiral, and no giant ellipticals.

population I (322)   Stars rich in atoms heavier than helium; nearly always relatively young stars found in the disk of the galaxy.

population II (322)   Stars poor in atoms heavier than helium; nearly always relatively old stars found in the halo, globular clusters, or the nuclear bulge.

position angle (191)   The angular direction of one body with respect to another; measured from north toward the east; typically used in the study of visual binaries.

precession (18)   The slow change in the direction of Earth's axis of rotation; one cycle takes nearly 26,000 years.

pressure (*P*) waves (442)   In geophysics, mechanical waves of compression and rarefaction that travel through Earth's interior.

primary minimum (200)   In the light curve of an eclipsing binary, the deeper eclipse.

prime focus (103)   The point at which the objective mirror forms an image in a reflecting telescope.

primeval atmosphere (450)   Earth's first air, composed of gases from the solar nebula.

primordial soup (599)   The rich solution of organic molecules in the earth's first oceans.

prominences (44, 159)   Eruptions on the solar surface; visible during total solar eclipses.

proper motion (175)   The rate at which a star moves across the sky; measured in seconds of arc per year.

proteins (597)   Complex molecules composed of amino acid units.

proton (128) A positively charged atomic particle contained in the nucleus of atoms; the nucleus of a hydrogen atom.

proton–proton chain (165) A series of three nuclear reactions that build a helium atom by adding together protons; the main energy source in the sun.

protoplanet (430) Massive object resulting from the coalescence of planetesimals in the solar nebula and destined to become a planet.

protostar (227) A collapsing cloud of gas and dust destined to become a star.

protostellar disk (227) A gas cloud around a forming star flattened by its rotation.

pulsar (293) A source of short, precisely timed radio bursts; believed to be spinning neutron stars.

quantum mechanics (129) The study of the behavior of atoms and atomic particles.

quasar (quasi-stellar object, or QSO) (377) Small, powerful source of energy believed to be the active core of very distant galaxies.

radial velocity (V_r) (140) That component of an object's velocity directed away from or toward Earth.

radial velocity curve (196) A graph of the velocity of recession or approach of the stars in a spectroscopic binary.

radiant (567) The point in the sky from which meteors in a shower seem to come.

radiation pressure (433) The force exerted on the surface of a body by its absorption of light. Small particles floating in the solar system can be blown outward by the pressure of the sunlight.

radio galaxy (365) A galaxy that is a strong source of radio signals.

radio interferometer (114) Two or more radio telescopes that combine their signals to achieve the resolving power of a larger telescope.

rays (461) Ejecta from meteorite impacts forming white streamers radiating from some lunar craters.

recombination (399) The stage within 10^6 years of the big bang when the gas became transparent to radiation.

reconnection (160) The process in the sun's atmosphere by which opposing magnetic fields combine and release energy to power solar flares.

red dwarf (181) Cool, low-mass star on the lower main sequence.

red shift (140) The lengthening of the wavelengths of light seen when the source and observer are receding from each other.

reflecting telescope (102) A telescope that uses a concave mirror to focus light into an image.

reflection nebula (209) A nebula produced by starlight reflecting off of dust particles in the interstellar medium.

refracting telescope (102) A telescope that forms images by bending (refracting) light with a lens.

regolith (464) A soil made up of crushed rock fragments.

relative age (460) The age of a geological feature referred to other features. For example, relative ages tell us the lunar maria are younger than the highlands.

relativistic Doppler formula (379) The formula for the Doppler shift produced by an object moving at a significant fraction of the speed of light.

relativistic jet model (382) An explanation of superluminal expansion based on a high-velocity jet from a quasar directed approximately toward Earth.

resolving power (106) The ability of a telescope to reveal fine detail; depends on the diameter of the telescope objective.

resonance (472) The coincidental agreement between two periodic phenomena; commonly applied to agreements between orbital periods, which can make orbits more or less stable.

retrograde loop (59) The apparent backward (westward) motion of planets as seen against the background of stars.

revolution (20) The motion of an object in a closed path about a point outside its volume; Earth revolves around the sun.

rich galaxy cluster (357) A cluster containing over 1000 galaxies, mostly elliptical, scattered over a volume about 3 Mpc in diameter.

rift valley (447) A long, straight, deep valley produced by the separation of crustal plates.

ring galaxy (354) A galaxy that resembles a ring around a bright nucleus; believed to be the result of a head-on collision of two galaxies.

RNA (ribonucleic acid) (597) Long carbon-chain molecules that use the information stored in DNA to manufacture complex molecules necessary to the organism.

Roche limit (516) The minimum distance between a planet and a satellite that holds itself together by its own gravity. If a satellite's orbit brings it within its planet's Roche limit, tidal forces will pull the satellite apart.

Roche surface (284) The outer boundary of the volume of space that a star's gravity can control within a binary system.

rolling plains (484) The most common type of terrain on Venus.

rotation (20) The turning of a body about an axis that passes through its volume; Earth rotates on its axis.

rotation curve (321, 351) A graph of orbital velocity versus radius in the disk of a galaxy.

RR Lyrae variable star (263) Variable star with a period of 12 to 24 hours; common in some globular clusters.

runoff channel (499) Geological feature on Mars that appears to have been cut by running water.

Sagittarius A* (334) The powerful radio source located at the core of the Milky Way galaxy.

saros cycle (48) An 18-year, $11\frac{1}{3}$-day period after which the pattern of lunar and solar eclipses repeats.

Schmidt-Cassegrain telescope (103) A Cassegrain telescope that uses a thin correcting lens as in a Schmidt camera.

Schwarzschild radius (R_s) (304) The radius of the event horizon around a black hole.

scientific model (14) An intellectual concept designed to help us think about a natural process without necessarily being a conjecture of truth.

scientific notation (4) The system of recording very large or very small numbers by using powers of 10.

secondary atmosphere (450) The gases outgassed from a planet's interior; rich in carbon dioxide.

secondary crater (461) A crater formed by the impact of debris ejected from a larger crater.

secondary minimum (200) In the light curve of an eclipsing binary, the shallower eclipse.

secondary mirror (103) In a reflecting telescope, the mirror that reflects the light to a point of easy observation.

second of arc (16) An angular measure; each minute of arc is divided into 60 seconds of arc.

seeing (106) Atmospheric conditions on a given night. When the atmosphere is unsteady, producing blurred images, the seeing is said to be poor.

seismic wave (442) A mechanical vibration that travels through Earth; usually caused by an earthquake.

seismograph (442) An instrument that records seismic waves.

selection effect (570) An influence on the probability that certain phenomena will be detected or selected, which can alter the outcome of a survey.

self-sustaining star formation (331) The process by which the birth of stars compresses the surrounding gas clouds and triggers the formation of more stars; proposed to explain spiral arms.

semimajor axis (a) (70) Half of the longest axis of an ellipse.

SETI (610) Search for Extra-Terrestrial Intelligence.

Seyfert galaxy (365) An otherwise normal spiral galaxy with an unusually bright, small core that fluctuates in brightness; believed to indicate the core is erupting.

Shapley-Curtis Debate (341) The 1920 debate between Harlow Shapley and Heber Curtis over the nature of the spiral nebulae.

shear (S) waves (442) Mechanical waves that travel through Earth's interior by the vibration of particles perpendicular to the direction of wave travel.

shepherd satellite (530) A satellite that, by its gravitational field, confines particles to a planetary ring.

shield volcano (487) Wide, low-profile volcanic cone produced by highly liquid lava.

shock wave (160) A sudden change in pressure that travels as an intense sound wave.

sidereal drive (104) The motor and gears on a telescope that turn it westward to keep it pointed at a star.

sidereal period (33) The period of rotation or revolution of an astronomical body referred to the stars.

single-line spectroscopic binary (198) A spectroscopic binary in which lines of one star are visible in the spectrum.

singularity (303) The object of zero radius into which the matter in a black hole is believed to fall.

sinuous rille (458) A narrow, winding valley on the moon caused by ancient lava flows along narrow channels.

small-angle formula (42) The mathematical formula that relates an object's linear diameter and distance to its angular diameter.

smooth plains (474) Apparently young plains on Mercury formed by lava flows at or soon after the formation of the Caloris Basin.

soft gamma-ray repeater (310) An object that produces repeated bursts of lower-energy gamma rays; thought to be produced by magnetars.

solar constant (163) A measure of the energy output of the sun; the total solar energy striking 1 m² just above Earth's atmosphere in 1 second.

solar eclipse (41) The event that occurs when the moon passes directly between Earth and the sun, blocking our view of the sun.

solar nebula hypothesis (416) The theory that the planets formed from the same cloud of gas and dust that formed the sun.

solar system (4) The sun and the nonluminous objects that orbit it, including the planets, comets, and asteroids.

solar wind (152) Rapidly moving atoms and ions that escape from the solar corona and blow outward through the solar system.

south celestial pole (15) The point of the celestial sphere directly above Earth's South Pole.

south circumpolar constellation (17) One of the consellations near the south celestial pole such that it is not seen to rise or set.

south point (16) The point on the horizon directly above the south celestial pole; exactly south.

special relativity (91) The first of Einstein's theories of relativity, which dealt with uniform motion.

spectral class or type (136) A star's position in the temperature classification system O, B, A F, G, K, and M. Based on the appearance of the star's spectrum.

spectral sequence (136) The arrangement of spectral classes (O, B, A, F, G, K, M) ranging from hot to cool.

spectrograph (112) A device that separates light by wavelength to produce a spectrum.

spectroscopic binary (196) A star system in which the stars are too close together to be visible separately. We see a single point of light, and only by taking a spectrum can we determine that there are two stars.

spectroscopic parallax (183) The method of determining a star's distance by comparing its apparent magnitude with its absolute magnitude as estimated from its spectrum.

spherical component (319) The part of the galaxy including all matter in a spherical distribution around the center (the halo and nuclear bulge).

spicule (151) Small, flamelike projection in the chromosphere of the sun.

spiral arms (319) Long, spiral patterns of bright stars, star clusters, gas, and dust that extend from the center to the edge of the disk of spiral galaxies.

spiral galaxy (344) A galaxy with an obvious disk component containing gas; dust; hot, bright stars; and spiral arms.

spiral nebulae (340) Nebulous objects with a spiral appearance observed in early telescopes; later recognized as spiral galaxies.

spiral tracers (327) Objects used to map the spiral arms (e.g., O and B associations, open clusters, clouds of ionized hydrogen, and some types of variable stars).

spoke (529) A radial feature in the rings of Saturn.

sporadic meteor (568) A meteor not part of a meteor shower.

spring tides (37) Ocean tides of high amplitude that occur at full and new moon.

standard candles (347) Objects of known brightness which astronomers use to find distance—for example, Cepheid variable stars and supernovae.

star (4) A celestial object composed of gas held together by its own gravity and supported by nuclear fusion occurring in its interior.

starburst galaxy (359) A bright blue galaxy in which many new stars are forming, believed caused by collisions between galaxies.

steady state theory (398) The theory (now generally abandoned) that the universe does not evolve.

stellar density function (184) A description of the abundance of stars of different types in space.

stellar model (246) A table of numbers representing the conditions in various layers within a star.

stellar parallax (p) (173) A measure of stellar distance. (See **parallax**.)

stony-iron meteorite (572) A meteorite that is a mixture of stone and iron.

stony meteorite (571) A meteorite composed of silicate (rocky) material.

stromatolite (601) A layered fossil formation caused by ancient mats of algae or bacteria that build up mineral deposits season after season.

strong force (164) One of the four forces of nature; the strong force binds protons and neutrons together in atomic nuclei.

subduction zone (445) A region of a planetary crust where a tectonic plate slides downward.

subsolar point (481) The point on a planet that is directly below the sun.

summer solstice (21) The point on the celestial sphere where the sun is at its most northerly point; also, the time when the sun passes this point, about June 22, and summer begins in the Northern Hemisphere.

sunspot (154) Relatively dark spot on the sun that contains intense magnetic fields.

supercluster (359, 406) A cluster of galaxy clusters.

supergiant (181) Exceptionally luminous star 10 to 1000 times the sun's diameter.

supergranule (149) A large granule on the sun's surface including many smaller granules.

superluminal expansion (382) The apparent expansion of parts of a quasar at speeds greater than the speed of light.

supernova remnant (279) The expanding shell of gas marking the site of a supernova explosion.

supernova (Type I) (278) The explosion of a star, believed caused by the transfer of matter to a white dwarf.

supernova (Type II) (278) The explosion of a star, believed caused by the collapse of a massive star.

synchrotron radiation (281) Radiation emitted when high-speed electrons move through a magnetic field.

synodic period (34) The period of rotation or revolution of a celestial body with respect to the sun.

T association (229) A large, loosely bound group of T Tauri stars.

temperature (125) A measure of the velocity of random motions among the atoms or molecules in a material.

terminator (458) The dividing line between daylight and darkness on a planet or moon.

terrestrial planet (422) Earthlike planet—small, dense, rocky.

theory (71) A system of assumptions and principles applicable to a wide range of phenomena that have been repeatedly verified.

thermal energy (125) The energy stored in an object as agitation among its atoms and molecules.

thermal pulse (270) Periodic eruptions in the helium fusion shell in an aging giant star; thought to aid in ejecting the surface layers of the stars to form planetary nebulae.

tidal coupling (457) The locking of the rotation of a body to its revolution around another body.

tidal heating (521) The heating of a planet or satellite because of friction caused by tides.

time dilation (305) The slowing of moving clocks or clocks in strong gravitational fields.

total eclipse (39, 41) A solar eclipse in which the moon completely covers the bright surface of the sun; a lunar eclipse in which the moon completely enters Earth's dark shadow.

transit (561) The crossing of a smaller object in front of a larger object.

transition (134) The movement of an electron from one atomic orbit to another.

transition region (150) The layer in the solar atmosphere between the chromosphere and the corona.

transverse velocity (140) The velocity of a star perpendicular to the line of sight.

triple alpha process (254) The nuclear fusion process that combines three helium nuclei (alpha particles) to make one carbon nucleus.

Trojan asteroid (577) Small, rocky body caught in Jupiter's orbit at the Lagrangian points, 60° ahead of and behind the planet.

true relative orbit (191) The orbit of one star in a visual binary with respect to the other star after correction for orbital inclination. (See **apparent relative orbit.**)

T Tauri stars (228) Young stars surrounded by gas and dust. Believed to be contracting toward the main sequence.

tuning fork diagram (342) A system of classification for elliptical, spiral, and irregular galaxies.

turnoff point (259) The point in an H-R diagram where a cluster's stars turn off of the main sequence and move toward the red giant region, revealing the approximate age of the cluster.

21-cm radiation (214) Radio emission produced by cold, low-density hydrogen in interstellar space.

ultraluminous infrared galaxy (356) A highly luminous galaxy so filled with dust that most of its energy escapes as infrared photons emitted by warmed dust.

ultraviolet radiation (100) Electromagnetic radiation with wavelengths shorter than visible light but longer than X rays.

umbra (38) The region of a shadow that is totally shaded.

uncompressed density (429) The density a planet would have if its gravity did not compress it.

unified model (374) The attempt to explain the different kinds of active galaxies and quasars by a single model.

uniform circular motion (56) The classical belief that the perfect heavens could move only by the combination of uniform motion along circular orbits.

universality (391) The assumption that the physical laws observed on Earth apply everywhere in the universe.

Van Allen belts (444) Radiation belts of high-energy particles trapped in Earth's magnetosphere.

variable star (262) A star whose brightness changes periodically.

velocity (81) A rate of travel that specifies both speed and direction.

velocity dispersion method (352) A method of finding a galaxy's mass by observing the range of velocities within the galaxy.

vernal equinox (21) The place on the celestial sphere where the sun crosses the celestial equator moving northward; also, the time of year when the sun crosses this point, about March 21, and spring begins in the Northern Hemisphere.

vesicular basalt (463) A porous rock formed by solidified lava with trapped bubbles.

violent motion (78) In Aristotelian physics, motion other than natural motion. (See **natural motion.**)

visual binary (190) A binary star system in which the two stars are separately visible in the telescope.

water hole (610) The interval of the radio spectrum between the 21-cm hydrogen radiation and the 18-cm OH radiation; likely wavelengths to use in the search for extraterrestrial life.

wavelength (99) The distance between successive peaks or troughs of a wave; usually represented by λ.

wavelength of maximum intensity (λ_{max}) (126) The wavelength at which a perfect radiator emits the maximum amount of energy; depends only on the object's temperature.

weak force (164) One of the four forces of nature; the weak force is responsible for some forms of radioactive decay.

west point (16) The point on the western horizon exactly halfway between the north point and the south point; exactly west.

white dwarf (181) Dying star that has collapsed to the size of Earth and is slowly cooling off; at the lower left of the H-R diagram.

Widmanstätten patterns (570) Bands in iron meteorites due to large crystals of nickel–iron alloys.

winter solstice (21) The point on the celestial sphere where the sun is farthest south; also, the time of year when the sun passes this point, about December 22, and winter begins in the Northern Hemisphere.

X-ray burster (300) An object that produces repeated bursts of x rays.

Zeeman effect (154) The splitting of spectral lines into multiple components when the atoms are in a magnetic field.

zenith (14) The point on the sky directly overhead.

zero-age main sequence (ZAMS) (250) The locus in the H-R diagram where stars first reach stability as hydrogen-burning stars.

zodiac (25) The band around the sky centered on the ecliptic within which the planets move.

zones (513) Yellow-white regions that circle Jupiter parallel to its equator; believed to be areas of rising gas.

Answers to Even-Numbered Problems

Chapter 1: 2. 1.28 s; **4.** 1.8×10^7

Chapter 2: 2. 4; **4.** 2800; **6.** A is brighter than B by a factor of 170; **8.** 66.5°; 113.5°; **10.** The sun would follow a path that coincides with the location of the celestial equator, rising directly east and setting directly west.

Chapter 3: 2. a) full; b) first quarter; c) waxing gibbous; d) waxing crescent; **4.** 12; 12.4. The moon orbits Earth and moves eastward about 13° per day; **6.** 6850 arc seconds or about 1.9°; **8.** a) The moon won't be full until Oct. 17; b) The moon will no longer be near the node of its orbit; **10.** August 12, 2026 [July 10, 1972 + 3 × ($6585\frac{1}{3}$ days). In order to get the Aug. 12 instead of Aug. 11, you must take into account the number of leap days in the interval.

Chapter 4: 2. Retrograde loops: Jupiter, Saturn, Uranus, Neptune, and Pluto; Never seen as crescents: Jupiter, Saturn, Uranus, Neptune, and Pluto; **4.** Mars, about 18 seconds of arc; Saturn's rings, about 44 seconds of arc; **6.** $\sqrt{27} = 5.2$ years

Chapter 5: 2. The force of gravity on the moon is about $\frac{1}{6}$ the force of gravity on Earth; **4.** 7350 m/s; **6.** 5070 s (1 hr and 25 min); **8.** The cannonball would move in an elliptical orbit with Earth's center at one focus of the ellipse; **10.** 6320 s (1 hr and 45 min)

Chapter 6: 2. short radio waves; **4.** The 10-m Keck telescope has a light-gathering power that is 1.56 million times greater than the human eye; **6.** No, his resolving power should have been about 5.8 seconds of arc at best; **8.** 0.013 m (1.3 cm or about 0.5 inches); **10.** about 50 cm (From 400 km above, a human is about 0.25 seconds of arc from shoulder to shoulder.)

Chapter 7: 2. 150 nm; **4.** by a factor of 16; **6.** 250 nm; **8.** a. B; b. F; c. M; d. K; **10.** about 0.58 nm

Chapter 8: 2. 730 km; **4.** about 3.6 times; **6.** 400,000 years; **8.** 9×10^{16} joules; **10.** 0.222 kg

Chapter 20: 2. It will look $206,265^2 = 4.3 \times 10^{10}$ times fainter, which is about 26.6 magnitudes fainter; 22.6 mag; **4.** about 3.3 times the half-life, or 4.3 billion years; **6.** large amounts of methane and water ices; **8.** about 1400

Chapter 21: 2. about 17%; **4.** 81×10^6 yr; they have been subducted; **6.** 0.22%

Chapter 22: 2. The rate at which an object radiates energy is proportional to its surface area (e.g., for a black body the rate of emission of energy is given by $\frac{4}{3}\pi r^2 \sigma T^4$). However, the energy an object has stored as heat is proportional to its mass and hence to its volume [i.e., mass = $\frac{4}{3}\pi r^3$ (density)]. The cooling time will therefore be proportional to the mass divided by the surface area:

$$\text{cooling time} = \frac{\text{mass}}{\text{surface area}} = \frac{\frac{4}{3}\pi r^3 \times \text{density}}{4\pi r^2}$$

$$= \tfrac{1}{3}\, r \times \text{density}$$

and we find that for planets of nearly the same density, the one with the largest radius will take the longest time to cool; **4.** No. Their angular diameter would be only 0.5 second of arc. They would be visible in photos taken from orbit around the moon; **6.** 0.5 second of arc (assuming an astronaut seen from above is 0.5 meters in diameter); no; **8.** 10.0016 cm; **10.** Mercury, $v_e = 4250$ m/s; moon, $v_e = 2380$ m/s; Earth, $v_e = 11,200$ m/s

Chapter 23: 2. 33,400 km (39,500 km from the center of Venus); **4.** 3.5×10^4 m/s; **6.** 260 km; **8.** 965″ (about 1/4°)

Chapter 24: 2. 35 Earth days; **4.** 4.4°; **6.** about 0.056 nm; **8.** 5.2 m/s

Chapter 25: 2. 42 seconds of arc; **4.** 256 m/s; **6.** 8.8 s; yes; **8.** 12.3 km/s; **10.** 1.04×10^{26} kg (17.2 Earth masses)

Chapter 26: 2. one billion; **4.** 80 km in diameter; **6.** 0.18 km/s; **8.** 3.28 AU and 2.5 AU; **10.** 9 million km (0.06 AU); too small; **12.** 33 Earth masses

Chapter 27: 2. 8.9 cm; 0.67 mm; **4.** about 1.3 solar masses; **6.** 380 km; **8.** pessimistic, 2×10^{-5}; optimistic, 10^7

Index

A ring, 528, 530, 531, 535
  shepherd satellites and, 530
A stars, 136, 137, 138
Aberration. *See also* Chromatic aberration
  in refracting telescopes, 103
Absolute ages, of impact craters, 460
Absolute zero, 125
Absorbers, perfect, 126
Absorption lines, 133
  Doppler effect on, 141–142
  stellar temperatures and, 135–136
Absorption spectrum, 132, 133–134
  of chromosphere and photosphere,
    150–151
  of corona, 152
  of sun, 148–149
Accelerated motion, 91
  in theory of relativity, 92–95
Acceleration, 80, 81
  curvature of space-time and, 93
  of falling bodies, 78–80
  in Newton's second law, 81, 82
  in orbital motion, 84
Acceleration of gravity, 78–80, 82–83
Accretion, 430
Achernar, Caracol temple and, 55
Achondrites, 572
  origin of, 573
Achromatic lenses, 103
Action and reaction, in Newton's third law,
  81–82
Action at a distance, gravity as, 83
Active optics, 109
Active regions, 153–154, 163
Adams, John Couch, 552
Adaptive optics, 106
Adenine, in DNA, 596–597
Adhesion, in formation of protoplanets,
  430–431
Adrastea, Jovian ring system and, 517
Advanced X-ray Astrophysics Facility
  (AXAF), 119
Air. *See also* Atmosphere
  as element, 78
  sphere of, 57
Air resistance, 79, 80
  in heating of meteors, 590
Aircraft
  in cosmic-ray astronomy, 119
  in infrared astronomy, 116, 117
Airy, George, 552
Aitkin Basin, 468
Al Jabbar, Orion as, 10
Albedo, 439n, 451
  in asteroid classification, 578–579
Aldebaran, 11
Aldrin, Edwin E., Jr., 456, 457, 462–463
Alexandria, 58
Alfonsine Tables, 60, 62, 72–73
  errors in, 67
Alfonso X, King, 60

ALH84001 meteorite, 501, 605–606
ALHA81005 meteorite, 461
Allegheny Mountains, 447, 448
Allende meteorite, 572
Almagest (Ptolemy), 59
Alne Corona, 488
α Aurigae. *See* Capella
α Bootis. *See* Arcturus
α Canis Majoris. *See* Sirius
α Draconis. *See* Thuban
α Eridani. *See* Achernar
α Lyrae. *See* Vega
α Orionis. *See* Betelgeuse
α Regio, 485
α Tauri. *See* Aldebaran
α Ursae Minoris. *See* Polaris
Alpheratz, constellation for, 10, 11
Alphonsine Tables, Tycho's revisions to, 68
Alt-azimuth mountings, 104–105
  for Gemini telescopes, 110
Altitude, in telescope mountings, 104–105
Aluminum (Al)
  in meteorites, 572, 573
  in Vesta, 579
Alvarez, Luis, 587
Alvarez, Walter, 587
AM radio, wavelengths of, 101
Amalthea
  Io and, 522
  Jovian ring system and, 517
American Indians, astronomy of, 53–55
Amino acids, 597
  in meteorites, 600
  in Miller experiment, 599
  in proteins, 600, 601
Ammonia (NH_3)
  in comets, 582
  in Earth's secondary atmosphere, 451
  in Jovian atmosphere, 509, 512, 513
  in Miller experiment, 599
  in Saturnian atmosphere, 527
  in Titan's atmosphere, 533–534
  on Triton, 556–557
  in Uranian atmosphere, 542, 543
  Uranian magnetic field and, 544
Ammonia hydrosulfide (NH_4SH)
  in Jovian atmosphere, 512, 513
  in Saturnian atmosphere, 527
Amplifier, in radio telescope, 113, 114
Anasazi Indians, astronomy by, 53–55
Anaximander, 56
Andes Mountains, 446–447
Andromeda, 10, 11
Angles. *See also* Angular diameter; Angular
  distance; Parallax (p); Small-angle
  formula
  on sky, 16, 68
Angstrom (Å), 100
Angular diameter, 16, 17
  of moon, 42, 45, 46

  of sun, 42, 45, 46
  of Uranus, 541
Angular distance, 16, 17
Angular momentum, 87. *See also* Rotation
  conservation of, 87
  in solar system origin, 414, 415
Angular momentum problem, 415, 417, 433
Annular eclipses, 45–46
  of 1994, 45, 46
  table of, 43
Anorthosite, 464
  in lunar crust, 465–466
Antarctica, meteorites found in, 570,
    605–606
Antares, 11
Antenna, in radio telescope, 113, 114
Apennine Mountains (lunar), Apollo 15
  mission to, 463
Aphelion, 23, 24
  changes in, 28
Aphrodite Terra, 485
Apian, Peter, 78
Apogee, 45, 86
Apollo 11, 462–463
Apollo 12, 463
Apollo 13, 463, 463n
Apollo 14, 463, 466
Apollo 15, 79, 80, 463
Apollo 16, 463
Apollo 17, 463
Apollo missions, 462–463, 466
  age of moon and, 427
  moon rocks retrieved by, 463–465
Apollo–Amor objects, 576
  orbits of, 576–577
  in Tunguska event, 590
Apparent visual magnitude (m_v), 13
Aquarius, 20
  as constellation along ecliptic, 21
Archaeoastronomy, 53–56
Arcturus
  Doppler effect in spectrum of, 140, 141
Arecibo, Puerto Rico
  radio telescope at, 115, 116, 472, 609
Argon (Ar)
  in determining age of solar system, 427
  in Martian atmosphere, 495
  in neutrino detectors, 166, 167
Argument, in science, 372
Argyre, 496, 501
Ariel, 548, 549
  visual appearance of, 550
Aries, as constellation along ecliptic, 21
Aristarchus, 57
Aristotle, 40–41, 53, 58, 60, 62, 64, 67, 82
  Copernican challenge to, 61
  cosmology of, 78
  Galileo on, 65
  and Galileo's experiments with motion,
    78–80
  gravity and, 83

Louis XIV, King, 163*n*
Low tide, 37
Lowell, Percival, 492, 558, 559
Lowell Observatory, 558
Lunar eclipses, 38–41, 50
　as seen from moon, 46
　Stonehenge and, 53
Lunar landing module (LM), 462–463
Lunar landings (Apollo), 462–463
Lunatic, etymology of, 32
Luther, Martin, 62, 73
Lyman series, 134–135

M stars, 136, 137, 138, 139
　abundance of, 184
　life zones around, 606–607
M13 globular cluster, message sent to, 609
Maat Mons, 486
Magellan spacecraft, radar images of Venus
　by, 483–484, 485–487, 488
Magma, 446
　moon rocks and, 463, 466
　Tharsis bulge and, 497
Magnesium (Mg)
　in meteorites, 573
　solar abundance of, 140
Magnetic cycle
　on stars, 159
　of sun, 157–159
Magnetic field reversals, 445
　plate tectonics and, 445–446
Magnetic fields
　angular momentum problem and, 433
　of Callisto, 519
　comet tails and, 582–583
　in corona, 152
　coronal activity and, 162
　of Earth, 444–445, 450
　in electromagnetic radiation, 99
　of Ganymede, 521
　of Jupiter, 510–512, 517, 545
　of Mars, 495, 501
　of Neptune, 545, 554
　in prominences, 160
　of Saturn, 525–526, 529–530, 545
　solar activity and, 152, 153–162, 164
　solar flares and, 160–161
　of Uranus, 543–545
　of Venus, 487–489
Magnetic force, electric force and, 403
Magnetic poles, solar activity and, 161
Magnetosphere, 444–445
　of Ganymede, 521
　of Jupiter, 510–512
　of Neptune, 554
　of Uranus, 545
Magnifying power, 106
　in purchasing telescopes, 108
Magnitude scale, 12
　awkwardnesses in, 14
Magnitudes. *See also* Apparent visual
　magnitude (m_v)
　of stars, 12–14
Main sequence, life zones around stars in,
　606–607
Mammoth tusk, lunar phases carved on, 55
"Man in the moon," 458
Mantle
　of Earth, 442, 443, 444, 446
　of moon, 466
　of Uranus, 543–544
Mare Crisium, 458, 467
Mare Foecunditatis, 458
Mare Humorum, 458

Mare Imbrium, 458, 466, 467
　lunar landing in, 463, 466
　origin of, 466, 468
Mare Nectaris, 458
Mare Nubium, 458
Mare Orientale, 466, 467–468
Mare Serenitatis, 458, 467
Mare Tranquillitatis, 458
　lunar landing in, 463
Maria (mare), 458–459
　formation of, 467–468, 470
Mariner 10 spacecraft, Mercury observed by,
　471, 473, 474
Mariner 4 spacecraft, Martian flyby of, 492,
　495
Mars, 479, 480, 490–505, 507, 558. *See also*
　Deimos; Phobos
　age of, 428
　apparent motion of, 59
　asteroid belt between Jupiter and, 575
　asteroid orbits and, 425
　atmosphere of, 453, 480, 490–491,
　　492–495, 502
　canals on, 491–492
　in Copernican cosmology, 61, 62
　crust of, 497
　data file for, 491
　density of, 429, 491
　diameter of, 490–491
　Earth versus, 491
　features of, 423
　history of, 501–503
　Hubble Space Telescope image of, 120
　impact craters on, 434
　inclination of, 491, 502
　intelligent life on, 491–492
　in Kepler's cosmology, 70
　life on, 604–606
　map of canals on, 492
　mass of, 491, 493–494
　meteorites from, 500–501, 605–606
　naked-eye visibility of, 24
　oceans of, 496–497, 499
　orbit of, 5, 24, 25, 421, 491, 577
　planetary science and, 413
　polar caps on, 491, 499–500
　in Ptolemaic cosmology, 60
　rotation of, 432, 491
　satellites of, 503–505
　scale of, 420, 422, 423, 490
　surface features of, 495–499, 503
　temperature of, 480, 493–494
　as terrestrial planet, 422, 423
　in Tychonic cosmology, 69
　visual appearance of, 423, 495
　water on, 499–501
Mars Global Surveyor spacecraft, 499, 500,
　501, 504
Mass, 81
　and angular momentum, 87
　of Callisto, 519
　center of, 84, 85
　of Charon, 561
　of comet nuclei, 583
　conversion to energy of, 165–166
　curvature of space-time and, 93–95
　distribution within atoms, 128
　of Earth, 439
　escape velocity and, 86–87
　as fundamental parameter, 81, 88, 125, 143
　gravity and, 83
　of Jupiter, 413, 508, 509
　in Kepler's third law, 88

of Mars, 491, 493–494
of Mercury, 471
momentum and, 80–81
of moon, 457
of Neptune, 553
of neutrinos, 168
in Newton's second law, 82
and orbital velocity, 84–85
of Pluto, 559, 561
of Saturn, 525
in theory of relativity, 91–92
of Uranus, 544
of Venus, 480
Mass extinctions, periodicity of, 588
Massive stars, life on planets of, 606
Mathematical Syntaxis (Ptolemy), 59
Mathematics. *See also* Computers
　Galileo and, 63
　Kepler and, 69
　in models, 14
　nature of universe and, 70
　Newton and, 88–90
　of Pythagoras, 56
Mathilde, visual appearance of, 578
Matter
　gravitation and, 33
　light and, 123, 130–135
　origin of, 415–416
Mauna Loa, Olympus Mons versus, 497
Maunder, E. Walter, 157
Maunder butterfly diagrams, 157, 158–159
Maunder Minimum, 163
Maxwell, James Clerk, 484–485, 528
Maxwell Montes, 484, 485, 489
Mayall telescope, 104
Mayan civilization
　archaeoastronomy in, 55–56
　eclipses and, 32, 46
Measurement, in science, 107
Mechanical universe, 89–90
Megachannel Extra-Terrestrial Assay
　(META), 610
Mercury, 456, 457, 461, 470, 471–476
　atmosphere of, 453, 456, 472
　in Copernican cosmology, 61, 62
　crust of, 472–474, 475
　data file for, 471
　density of, 429, 471, 474
　diameter of, 471
　history of, 475–476
　impact craters on, 434
　internal structure of, 474
　in Kepler's cosmology, 70, 72
　life on, 604
　mass of, 471
　Mimas versus, 534
　moon and, 473, 474, 475
　naked-eye visibility of, 24, 25
　orbit of, 4, 24, 25, 421, 471–472
　orbital inclination of, 420, 428
　in Ptolemaic cosmology, 60
　relativistic precession of orbit of, 93–94
　rotation of, 471–472
　scale of, 4, 5, 420, 422, 471
　surface features of, 472–474, 475
　temperature of, 471, 472, 475
　as terrestrial planet, 422–423
　in Tychonic cosmology, 69
　visual appearance of, 471
Mercury-vapor lights, emission spectra of,
　133
Merging galaxies, 355–357
　in galactic evolution, 359–360
　quasars and, 384–385
Messages, to extraterrestrial civilizations,
　608–611

Table A-6 Temperature Scales

	Kelvin (K)	Centigrade (°C)	Fahrenheit (°F)
Absolute zero	0 K	−273°C	−459°F
Freezing point of water	273 K	0°C	32°F
Boiling point of water	373 K	100°C	212°F

Conversions:

$K = °C + 273$

$°C = \frac{5}{9}(°F - 32)$

$°F = \frac{9}{5}°C + 32$

Table A-7 Units Used in Astronomy

1 angstrom (Å)	$= 10^{-8}$ cm
	$= 10^{-10}$ m
1 astronomical unit (AU)	$= 1.495979 \times 10^{11}$ m
	$= 92.95582 \times 10^6$ miles
1 light-year (ly)	$= 6.3240 \times 10^4$ AU
	$= 9.46053 \times 10^{15}$ m
	$= 5.9 \times 10^{12}$ miles
1 parsec (pc)	$= 206,265$ AU
	$= 3.085678 \times 10^{16}$ m
	$= 3.261633$ ly
1 kiloparsec (kpc)	$= 1000$ pc
1 megaparsec (Mpc)	$= 1,000,000$ pc

Table A-8 The Nearest Stars

Name	Absolute Magnitude (M_v)	Distance (ly)	Spectral Type	Apparent Visual Magnitude (m_v)
Sun	4.83		G2	−26.8
Proxima Cen	15.45	4.28	M5	11.05
α Cen A	4.38	4.3	G2	0.1
B	5.76	4.3	K5	1.5
Barnard's Star	13.21	5.9	M5	9.5
Wolf 359	16.80	7.6	M6	13.5
Lalande 21185	10.42	8.1	M2	7.5
Sirius A	1.41	8.6	A1	−1.5
B	11.54	8.6	white dwarf	7.2
Luyten 726-8A	15.27	8.9	M5	12.5
B (UV Cet)	15.8	8.9	M6	13.0
Ross 154	13.3	9.4	M5	10.6
Ross 248	14.8	10.3	M6	12.2
ε Eri	6.13	10.7	K2	3.7
Luyten 789–6	14.6	10.8	M7	12.2
Ross 128	13.5	10.8	M5	11.1
61 CYG A	7.58	11.2	K5	5.2
B	8.39	11.2	K7	6.0
ε Ind	7.0	11.2	K5	4.7
Procyon A	2.64	11.4	F5	0.3
B	13.1	11.4	white dwarf	10.8
Σ 2398 A	11.15	11.5	M4	8.9
B	11.94	11.5	M5	9.7
Groombridge 34 A	10.32	11.6	M1	8.1
B	13.29	11.6	M6	11.0
Lacaille 9352	9.59	11.7	M2	7.4
τ Ceti	5.72	11.9	G8	3.5
BD + 5° 1668	11.98	12.2	M5	9.8
L 725–32	15.27	12.4	M5	11.5
Lacaille 8760	8.75	12.5	M0	6.7
Kapteyn's Star	10.85	12.7	M0	8.8
Kruger 60 A	11.87	12.8	M3	9.7
B	13.3	12.8	M4	11.2